Applied Robotic Analysis

PRENTICE HALL INDUSTRIAL ROBOTS SERIES

W. E. Snyder, Series Editor

Parkin, APPLIED ROBOTIC ANALYSIS

Snyder, INDUSTRIAL ROBOTS: COMPUTER INTERFACING AND CONTROL

Applied Robotic Analysis

ROBERT E. PARKIN

University of Lowell

PRENTICE HALL, Englewood Cliffs, New Jersey 07632

Library of Congress Cataloging-in-Publication Data

PARKIN, ROBERT E.
Applied robotic analysis / Robert E. Parkin.
p. cm.—(Prentice Hall industrial robots series)
Includes bibliographical references and index.
ISBN 0-13-773391-7
1. Robots—Motion. 2. Machinery, Kinematics of. I. Title.
II. Series.
TJ211.4.P37 1991
629.8′92—dc20 90-42773
CIP

Acquisitions editor: Bernard Goodwin
Editorial/production supervision: Merrill Peterson
Interior design: Joan Stone
Cover design: Wanda Lubelska
Prepress buyer: Kelly Behr
Manufacturing buyer: Susan Brunke

The publisher offers discounts on this book when ordered in bulk quantities. For more information, write:
Special Sales/College Marketing
Prentice Hall
College Technical and Reference Division
Englewood Cliffs, NJ 07632

Printed in the United States of America
10 9 8 7 6 5 4 3 2 1

ISBN 0-13-773391-7

Prentice-Hall International (UK) Limited, *London*
Prentice-Hall of Australia Pty. Limited, *Sydney*
Prentice-Hall Canada Inc., *Toronto*
Prentice-Hall Hispanoamericana, S.A., *Mexico*
Prentice-Hall of India Private Limited, *New Delhi*
Prentice-Hall of Japan, Inc., *Tokyo*
Simon & Schuster Asia Pte. Ltd., *Singapore*
Editora Prentice-Hall do Brasil, Ltda., *Rio de Janeiro*

Contents

Preface

There is a considerable body of sophisticated analyses in existing books and journals, but all the analysis in the world does not readily seem to permit the remote control of a robot or the automated control of two robots in the same workspace. The heart of the problem is the reliance on kinematic notations devised in the nineteenth century that are inappropriate for robots.

Kinematics is the glue that holds all aspects of manipulator analysis and control together. As shown in the accompanying figure, most aspects of robotics rely on kinematics.

This book gives a thorough grounding in kinematics, permitting such subjects as path control, dynamics, and control to be properly addressed. A conceptually more important result is that the promise of fully automated systems of manipulators can be realized. For example, the techniques developed in this book permit a trajectory to be planned at a remote terminal and downloaded for real-time execution to a robot controller.

The kinematic methodology developed is called the point-plane method. The first three chapters of this book lay the foundation for this method. The material is developed logically, from the relationships between points, lines, and planes, through the algebra of movement in Cartesian space, to the classification of manipulators and their capabilities. Six kinematically simple robotic joints are identified and shown to model all the joints in common commercial robots.

The capabilities of a robot can be determined from its plane(s) of motion alone. For example, the three degree of freedom (3 dof) revolute, cylin-

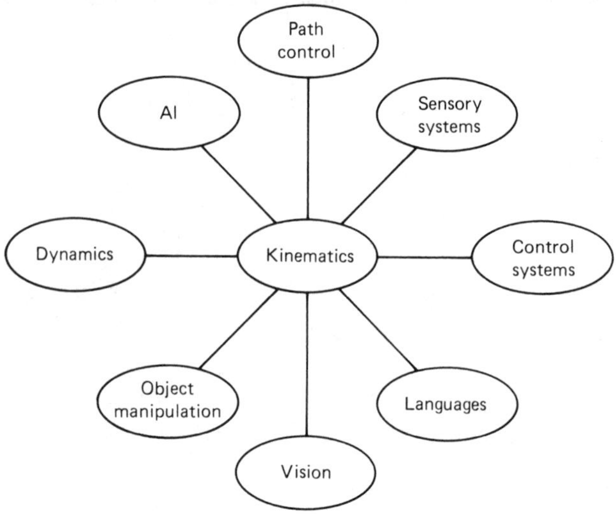

drical, and polar robots can be described by plane of motion $\overline{\mathbf{p}}^2$, and this single entity defines the capability of these three types of robot. The motion of all manipulators can be presented geometrically without ambiguity in two dimensions, and all calculations are performed in the plane(s) of motion.

The kinematic equations, derived for general orientations, are canonical, ensuring the best possible chance at real-time control. The methods for obtaining inverse kinematic solutions are simple and more powerful than traditional methods. Trajectory planning emphasizes polynomial methods, particularly Bezier function techniques, for effective velocity control.

The fact that the point-plane method may appear novel to some does not mean that this work is advanced. The work is intended as an introductory text and is a study of robotics starting at the basics. The book fulfills two functions: It serves as a reference book for working engineers involved in automation and as a textbook. The text is for swing students: seniors or first-year graduate students in engineering, physics, mathematics, or computer science; it has been used successfully as such for four years. The reader is assumed to have a solid background in calculus and college-level physics and familiarity with matrix notation and algebra.

HOW TO USE THE TEXT

The point-plane method is different to standard notations, and a primer on how to use the book is in order. Remembering formulae is not considered important, but understanding concepts and knowing where to find formulae is all-important; this author gives open book examinations. Whether the reader is a student taking a college course, a working engineer with a need to analyze some robotic system, or a professional seeking an understanding of robotics, the following sequence is recommended.

Read Chapter 1 until comfortable with the concepts of points and planes as 4-tuples, vectors as 3 or 4-tuples, lines, and the relationships between these entities. Read carefully Section 1.2 of Chapter 2; the rest of the chapter can be skipped until an understanding of graphics is needed. Read Chapter 3 in its entirety, remembering the Greek symbols and geometric symbols for the kinematically simple joints. Understand the geometry of each joint and the spatial meaning of the fold lines. The reader is encouraged to draw the spatial positioning of the joints and links on a plain sheet of paper and demonstrate the true spatial positioning with the fold lines.

Read all of Chapter 4 and remember that joints λ, ψ, ϕ and $\tilde{\phi}$ produce independent planes $\bar{\mathbf{p}}^0$, whereas joints γ and $\tilde{\psi}$ produce dependent planes $\mathbf{p}^1$. Understand the main robot types (placeable, directable and orientable) and mark the page containing Table 1 showing the planes of motion of the various types of robot. Feel comfortable with the meaning of position space.

Chapter 5 is heavy on equations and light on concept, and most of the chapter should be used for reference. The discussion on working envelopes can be omitted at a first reading. Chapter 6 is also heavy on equations, but also presents geometric inverse kinematic solutions; such solutions should be worked and understood by the reader. Appendix C is reference material of interest to the working engineer, and should be considered an appendix of Chapter 6.

The polynomial interpolation material in Chapter 7 is available elsewhere, but some of the Bezier curve applications may be less familiar. The reader is advised to read the complete chapter and remember what is in the various sections without committing their particulars to memory. The reader can refer back to Chapter 7 when studying Sections III, V and VI of Chapter 8. All of Chapter 8 is important.

The reader interested in robot dynamics will find Chapter 9 to be both satisfying and unsatisfying. The satisfaction will come in finding two explicit equations that determine the joint torques for any robot configuration. The dissatisfaction comes when the reader realizes that torque space task planning, the mapping from torque space to joint space or position space, is open or iterative. Chapter 10 discusses robotic drive systems and their control, and the reader may find the worked examples.

Chapters 1 through 6 provide a comprehensive and complete presentation of the kinematics of common robots. The ability to obtain the inverse kinematic solution for almost every commercial robot is an important milestone. Some may consider the material up to and including Chapter 6 a complete course, and the later chapters a follow up course.

ONE-SEMESTER ROBOTICS COURSE

Three-credit-hour senior level undergraduate or first year graduate level course that can be offered to electrical or mechanical engineers, computer science, mathematics or physics students.

Prerequisites: 1 year calculus. One year of college physics. Familiarity with matrix algebra. Some facility in transform theory is useful.

Goals: This course is intended to give students a basic grounding in the mathematics of robotic control, enabling them to analyze, design and remotely control the robotic workplace.

Week 1: The homogeneous representation. Relations between vectors, points, lines and planes in homogeneous representation.

Week 2: More work in the relations involving planes. Rotation of a vector about a point in a plane.

Week 3: Common robotic joints. Robotic classification. Degrees of freedom.

Week 4: Mathematical classification of robots, and some simple three degree of freedom robots. Fold lines. Planes of motion.

Week 5: Capability of manipulators in terms of the planes of motion. Position space.

Week 6: Base orientation. Forward kinematic solutions.

Week 7: Working envelopes. Inverse kinematic solutions.

Week 8: Common commercial robots, their characteristics and applications. Midterm test during the eighth week of material up to and including week #7.

Week 9: Parametric Description of Curves. Bezier curves.

Week 10: Slew solutions. Straight line and polynomial trajectories.

Week 11: Spline curves, velocity control and the Bezier function.

Week 12: Dynamics of loaded kinematic chains.

Week 13: Dynamics of common robots.

Week 14: Actuators and their control.

Week 15: Task planning and control. Splined Bezier paths.

Week 16: Final exams.

ACKNOWLEDGMENTS

The manuscript was prepared on a word processor, and typographic and all other errors are mine. Although most perspectives are mine, it could not have been produced without help. First, my wife has put up with a lot during its preparation and has permitted notes, documents, printouts and ill humor to spread throughout the house. For about three years my wife, daughter and son have had a part-time husband and father.

Many classes of students have assisted in this work. Typos were found by students with the impetus of a point for every error found. Conceptual errors were awarded 5 points, but only one student collected; I hope it is the only one. When students at the University of Lowell do not understand some material, they let you know. One class vehemently protested during a class on robotic movement, telling me I was wrong. In an attempt to explain the motion the concept of fold lines was developed.

I owe a great debt to my colleagues in the Electrical Engineering Department at the University of Lowell and the IEEE Robotics Chapter in Boston. They have given support and encouragement when needed. Several must be named. Ross Holmstrom hired me from the business world in 1982 and was instrumental in my getting tenure in 1984. Ron Brunelle has a sense of perspective and a "How can I help" attitude that smooths the roughest road. Bob Lemieux is a most generous and thoughtful person. Dave Wade is constant, as a friend should be.

Nomenclature

GEOMETRIC ELEMENTS

$\alpha, \beta, \gamma, \eta, \mu, \nu, \rho$	Vectors
β_1, η_y, ν_z	Elements of vectors
$\mathbf{u}, \mathbf{v}, \mathbf{w}$	Points
$x, y, z, x_3, z_1, y_i, u_z, v_x, w_y$	Elements of points
$\mathbf{p}, \mathbf{q}, \mathbf{p}_i$	Planes
$\bar{\mathbf{p}}, \bar{\mathbf{q}}_j, \bar{\mathbf{p}}_2^1$	Independent planes
a through h, b_3, d_i	Elements of planes
J_i	Length of link i
$\boldsymbol{\mu} \times \boldsymbol{\eta}$	Cross product of $\boldsymbol{\mu}$ and $\boldsymbol{\eta}$
$\boldsymbol{\mu} \cdot \boldsymbol{\eta}$	Dot product of $\boldsymbol{\mu}$ and $\boldsymbol{\eta}$
$\lvert\boldsymbol{\nu}\rvert$	Magnitude of the vector $\boldsymbol{\nu}$
ζ	Vector defining position space
ξ	Vector defining joint space

OBJECTS

C $4 \times n$ matrix, where the columns comprise the n corner points of an object

S	$m \times 4$ matrix, where the rows comprise the m planes defining the surfaces of the object

MATRICES

A through Z	Matrices
$\mathbf{w}^{\#}$, $\boldsymbol{\nu}^{\#}$, $B^{\#}$	Transposition of rows and columns of $\mathbf{w}$, $\boldsymbol{\nu}$ and B
T	Translational matrix (nonsingular)
R	Rotational matrix (nonsingular)
G	Scaling matrix (nonsingular)
L	Three-dimensional imaging (nonsingular)

JOINTS

θ	Pitch joint
λ	Roll joint
γ	Yaw joint
ϕ	Forward cylindrical joint
$\tilde{\phi}$	Reverse cylindrical joint
ψ	Forward crank joint
$\tilde{\psi}$	Reverse crank joint
$\underline{\psi}$ or $\underline{\phi}$	Crank or cylindrical joint whose uplink has a negative length (it points the opposite way to normal)
Θ	Any revolute joint
σ	Sliding link

MISCELLANEOUS SYMBOLS

t, α, β	Parameters
$\mathbf{\Omega}$, $\mathbf{\Omega}_i$	Axis of rotation
τ_3	The torque at joint three
or	
τ_1	The arithmetic time to perform multiplication or division
τ_2	The arithmetic time to perform addition or subtraction
τ_3	The arithmetic time to perform functional calculations such as exponentiation or trigonometric evaluation.

SYMBOLS FOR ROBOTIC JOINTS

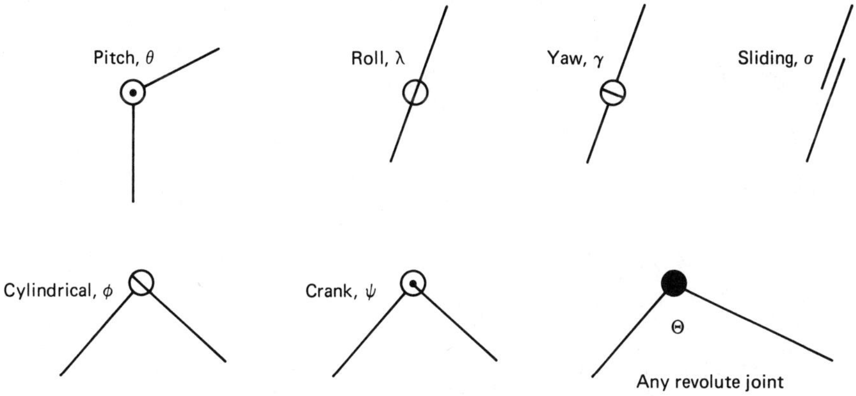

Useful Trigonometric Formulas

$\cos^2\theta + \sin^2\theta = 1 \qquad \cos^2\theta = \frac{1}{2}(1 + \cos 2\theta) \qquad \sin^2\theta = \frac{1}{2}(1 - \cos 2\theta)$

$\sin 2\theta = 2 \sin\theta \cos\theta \qquad \cos 2\theta = 2\cos^2\theta - 1 = 1 - 2\sin^2\theta = \cos^2\theta - \sin^2\theta$

$\sin(\alpha \pm \beta) = \sin\alpha \cos\beta \pm \cos\alpha \sin\beta$

$\cos(\alpha \pm \beta) = \cos\alpha \cos\beta \mp \sin\alpha \sin\beta$

$\cos\theta + \sin\theta = \sqrt{2}\sin(45^\circ + \theta) = \sqrt{2}\cos(45^\circ - \theta)$

$\cos\theta - \sin\theta = \sqrt{2}\cos(45^\circ + \theta) = \sqrt{2}\sin(45^\circ - \theta)$

$a\sin\theta + b\cos\theta = a\sqrt{1 + \frac{b^2}{a^2}}\sin\left(\theta + \tan^{-1}\frac{b}{a}\right)$, which is valid for all $a \neq 0$

or

$a\sin\theta + b\cos\theta = b\sqrt{1 + \frac{a^2}{b^2}}\cos\left(\theta - \tan^{-1}\frac{a}{b}\right)$, which is valid for all $b \neq 0$

$\sin(\alpha + \beta)\sin(\alpha - \beta) = \cos^2\beta - \cos^2\alpha = \sin^2\alpha - \sin^2\beta$

$\cos(\alpha + \beta)\cos(\alpha - \beta) = \cos^2\beta - \sin^2\alpha$

$$\sin\alpha\sin\beta = \frac{1}{2}\{\cos(\alpha - \beta) - \cos(\alpha + \beta)\}$$

$$\cos\alpha\cos\beta = \frac{1}{2}\{\cos(\alpha - \beta) + \cos(\alpha + \beta)\}$$

$$\sin\alpha\cos\beta = \frac{1}{2}\{\sin(\alpha - \beta) + \sin(\alpha + \beta)\}$$

$$\cos\alpha\sin\beta = \frac{1}{2}\{\sin(\alpha + \beta) - \sin(\alpha - \beta)\}$$

$$\sin(\alpha + \beta)\sin\beta = \frac{1}{2}\{\cos\alpha - \cos(\alpha + 2\beta)\}$$

$$\cos(\alpha + \beta)\cos\beta = \frac{1}{2}\{\cos\alpha + \cos(\alpha + 2\beta)\}$$

$$\sin(\alpha + \beta)\cos\beta = \frac{1}{2}\{\sin\alpha + \sin(\alpha + 2\beta)\}$$

$$\cos(\alpha + \beta)\sin\beta = \frac{1}{2}\{\sin(\alpha + 2\beta) - \sin\alpha\}$$

$$\sin\theta = \frac{2\tan\theta/2}{1 + \tan^2\theta/2} \qquad \cos\theta = \frac{1 - \tan^2\theta/2}{1 + \tan^2\theta/2}$$

$$\sin\theta = \pm\sqrt{\frac{1 - \cos 2\theta}{2}} \qquad \cos\theta = \pm\sqrt{\frac{1 + \cos 2\theta}{2}}$$

Given a triangle with angles θ_a, θ_b, θ_c and length of opposite sides a, b, and c, $\theta_a + \theta_b + \theta_c = 180°$. If $\theta_c = 90°$, $a^2 + b^2 = c^2$. For any angles, the following rules hold:

Law of sines $\qquad \dfrac{a}{\sin\theta_a} = \dfrac{b}{\sin\theta_b} = \dfrac{c}{\sin\theta_c}$

Law of cosines $\qquad a^2 = b^2 + c^2 - 2bc\cos\theta_a$

$a = b\cos\theta_c + c\cos\theta_b$

Applied Robotic Analysis

CHAPTER ONE

Relationship of Vectors, Points, Lines, Planes, Spheres, and Circles

This chapter provides the basis for position in Cartesian space and derives the interrelationships among the entities used to describe position and orientation in space. A full understanding of position is required before movement in general can be considered and before manipulator position and movement in particular can be described. A prerequisite for reading this chapter is a thorough understanding of vector notation. The algebraic and geometric meanings of the scalar or dot product and the vector or cross product must be fully understood. Standard linear algebraic notation is used throughout.

A plane is defined in homogeneous representation as $\mathbf{p} = [a \quad b \quad c \quad d]$, and its position in space is fixed. Most roboticists use planes sparingly, instead relying on a system of frames (a frame is a set of three mutually orthogonal unit vectors). It will be shown that the extensive use of planes in this work leads to simpler methods of defining rotation (Chapter 2) and to simpler specifications for robotic joints (Chapter 3). The extensive use of planes permits the mathematics of imaging (Chapter 2) of robots and objects to be greatly simplified.

1.1 THE HOMOGENEOUS REPRESENTATION

A vector $\boldsymbol{\nu} = [\nu_1 \quad \nu_2 \quad \nu_3]^{\#}$ is a 3-tuple (an ordered set of three quantities) and has direction and length but not location; we use the superscript # to denote matrix transposition. To define the position of a point in space we could use the 3-tuple $[x \quad y \quad z]^{\#}$; since there is no distinction between the 3-tuples, it is possible to confuse vectors and points. In homogeneous representation form, a point is defined by four quantities $[x \quad y \quad z \quad 1]^{\#}$ [Paul 1981b]. A vector can be defined by a 3-tuple or in homogeneous representation form as the 4-tuple $[v_1 \quad v_2 \quad v_3 \quad 0]^{\#}$, where the fourth quantity is zero. The homogeneous form is essential in treating points in their relationship to planes.

In conventional vector analysis, a plane within which two vectors lie is defined as the cross product of these two vectors: The vector $[a \quad b \quad c]^{\#}$ formed by this cross product is orthogonal to the surface of the plane. A vector has no position in Cartesian space, and neither does a plane in this definition. However, the homogeneous representation of a plane is $\mathbf{p} = [a \quad b \quad c \quad d]$, where d provides position in Cartesian space. When using $\mathbf{p}$ in rotations and robotic joints, we restrict vector $[a \quad b \quad c]^{\#}$ to be a unit vector. Thus the forms we use are

$$\begin{bmatrix} v_1 \\ v_2 \\ v_3 \end{bmatrix} \text{ or } \begin{bmatrix} v_1 \\ v_2 \\ v_3 \\ 0 \end{bmatrix} \text{ for vectors; } \quad \begin{bmatrix} x \\ y \\ z \\ 1 \end{bmatrix} \text{ for points;}$$

$$[a \quad b \quad c \quad d] \text{ for planes, } \qquad \text{where } a^2 + b^2 + c^2 = 1$$

A transformation that uses homogeneous representation form is called a *homogeneous transformation*. The homogeneous transformation has the ability to provide for every kind of transformation (translation, rotation, scaling, and imaging) within a single matrix [Paul 1981b], although this is sometimes inconvenient and is avoided. This is seen in Chapter 2.

1.2 VECTORS AND POINTS

A vector or a point in three-dimensional space can be written as

$$\mathbf{u} = \begin{bmatrix} x \\ y \\ z \\ s \end{bmatrix} \equiv \begin{bmatrix} \alpha x \\ \alpha y \\ \alpha z \\ \alpha s \end{bmatrix} \equiv - \begin{bmatrix} \alpha x \\ \alpha y \\ \alpha z \\ \alpha s \end{bmatrix}, \qquad \text{where } -\infty \leq s \leq \infty$$

The distance of a point from the origin is $\sqrt{x^2 + y^2 + z^2}/s$, so we see $[x \quad y \quad z \quad s/2]^{\#}$ is twice the distance from the origin as $[x \quad y \quad z \quad s]^{\#}$ but has the same direction with respect to the origin: if $s = \infty$, the point is at the origin. We have two problems with this notation: First, if $s = 0$, the distance from the origin becomes unbounded; second, there is no way to distinguish between points and vectors. The problem is resolved without losing generality by setting $s = 1$ for points and $s = 0$ for vectors.

We use lowercase Greek symbols for vectors and sometimes reduce a vector to a 3-tuple. For example, we write

$$\boldsymbol{\nu} = \begin{bmatrix} \nu_1 \\ \nu_2 \\ \nu_3 \\ 0 \end{bmatrix} \quad \text{or} \quad \boldsymbol{\nu} = \begin{bmatrix} \nu_1 \\ \nu_2 \\ \nu_3 \end{bmatrix}$$

and use either form, depending on the situation. The 3-tuple is always used when forming dot or cross products, but the 4-tuple is always used with the homogeneous representation.

The distance between two points

$$\begin{bmatrix} x_1 \\ y_1 \\ z_1 \\ 1 \end{bmatrix} \quad \text{and} \quad \begin{bmatrix} x_2 \\ y_2 \\ z_2 \\ 1 \end{bmatrix}$$

is

$$D = \sqrt{(x_1 - x_2)^2 + (y_1 - y_2)^2 + (z_1 - z_2)^2} \tag{1.1}$$

The vector joining the same two points is

$$\mathbf{v} = \begin{bmatrix} x_1 - x_2 \\ y_1 - y_2 \\ z_1 - z_2 \end{bmatrix}$$

and the length of this vector is the same as D in equation (1.1). The addition of two or more vectors produces another vector, and the addition of a vector to a point produces another point. However, the addition of two points produces a nonstandard form for a point, which is the average of the two points—that is,

$$\mathbf{u} + \mathbf{v} = \begin{bmatrix} u_x \\ u_y \\ u_z \\ 1 \end{bmatrix} + \begin{bmatrix} v_x \\ v_y \\ v_z \\ 1 \end{bmatrix} = \begin{bmatrix} u_x + v_x \\ u_y + v_y \\ u_z + v_z \\ 2 \end{bmatrix} \equiv \begin{bmatrix} \dfrac{u_x + v_x}{2} \\ \dfrac{u_y + v_y}{2} \\ \dfrac{u_z + v_z}{2} \\ 1 \end{bmatrix}$$

1.3 LINES

A straight line in three-dimensional space can be specified by two separate points, a point and a vector, a point and the partial derivatives of y and z with respect to x [Parkin 1986], and so on. A convenient way to define a line is parametrically as the addition of a point on the line and a vector whose length depends on some parameter and whose direction is the same as the line—that is,

$$\mathbf{u}_1 + \alpha(\mathbf{u}_2 - \mathbf{u}_1) = \begin{bmatrix} x_1 \\ y_1 \\ z_1 \\ 1 \end{bmatrix} + \alpha \begin{bmatrix} x_2 - x_1 \\ y_2 - y_1 \\ z_2 - z_1 \\ 0 \end{bmatrix} \equiv \begin{bmatrix} x_1 \\ y_1 \\ z_1 \\ 1 \end{bmatrix} + \beta \begin{bmatrix} 1 \\ \partial y/\partial x \\ \partial z/\partial x \\ 0 \end{bmatrix}$$

$$\equiv \begin{bmatrix} x \\ a_1x + b_1 \\ a_2x + b_2 \\ 1 \end{bmatrix}, \qquad \begin{matrix} -\infty \le \alpha \le \infty \\ -\infty \le \beta \le \infty \\ -\infty \le x \le \infty \end{matrix} \tag{1.2}$$

where $\mathbf{u}_1 = [x_1 \quad y_1 \quad z_1 \quad 1]^\#$ and $\mathbf{u}_2 = [x_2 \quad y_2 \quad z_2 \quad 1]^\#$ are points on the line and the partial derivatives are defined by

$$\frac{\partial y}{\partial x} = \frac{y_2 - y_1}{x_2 - x_1} = a_1 \quad \text{and} \quad \frac{\partial z}{\partial x} = \frac{z_2 - z_1}{x_2 - x_1} = a_2$$

$$\beta = \alpha(x_2 - x_1), \qquad b_1 = y \quad \text{at} \quad x = 0, \quad \text{and} \quad b_2 = z \quad \text{at} \quad x = 0$$

Often the equivalence of two lines is not obvious and should be proven or disproven by testing that both lines pass through the same two points. For example, we can show that the two lines

$$\begin{bmatrix} 3 \\ 0 \\ 0 \\ 1 \end{bmatrix} + \alpha \begin{bmatrix} -3 \\ \frac{3}{2} \\ 0 \\ 0 \end{bmatrix} \quad \text{and} \quad \begin{bmatrix} 0 \\ \frac{3}{2} \\ 0 \\ 1 \end{bmatrix} + \beta \begin{bmatrix} -3 \\ \frac{3}{2} \\ 0 \\ 0 \end{bmatrix}$$

are equivalent and points on the lines are the same if $\beta = \alpha - 1$.

1.3.1 The Intersection of Two Lines

The intersection of two lines $y = a_1x + b_1$ and $y = a_2x + b_2$ in two-dimensional space occurs at

$$x = \frac{b_2 - b_1}{a_1 - a_2} \tag{1.3}$$

Provided $a_1 \neq a_2$, the lines will always intersect. In three-dimensional space it is unlikely that two lines will intersect due to rounding or experimental error. Suppose two lines are defined by $y = a_1x + b_1$, $z = a_2x + b_2$ and $y = c_1x + d_1$, $z = c_2x + d_2$. Then the intersection occurs at $x = (b_1 - d_1)/(c_1 - a_1)$ or $x = (b_2 - d_2)/(c_2 - a_2)$. If the coefficients were obtained without error, the solutions will be the same. In practice, errors will occur, so we use the average value

$$x = \frac{1}{2}\left[\frac{b_1 - d_1}{c_1 - a_1} + \frac{b_2 - d_2}{c_2 - a_2}\right] \tag{1.4}$$

The difference between the two values of x (or y or z) can be used as a measure of accuracy and to determine if the two lines are close enough to be considered to intersect. To find the actual minimum distance between the two lines, refer to Section 1.3.3.

1.3.2 The Minimum Distance of a Line from a Point

Given line $\begin{bmatrix} \alpha \\ a_1\alpha + b_1 \\ a_2\alpha + b_2 \\ 1 \end{bmatrix}$ and point $\begin{bmatrix} x \\ y \\ z \\ 1 \end{bmatrix}$, the distance from this point to a point on the line is D, where $D^2 = (\alpha - x)^2 + (a_1\alpha + b_1 - y)^2 + (a_2\alpha + b_2 - z)^2$. At a maximum or minimum,

$$\frac{\partial D^2}{\partial \alpha} = 2(\alpha - x) + 2a_1(a_1\alpha + b_1 - y) + 2a_2(a_2\alpha + b_2 - z) = 0$$

However, the maximum is not physically meaningful. Thus

$$\alpha = \frac{x - a_1(b_1 - y) - a_2(b_2 - z)}{1 + a_1^2 + a_2^2} \tag{1.5}$$

To find the minimum distance of a line from the origin we set $x = y = z = 0$, so

$$\alpha = -\frac{a_1b_1 + a_2b_2}{1 + a_1^2 + a_2^2} \tag{1.6}$$

Alternatively, if we express a line in terms of two points as $\mathbf{u}_1 + \alpha(\mathbf{u}_2 - \mathbf{u}_1)$, then distance D shown in Figure 1.1 is given by

$$D^2 = \{\alpha(x_2 - x_1) + x_1 - x\}^2 + \{\alpha(y_2 - y_1) + y_1 - y\}^2 + \{\alpha(z_2 - z_1) + z_1 - z\}^2$$

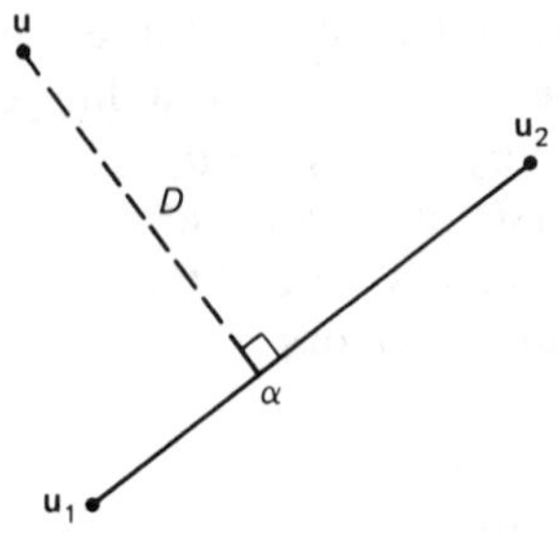

Figure 1.1 Normal from **u** to a line

and

$$\frac{\partial D^2}{\partial \alpha} = 2(x_2 - x_1)\{\alpha(x_2 - x_1) + x_1 - x\} + 2(y_2 - y_1)\{\alpha(y_2 - y_1) + y_1 - y\}$$
$$+ 2(z_2 - z_1)\{\alpha(z_2 - z_1) + z_1 - z\} = 0$$

at a maximum or a minimum. Thus

$$\alpha = -\frac{(x_2 - x_1)(x_1 - x) + (y_2 - y_1)(y_1 - y) + (z_2 - z_1)(z_1 - z)}{(x_2 - x_1)^2 + (y_2 - y_1)^2 + (z_2 - z_1)^2} \tag{1.7}$$

The location α along the line $\mathbf{u}_1 + \alpha(\mathbf{u}_2 - \mathbf{u}_1)$ at minimum distance from $\mathbf{u}$ is shown in Figure 1.1. When computations are to be carried out manually, it is usually easier to calculate D^2 in terms of the specific case—that is, form the specific equations and differentiate and solve rather than using equations (1.5), (1.6), and (1.7). These equations are programmed when a computer is used.

Alternatively, we can find this minimum distance from cross products. The cross product of the two vectors $(\mathbf{u} - \mathbf{u}_1)$ and $(\mathbf{u} - \mathbf{u}_2)$ produces a mutually orthogonal third vector, and the cross product of this third vector with $(\mathbf{u}_2 - \mathbf{u}_1)$ produces a vector in the direction of vector $(\bar{\mathbf{u}} - u)$, i.e.,

$$\bar{\mathbf{u}} - \mathbf{u} = \left\{\frac{(\mathbf{u} - \mathbf{u}_1) \times (\mathbf{u}_2 - \mathbf{u}_1)}{|\mathbf{u}_2 - \mathbf{u}_1|}\right\} \times \left\{\frac{\mathbf{u}_2 - \mathbf{u}_1}{|\mathbf{u}_2 - \mathbf{u}_1|}\right\}$$
$$= \frac{\{(\mathbf{u} - \mathbf{u}_1) \times (\mathbf{u}_2 - \mathbf{u}_1)\} \times (\mathbf{u}_2 - \mathbf{u}_1)}{|\mathbf{u}_2 - \mathbf{u}_1|^2}$$

and $D = |\bar{\mathbf{u}} - \mathbf{u}|$. This method is computationally more expensive than using equation (1.7).

1.3.3 The Minimum Distance Between Two Lines

Suppose two lines are given by

$$\begin{bmatrix} x_1 + \alpha(x_2 - x_1) \\ y_1 + \alpha(y_2 - y_1) \\ z_1 + \alpha(z_2 - z_1) \\ 1 \end{bmatrix} \quad \text{and} \quad \begin{bmatrix} x_3 + \beta(x_4 - x_3) \\ y_3 + \beta(y_4 - y_3) \\ z_3 + \beta(z_4 - z_3) \\ 1 \end{bmatrix}$$

The distance D between two points on these lines is given by

$$\begin{aligned} D^2 = {} & \{x_1 + \alpha(x_2 - x_1) - x_3 - \beta(x_4 - x_3)\}^2 \\ & + \{y_1 + \alpha(y_2 - y_1) - y_3 - \beta(y_4 - y_3)\}^2 \\ & + \{z_1 + \alpha(z_2 - z_1) - z_3 - \beta(z_4 - z_3)\}^2 \end{aligned}$$

For this distance to be a minimum, we require $\partial D^2/\partial\alpha = \partial D^2/\partial\beta = 0$, so

$$\begin{aligned} & \{x_1 + \alpha(x_2 - x_1) - x_3 - \beta(x_4 - x_3)\}(x_2 - x_1) \\ & \qquad + \{y_1 + \alpha(y_2 - y_1) - y_3 - \beta(y_4 - y_3)\}(y_2 - y_1) \\ & \qquad\qquad + \{z_1 + \alpha(z_2 - z_1) - z_3 - \beta(z_4 - z_3)\}(z_2 - z_1) = 0 \end{aligned}$$

and

$$\begin{aligned} & \{x_1 + \alpha(x_2 - x_1) - x_3 - \beta(x_4 - x_3)\}(x_4 - x_3) \\ & \qquad + \{y_1 + \alpha(y_2 - y_1) - y_3 - \beta(y_4 - y_3)\}(y_4 - y_3) \\ & \qquad\qquad + \{z_1 + \alpha(z_2 - z_1) - z_3 - \beta(z_4 - z_3)\}(z_4 - z_3) = 0 \end{aligned}$$

The preceding two equations can be rewritten as

$$\begin{aligned} a_1\alpha + a_2\beta &= a_3 \\ b_1\alpha + b_2\beta &= b_3 \end{aligned}$$

with solution

$$\alpha = \frac{a_3 b_2 - a_2 b_3}{a_1 b_2 - a_2 b_1} \quad \text{and} \quad \beta = \frac{a_1 b_3 - a_3 b_1}{a_1 b_2 - a_2 b_1} \tag{1.8}$$

where

$$\left.\begin{aligned} a_1 &= (x_2 - x_1)^2 + (y_2 - y_1)^2 + (z_2 - z_1)^2, \\ a_2 &= -(x_4 - x_3)(x_2 - x_1) - (y_4 - y_3)(y_2 - y_1) - (z_4 - z_3)(z_2 - z_1) \\ &= -b_1, \\ a_3 &= -(x_1 - x_3)(x_2 - x_1) - (y_1 - y_3)(y_2 - y_1) - (z_1 - z_3)(z_2 - z_1), \\ b_2 &= -(x_4 - x_3)^2 - (y_4 - y_3)^2 - (z_4 - z_3)^2, \\ b_3 &= -(x_1 - x_3)(x_4 - x_3) - (y_1 - y_3)(y_4 - y_3) - (z_1 - z_3)(z_4 - z_3) \end{aligned}\right\} \tag{1.9}$$

As in the prior section, it is usually more convenient to calculate D^2 and differentiate for the specific case considered; see the first example in Section 1.6.

1.4 PLANES

A plane is represented as a row vector—that is, $\mathbf{p} = [a \quad b \quad c \quad d]$. Planes $\mathbf{p}_1 = [a_1 \quad b_1 \quad c_1 \quad d_1]$ and $\mathbf{p}_2 = [a_2 \quad b_2 \quad c_2 \quad d_2]$ are parallel if $a_1 = \alpha a_2$, $b_1 = \alpha b_2$, $c_1 = \alpha c_2$ for any nonzero α. Planes $\mathbf{p}_1$ and $\mathbf{p}_2$ are the same if in addition $d_1 = \alpha d_2$. Although four quantities define a plane, only three and any three of them are independent [Parkin 1986]. For convenience in dealing with rotation of a vector (in Chapter 2) and robot joints (in Chapter 3), we usually normalize the length of the vector partially defining the plane to unity—that is, $a^2 + b^2 + c^2 = 1$—and then refer to the plane as normalized.

The unit vector $[a \quad b \quad c]^\#$ is orthogonal to the plane $[a \quad b \quad c \quad d]$, so the angle θ between two normalized planes $[a_1 \quad b_1 \quad c_1 \quad d_1]$ and $[a_2 \quad b_2 \quad c_2 \quad d_2]$ is given by the dot product

$$\cos\theta = \begin{bmatrix} a_1 \\ b_1 \\ c_1 \end{bmatrix} \cdot \begin{bmatrix} a_2 \\ b_2 \\ c_2 \end{bmatrix} = [a_1 \quad b_1 \quad c_1] \begin{bmatrix} a_2 \\ b_2 \\ c_2 \end{bmatrix} = a_1a_2 + b_1b_2 + c_1c_2 \quad (1.10)$$

The cross product is

$$\begin{bmatrix} a_3 \\ b_3 \\ c_3 \end{bmatrix} = \begin{bmatrix} a_2 \\ b_2 \\ c_2 \end{bmatrix} \times \begin{bmatrix} a_1 \\ b_1 \\ c_1 \end{bmatrix} = \begin{bmatrix} 0 & -c_2 & b_2 \\ c_2 & 0 & -a_2 \\ -b_2 & a_2 & 0 \end{bmatrix} \begin{bmatrix} a_1 \\ b_1 \\ c_1 \end{bmatrix}$$
$$= -\begin{bmatrix} 0 & -c_1 & b_1 \\ c_1 & 0 & -a_1 \\ -b_1 & a_1 & 0 \end{bmatrix} \begin{bmatrix} a_2 \\ b_2 \\ c_2 \end{bmatrix} = \begin{bmatrix} b_2c_1 - c_2b_1 \\ c_2a_1 - a_2c_1 \\ a_2b_1 - b_2a_1 \end{bmatrix} \quad (1.11)$$

and

$$\sin\theta = \sqrt{a_3^2 + b_3^2 + c_3^2} \quad (1.12)$$

1.4.1 Points and Planes

A point $\mathbf{u} = [x \quad y \quad z \quad 1]^\#$ lies in the plane $\mathbf{p} = [a \quad b \quad c \quad d]$ if $\mathbf{pu} = 0$—that is, $ax + by + cz + d = 0$. The plane will intersect the z-axis at $x = 0$, $y = 0$, for which $cz + d = 0$, or $z = -d/c$. Similarly, the plane will intersect the x- and y-axes at $-d/a$ and $-d/b$, respectively. If $d = 0$ the plane intersects the origin. Plane $[1 \quad 0 \quad 0 \quad 0]$ is the yz-plane passing through the origin.

The points where $\mathbf{p}$ intersects the x-axis (at $y = z = 0$), y-axis, and z-axis are

$$\begin{bmatrix} \frac{-d}{a} \\ 0 \\ 0 \\ 1 \end{bmatrix}, \quad \begin{bmatrix} 0 \\ \frac{-d}{b} \\ 0 \\ 1 \end{bmatrix}, \quad \text{and} \quad \begin{bmatrix} 0 \\ 0 \\ \frac{-d}{c} \\ 1 \end{bmatrix}$$

respectively.

1.4.2 Use of Three Points to Define a Plane

Three points can define a plane. If the three points are

$$\begin{bmatrix} x_1 \\ y_1 \\ z_1 \\ 1 \end{bmatrix}, \quad \begin{bmatrix} x_2 \\ y_2 \\ z_2 \\ 1 \end{bmatrix}, \quad \text{and} \quad \begin{bmatrix} x_3 \\ y_3 \\ z_3 \\ 1 \end{bmatrix}$$

then $\mathbf{p} = [a \quad b \quad c \quad d]$ can be found from

$$[a \quad b \quad c \quad d] \begin{bmatrix} x_1 & x_2 & x_3 \\ y_1 & y_2 & y_3 \\ z_1 & z_2 & z_3 \\ 1 & 1 & 1 \end{bmatrix} = [0 \quad 0 \quad 0]$$

We have only three unknowns in this equation, since a linear relationship links a, b, and c, and the matrix equation yields three algebraic equations, which can be solved. Suppose $\mathbf{p} = [a \quad b \quad c \quad 1]$; then

$$[a \quad b \quad c] \begin{bmatrix} x_1 & x_2 & x_3 \\ y_1 & y_2 & y_3 \\ z_1 & z_2 & z_3 \end{bmatrix} = [a \quad b \quad c]\mathbf{X} = -[1 \quad 1 \quad 1]$$

Provided

$$\begin{bmatrix} x_1 \\ y_1 \\ z_1 \end{bmatrix}, \quad \begin{bmatrix} x_2 \\ y_2 \\ z_2 \end{bmatrix}, \quad \text{and} \quad \begin{bmatrix} x_3 \\ y_3 \\ z_3 \end{bmatrix}$$

are linearly independent, matrix $\mathbf{X}$ is nonsingular. Thus $[a \quad b \quad c] = -[1 \quad 1 \quad 1]\mathbf{X}^{-1}$.

Example:

If

$$\mathbf{v}_1 = \begin{bmatrix} 1 \\ 0 \\ 0 \\ 1 \end{bmatrix}, \quad \mathbf{v}_2 = \begin{bmatrix} 0 \\ 1 \\ 0 \\ 1 \end{bmatrix}, \quad \text{and} \quad \mathbf{v}_3 = \begin{bmatrix} 0 \\ 0 \\ 1 \\ 1 \end{bmatrix}$$

then

$$[a \quad b \quad c \quad 1] \begin{bmatrix} 1 & 0 & 0 \\ 0 & 1 & 0 \\ 0 & 0 & 1 \\ 1 & 1 & 1 \end{bmatrix} = [0 \quad 0 \quad 0]$$

and

$$[a \quad b \quad c] \begin{bmatrix} 1 & 0 & 0 \\ 0 & 1 & 0 \\ 0 & 0 & 1 \end{bmatrix} = -[1 \quad 1 \quad 1]$$

so $[a \quad b \quad c] = [-1 \quad -1 \quad -1]$ and $\mathbf{p} = [a \quad b \quad c \quad 1]$.

In this simple example, matrix $\mathbf{X}$ is the identity matrix. Usually inverting $\mathbf{X}$ is computationally expensive and unnecessary. Furthermore, the method fails if the plane passes through the origin, and so $d = 0$. It is more efficient to find $[a \quad b \quad c]$ from the cross product of the two vectors formed as the difference between two different pairs of points, i.e.,

$$\begin{aligned} [a \quad b \quad c]^{\#} &= \begin{bmatrix} x_1 - x_2 \\ y_1 - y_2 \\ z_1 - z_2 \end{bmatrix} \times \begin{bmatrix} x_1 - x_3 \\ y_1 - y_3 \\ z_1 - z_3 \end{bmatrix} \\ &= \begin{bmatrix} (y_1 - y_2)(z_1 - z_3) - (z_1 - z_2)(y_1 - y_3) \\ (z_1 - z_2)(x_1 - x_3) - (x_1 - x_2)(z_1 - z_3) \\ (x_1 - x_2)(y_1 - y_3) - (y_1 - y_2)(x_1 - x_3) \end{bmatrix} \end{aligned} \tag{1.13}$$

where the three points are $[x_i \quad y_i \quad z_i \quad 1]^{\#}$, $i = 1, 2, 3$.

Geometrically, we notice that if the angle θ between the two vectors of the cross product is zero, then $a = b = c = 0$, but this is not permitted in the specification of a plane. This requirement means that the vectors in the cross product must be linearly independent; that is, the three points must not lie on a straight line in three-dimensional space if these three points are to be used to define a plane. The specification of the plane is completed by setting $d = -(ax_i + by_i + cz_i)$ for $i = 1, 2,$ or 3.

Repeating the prior example using the cross-product method of determination, we have

$$\begin{bmatrix} a \\ b \\ c \end{bmatrix} = \begin{bmatrix} 1 \\ -1 \\ 0 \end{bmatrix} \times \begin{bmatrix} 1 \\ 0 \\ -1 \end{bmatrix} = \begin{bmatrix} 1 - 0 \\ 0 + 1 \\ 0 + 1 \end{bmatrix} = \begin{bmatrix} 1 \\ 1 \\ 1 \end{bmatrix}$$

This result is the negative of the prior result, but is not incorrect. Reversing any two points in forming one of the two vectors changes the sign in the cross product and produces the prior result. Recall that a plane is partially specified by the vector $[a \quad b \quad c]^{\#}$ that is orthogonal to the surface of the plane, and this orthogonal vector can be directed from this surface in two opposite directions. Finally, the plane specification is completed using $\mathbf{v}_1$ to give $d = -[1 + 0 + 0] = -1$, so the plane is $\mathbf{p} = [1 \quad 1 \quad 1 \quad -1]$.

1.4.3 Use of a Vector and a Point to Define a Plane

In the prior section we saw that a family of parallel planes can be described by a unit vector $\boldsymbol{\mu} = [a \quad b \quad c]^{\#}$, which is orthogonal to the planes in this family. The family of planes is given by $\mathbf{p} = [a \quad b \quad c \quad d]$, where d is arbitrary. A specific d produces a unique plane. Suppose $\mathbf{p}$ is required to pass through point $[x \quad y \quad z \quad 1]^{\#}$; then, as shown previously, $d = -ax - by - cz$.

1.4.4 Use of a Point and a Line to Define a Plane

A point $\begin{bmatrix} x_1 \\ y_1 \\ z_1 \\ 1 \end{bmatrix}$ and a line $\begin{bmatrix} x \\ a_1x + b_1 \\ a_2x + b_2 \\ 1 \end{bmatrix}$ uniquely define a plane $\mathbf{p} = [a \quad b \quad c \quad d]$ provided the point is not on the line. We convert the problem to one of finding the plane from either of the following:

1. Three points by using the point given and two points on the line, as discussed in Section 1.4.2.
2. Vector $\begin{bmatrix} x_1 \\ y_1 - b_1 \\ z_1 - b_2 \end{bmatrix} \times \begin{bmatrix} 1 \\ a_1 \\ a_2 \end{bmatrix}$ and point $\begin{bmatrix} x_1 \\ y_1 \\ z_1 \\ 1 \end{bmatrix}$, as discussed in section 1.4.3, where $\begin{bmatrix} 1 \\ a_1 \\ a_2 \end{bmatrix}$ is the vector on the line defined by $x = 0$ and

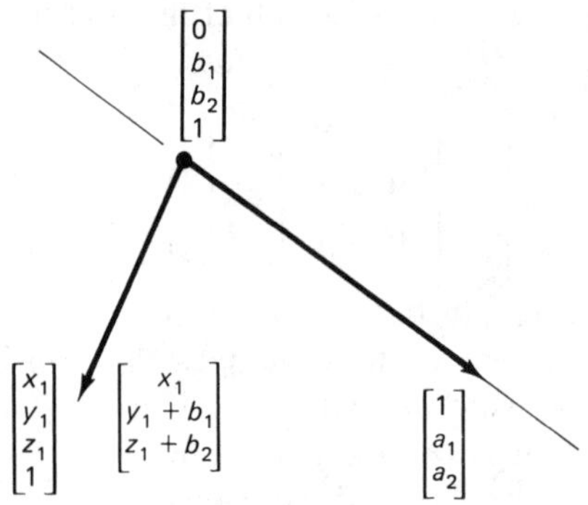

Figure 1.2 Formation of two vectors from a point and a line

$x = 1$ and $\begin{bmatrix} x_1 \\ y_1 - b_1 \\ z_1 - b_2 \end{bmatrix}$ is the vector from the point on the line at $x = 0$ to the point $\begin{bmatrix} x_1 \\ y_1 \\ z_1 \\ 1 \end{bmatrix}$. Thus the cross product produces a vector partially defining the required plane. The situation is shown in Figure 1.2.

1.4.5 Family of Planes About a Line

A line can define a family of planes (sometimes known as a *pencil of planes*) such that the line lies on all the planes [Gellert 1977]. Thus

$$[a \quad b \quad c \quad d]\begin{bmatrix} x \\ a_1x + b_1 \\ a_2x + b_2 \\ 1 \end{bmatrix} = ax + b(a_1x + b_1) + c(a_2x + b_2) + d = 0 \qquad \text{for all } x$$

At $x = 0$, $bb_1 + cb_2 + d = 0$; at $x = -b_1/a_1$, $-ab_1/a_1 + c(-a_2b_1/a_1 + b_2) + d = 0$; these equations can be rewritten as

$$b = \frac{-cb_2}{b_1} - \frac{d}{b_1} \quad \text{and} \quad a = c\left(-a_2 + \frac{b_2a_1}{b_1}\right) + \frac{da_1}{b_1} \tag{1.14}$$

and together with the constraint $a^2 + b^2 + c^2 = 1$, the normalized family of planes is written in terms of parameter d. When $b_1 = 0$, choose $x = -b_2/a_2$ or some other point as the second point.

1.4.6 Location of a Plane with Respect to the Origin

Suppose point $[x \quad y \quad z \quad 1]^\#$ is on plane $\mathbf{p} = [a \quad b \quad c \quad d]$ at a minimum distance from the origin. Then

$$[a \quad b \quad c \quad d]\begin{bmatrix} x \\ y \\ z \\ 1 \end{bmatrix} = 0$$

so

$$ax + by + cz = -d, \quad \text{or} \quad z = -\frac{1}{c}\{d + ax + by\}$$

The distance D from the origin is defined by

$$D^2 = x^2 + y^2 + z^2 = x^2 + y^2 + \frac{1}{c^2}(d + ax + by)^2$$

For this distance to be a minimum,

$$\frac{\partial}{\partial x} D^2 = 0 = 2x + \frac{2}{c^2}(d + ax + by)a \quad \text{and}$$

$$\frac{\partial}{\partial y} D^2 = 0 = 2y + \frac{2}{c^2}(d + ax + by)b$$

so $(c^2 + a^2)x + aby + ad = 0$ and $abx + (c^2 + b^2)y + bd = 0$. The pair of simultaneous equations can be written in matrix form as

$$\begin{bmatrix} a^2 + c^2 & ab \\ ab & b^2 + c^2 \end{bmatrix}\begin{bmatrix} x \\ y \end{bmatrix} = -\begin{bmatrix} da \\ db \end{bmatrix}$$

which can be solved as

$$\begin{bmatrix} x \\ y \end{bmatrix} = \frac{-1}{c^2}\begin{bmatrix} b^2 + c^2 & -ab \\ -ab & a^2 + c^2 \end{bmatrix}\begin{bmatrix} da \\ db \end{bmatrix} = -d\begin{bmatrix} a \\ b \end{bmatrix}$$

so $z = -dc$ and $D = \sqrt{x^2 + y^2 + z^2} = \sqrt{d^2a^2 + d^2b^2 + d^2c^2} = |\, d \,|$. This same result can be derived more simply by noting that the line from the origin to the point on the plane at a minimum distance from the origin must be normal to the plane and so has a direction the same as the vector $[a \quad b \quad c]^\#$ partially defining the plane. The line from the origin is

$$\begin{bmatrix} 0 \\ 0 \\ 0 \\ 1 \end{bmatrix} + \alpha\begin{bmatrix} a \\ b \\ c \\ 0 \end{bmatrix}$$

which intersects plane $[a \quad b \quad c \quad d]$ at $\alpha a^2 + \alpha b^2 + \alpha c^2 + d = 0$, or $\alpha = -d$. The distance of point $\begin{bmatrix} -da \\ -db \\ -dc \\ 1 \end{bmatrix}$ from the origin is $\sqrt{d^2a^2 + d^2b^2 + d^2c^2} = |\, d \,|$.

1.4.7 Minimum Distance of a Point from a Plane

Given a plane $\mathbf{p} = [a \quad b \quad c \quad d]$ and a point $\mathbf{u} = [x \quad y \quad z \quad 1]^{\#}$, the line orthogonal to the plane that passes through point $[x \quad y \quad z \quad 1]^{\#}$ is

$$\begin{bmatrix} x \\ y \\ z \\ 1 \end{bmatrix} + \alpha \begin{bmatrix} a \\ b \\ c \\ 0 \end{bmatrix}$$

and the point of intersection of this line with the normalized plane can be found by solving α in the equation.

$$[a \quad b \quad c \quad d]\left\{\begin{bmatrix} x \\ y \\ z \\ 1 \end{bmatrix} + \alpha \begin{bmatrix} a \\ b \\ c \\ 0 \end{bmatrix}\right\} = ax + by + cz + d + \alpha = 0$$

so $\alpha = -\mathbf{pu}$. The point is

$$\begin{bmatrix} x \\ y \\ z \\ 1 \end{bmatrix} - (ax + by + cz + d)\begin{bmatrix} a \\ b \\ c \\ 0 \end{bmatrix} \tag{1.15}$$

1.4.8 Intersection of a Line and a Plane

With a line defined parametrically in terms of two points $\mathbf{u}$ and $\mathbf{v}$ on the line as $\mathbf{u} + \alpha(\mathbf{v} - \mathbf{u})$, the point of intersection of the line with the plane $\mathbf{p}$ is given by

$$\mathbf{p}\{\mathbf{u} + \alpha(\mathbf{v} - \mathbf{u})\} = 0, \quad \text{so} \quad \alpha = -\frac{\mathbf{pu}}{\mathbf{p}(\mathbf{v} - \mathbf{u})} \tag{1.16}$$

1.4.9 Position of a Plane with Respect to a Point and the Origin

The line joining a point $\mathbf{u} = [x \quad y \quad z \quad 1]^{\#}$ to the origin is $(1 - \alpha)\mathbf{u} + \alpha[0 \quad 0 \quad 0 \quad 1]^{\#} = [(1 - \alpha)x \quad (1 - \alpha)y \quad (1 - \alpha)z \quad 1]$, as shown in Figure 1.3. This line intersects plane $\mathbf{p} = [a \quad b \quad c \quad d]$ at $(1 - \alpha)\mathbf{pu} + \alpha d = 0$. The location of the point of intersection of the plane on the line is given by

$$\alpha = \frac{\mathbf{pu}}{\mathbf{pu} - d} \tag{1.17}$$

If $\alpha < 0$, the plane is outside u with respect to the origin.
If $\alpha = 0$, the plane passes through $\mathbf{u}$ ($\mathbf{pu} = 0$).
If $0 < \alpha < 1$, the plane is between the point and the origin.
If $\alpha = 1$, the plane passes through the origin ($d = 0$).
If $\alpha > 1$, the point is on the other side of the origin with respect to $\mathbf{p}$.

If the vector $[x \quad y \quad z]^{\#}$ is orthogonal to $[a \quad b \quad c]^{\#}$, then $\alpha = \infty$ for all d except $d = 0$, at which α is indeterminate. Provided $[x \quad y \quad z]^{\#}$ is not orthogonal to $[a \quad b \quad c]^{\#}$, α will always be bounded.

Example:

Find the location of the following points with respect to the origin and the plane $\mathbf{p} = [1 \quad 2 \quad 2 \quad 1]$:

$$\begin{bmatrix} 1 \\ 2 \\ 1 \\ 1 \end{bmatrix}, \begin{bmatrix} -1 \\ -2 \\ -1 \\ 1 \end{bmatrix}, \begin{bmatrix} 3 \\ -1 \\ -1 \\ 1 \end{bmatrix}, \begin{bmatrix} 1 \\ -2 \\ 1 \\ 1 \end{bmatrix}, \text{ and } \begin{bmatrix} -1 \\ 0 \\ 1 \\ 1 \end{bmatrix}$$

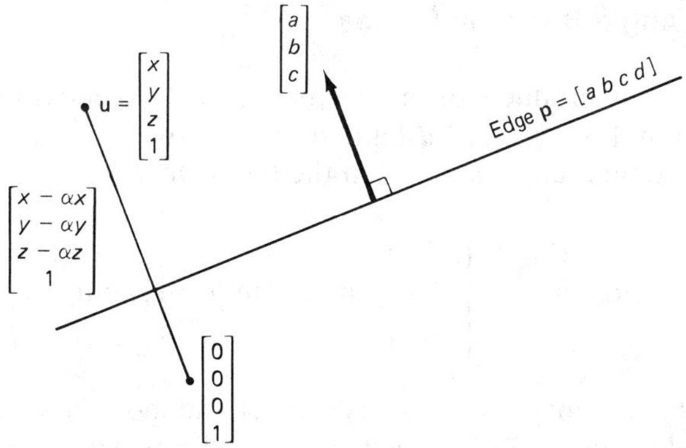

Figure 1.3 A point with respect to **p**

$\mathbf{pu} = [1 \quad 2 \quad 2 \quad 1]\begin{bmatrix} 1 \\ 2 \\ 1 \\ 1 \end{bmatrix} = 8$, and from equation (1.17) $\alpha = \frac{8}{7}$, so point $\begin{bmatrix} 1 \\ 2 \\ 1 \\ 1 \end{bmatrix}$ and the plane are on opposite sides of the origin.

$\mathbf{pu} = [1 \quad 2 \quad 2 \quad 1]\begin{bmatrix} -1 \\ -2 \\ -1 \\ 1 \end{bmatrix} = -6$, so $\alpha = \frac{6}{7}$; the plane is between the point and the origin.

$\mathbf{pu} = [1 \quad 2 \quad 2 \quad 1]\begin{bmatrix} 3 \\ -1 \\ -1 \\ 1 \end{bmatrix} = 0$ and point $\begin{bmatrix} 3 \\ -1 \\ -1 \\ 1 \end{bmatrix}$ is on the plane.

$\mathbf{pu} = [1 \quad 2 \quad 2 \quad 1]\begin{bmatrix} 1 \\ -2 \\ 1 \\ 1 \end{bmatrix} = 0$ and point $\begin{bmatrix} 1 \\ -2 \\ 1 \\ 1 \end{bmatrix}$ is on the plane.

$\mathbf{pu} = [1 \quad 2 \quad 2 \quad 1]\begin{bmatrix} -1 \\ 0 \\ 1 \\ 1 \end{bmatrix} = 2$, so $\alpha = \frac{2}{1}$; thus point $\begin{bmatrix} -1 \\ 0 \\ 1 \\ 1 \end{bmatrix}$ and the plane are on opposite sides of the origin.

1.4.10 Angle Between Planes

The dot and cross products of the vectors partially specifying two normalized planes $\mathbf{p}_1 = [a_1 \quad b_1 \quad c_1 \quad d_1]$ and $\mathbf{p}_2 = [a_2 \quad b_2 \quad c_2 \quad d_2]$ can be used to define the maximum angle θ between the two planes:

$$\cos\theta = [a_1 \quad b_1 \quad c_1]\begin{bmatrix} a_2 \\ b_2 \\ c_2 \end{bmatrix} \quad \text{and} \quad \sin\theta = \left\| \begin{bmatrix} b_1c_2 - c_1b_2 \\ c_1a_2 - a_1c_2 \\ a_1b_2 - b_1a_2 \end{bmatrix} \right\|$$

The first evaluation involving $\cos\theta$ is simpler and more efficient and should be the method of choice for finding θ. If $\theta = 0$, the planes are parallel. If $\theta = 90°$, the planes are orthogonal.

1.4.11 Line of Intersection of Two Planes

Two planes will intersect at a line provided these two planes are not parallel (as in Section 1.4.10 if $\theta = 0$). A point $[x \quad y \quad z \quad 1]^{\#}$ on the line of intersection satisfies

$$a_1x + b_1y + c_1z + d_1 = 0$$

and

$$a_2x + b_2y + c_2z + d_2 = 0$$

If the line of intersection has $x \equiv 0$, then the two equations can be combined as

$$y = \frac{c_2 - c_1}{b_1 - b_2}z + \frac{d_2 - d_1}{b_1 - b_2} \quad \text{or} \quad z = \frac{b_2 - b_1}{c_1 - c_2}y + \frac{d_2 - d_1}{c_2 - c_1}$$

and any two pairs of values for y and z that satisfy one of these equations defines the line of intersection with $x = 0$.

Provided x is not identically zero on the line of intersection, we can eliminate x from these two equations to give

$$(a_2b_1 - a_1b_2)y + (a_2c_1 - a_1c_2)z + a_2d_1 - a_1d_2 = 0$$

At least one of the two conditions $y = 0$ or $z = 0$ must lie on the line of intersection: At $y = 0$, $z = (a_1d_2 - a_2d_1)/(a_2c_1 - a_1c_2)$ and at $z = 0$, $y = (a_1d_2 - a_2d_1)/(a_2b_1 - a_1b_2)$. Given y and z, the value of x is found by substituting in one of the original two equations.

If $x = 0$ on the line of intersection, then at least one of the two conditions $y = 0$ or $z = 0$ apply:

$$\text{At } y = 0, \quad z = \frac{b_1d_2 - b_2d_1}{b_2c_1 - b_1c_2}; \qquad \text{at } z = 0, \quad y = \frac{c_1d_2 - c_2d_1}{b_1c_2 - b_2c_1}$$

The three separate points are

$$\begin{bmatrix} \dfrac{b_1d_2 - b_2d_1}{a_1b_2 - a_2b_1} \\ \dfrac{a_2d_1 - a_1d_2}{a_1b_2 - a_2b_1} \\ 0 \\ 1 \end{bmatrix}, \quad \begin{bmatrix} \dfrac{c_1d_2 - c_2d_1}{a_1c_2 - a_2c_1} \\ 0 \\ \dfrac{a_2d_1 - a_1d_2}{a_1c_2 - a_2c_1} \\ 1 \end{bmatrix}, \quad \text{and} \quad \begin{bmatrix} 0 \\ \dfrac{c_1d_2 - c_2d_1}{b_1c_2 - b_2c_1} \\ \dfrac{b_1d_2 - b_2d_1}{b_1c_2 - b_2c_1} \\ 1 \end{bmatrix} \qquad (1.20)$$

These points are the origin if $d_1 = d_2 = 0$, in which case we use the origin and the cross product discussed next to define the line.

The direction of line of intersection of the two planes $[a_1 \quad b_1 \quad c_1 \quad d_1]$ and $[a_2 \quad b_2 \quad c_2 \quad d_2]$ can be found from the cross product

$$\begin{bmatrix} a_1 \\ b_1 \\ c_1 \end{bmatrix} \times \begin{bmatrix} a_2 \\ b_2 \\ c_2 \end{bmatrix} = \begin{bmatrix} b_1c_2 - c_1b_2 \\ c_1a_2 - a_1c_2 \\ a_1b_2 - b_1a_2 \end{bmatrix}$$

since the cross product of two vectors is a vector orthogonal to both these vectors; the line of intersection must lie in both planes (and so must be orthogonal to the vectors partially specifying the planes). It is usually more efficient to use this direction with one point on the line to define the line rather than finding two points on the line.

1.4.12 Point of Intersection of Three Planes

For there to be a point of intersection of three planes $\mathbf{p}_i = [a_i \quad b_i \quad c_i \quad d_i]$, $i = 1, 2, 3$, we require vectors

$$\begin{bmatrix} a_1 \\ b_1 \\ c_1 \end{bmatrix} \times \begin{bmatrix} a_2 \\ b_2 \\ c_2 \end{bmatrix}, \quad \begin{bmatrix} a_1 \\ b_1 \\ c_1 \end{bmatrix} \times \begin{bmatrix} a_3 \\ b_3 \\ c_3 \end{bmatrix}, \quad \text{and} \quad \begin{bmatrix} a_2 \\ b_2 \\ c_2 \end{bmatrix} \times \begin{bmatrix} a_3 \\ b_3 \\ c_3 \end{bmatrix}$$

to be nonzero and not to be colinear. If these conditions are met, the point of intersection of the three planes will occur at point

$$\begin{bmatrix} x \\ y \\ z \\ 1 \end{bmatrix}, \quad \text{where} \quad \begin{bmatrix} a_1 & b_1 & c_1 & d_1 \\ a_2 & b_2 & c_2 & d_2 \\ a_3 & b_3 & c_3 & d_3 \end{bmatrix} \begin{bmatrix} x \\ y \\ z \\ 1 \end{bmatrix} = \begin{bmatrix} 0 \\ 0 \\ 0 \end{bmatrix},$$

$$\text{or} \quad \begin{bmatrix} x \\ y \\ z \end{bmatrix} = - \begin{bmatrix} a_1 & b_1 & c_1 \\ a_2 & b_2 & c_2 \\ a_3 & b_3 & c_3 \end{bmatrix}^{-1} \begin{bmatrix} d_1 \\ d_2 \\ d_3 \end{bmatrix}$$

This is not the method of choice, since it is computationally expensive to form the matrix inverse. Instead, we notice that expressions (1.21) give three points of intersection of $\mathbf{p}_1$ and $\mathbf{p}_2$, and two of these three points $\mathbf{u}_1$, $\mathbf{u}_2$, and $\mathbf{u}_3$ are used to define the line of intersection—that is, $\mathbf{u}_1 + \alpha(\mathbf{u}_2 - \mathbf{u}_1)$. The line will intersect $\mathbf{p}_3$ when $[a_3 \quad b_3 \quad c_3 \quad d_3][\mathbf{u}_1 + \alpha(\mathbf{u}_2 - \mathbf{u}_1)] = 0$, or

$$\alpha = \frac{a_3x_1 + b_3y_1 + c_3z_1 + d_3}{a_3(x_1 - x_2) + b_3(y_1 - y_2) + c_3(z_1 - z_2)} \tag{1.21}$$

where $\mathbf{u}_1 = [x_1 \quad y_1 \quad z_1 \quad 1]^{\#}$. The point of intersection can also be found using pairs of points $\mathbf{u}_1$ and $\mathbf{u}_3$, and $\mathbf{u}_2$ and $\mathbf{u}_3$.

1.4.13 Distance Between Parallel Planes

Given normalized parallel planes $\mathbf{p}_1 = [a \quad b \quad c \quad d_1]$ and $\mathbf{p}_2 = [a \quad b \quad c \quad d_2]$, a point on $\mathbf{p}_1$ is (except for the degenerate case $a = 0$) $u = [x \quad 0 \quad 0 \quad 1]^{\#}$, so from $\mathbf{p}_1 u = 0$ we have $x = -d_1/a$. The line between $\mathbf{p}_1$ and $\mathbf{p}_2$, normal to the two planes,

$$\begin{bmatrix} x \\ 0 \\ 0 \\ 1 \end{bmatrix} + \alpha \begin{bmatrix} a \\ b \\ c \\ 0 \end{bmatrix} = \begin{bmatrix} \dfrac{\alpha a - d_1}{a} \\ \alpha b \\ \alpha c \\ 1 \end{bmatrix}$$

If α is chosen so that it defines the point of intersection with $\mathbf{p}_2$, then

$$[a \quad b \quad c \quad d_2] \begin{bmatrix} \dfrac{\alpha a - d_1}{a} \\ \alpha b \\ \alpha c \\ 1 \end{bmatrix} = 0$$

which yields $\alpha = d_1 - d_2$. This result could have been deduced more simply using the fact that the minimum distances of $\mathbf{p}_1$ and $\mathbf{p}_2$ from the origin are d_1 and d_2, respectively, and since the planes are parallel, the same line connects the points on $\mathbf{p}_1$ and $\mathbf{p}_2$ nearest to the origin.

1.5 SPHERES AND CIRCLES

Given the location $\mathbf{u}$ of one end of a rigid link of length r, the motion of the other end of the link can be described as the surface of a sphere of radius r centered at u. The relationship between spheres and points and planes is developed in this section.

The surface of a sphere of radius r_0 centered at

$$\mathbf{u}_0 = \begin{bmatrix} x_0 \\ y_0 \\ z_0 \\ 1 \end{bmatrix}$$

is given by the set of points

$$\mathbf{u} = \begin{bmatrix} x \\ y \\ z \\ 1 \end{bmatrix}$$

such that $(x - x_0)^2 + (y - y_0)^2 + (z - z_0)^2 = r_0^2$, so

$$x^2 + y^2 + z^2 - 2x_0x - 2y_0y - 2z_0z = r_0^2 - x_0^2 - y_0^2 - z_0^2 \tag{1.22}$$

or in point-plane notation,

$$[x - 2x_0 \quad y - 2y_0 \quad z - 2z_0 \quad x_0^2 + y_0^2 + z_0^2 - r_0^2]\begin{bmatrix} x \\ y \\ z \\ 1 \end{bmatrix} = \mathbf{p}(\mathbf{u}_0, \mathbf{r}_0)\mathbf{u} = 0 \tag{1.23}$$

That is, the surface of the sphere is the set of points $\mathbf{u}$ on the parametric plane $\mathbf{p}(\mathbf{u}_0, \mathbf{r}_0)$. More simply, we say that $\mathbf{p}(\mathbf{u}_0, \mathbf{r}_0)$ defines the surface of the sphere.

1.5.1 The Intersection of Spheres

Given the spheres defined by surfaces $\mathbf{p}(\mathbf{u}_1, r_1)$ and $\mathbf{p}(\mathbf{u}_2, r_2)$, we notice from equation (1.22) that the quadratic terms in both equations are identical, so the difference between these two equations in point-plane form is

$$[2(x_2 - x_1) \quad 2(y_2 - y_1) \quad 2(z_2 - z_1) \quad r_2^2 - r_1^2 + x_1^2 + y_1^2 + z_1^2 - x_2^2 - y_2^2 - z_2^2]\begin{bmatrix} x \\ y \\ z \\ 1 \end{bmatrix} = 0 \tag{1.24}$$

That is, the intersection of two spheres lies in a plane. The intersection of this plane with either of the two spheres is a unique circle. The equations with symbols for center points become messy, and it is easier to treat specific examples; see Example 10 in Section 1.6.

The intersection of the surface of two spheres is a circle, and the intersection of this circle with the surface of another sphere produces two points. The worked example in Section 1.6 demonstrates the method of finding these two points better than a symbolic derivation.

1.5.2 Points, Lines, and Spheres

The distance of $\mathbf{u}$ from the center of sphere $\mathbf{p}(\mathbf{u}_0, \mathbf{r}_0)$ is $|\mathbf{u} - \mathbf{u}_0|$, so the minimum distance of $\mathbf{u}$ from the surface of the sphere is $|\mathbf{u} - \mathbf{u}_0| - r_0$. The line joining $\mathbf{u}$ and $\mathbf{u}_0$ intersects the sphere at

$$\mathbf{u}_0 \pm \frac{r_0(\mathbf{u} - \mathbf{u}_0)}{|\mathbf{u} - \mathbf{u}_0|} = 0 \tag{1.25}$$

The line $\mathbf{u}_1 + \alpha(\mathbf{u}_2 - \mathbf{u}_1)$ intersects $\mathbf{p}(\mathbf{u}_0, r_0)$ at the two values of α that satisfy

$$[x_1 + \alpha(x_2 - x_1) - 2x_0 \quad y_1 + \alpha(y_2 - y_1) - 2y_0 \quad z_1 + \alpha(z_2 - z_1) - 2z_0$$

$$x_0^2 + y_0^2 + z_0^2 - r_0^2] \begin{bmatrix} x_1 + \alpha(x_2 - x_1) \\ y_1 + \alpha(y_2 - y_1) \\ z_1 + \alpha(z_2 - z_1) \\ 1 \end{bmatrix} = 0 \quad (1.26)$$

1.5.3 Planes and Spheres

The intersection of plane $[a \quad b \quad c \quad d]$ with $\mathbf{p}(\mathbf{u}_0, r_0)$ is a circle, as discussed in Section 1.5.1. To find this circle of intersection, we can solve equation (1.23) with the restriction that $ax + by + cz + d = 0$. Alternatively, since the circle lies in $[a \quad b \quad c \quad d]$ with radius r_0 and center $\mathbf{u}_0$, we can use the rotation of a vector in a plane about a point method discussed in Chapter 2.

The intersection of two planes and a sphere is two points, since the intersection of two planes is a line and the intersection of a line and a sphere is two points.

1.5.4 Circles

A circle can be defined by a sphere with the same center and radius as the circle and the plane within which the circle lies. Thus, a circle can be defined by

$$(x - x_0)^2 + (y - y_0)^2 + (z - z_0)^2 = r_0^2$$

and

$$ax + by + cz + d = 0$$

Provided $a \neq 0$, we eliminate x:

$$\{(by + cz + d)/a + x_0\}^2 + (y - y_0)^2 + (z - z_0)^2 = r_0^2$$

which can be resorted as

$$(1 + b^2)y^2 + (1 + c^2)z^2 + \alpha yz + \beta = 0 \quad (1.26)$$

where α and β are constants. The specification of a circle by equation (1.26) has less application in the field of robotics than the circles defined as the rotation of a vector about a point in a plane, as discussed in Section 2.1.2.

The intersection of the circle defined by equation (1.26) with the plane $[a_1 \quad b_1 \quad c_1 \quad d_1]$ imposes the restriction $a_1x + b_1y + c_1z + d_1 = 0$ on equation (1.26), or

$$(ab_1 - a_1b)y + (ac_1 - a_1c)z + ad_1 - a_1d = 0$$

which combines with equation (1.26) to give

$$(1 + b^2)\{(a_1c - ac_1) + a_1d - ad_1\}^2 + (1 + c^2)z^2 + \alpha\{(a_1c - ac_1) + a_1d - ad_1\}z + \beta = 0 \quad (1.27)$$

which is a quadratic in z with two solutions.

1.6 EXAMPLES

1. Given points $\mathbf{u}_1 = \begin{bmatrix} -1 \\ 2 \\ 1 \\ 1 \end{bmatrix}$, $\mathbf{u}_2 = \begin{bmatrix} 1 \\ -2 \\ -1 \\ 1 \end{bmatrix}$, $\mathbf{u}_3 = \begin{bmatrix} 1 \\ 1 \\ 0 \\ 1 \end{bmatrix}$ and $\mathbf{u}_4 = \begin{bmatrix} 0 \\ -1 \\ 2 \\ 1 \end{bmatrix}$.

(a) Calculate the distance from each point to the origin.
(b) Find the distance between pairs of points.
(c) Assuming the first pair of points defines a line, determine the minimum distances from the third and fourth points to the line.
(d) Assuming the third and fourth points define a line, find the minimum distance between the two lines.
(e) Find the minimum distance from each line to the origin.

(a) For $\mathbf{u}_1$, $D^2 = x_1^2 + y_1^2 + z_1^2 = 1^2 + 2^2 + 1^2 = 6$, so $D = 2.449$.
For $\mathbf{u}_2$, $D^2 = 6$, so $D = 2.449$.
For $\mathbf{u}_3$, $D^2 = 2$, so $D = 1.414$.
For $\mathbf{u}_4$, $D^2 = 5$, so $D = 2.236$.

(b) The distance between $\mathbf{u}_1$ and $\mathbf{u}_2$ is $D^2 = (x_1 - x_2)^2 + (y_1 - y_2)^2 + (z_1 - z_2)^2 = 2^2 + 4^2 + 2^2 = 24$, so $D = 4.899$.
$\mathbf{u}_1$ to $\mathbf{u}_3$: $D^2 = 2^2 + 1^2 + 1^2 = 6$, so $D = 2.449$.
$\mathbf{u}_1$ to $\mathbf{u}_4$: $D^2 = 1^2 + 3^2 + 1^2 = 11$, so $D = 3.317$.
$\mathbf{u}_2$ to $\mathbf{u}_3$: $D^2 = 0 + 3^2 + 1^2 = 10$, so $D = 3.162$.
$\mathbf{u}_2$ to $\mathbf{u}_4$: $D^2 = 1^2 + 1^2 + 3^2 = 11$, so $D = 3.317$.
$\mathbf{u}_3$ to $\mathbf{u}_4$: $D^2 = 1^2 + 2^2 + 2^2 = 9$, so $D = 3$.

(c) The line between $\mathbf{u}_1$ and $\mathbf{u}_2$ is $\mathbf{u}_1 + \alpha(\mathbf{u}_2 - \mathbf{u}_1) = \begin{bmatrix} -1 + 2\alpha \\ 2 - 4\alpha \\ 1 - 2\alpha \\ 1 \end{bmatrix}$.

The distance between this line and $\mathbf{u}_3$ is

$$\begin{aligned} D^2 &= (x_1 + \alpha(x_2 - x_1) - x_3)^2 + (y_1 + \alpha(y_2 - y_1) - y_3)^2 + (z_1 + \alpha(z_2 - z_1) - z_3)^2 \\ &= (-1 + 2\alpha - 1)^2 + (2 - 4\alpha - 1)^2 + (1 - 2\alpha)^2 \\ &= 4(\alpha - 1)^2 + (1 - 4\alpha)^2 + (1 - 2\alpha)^2 = 24\alpha^2 - 20\alpha + 6 \end{aligned}$$

We differentiate this line with respect to α to find the minimum distance, so $\partial D^2/\partial\alpha = 48\alpha - 20 = 0$ at a max or min; thus $\alpha = 0.417$, so $D^2 = 1.833$ and $D = 1.354$. The distance between the line and $\mathbf{u}_4$ is

$$D^2 = (-1 + 2\alpha - 0)^2 + (2 - 4\alpha + 1)^2 + (1 - 2\alpha - 2)^2$$

$$= (2\alpha - 1)^2 + (3 - 4\alpha)^2 + (1 + 2\alpha)^2 = 24\alpha^2 - 24\alpha + 11$$

$\partial D^2/\partial\alpha = 48\alpha - 24 = 0$ at a max or min, giving $\alpha = 0.500$. Thus $D^2 = 5.000$ and $D = 2.236$.

(d) The line between $\mathbf{u}_3$ and $\mathbf{u}_4$ is $\begin{bmatrix} 1-\beta \\ 1-2\beta \\ 2\beta \\ 1 \end{bmatrix}$.

The distance between the two lines is given by the distance between two points on the lines:

$$D^2 = (-1 + 2\alpha - 1 + \beta)^2 + (2 - 4\alpha - 1 + 2\beta)^2 + (1 - 2\alpha - 2\beta)^2$$

$$= (-2 + 2\alpha + \beta)^2 + (1 - 4\alpha + 2\beta)^2 + (1 - 2\alpha - 2\beta)^2$$

The minimum distance is found by differentiating with respect to the two parameters α and β and equating the result to zero:

$$\frac{\partial D^2}{\partial\alpha} = 4(-2 + 2\alpha + \beta) - 8(1 - 4\alpha + 2\beta) - 4(1 - 2\alpha - 2\beta)$$

$$= -20 + 48\alpha - 4\beta = 0$$

$$\frac{\partial D^2}{\partial\beta} = 2(-2 + 2\alpha + \beta) + 4(1 - 4\alpha + 2\beta) - 4(1 - 2\alpha - 2\beta)$$

$$= -4 - 4\alpha + 18\beta = 0$$

We rewrite this pair of simultaneous equations as

$$12\alpha - \beta = 5$$

$$-\alpha + 4.5\beta = 1$$

which have solution $\alpha = 0.443$ and $\beta = 0.321$. Thus, the two points on the line at a minimum distance apart are

$$\begin{bmatrix} -0.114 \\ 0.228 \\ 0.114 \\ 1 \end{bmatrix} \text{ and } \begin{bmatrix} 0.679 \\ 0.358 \\ 0.642 \\ 1 \end{bmatrix}$$

and the minimum distance is given by $D^2 = 0.9245$, so $D = 0.9615$.

(e) The distance of the first line from the origin is given by

$$D^2 = (-1 + 2\alpha)^2 + (2 - 4\alpha)^2 + (1 - 2\alpha)^2$$

$$\frac{\partial D^2}{\partial\alpha} = 4(-1 + 2\alpha) - 8(2 - 4\alpha) - 4(1 - 2\alpha) = -24 + 48\alpha = 0$$

so $\alpha = \frac{1}{2}$ and $D = 0$.

Hence the line passes through the origin. The distance of the second line from the origin is given by

$$D^2 = (1 - \beta)^2 + (1 - 2\beta)^2 + (2\beta)^2$$

$$\frac{\partial D^2}{\partial \beta} = -2(1 - \beta) - 4(1 - 2\beta) + 4(2\beta) = -6 + 18\beta = 0$$

so $\beta = \frac{1}{3}$ and $D = 1$.

---

2. Find the point of intersection of the line passing through points $\begin{bmatrix} 1 \\ 1 \\ 1 \\ 1 \end{bmatrix}$ and $\begin{bmatrix} 1 \\ 2 \\ 3 \\ 1 \end{bmatrix}$ and the xy-plane, the xz-plane, and the yz-plane.

The line is $\begin{bmatrix} 1 \\ 1 \\ 1 \\ 1 \end{bmatrix} + \alpha \begin{bmatrix} 1 - 1 \\ 2 - 1 \\ 3 - 1 \\ 1 - 1 \end{bmatrix}$, and at the point of intersection $\mathbf{w}$ with $\mathbf{p} = [a \quad b \quad c \quad d]$, as given by equation (1.16),

$$\alpha = -\frac{a + b + c + d}{b + 2c}$$

For the xy-plane, $\mathbf{p} = [0 \quad 0 \quad 1 \quad 0]$ and $\alpha = -\frac{1}{2}$, so $\mathbf{w} = \begin{bmatrix} 1 \\ \frac{1}{2} \\ 0 \\ 1 \end{bmatrix}$.

For the xz-plane, $\mathbf{p} = [0 \quad 1 \quad 0 \quad 0]$, $\alpha = -1$, and $\mathbf{w} = \begin{bmatrix} 1 \\ 0 \\ -1 \\ 1 \end{bmatrix}$.

For the yz-plane, $\mathbf{p} = [1 \quad 0 \quad 0 \quad 0]$ and $\alpha = \infty$, which means that the line is parallel to the yz-plane and can never intersect it.

---

3. Find the minimum distance between the two lines $\begin{bmatrix} 1 + \alpha \\ -\alpha \\ 1 \\ 1 \end{bmatrix}$ and $\begin{bmatrix} 2 - \beta \\ 1 \\ \beta \\ 1 \end{bmatrix}$.

$D^2 = (1 + \alpha - 2 + \beta)^2 + (-\alpha - 1)^2 + (1 - \beta)^2$,
$\partial D^2/\partial \alpha = 2(-1 + \alpha + \beta) + 2(\alpha + 1) + 0 = 4\alpha + 2\beta = 0$ at a max or min, and
$\partial D^2/\partial \beta = 2(-1 + \alpha + \beta) + 0 + -2(1 - \beta) = -4 + 2\alpha + 4\beta = 0$ at a max or min.

$$2\alpha + \beta = 0$$

$$\alpha + 2\beta = 2$$

so $\alpha = -\frac{2}{3}$, $\beta = \frac{4}{3}$, and $D^2 = (-1 - \frac{2}{3} + \frac{4}{3})^2 + (\frac{2}{3} - 1)^2 + (1 - \frac{4}{3})^2 = 0.333$; hence $D = 0.577$.

---

4. A line passes through the origin and intersects $\mathbf{u}_3$ and a line on which points $\mathbf{u}_1$ and $\mathbf{u}_2$ lie. What restrictions must be placed on $\mathbf{u}_1$, $\mathbf{u}_2$, and $\mathbf{u}_3$ for this to happen?

The line between $\mathbf{u}_1$ and $\mathbf{u}_2$ is

$$\begin{bmatrix} x_1 + \alpha(x_2 - x_1) \\ y_1 + \alpha(y_2 - y_1) \\ z_1 + \alpha(z_2 - z_1) \\ 1 \end{bmatrix}$$

The line between $\mathbf{u}_3$ and a point on this line passes through the origin when

$$x_3 + \beta\{x_1 + \alpha(x_2 - x_1) - x_3\} = 0$$

$$y_3 + \beta\{y_1 + \alpha(y_2 - y_1) - y_3\} = 0$$

$$z_3 + \beta\{z_1 + \alpha(z_2 - z_1) - z_3\} = 0$$

so $\beta = -z_3/(z_1 + \alpha(z_2 - z_1) - z_3)$, which is eliminated from the first two equations as

$$x_3\{z_1 + \alpha(z_2 - z_1) - z_3\} - z_3\{x_1 + \alpha(x_2 - x_1) - x_3\} = 0$$

$$y_3\{z_1 + \alpha(z_2 - z_1) - z_3\} - z_3\{y_1 + \alpha(y_2 - y_1) - y_3\} = 0$$

That is,

$$\alpha\{x_3(z_2 - z_1) - z_3(x_2 - x_1)\} + x_3 z_1 - z_3 x_1 = 0$$

$$\alpha\{y_3(z_2 - z_1) - z_3(y_2 - y_1)\} + y_3 z_1 - z_3 y_1 = 0$$

Eliminating α gives

$$\{y_3 z_1 - z_3 y_1\}\{x_3(z_2 - z_1) - z_3(x_2 - x_1)\} - \{x_3 z_1 - z_3 x_1\}\{y_3(z_2 - z_1) - z_3(y_2 - y_1)\} = 0$$

which is a restriction on points $\mathbf{u}_1$, $\mathbf{u}_2$, and $\mathbf{u}_3$ in order for the required line to pass through the origin. That is, we can arbitrarily choose five of the six parameters in the preceding equation, and the equation determines the value of the sixth.

---

5. Find the minimum distance of point $[1 \ \ 0 \ \ 0 \ \ 1]^\#$ from plane $[1 \ \ 2 \ \ 3 \ \ 1]/\sqrt{14}$.

From Section 1.4.7, the point of intersection is given by

$$\begin{bmatrix} x \\ y \\ z \\ 1 \end{bmatrix} + \alpha \begin{bmatrix} a \\ b \\ c \\ 0 \end{bmatrix}$$

where $\alpha = -(ax + by + cz + d) = -2/\sqrt{14}$. Thus the point of intersection is

$$\begin{bmatrix} 1 - 2/14 \\ -4/14 \\ -6/14 \\ 1 \end{bmatrix} = \begin{bmatrix} 6/7 \\ -2/7 \\ -3/7 \\ 1 \end{bmatrix}$$

and the distance between the two points is

$$\sqrt{(1-\tfrac{6}{7})^2+(\tfrac{2}{7})^2+(\tfrac{3}{7})^2}=\sqrt{\tfrac{2}{7}}.$$

---

6. Find all the planes passing through all sets of three points of

$$\begin{bmatrix}1\\2\\1\\1\end{bmatrix},\quad \begin{bmatrix}1\\-1\\1\\1\end{bmatrix},\quad \begin{bmatrix}1\\0\\1\\1\end{bmatrix},\quad \text{and} \quad \begin{bmatrix}0\\1\\2\\1\end{bmatrix}$$

If the planes obtained for two separate cases are the same, indicate why. If the procedure fails to produce a valid plane, indicate why.

Numbering the points 1–4, the plane $[a \quad b \quad c \quad d]$ passing through each group of indicated points is as follows.

Points 1, 2, 3:

$$\begin{bmatrix}a\\b\\c\end{bmatrix}=\begin{bmatrix}0\\3\\0\end{bmatrix}\times\begin{bmatrix}0\\2\\0\end{bmatrix}=\begin{bmatrix}0\\0\\0\end{bmatrix}$$

which is not valid, since vector $[a \quad b \quad c]^{\#}$ must be nonzero. The problem is caused because the three points lie on the same line.

Points 1, 2, 4:

$$\begin{bmatrix}a\\b\\c\end{bmatrix}=\begin{bmatrix}0\\3\\0\end{bmatrix}\times\begin{bmatrix}1\\1\\-1\end{bmatrix}=\begin{bmatrix}-3\\0\\-3\end{bmatrix}$$

$d=-(-3+0-3)=6$, or $\mathbf{p}=[1 \quad 0 \quad 1 \quad -2]$.

Points 1, 3, 4:

$$\begin{bmatrix}a\\b\\c\end{bmatrix}=\begin{bmatrix}0\\2\\0\end{bmatrix}\times\begin{bmatrix}1\\1\\-1\end{bmatrix}=\begin{bmatrix}-2\\0\\-2\end{bmatrix}$$

$d=-(-2+0-2)=4$, or $\mathbf{p}=[1 \quad 0 \quad 1 \quad -2]$, which is the same as for points 1, 2, and 4. Recall points 1, 2, and 3 lie on the same line, so a plane formed using point 4 in association with points 1 and 2 is the same as a plane formed using point 4 in association with points 1 and 3.

Points 2, 3, 4:

$$\begin{bmatrix}a\\b\\c\end{bmatrix}=\begin{bmatrix}0\\-1\\0\end{bmatrix}\times\begin{bmatrix}1\\-2\\-1\end{bmatrix}=\begin{bmatrix}1\\0\\1\end{bmatrix}$$

$d=-(1+0+1)=-2$, or $\mathbf{p}=[1 \quad 0 \quad 1 \quad -2]$, which is the same plane as for points 1, 2, and 3 or 1, 3, and 4 for the reasons just given.

---

7. Find all the points of intersection for the five planes

$$\mathbf{p}_1 = [1 \quad 0 \quad -1 \quad 1]$$
$$\mathbf{p}_2 = [1 \quad 1 \quad 0 \quad 0]$$
$$\mathbf{p}_3 = [-1 \quad 0 \quad 1 \quad 0]$$
$$\mathbf{p}_4 = [1 \quad 1 \quad 0 \quad -1]$$
$$\mathbf{p}_5 = [-1 \quad -1 \quad 0 \quad 2]$$

The number of ways that five items can be taken three at a time is 10. Why are less than 10 points of intersection found?

The number of times that n items can be taken m at a time is given by the binomial function

$$\begin{bmatrix} n \\ m \end{bmatrix} = \frac{n!}{m!(n-m)!}$$

and for $n = 5$ and $m = 3$, $\begin{bmatrix} 5 \\ 3 \end{bmatrix} = \dfrac{5!}{3!(5-3)!} = 10$. We see that $\mathbf{p}_1$ and $\mathbf{p}_3$ are parallel planes; the cross product of the vectors partially defining the planes is zero. Similarly, planes $\mathbf{p}_2$, $\mathbf{p}_4$, and $\mathbf{p}_5$ are parallel. Thus, no point of intersection can be found for any three of the five planes.

---

8. Find the line defining the intersection of the two planes $[1 \quad 2 \quad 3 \quad 1]$ and $[2 \quad 1 \quad 1 \quad 1]$.

$[1 \quad 2 \quad 3 \quad 1]\begin{bmatrix} x \\ y \\ z \\ 1 \end{bmatrix} = 0$ and $[2 \quad 1 \quad 1 \quad 1]\begin{bmatrix} x \\ y \\ z \\ 1 \end{bmatrix}$, so

$$x + 2y + 3z = -1$$
$$2x + y + z = -1$$

Eliminating x gives $3y + 5z = -1$, and for the point on the line of intersection $y = 0$, we get $z = -\frac{1}{5}$ with a corresponding value of $x = -\frac{2}{5}$. Similarly, for $z = 0$, we get $y = -\frac{1}{3}$ and $x = -\frac{1}{3}$. The line of intersection is

$$\begin{bmatrix} -\frac{2}{5} \\ 0 \\ -\frac{1}{5} \\ 1 \end{bmatrix} + \alpha \begin{bmatrix} -\frac{1}{3} + \frac{2}{5} \\ -\frac{1}{3} + \frac{1}{5} \\ 0 - \frac{1}{5} \\ 1 - 1 \end{bmatrix} = \begin{bmatrix} -\frac{2}{5} \\ 0 \\ -\frac{1}{5} \\ 1 \end{bmatrix} + \alpha \begin{bmatrix} \frac{1}{15} \\ -\frac{2}{15} \\ -\frac{1}{5} \\ 0 \end{bmatrix}$$

Alternatively, the line of intersection will intersect another plane provided the angle between the vector specifying the direction of the line and the vector partially specifying the plane are not orthogonal. Thus, for three planes $\mathbf{p}_i = [a_i \quad b_i \quad c_i \quad d_i]$,

$i = 1, 2, 3$, to intersect at a point, we require

$$[a_1 \quad b_1 \quad c_1]\begin{bmatrix} a_2 \\ b_2 \\ c_2 \end{bmatrix} \neq \pm 1 \quad \text{and} \quad [a_3 \quad b_3 \quad c_3]\left\|\begin{bmatrix} b_1c_2 - c_1b_2 \\ c_1a_2 - a_1c_2 \\ a_1b_2 - b_1a_2 \end{bmatrix}\right\| \neq 0$$

Except for the degenerate case, the line of intersection of the two planes $\mathbf{p}_1 = [a_1 \quad b_1 \quad c_1 \quad d_1]$ and $\mathbf{p}_2 = [a_2 \quad b_2 \quad c_2 \quad d_2]$ must cross the xy-plane at $z = 0$; the degenerate case occurs when $\mathbf{p}_1$ or $\mathbf{p}_2$ is parallel to the xy-plane (so $a_1b_2 = a_2b_1$ and the xy-plane will not be crossed). If this occurs, choose the xz-plane or the yz-plane to find a point. At the xy-plane,

$$[a_1 \quad b_1 \quad c_1 \quad d_1]\begin{bmatrix} x_1 \\ y_1 \\ 0 \\ 1 \end{bmatrix} = 0 \quad \text{and} \quad [a_2 \quad b_2 \quad c_2 \quad d_2]\begin{bmatrix} x_1 \\ y_1 \\ 0 \\ 1 \end{bmatrix} = 0$$

so

$$\begin{aligned} a_1x_1 + b_1y_1 &= -d_1 \\ a_2x_1 + b_2y_1 &= -d_2 \end{aligned} \quad \text{or} \quad \begin{bmatrix} a_1 & b_1 \\ a_2 & b_2 \end{bmatrix}\begin{bmatrix} x_1 \\ y_1 \end{bmatrix} = -\begin{bmatrix} d_1 \\ d_2 \end{bmatrix}$$

and

$$\begin{bmatrix} x_1 \\ y_1 \end{bmatrix} = \frac{-1}{a_1b_2 - b_1a_2}\begin{bmatrix} b_2 & -b_1 \\ -a_2 & a_1 \end{bmatrix}\begin{bmatrix} d_1 \\ d_2 \end{bmatrix}$$

If the determinant $(a_1b_2 - b_1a_2)$ is zero, x_1 and y_1 cannot be found, since the line of intersection is parallel to the xy-plane. Similarly, the line must cross (except for degenerate cases) the xz-plane at point $[x_2 \quad 0 \quad z_2 \quad 1]^\#$ and the yz-plane at point $[0 \quad y_3 \quad z_3 \quad 1]^\#$, where

$$\begin{bmatrix} x_2 \\ z_2 \end{bmatrix} = \frac{-1}{a_1c_2 - c_1a_2}\begin{bmatrix} c_2 & -c_1 \\ -a_2 & a_1 \end{bmatrix}\begin{bmatrix} d_1 \\ d_2 \end{bmatrix}$$

$$\text{and} \quad \begin{bmatrix} y_3 \\ z_3 \end{bmatrix} = \frac{-1}{b_1c_2 - c_1b_2}\begin{bmatrix} c_2 & -c_1 \\ -b_2 & b_1 \end{bmatrix}\begin{bmatrix} d_1 \\ d_2 \end{bmatrix}$$

Thus, three points on the line of intersection are

$$\begin{bmatrix} -\frac{1}{3} \\ -\frac{1}{3} \\ 0 \\ 1 \end{bmatrix}, \quad \begin{bmatrix} -\frac{2}{5} \\ 0 \\ -\frac{1}{5} \\ 1 \end{bmatrix}, \quad \text{and} \quad \begin{bmatrix} 0 \\ -2 \\ 1 \\ 1 \end{bmatrix}$$

We see that these points are correct by showing that $\mathbf{pu} = 0$ in every case. Only two and any two of these can be used to define the line. The line is then defined parametrically as

$$\begin{bmatrix} -\frac{1}{3} - \frac{\alpha}{15} \\ -\frac{1}{3} + \frac{\alpha}{3} \\ -\frac{\alpha}{5} \\ 1 \end{bmatrix} \quad \text{or} \quad \begin{bmatrix} -\frac{1}{3}(1-\alpha) \\ -\frac{1}{3} - \frac{5\alpha}{3} \\ \alpha \\ 1 \end{bmatrix} \quad \text{or} \quad \begin{bmatrix} -\frac{2}{5}(1-\alpha) \\ -2\alpha \\ -\frac{1}{5} + \frac{6\alpha}{5} \\ 1 \end{bmatrix}$$

---

9. Given the three spheres whose surfaces are defined by $\mathbf{p}(\mathbf{u}_1, 1)$, $\mathbf{p}(\mathbf{u}_2, 2)$, and $\mathbf{p}(\mathbf{u}_3, 2)$, where

$$\mathbf{u}_1 = \begin{bmatrix} 1 \\ 0 \\ 0 \\ 1 \end{bmatrix}; \quad \mathbf{u}_2 = \begin{bmatrix} 2 \\ 1 \\ 0 \\ 1 \end{bmatrix}, \quad \mathbf{u}_3 = \begin{bmatrix} 0 \\ 0 \\ 2 \\ 1 \end{bmatrix}$$

and planes $\mathbf{p}_1 = [1 \quad 1 \quad 1 \quad -4]$ and $\mathbf{p}_2 = [1 \quad -2 \quad 1 \quad -1]$.
Find each of the following.
(a) The circle of intersection of $\mathbf{p}(\mathbf{u}_1, 1)$ and $\mathbf{p}(\mathbf{u}_2, 2)$
(b) The points of intersection of $\mathbf{p}(\mathbf{u}_1, 1)$, $\mathbf{p}(\mathbf{u}_2, 2)$ and $\mathbf{p}(\mathbf{u}_3, 2)$
(c) The circle of intersection of $\mathbf{p}(\mathbf{u}_1, 1)$ and $\mathbf{p}_1$
(d) The points of intersection of $\mathbf{p}(\mathbf{u}_2, 2)$, $\mathbf{p}_1$, and $\mathbf{p}_2$

The plane of intersection of planes $\mathbf{p}(\mathbf{u}_1, 1)$ and $\mathbf{p}(\mathbf{u}_2, 2)$ is $[2 \quad 2 \quad 0 \quad -1]$, and that of planes $\mathbf{p}(\mathbf{u}_1, 1)$ and $\mathbf{p}(\mathbf{u}_3, 2)$ is $[-2 \quad 0 \quad 4 \quad 0]$.
(a) Sphere $\mathbf{p}(\mathbf{u}_1, 1)$ is

$$[x-2 \quad y \quad z \quad 0]\begin{bmatrix} x \\ y \\ z \\ 1 \end{bmatrix}$$

and sphere $\mathbf{p}(\mathbf{u}_2, 2)$ is

$$[x-4 \quad y-2 \quad z \quad 1]\begin{bmatrix} x \\ y \\ z \\ 1 \end{bmatrix}$$

The plane defining the intersection of the two spheres satisfies

$$[2 \quad 2 \quad 0 \quad 1]\begin{bmatrix} x \\ y \\ z \\ 1 \end{bmatrix} = 0, \quad \text{or} \quad y = -x - \tfrac{1}{2}$$

With this condition applied to $\mathbf{p}(\mathbf{u}_1, 1)$ we have

$$x(x-2) + (x+\tfrac{1}{2})^2 + z^2 = 0$$

Thus, the circle is defined by $2x^2 - x + z^2 = -\frac{1}{4}$ and $y = -x - \frac{1}{2}$.

(b) The three spheres intersect at the two points given by the solution of the three equations

$$x^2 + y^2 + (z-2)^2 = 4$$
$$2x^2 - x + z^2 = -\tfrac{1}{4}$$
$$y = -x - \tfrac{1}{2}$$

The second equation is $z^2 = x - 2x^2 - \frac{1}{4}$, but this causes difficulty in solution.

Alternatively, we have the two planes of intersection as $[2 \quad 2 \quad 0 \quad -1]$ and $[-2 \quad 0 \quad 4 \quad 0]$. Two points on the line of intersection of the two planes are

$$\begin{bmatrix} 0 \\ 0.5 \\ 0 \\ 1 \end{bmatrix} \quad \text{and} \quad \begin{bmatrix} 8.0 \\ -7.5 \\ 4.0 \\ 1 \end{bmatrix}$$

From equation (1.26) we get the quadratic $144\alpha^2 - 24\alpha + 0.25 = 0$, with solutions $\alpha = 0.156$ and $\alpha = 0.011$. With these values for α, the points on the line (which are the points of intersection of the spheres) are

$$\begin{bmatrix} 1.244 \\ -0.744 \\ 0.622 \\ 1 \end{bmatrix} \quad \text{and} \quad \begin{bmatrix} 0.089 \\ 0.411 \\ 0.045 \\ 1 \end{bmatrix}$$

We verify that these points are correct by calculating the distance from each point to the center of each sphere as the radii of the spheres.

(c) The two conditions that define the required circle of intersection are

$$[x-2 \quad y \quad z \quad 1-1]\begin{bmatrix} x \\ y \\ z \\ 1 \end{bmatrix} = 0 \quad \text{and} \quad [1 \quad 1 \quad 1 \quad -4]\begin{bmatrix} x \\ y \\ z \\ 1 \end{bmatrix} = 0$$

so $x^2 - 2x + y^2 + z^2 = 0$ and $z = 4 - x - y$.

Eliminating z in the first equation gives the circle $x^2 + y^2 - 5x - 4y + 8 = 0$, with $z = 4 - x - y$.

(d) The line of intersection of the two planes is determined from

$$[1 \quad 1 \quad 1 \quad -4]\begin{bmatrix} x \\ y \\ z \\ 1 \end{bmatrix} = 0 \quad \text{and} \quad [1 \quad -2 \quad 1 \quad -1]\begin{bmatrix} x \\ y \\ z \\ 1 \end{bmatrix} = 0$$

so

$$x + y + z = 4$$

$$x - 2y + z = 1$$

Subtracting the second equation from the first gives $y = 1$ for all x and z. The line of intersection is

$$\begin{bmatrix} \alpha \\ 1 \\ 3 - \alpha \\ 1 \end{bmatrix}$$

Applying equation (1.26) gives

$$[\alpha - 4 \quad 1 - 2 \quad 3 - \alpha \quad 5 - 4] \begin{bmatrix} \alpha \\ 1 \\ 3 - \alpha \\ 1 \end{bmatrix} = 0$$

so $2\alpha^2 - 10\alpha + 9$, with solutions $\alpha = 3.829$ and $\alpha = 1.177$.

The points of intersection are

$$\begin{bmatrix} 3.829 \\ 1 \\ -0.829 \\ 1 \end{bmatrix} \quad \text{and} \quad \begin{bmatrix} 1.177 \\ 1 \\ 1.823 \\ 1 \end{bmatrix}$$

---

10. Find the intersection of the locus of a point at distance l from point $\mathbf{u}$ and the plane $\mathbf{p}$.

The locus of a point at distance l from $\mathbf{u}$ is the surface of a sphere of radius l centered at $\mathbf{u}$. The intersection of the surface of a sphere and plane $\mathbf{p} = [a \quad b \quad c \quad d]$ (if it occurs) is a circle. We have the geometry shown in Figure 1.4. The line from $\mathbf{u}$ that is normal to $\mathbf{p}$ is given by

$$\mathbf{u} + \alpha \begin{bmatrix} a \\ b \\ c \\ 0 \end{bmatrix}$$

This line intersects $\mathbf{p}$ at $\mathbf{v}$, so

$$\mathbf{p} \left\{ \mathbf{u} + \alpha \begin{bmatrix} a \\ b \\ c \\ 0 \end{bmatrix} \right\} = 0$$

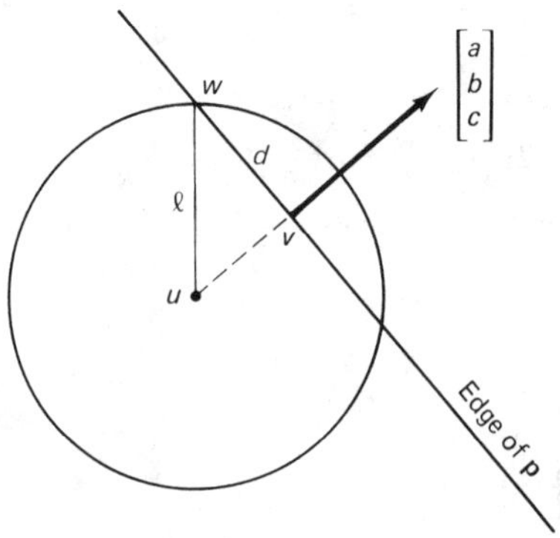

Figure 1.4 Locus of point w

or $\alpha = -\mathbf{pu}$, provided $\mathbf{p}$ is normalized. We can see from Figure 1.2 that the distance d of the circle of intersection from $\mathbf{v}$ is given by $d^2 = l^2 - \alpha^2$. Suppose a unit vector $[e \quad f \quad 0]^\#$ is orthogonal to $[a \quad b \quad c]^\#$; then $ae + bf = 0$, or $a\sqrt{1 - f^2} + bf = 0$, or

$$f = \frac{a}{\sqrt{a^2 + b^2}} \quad \text{and} \quad e = \frac{b}{\sqrt{a^2 + b^2}}$$

The cross product of $[a \quad b \quad c]^\#$ and $[e \quad f \quad 0]^\#$ is a unit vector that is orthogonal to the two vectors in its formation and so must lie on the surface of the plane $\mathbf{p}$; that is,

$$\begin{bmatrix} a \\ b \\ c \end{bmatrix} \times \begin{bmatrix} e \\ f \\ 0 \end{bmatrix} = \begin{bmatrix} -cf \\ ce \\ af - be \end{bmatrix} = \begin{bmatrix} -ac \\ bc \\ a^2 + b^2 \end{bmatrix} \div \sqrt{a^2 + b^2} = \boldsymbol{\nu}$$

Thus, one point on the circle of intersection is $\mathbf{v} + d\boldsymbol{\nu}$, and the circle of intersection is the rotation of vector $d\boldsymbol{\nu}$ in plane $\mathbf{p}$ about point $\mathbf{v}$.

EXERCISES FOR CHAPTER 1

1. Find the distance between the point $\begin{bmatrix} 1 \\ 0 \\ 1 \\ 1 \end{bmatrix}$ and the line $\begin{bmatrix} 2 \\ 1 \\ 1 \\ 1 \end{bmatrix} + \alpha \begin{bmatrix} -1 \\ -1 \\ 1 \\ 0 \end{bmatrix}$.

---

2. Find the distance between the point $\begin{bmatrix} 3 \\ 2 \\ 1 \\ 1 \end{bmatrix}$ and the line $\begin{bmatrix} 2 - \alpha \\ 1 + \alpha \\ \alpha \\ 1 \end{bmatrix}$.

---

3. Find the minimum distance between lines $\begin{bmatrix} \alpha \\ 2+\alpha \\ 2+\alpha \\ 1 \end{bmatrix}$ and $\begin{bmatrix} 1-\beta \\ 1 \\ 2+\beta \\ 1 \end{bmatrix}$.

---

4. Find the minimum distance between lines $\begin{bmatrix} 0 \\ 1 \\ 1 \\ 1 \end{bmatrix} + \alpha \begin{bmatrix} -1 \\ 0 \\ 2 \\ 0 \end{bmatrix}$ and $\begin{bmatrix} 2 \\ 1 \\ 0 \\ 1 \end{bmatrix} + \beta \begin{bmatrix} 1 \\ 0 \\ 2 \\ 1 \end{bmatrix}$. Determine the points on each line that are closest to the other line.

---

5. Find the plane that intersects the points $\begin{bmatrix} 1 \\ 1 \\ 2 \\ 1 \end{bmatrix}$, $\begin{bmatrix} 2 \\ 0 \\ 0 \\ 1 \end{bmatrix}$, and $\begin{bmatrix} 1 \\ 1 \\ 0 \\ 1 \end{bmatrix}$. Where does this plane intersect the x-, y-, and z-axes. What is the minimum distance of the plane from the origin.

---

6. **(a)** Find the line of intersection of the two planes $[1 \quad 2 \quad 2 \quad 1]$ and $[2 \quad 0 \quad 1 \quad 2]$.
(b) Find the distance of both planes from the origin.
(c) A line passes through points $[1 \quad 0 \quad 0 \quad 1]^{\#}$ and $[0 \quad 1 \quad 1 \quad 1]^{\#}$. Find where this line intersects the plane $[1 \quad 2 \quad 2 \quad 1]$.

---

7. Determine the family of planes that contain the vectors $\begin{bmatrix} 2 \\ 2 \\ 1 \end{bmatrix}$ and $\begin{bmatrix} 1 \\ 1 \\ 2 \end{bmatrix}$. If the plane passes through the point $\begin{bmatrix} 1 \\ 2 \\ 3 \\ 1 \end{bmatrix}$, what is the plane? If the plane passes through the point $\begin{bmatrix} 2 \\ 1 \\ 0 \\ 1 \end{bmatrix}$, what is the plane? Calculate the distance between the two points and the distance between the two planes. What is the requirement for these two quantities to be the same?

---

8. Given point $\begin{bmatrix} 1 \\ 1 \\ 2 \\ 1 \end{bmatrix}$, line $\begin{bmatrix} x \\ 1 + x \\ -1 + x \\ 1 \end{bmatrix}$, and planes $\mathbf{p} = [-2 \quad -1 \quad 1 \quad 1]$ and $\mathbf{q} = [1 \quad 0 \quad 1 \quad 1]$. Find:

(a) The minimum distance of the point from the line

(b) The minimum distance of the point from the plane p

(c) The point of intersection of the line on the plane p

(d) The line of intersection of the two planes

---

9. **(a)** Find the line of intersection of the two planes $[\frac{1}{3} \quad \frac{2}{3} \quad \frac{2}{3} \quad \frac{1}{3}]$ and $[2/\sqrt{5} \quad 0 \quad 1/\sqrt{5} \quad 2/\sqrt{5}]$. Find the distance of both planes from the origin.

(b) A line passes through points $[1 \quad 0 \quad 0 \quad 1]^{\#}$ and $[0 \quad 1 \quad 1 \quad 1]^{\#}$. Find where this line intersects the plane $[\frac{1}{3} \quad \frac{2}{3} \quad \frac{2}{3} \quad \frac{1}{3}]$.

---

10. Find the relative location of planes $[1 \quad 1 \quad -2 \quad 1]$, $[-1 \quad -2 \quad 1 \quad 1]$, and $[-2 \quad 2 \quad 0 \quad -1]$ with respect to point $[1 \quad 1 \quad 1 \quad 1]^{\#}$ and the origin.

---

11. Find the relative location of points $\begin{bmatrix} 1 \\ 2 \\ 3 \\ 1 \end{bmatrix}$, $\begin{bmatrix} -1 \\ -2 \\ -3 \\ 1 \end{bmatrix}$, $\begin{bmatrix} 0 \\ 1 \\ -1 \\ 1 \end{bmatrix}$, and $\begin{bmatrix} -2 \\ 1 \\ -2 \\ 1 \end{bmatrix}$ with respect to plane $[-1 \quad -1 \quad 2 \quad 1]$ and the origin.

---

12. **(a)** Determine the relative positions of the points $\mathbf{u}_1 = \begin{bmatrix} 2 \\ 2 \\ 1 \\ 1 \end{bmatrix}$ and $\mathbf{u}_2 = \begin{bmatrix} 0 \\ 0 \\ 2 \\ 1 \end{bmatrix}$ with respect to plane $\mathbf{p} = [1 \quad 2 \quad -1 \quad 1]$ and the origin.

(b) Find the point of intersection of the line joining $\mathbf{u}_1$ and $\mathbf{u}_2$ and $\mathbf{p}$.

(c) What is the distance of $\mathbf{u}_1$ from $\mathbf{p}$?

---

13. **(a)** What is the distance D of point $\begin{bmatrix} -1 \\ 2 \\ 1 \\ 1 \end{bmatrix}$ from the origin?

(b) Find the two values of d such that the plane $[2 \quad 2 \quad 0 \quad d]$ is at distance D from the point.

(c) For the values of d obtained, what is the distance of the plane from the origin?

---

14. Given $\mathbf{p} = [a \quad b \quad c \quad d] = [1/\sqrt{2} \quad 1/\sqrt{2} \quad 0 \quad -2]$ and $\mathbf{u}$ is a point on the line passing through $\begin{bmatrix} 1 \\ 2 \\ 1 \\ 1 \end{bmatrix}$ and the origin. If $\alpha = \dfrac{\mathbf{pu}}{\mathbf{pu} - \mathbf{d}}$, explain what $\alpha = 0$ and $\alpha = \infty$ mean.

---

15. Given three points $\begin{bmatrix} 2 \\ 1 \\ 1 \\ 1 \end{bmatrix}$, $\begin{bmatrix} 1 \\ 2 \\ 2 \\ 1 \end{bmatrix}$, and $\begin{bmatrix} 1 \\ 1 \\ 0 \\ 1 \end{bmatrix}$. Find:

(a) Plane $\mathbf{p}$ that passes through all three points
(b) The line of intersection of $\mathbf{p}$ with the xy-plane
(c) The line of intersection of $\mathbf{p}$ with the yz-plane
(d) The point of intersection of $\mathbf{p}$ with the xy- and yz-planes.

---

16. Given the four points $\begin{bmatrix} 1 \\ 1 \\ 0 \\ 1 \end{bmatrix}$, $\begin{bmatrix} 1 \\ 0 \\ 2 \\ 1 \end{bmatrix}$, $\begin{bmatrix} 0 \\ 2 \\ 2 \\ 1 \end{bmatrix}$, and $\begin{bmatrix} 2 \\ 0 \\ 0 \\ 1 \end{bmatrix}$, find each of the following.

(a) The plane containing points $\begin{bmatrix} 1 \\ 1 \\ 0 \\ 1 \end{bmatrix}$, $\begin{bmatrix} 1 \\ 0 \\ 2 \\ 1 \end{bmatrix}$, and $\begin{bmatrix} 0 \\ 2 \\ 2 \\ 1 \end{bmatrix}$

(b) The line connecting points $\begin{bmatrix} 1 \\ 1 \\ 0 \\ 1 \end{bmatrix}$ and $\begin{bmatrix} 1 \\ 0 \\ 2 \\ 1 \end{bmatrix}$ and the line connecting $\begin{bmatrix} 0 \\ 2 \\ 2 \\ 1 \end{bmatrix}$ and $\begin{bmatrix} 2 \\ 0 \\ 0 \\ 1 \end{bmatrix}$

(c) The minimum distance between the two lines and the points on these lines at which it occurs

(d) The minimum distance of point $\begin{bmatrix} 2 \\ 0 \\ 0 \\ 1 \end{bmatrix}$ from the plane.

---

17. Given point $\mathbf{u} = \begin{bmatrix} 1 \\ 0 \\ 0 \\ 1 \end{bmatrix}$, vectors $\boldsymbol{\nu} = \begin{bmatrix} 0 \\ 1 \\ -1 \end{bmatrix}$ and $\boldsymbol{\eta} = \begin{bmatrix} 1 \\ 1 \\ 0 \end{bmatrix}$, and plane $\mathbf{p} = [-1 \quad 1 \quad 1 \quad 1]$.

(a) Show that $\mathbf{u}$ is on $\mathbf{p}$.
(b) Show that $\boldsymbol{\nu}$ lies in $\mathbf{p}$.
(c) Show that $\boldsymbol{\eta}$ lies in $\mathbf{p}$.

---

18. Given the plane $\mathbf{p} = [-1 \quad -2 \quad -1 \quad 1]$ and points $\mathbf{u} = [-1 \quad 0 \quad 0 \quad 1]^{\#}$ and $\mathbf{v} = [0 \quad 4 \quad 0 \quad 1]^{\#}$. Find the point of intersection between the line joining the two points and the plane. Find the minimum distance of $\mathbf{u}$ and $\mathbf{v}$ from $\mathbf{p}$. Comment on the location of $\mathbf{u}$ and $\mathbf{v}$ with respect to $\mathbf{p}$ and the origin.

---

19. Given the four planes $\mathbf{p}_1 = [0 \quad 0 \quad 1 \quad 0]$, $\mathbf{p}_2 = [-1/\sqrt{2} \quad 1/\sqrt{2} \quad 0 \quad 2]$, $\mathbf{p}_3 = [1 \quad 0 \quad 0 \quad 1]$, and $\mathbf{p}_4 = [\frac{1}{2} \quad 0 \quad -\sqrt{3}/2 \quad 2]$. Find:

(a) The line of intersection of $\mathbf{p}_1$ and $\mathbf{p}_2$
(b) The line of intersection of $\mathbf{p}_3$ and $\mathbf{p}_4$
(c) The point of intersection $\mathbf{u}$ of the three planes $\mathbf{p}_1$, $\mathbf{p}_2$, and $\mathbf{p}_3$
(d) The minimum distance of $\mathbf{u}$ from $\mathbf{p}_4$

---

20. Find the line formed as the intersection of the two planes $[1 \quad 1 \quad 1 \quad 1]$ and $[1 \quad 2 \quad 2 \quad 1]$.

---

21. Given a cube of side unity that sits on a horizontal surface $[0 \quad 0 \quad 1 \quad 0]$ such that its centroid is at $[1/\sqrt{2} \quad 1/\sqrt{2} \quad 1/2 \quad 1]^{\#}$ and a corner point is at $[1/\sqrt{2} \quad 0 \quad 0 \quad 1]^{\#}$. Specify the cube by each of the following:

(a) Its corner points and associated edges
(b) The planes coincident with its surfaces and associated edges

---

22. Given a cube sits on a horizontal surface $[0 \quad 0 \quad 1 \quad 1]$ such that its centroid is at $[0 \quad 0 \quad 0 \quad 1]^{\#}$ and one corner point is at $[2\sqrt{2} \quad 0 \quad -1 \quad 1]^{\#}$. Specify the cube by:

(a) Its corner points and associated edges
(b) The planes coincident with its surfaces and associated edges

---

23. Find the distance between line $\begin{bmatrix} 1 \\ 2 + \alpha \\ 3 + \alpha \\ 1 \end{bmatrix}$ and the sphere of radius 2 centered at $\begin{bmatrix} -2 \\ 1 \\ -1 \\ 1 \end{bmatrix}$.

---

24. Find the circle of intersection of the two spheres $\mathbf{p}(\mathbf{u}_1, 2)$ and $\mathbf{p}(\mathbf{u}_2, 3)$, where

$$\mathbf{u}_1 = \begin{bmatrix} 2 \\ 1 \\ 0 \\ 1 \end{bmatrix} \quad \text{and} \quad \mathbf{u}_2 = \begin{bmatrix} 1 \\ 2 \\ 1 \\ 1 \end{bmatrix}$$

---

25. Find the points of intersection of the three spheres $\mathbf{p}(\mathbf{u}_1, 2)$, $\mathbf{p}(\mathbf{u}_2, 2)$ and $\mathbf{p}(\mathbf{u}_3, 1)$, where

$$\mathbf{u}_1 = \begin{bmatrix} 1 \\ 1 \\ 0 \\ 1 \end{bmatrix}, \quad \mathbf{u}_2 = \begin{bmatrix} 1 \\ 0 \\ 1 \\ 1 \end{bmatrix}, \quad \text{and} \quad \mathbf{u}_3 = \begin{bmatrix} 0 \\ 1 \\ 2 \\ 1 \end{bmatrix}$$

---

26. Find the points of intersection of the three spheres $\mathbf{p}(\mathbf{u}_1, 2)$, $\mathbf{p}(\mathbf{u}_2, 2)$ and $\mathbf{p}(\mathbf{u}_3, 2)$, where

$$\mathbf{u}_1 = \begin{bmatrix} 1 \\ 0 \\ 0 \\ 1 \end{bmatrix}, \quad \mathbf{u}_2 = \begin{bmatrix} 2 \\ 1 \\ 0 \\ 1 \end{bmatrix}, \quad \text{and} \quad \mathbf{u}_3 = \begin{bmatrix} 0 \\ 0 \\ 2 \\ 1 \end{bmatrix}$$

---

27. Find the points of intersection of the three spheres $\mathbf{p}(\mathbf{u}_1, 1)$, $\mathbf{p}(\mathbf{u}_2, 1)$ and $\mathbf{p}(\mathbf{u}_3, 1)$, where

$$\mathbf{u}_1 = \begin{bmatrix} 0 \\ 2 \\ 0 \\ 1 \end{bmatrix}, \quad \mathbf{u}_2 = \begin{bmatrix} 1 \\ 1 \\ 0 \\ 1 \end{bmatrix}, \quad \text{and} \quad \mathbf{u}_3 = \begin{bmatrix} 0 \\ 1 \\ 2 \\ 1 \end{bmatrix}$$

---

28. Determine the range of values of α such that spheres $\mathbf{p}(\mathbf{u}_1, 2)$, $\mathbf{p}(\mathbf{u}_2, 2)$ and $\mathbf{p}(\mathbf{u}_3, 2)$ intersect, where

$$\mathbf{u}_1 = \begin{bmatrix} 2 \\ 2 \\ 1 \\ 1 \end{bmatrix}, \quad \mathbf{u}_2 = \begin{bmatrix} 0 \\ \alpha \\ 3 \\ 1 \end{bmatrix}, \quad \text{and} \quad \mathbf{u}_3 = \begin{bmatrix} 1 \\ 0 \\ 2 \\ 1 \end{bmatrix}$$

---

29. Determine the points of intersection of the spheres $\mathbf{p}(\mathbf{u}_1, 2)$, $\mathbf{p}(\mathbf{u}_2, 4)$, and plane $[0.2 \quad 0.4 \quad 0.8944 \quad -1]$, where

$$\mathbf{u}_1 = \begin{bmatrix} 2 \\ -1 \\ 1 \\ 1 \end{bmatrix} \quad \text{and} \quad \mathbf{u}_2 = \begin{bmatrix} 1 \\ 3 \\ 2 \\ 1 \end{bmatrix}$$

CHAPTER TWO

Movement and Imaging

In this chapter transformations for the translation, rotation, scaling, imaging, and perspective views of vectors and points are developed. The rotation of a vector about a point in a plane (Section 2.1.2) is a precursor to a discussion of robotic joints in the next chapter. The reader can safely omit all but Section 2.1.2 from a first reading, since only this section is needed in the derivation of robotic joints. The sections on imaging, perspective projection, and cameras are needed when task planning and robotic graphics are studied.

The relationships between vectors, points, lines, and planes were developed in Chapter 1 without any assessment of positiveness or negativeness. We added a vector to a point without ambiguity. We found the unique point of intersection of a line and a plane without regard for the orientation of the vector partially defining the plane. Here counterclockwise angles are assumed positive under rotation, and a determination of the direction of the vector $[a \quad b \quad c]^{\#}$ (partially defining the plane $[a \quad b \quad c \quad d]$ within which rotation is performed) is important. When illustrating points and vectors in a plane **p**, we will always assume that $[a \quad b \quad c]^{\#}$ is oriented out of the paper.

There is ambiguity when assessing the angle between two vectors. For example, the dot and cross product of the two unit vectors

$$\begin{bmatrix} 1 \\ 0 \\ 0 \end{bmatrix} \quad \text{and} \quad \begin{bmatrix} -0.866 \\ 0.5 \\ 0 \end{bmatrix}$$

are

$$[1 \quad 0 \quad 0]\begin{bmatrix} -0.866 \\ 0.5 \\ 0 \end{bmatrix} = -0.866 \quad \text{and} \quad \begin{bmatrix} 1 \\ 0 \\ 0 \end{bmatrix} \times \begin{bmatrix} -0.866 \\ 0.5 \\ 0 \end{bmatrix} = \begin{bmatrix} 0 \\ 0 \\ 0.5 \end{bmatrix}$$

The dot product -0.866 is the cosine of the angle θ between the two vectors and the cross product 0.5 is the sine of the angle between the two vectors. A computer will calculate $\cos^{-1}(-0.866) = 150°$ and $\sin^{-1}0.5 = 30°$; that is, the inverse cosine assumes an angle in the range $0 \le \theta \le 180°$ and the inverse sine assumes an angle in the range $-90° \le \theta \le 90°$. The conclusion that can be drawn is that both dot and cross products are needed to determine the actual angle. Recall that $210° \equiv -150°$.

The cross product of two vectors is

$$\begin{bmatrix} a \\ b \\ c \end{bmatrix} \times \begin{bmatrix} d \\ e \\ f \end{bmatrix} = \begin{bmatrix} bf - ce \\ cd - af \\ ae - bd \end{bmatrix} \equiv \begin{bmatrix} 0 & -c & b \\ c & 0 & -a \\ -b & a & 0 \end{bmatrix}\begin{bmatrix} d \\ e \\ f \end{bmatrix}$$

$$\equiv \begin{bmatrix} 0 & f & -e \\ -f & 0 & d \\ e & -d & 0 \end{bmatrix}\begin{bmatrix} a \\ b \\ c \end{bmatrix}$$

Notice that the cross product is a singular transformation, since

$$\begin{vmatrix} 0 & -c & b \\ c & 0 & -a \\ -b & a & 0 \end{vmatrix} = 0 \quad \text{and} \quad \begin{vmatrix} 0 & f & -e \\ -f & 0 & d \\ e & -d & 0 \end{vmatrix} = 0$$

2.1 NONSINGULAR TRANSFORMATIONS

A nonsingular transformation has an inverse; that is, the transformation can be applied in reverse to return to the original condition. A singular transformation does not permit this. We consider both nonsingular and singular transformations.

Suppose a point **u** is transformed to point **v** by operation **A**. Then **v** = **Au**, where **A** is a nonsingular 4×4 matrix. **A** can transform [Bottema 1979; Duffy 1980; Hartenberg 1964; Lee 1983a; Paul 1981b; Pieper 1968; Pieper 1969] from **u** to **v** by adding a vector (translation), rotating **u** with respect to the origin, scaling the point **u** with respect to its distance from the origin, inverting, or imaging (as with a lens of a camera). Some movements are inconvenient to describe in terms of a single matrix (the homogeneous transformation) [Parkin 1986]. For example, the transformation involving the addition of a point to a rotated vector is better specified as the addition of two quantities. We will use the form that is most convenient in the application.

2.1.1 Translation of Points and Planes by a Vector

The translation of point $\mathbf{u} = [x \quad y \quad z \quad 1]^\#$ by vector $\boldsymbol{\nu} = [\nu_x \quad \nu_y \quad \nu_z]^\#$ is

$$\begin{bmatrix} x + \nu_x \\ y + \nu_y \\ z + \nu_z \\ 1 \end{bmatrix} = \mathbf{Tu} = \begin{bmatrix} 1 & 0 & 0 & \nu_x \\ 0 & 1 & 0 & \nu_y \\ 0 & 0 & 1 & \nu_z \\ 0 & 0 & 0 & 1 \end{bmatrix} \begin{bmatrix} x \\ y \\ z \\ 1 \end{bmatrix} \tag{2.1}$$

Matrix $\mathbf{T}$ has determinant $|\mathbf{T}| = 1$, and

$$\mathbf{T}^{-1} = \begin{bmatrix} 1 & 0 & 0 & -\nu_x \\ 0 & 1 & 0 & -\nu_y \\ 0 & 0 & 1 & -\nu_z \\ 0 & 0 & 0 & 1 \end{bmatrix} \tag{2.2}$$

which is the same as $\mathbf{T}$ but with the signs of the off-diagonal terms reversed.

To translate a plane $\mathbf{p}$, use $\mathbf{T}^{-1}$, as in $\mathbf{q} = \mathbf{pT}^{-1}$. Then $\mathbf{qv} = \mathbf{pT}^{-1}\mathbf{Tu} = \mathbf{pu}$, so point $\mathbf{v}$ is on plane $\mathbf{q}$ if point $\mathbf{u}$ is on plane $\mathbf{p}$.

$$\mathbf{q} = \mathbf{pT}^{-1} = [a \quad b \quad c \quad d] \begin{bmatrix} 1 & 0 & 0 & -\nu_x \\ 0 & 1 & 0 & -\nu_y \\ 0 & 0 & 1 & -\nu_z \\ 0 & 0 & 0 & 1 \end{bmatrix} = [a \quad b \quad c \quad d']$$

where $d' = d - a\nu_x - b\nu_y - c\nu_z$.

2.1.2 Rotation of a Vector by Angle θ in Plane p

We are given vector $\boldsymbol{\nu} = [\nu_x \quad \nu_y \quad \nu_z]^\#$ in normalized plane $\mathbf{p} = [a \quad b \quad c \quad d]$ that is to be rotated by angle θ about point $\mathbf{v}$ on plane $\mathbf{p}$ to become vector $\boldsymbol{\eta}$, as shown in Figure 2.1. Suppose we construct a normal to vector $\boldsymbol{\nu}$ from the end of $\boldsymbol{\eta}$. Then from Figure 2.1 we see

$$|\boldsymbol{\alpha}| = |\boldsymbol{\nu}|\cos\theta \quad \text{and} \quad |\boldsymbol{\beta}| = |\boldsymbol{\nu}|\sin\theta$$

Therefore, $\boldsymbol{\alpha} = \boldsymbol{\nu}\cos\theta$ and

$$\boldsymbol{\beta} = k \begin{bmatrix} a \\ b \\ c \end{bmatrix} \times \alpha$$

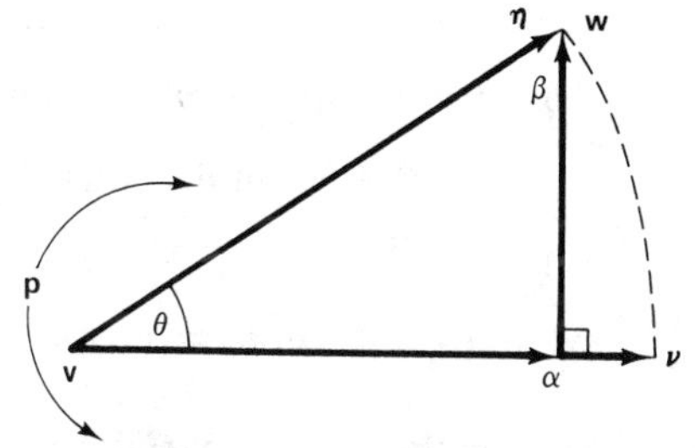

Figure 2.1 Rotation of vector $\boldsymbol{\nu}$ about point $\mathbf{v}$ in plane $\mathbf{p}$.

where k is a scalar. Vector $\boldsymbol{\alpha}$ and unit vector $[a \quad b \quad c]^{\#}$ are orthogonal, so $|\boldsymbol{\beta}| = |k||\boldsymbol{\alpha}|$, yielding $k = \tan\theta$. Thus

$$\boldsymbol{\eta} = \boldsymbol{\alpha} + \boldsymbol{\beta} = \boldsymbol{\alpha} + \tan\theta \begin{bmatrix} a \\ b \\ c \end{bmatrix} x\boldsymbol{\alpha} = \boldsymbol{\nu}\cos\theta + \sin\theta \begin{bmatrix} 0 & -c & b \\ c & 0 & -a \\ -b & a & 0 \end{bmatrix} \boldsymbol{\nu}$$

$$= \begin{bmatrix} \cos\theta & -c\sin\theta & b\sin\theta \\ c\sin\theta & \cos\theta & -a\sin\theta \\ -b\sin\theta & a\sin\theta & \cos\theta \end{bmatrix} \boldsymbol{\nu} = \mathrm{R}(\theta)\boldsymbol{\nu} \tag{2.3}$$

It may seem that the 3×3 matrix is incorrect, since rotation by angle $-\theta$ is

$$R(-\theta) = \begin{bmatrix} \cos\theta & c\sin\theta & -b\sin\theta \\ -c\sin\theta & \cos\theta & a\sin\theta \\ b\sin\theta & -a\sin\theta & \cos\theta \end{bmatrix} \tag{2.4}$$

and rotations by $-\theta$ followed by θ should produce a combined rotation matrix that is the identity matrix, but

$R(-\theta)\mathbf{R}(\theta)$

$$= \begin{bmatrix} \cos\theta & c\sin\theta & -b\sin\theta \\ -c\sin\theta & \cos\theta & a\sin\theta \\ b\sin\theta & -a\sin\theta & \cos\theta \end{bmatrix} \begin{bmatrix} \cos\theta & -c\sin\theta & b\sin\theta \\ c\sin\theta & \cos\theta & -a\sin\theta \\ -b\sin\theta & a\sin\theta & \cos\theta \end{bmatrix}$$

$$= \begin{bmatrix} 1 - a^2\sin^2\theta & -ab\sin^2\theta & -ac\sin^2\theta \\ -ab\sin^2\theta & 1 - b^2\sin^2\theta & -bc\sin^2\theta \\ -ac\sin^2\theta & -bc\sin^2\theta & 1 - c^2\sin^2\theta \end{bmatrix} \neq I$$

However,

$$\boldsymbol{\nu} = R(-\theta)\boldsymbol{\eta} = R(-\theta)R(\theta)\boldsymbol{\nu} = \begin{bmatrix} 1 - a^2\sin^2\theta & -ab\sin^2\theta & -ac\sin^2\theta \\ -ab\sin^2\theta & 1 - b^2\sin^2\theta & -bc\sin^2\theta \\ -ac\sin^2\theta & -bc\sin^2\theta & 1 - c^2\sin^2\theta \end{bmatrix} \boldsymbol{\nu}$$

$$= \begin{bmatrix} \nu_x - a\sin^2\theta\{a\nu_x + b\nu_y + c\nu_z\} \\ \nu_y - b\sin^2\theta\{a\nu_x + b\nu_y + c\nu_z\} \\ \nu_z - c\sin^2\theta\{a\nu_x + b\nu_y + c\nu_z\} \end{bmatrix}$$

which are all true, since $\boldsymbol{\nu}$ is in the plane $\mathbf{p}$ and so $a\nu_x + b\nu_y + c\nu_z = 0$.

$$\begin{aligned}|R(\theta)| &= \cos\theta(\cos^2\theta + a^2\sin^2\theta) - c\sin\theta(-c\sin\theta\cos\theta - ab\sin^2\theta)\\ &\quad - b\sin\theta(ac\sin^2\theta - b\sin\theta\cos\theta)\\ &= \cos\theta(\cos^2\theta + a^2\sin^2\theta + c^2\sin^2\theta + b^2\sin^2\theta) + abc\sin^3\theta\\ &\quad - abc\sin^3\theta\\ &= \cos\theta(\cos^2\theta + \sin^2\theta) = \cos\theta\end{aligned}$$

Using standard methods of linear algebra, we calculate

$$R(\theta)^{-1} = \frac{1}{\cos\theta}$$

$$\begin{bmatrix} a^2\sin^2\theta + \cos^2\theta & ab\sin^2\theta - c\cos\theta\sin\theta & ac\sin^2\theta + b\cos\theta\sin\theta \\ ab\sin^2\theta + c\cos\theta\sin\theta & b^2\sin^2\theta + \cos^2\theta & bc\sin^2\theta - a\cos\theta\sin\theta \\ ac\sin^2\theta - b\cos\theta\sin\theta & bc\sin^2\theta + a\cos\theta\sin\theta & c^2\sin^2\theta + \cos^2\theta \end{bmatrix} \tag{2.5}$$

which is singular when $\cos\theta = 0$, or $\theta = \pm(n + 1/2)\pi$, $n = 0, 1, \ldots$.

Since $|AB| = |A||B|$, then we expect $|R(\theta_1)R(\theta_2)|$ to equal $|R(\theta_1)||R(\theta_2)|$. But we have $|R(\pi/2)| = 0$ and $|R(\pi)| = -1$, so in general $|R(\theta_1)R(\theta_2)| \neq |R(\theta_1+\theta_2)|$, implying $R(\theta_1+\theta_2) \neq R(\theta_1)R(\theta_2)$. However, we have

$$R(\theta)R(\theta) = \begin{bmatrix} \cos\theta & -c\sin\theta & b\sin\theta \\ c\sin\theta & \cos\theta & -a\sin\theta \\ -b\sin\theta & a\sin\theta & \cos\theta \end{bmatrix}\begin{bmatrix} \cos\theta & -c\sin\theta & b\sin\theta \\ c\sin\theta & \cos\theta & -a\sin\theta \\ -b\sin\theta & a\sin\theta & \cos\theta \end{bmatrix}$$

$$= \begin{bmatrix} \cos^2\theta - c^2\sin^2\theta - b^2\sin^2\theta & -2c\cos\theta\sin\theta + ab\sin^2\theta & 2b\cos\theta\sin\theta + ac\sin^2\theta \\ 2c\cos\theta\sin\theta + ab\sin^2\theta & \cos^2\theta - a^2\sin^2\theta - c^2\sin^2\theta & -2a\cos\theta\sin\theta + bc\sin^2\theta \\ -2b\cos\theta\sin\theta + ac\sin^2\theta & 2a\cos\theta\sin\theta + bc\sin^2\theta & \cos^2\theta - a^2\sin^2\theta - b^2\sin^2\theta \end{bmatrix}$$

$$= \begin{bmatrix} \cos^2\theta - \sin^2\theta + a^2\sin^2\theta & -c\sin 2\theta + ab\sin^2\theta & b\sin 2\theta + ac\sin^2\theta \\ c\sin 2\theta + ab\sin^2\theta & \cos^2\theta - \sin^2\theta + b^2\sin^2\theta & -a\sin 2\theta + bc\sin^2\theta \\ -b\sin 2\theta + ac\sin^2\theta & a\sin 2\theta + bc\sin^2\theta & \cos^2\theta - \sin^2\theta + c^2\sin^2\theta \end{bmatrix}$$

$$= \begin{bmatrix} \cos 2\theta & -c\sin 2\theta & b\sin 2\theta \\ c\sin 2\theta & \cos 2\theta & -a\sin 2\theta \\ -b\sin 2\theta & a\sin 2\theta & \cos 2\theta \end{bmatrix} + \sin^2\theta\begin{bmatrix} a^2 & ab & ac \\ ab & b^2 & bc \\ ac & bc & c^2 \end{bmatrix} = R(2\theta) + \sin^2\theta\begin{bmatrix} a \\ b \\ c \end{bmatrix}[a \quad b \quad c]$$

and for any vector $\boldsymbol{\nu}$ in $\mathbf{p}$, $[a \quad b \quad c]\boldsymbol{\nu} = 0$, so the result is correct. It is easy to show that the product of n rotation matrices $R(\theta)$ produces

$$R(\theta)R(\theta)\cdots R(\theta) = R(n\theta) + A(\theta)[a \quad b \quad c]$$

where $A(\theta)$ is some matrix that could be a column vector.

Point **w** of Figure 2.1 is given by

$$\mathbf{w} = \begin{bmatrix} v_x \\ v_y \\ v_z \\ 1 \end{bmatrix} + \begin{bmatrix} \eta_x \\ \eta_y \\ \eta_z \\ 0 \end{bmatrix} = \begin{bmatrix} v_x \\ v_y \\ v_z \\ 1 \end{bmatrix} + \begin{bmatrix} \cos\theta & -c\sin\theta & b\sin\theta & 0 \\ c\sin\theta & \cos\theta & -a\sin\theta & 0 \\ -b\sin\theta & a\sin\theta & \cos\theta & 0 \\ 0 & 0 & 0 & 1 \end{bmatrix} \begin{bmatrix} \nu_x \\ \nu_y \\ \nu_z \\ 0 \end{bmatrix} \tag{2.6}$$

$$= \begin{bmatrix} 1 & 0 & 0 & \nu_x\cos\theta - \nu_y c\sin\theta + \nu_z b\sin\theta \\ 0 & 1 & 0 & \nu_x c\sin\theta + \nu_y\cos\theta - \nu_z a\sin\theta \\ 0 & 0 & 1 & -\nu_x b\sin\theta + \nu_y a\sin\theta + \nu_z\cos\theta \\ 0 & 0 & 0 & 1 \end{bmatrix} \begin{bmatrix} v_x \\ v_y \\ v_z \\ 1 \end{bmatrix} \tag{2.7}$$

If **v** is the origin, we can use the mathematics just presented to rotate a point $\boldsymbol{\nu}$ (which we rewrite as **v**) on plane **p** by angle θ about the origin to reach point **w**, where

$$\mathbf{w} = \begin{bmatrix} \cos\theta & -c\sin\theta & b\sin\theta & 0 \\ c\sin\theta & \cos\theta & -a\sin\theta & 0 \\ -b\sin\theta & a\sin\theta & \cos\theta & 0 \\ 0 & 0 & 0 & 1 \end{bmatrix} \begin{bmatrix} v_x \\ v_y \\ v_z \\ 1 \end{bmatrix} = R_\theta \mathbf{v} \tag{2.8}$$

and if $|R(\theta)| \neq 0$, $\mathbf{v} = R^{-1}(\theta)\mathbf{w}$.

Suppose we rotate in the yz-plane (or about the x-axis), so $\mathbf{p} = [1 \quad 0 \quad 0 \quad 0]$. Then

$$\begin{bmatrix} w_x \\ w_y \\ w_z \\ 1 \end{bmatrix} = \begin{bmatrix} \cos\theta & 0 & 0 & 0 \\ 0 & \cos\theta & -\sin\theta & 0 \\ 0 & \sin\theta & \cos\theta & 0 \\ 0 & 0 & 0 & 1 \end{bmatrix} \begin{bmatrix} v_x \\ v_y \\ v_z \\ 1 \end{bmatrix}$$

$$\text{and} \quad \begin{bmatrix} v_x \\ v_y \\ v_z \\ 1 \end{bmatrix} = \begin{bmatrix} \sec\theta & 0 & 0 & 0 \\ 0 & \cos\theta & \sin\theta & 0 \\ 0 & -\sin\theta & \cos\theta & 0 \\ 0 & 0 & 0 & 1 \end{bmatrix} \begin{bmatrix} w_x \\ w_y \\ w_z \\ 1 \end{bmatrix} \tag{2.9}$$

or $\mathbf{w} = R_x(\theta)\mathbf{v}$, $\mathbf{v} = R_x^{-1}(\theta)\mathbf{w}$. Rotating in the xz- and xy-planes produces

$$R_{y\theta} = \begin{bmatrix} \cos\theta & 0 & \sin\theta & 0 \\ 0 & \cos\theta & 0 & 0 \\ -\sin\theta & 0 & \cos\theta & 0 \\ 0 & 0 & 0 & 1 \end{bmatrix}$$

$$\text{and} \quad R_{z\theta} = \begin{bmatrix} \cos\theta & -\sin\theta & 0 & 0 \\ \sin\theta & \cos\theta & 0 & 0 \\ 0 & 0 & \cos\theta & 0 \\ 0 & 0 & 0 & 1 \end{bmatrix} \tag{2.10}$$

respectively.

2.1.3 Rotation of a Point about an Axis

Given point $\mathbf{w}$, which is to be rotated about an axis $\mathbf{u} + \alpha(\mathbf{v} - \mathbf{u})$, where $|\mathrm{v} - \mathrm{u}| = 1$. If the direction of the line is used to partially define a plane $\mathbf{p}$, which contains $\mathbf{w}$, then

$$\mathbf{p} = [v_x - u_x \quad v_y - u_y \quad v_z - u_z \quad d] \tag{2.11}$$

Rotation of $\mathbf{w} = [w_x \quad w_y \quad w_z \quad 1]^{\#}$ about the axis by angle θ is the same as rotation of $\mathbf{w}$ in $\mathbf{p}$ by angle θ about the axial point defined by equation (1.7) as

$$\alpha = -\frac{(v_x - u_x)(u_x - w_x) + (v_y - u_y)(u_y - w_y) + (v_z - u_z)(u_z - w_z)}{(v_x - u_x)^2 + (v_y - u_y)^2 + (v_z - u_z)^2} \tag{2.12}$$

This rotation will produce a new point $\mathbf{u} + \alpha(\mathbf{v} - \mathbf{u}) + \boldsymbol{\eta}$, where vector $\boldsymbol{\eta}$ is defined by equation

$$\begin{bmatrix} \eta_x \\ \eta_y \\ \eta_z \end{bmatrix} = \begin{bmatrix} \cos\theta & -(v_z - u_z)\sin\theta & (v_y - u_y)\sin\theta \\ (v_z - u_z)\sin\theta & \cos\theta & -(v_x - u_x)\sin\theta \\ -(v_y - u_y)\sin\theta & (v_x - u_x)\sin\theta & \cos\theta \end{bmatrix} \begin{bmatrix} w_x - u_x - \alpha(v_x - u_x) \\ w_y - u_y - \alpha(v_y - u_y) \\ w_z - u_z - \alpha(v_z - u_z) \end{bmatrix} \tag{2.13}$$

If $\mathbf{u}$ is the origin, the new point is given by

$$\begin{bmatrix} \alpha v_x \\ \alpha v_y \\ \alpha v_z \\ 1 \end{bmatrix} + \boldsymbol{\eta} = \begin{bmatrix} \cos\theta & -v_z\sin\theta & v_y\sin\theta & \alpha v_x(1 - \cos\theta) \\ v_z\sin\theta & \cos\theta & -v_x\sin\theta & \alpha v_y(1 - \cos\theta) \\ -v_y\sin\theta & v_x\sin\theta & \cos\theta & \alpha v_z(1 - \cos\theta) \\ 0 & 0 & 0 & 1 \end{bmatrix} \mathbf{w} \tag{2.14}$$

Suppose the line is the x-axis; then $v_x = 1$ and $v_y = v_z = 0$, and equation (2.14) becomes

$$\begin{bmatrix} 1 & 0 & 0 & 0 \\ 0 & \cos\theta & -\sin\theta & 0 \\ 0 & \sin\theta & \cos\theta & 0 \\ 0 & 0 & 0 & 1 \end{bmatrix} \begin{bmatrix} w_x \\ w_y \\ w_z \\ 1 \end{bmatrix} = R_x(\theta)\mathbf{w} \tag{2.15}$$

which is the result familiar to many [Paul 1981b]. Similarly, for rotations about the y- and z-axes, respectively,

$$R_y(\theta) = \begin{bmatrix} \cos\theta & 0 & \sin\theta & 0 \\ 0 & 1 & 0 & 0 \\ -\sin\theta & 0 & \cos\theta & 0 \\ 0 & 0 & 0 & 1 \end{bmatrix}$$

$$\text{and} \quad R_z(\theta) = \begin{bmatrix} \cos\theta & -\sin\theta & 0 & 0 \\ \sin\theta & \cos\theta & 0 & 0 \\ 0 & 0 & 1 & 0 \\ 0 & 0 & 0 & 1 \end{bmatrix} \tag{2.16}$$

The reader will have noticed that equations (2.9) and (2.10) are inconsistent with equations (2.15) and (2.16). However, there is no inconsistency in the context for which they apply. Rotation of a point $\mathbf{v}$ that lies in the yz-plane about the x-axis requires $\mathbf{v}_x = 0$, and the leading $\cos\theta$ or $\sec\theta$ terms in the matrices of equation (2.9) become multiplied by zero, thus having no effect. On the other hand, the rotation of $\mathbf{w}$ about the x-axis defined by equation (2.15) does not require w_x to be zero, and the x value of $\mathbf{R}_x(\theta)\mathbf{w}$ is w_x.

Rotation about a major axis such as $R_x(\theta)$ requires no restriction on the point to be rotated, and so the associative rule does not apply to operations such as $R_z(\theta_1)R_y(\theta_2)R_x(\theta_3)\mathbf{u}$, permitting a composite matrix to be calculated as in $R_T = R_z(\theta_1)R_y(\theta_2)R_x(\theta_3)$. If point $\mathbf{u}$ is rotated by $-\theta_1$ about the x-axis, by $-\theta_2$ about the y-axis, and by $-\theta_3$ about the z-axis to reach point $\mathbf{v}$ (for simplicity we write $\sin\theta_1$ as s_1, and so on), the transformations are

$$\begin{aligned}
\mathbf{v} &= \begin{bmatrix} c_3 & s_3 & 0 & 0 \\ -s_3 & c_3 & 0 & 0 \\ 0 & 0 & 1 & 0 \\ 0 & 0 & 0 & 1 \end{bmatrix} \begin{bmatrix} c_2 & 0 & -s_2 & 0 \\ 0 & 1 & 0 & 0 \\ s_2 & 0 & c_2 & 0 \\ 0 & 0 & 0 & 1 \end{bmatrix} \begin{bmatrix} 1 & 0 & 0 & 0 \\ 0 & c_1 & s_1 & 0 \\ 0 & -s_1 & c_1 & 0 \\ 0 & 0 & 0 & 1 \end{bmatrix} \mathbf{u} \\
&= \begin{bmatrix} c_3 & s_3 & 0 & 0 \\ -s_3 & c_3 & 0 & 0 \\ 0 & 0 & 1 & 0 \\ 0 & 0 & 0 & 1 \end{bmatrix} \begin{bmatrix} c_2 & s_2 s_1 & -s_2 c_1 & 0 \\ 0 & c_1 & s_1 & 0 \\ s_2 & -c_2 s_1 & c_2 c_1 & 0 \\ 0 & 0 & 0 & 1 \end{bmatrix} \mathbf{u} \\
&= \begin{bmatrix} c_3 c_2 & s_3 c_1 + c_3 s_2 s_1 & s_3 s_1 - c_3 s_2 c_1 & 0 \\ -s_3 c_2 & c_3 c_1 - s_3 s_2 s_1 & c_3 s_1 + s_3 s_2 c_1 & 0 \\ s_2 & -c_2 s_1 & c_2 c_1 & 0 \\ 0 & 0 & 0 & 1 \end{bmatrix} \mathbf{u}
\end{aligned} \tag{2.17}$$

Notice the asymmetry of this matrix, indicating that rotations about the major axes are not commutative. Also notice that the rotation about a major axis does not change the distance of a point from the origin, so we can use the origin as the point of rotation by the solid angle θ as discussed in the next paragraph. As a point of interest, there is an infinity of such points of rotation contained in the plane defined by $[u_x - v_x \quad u_y - v_y \quad u_z - v_z \quad 0]$ that could be chosen, and the freedom to choose this point is the reason for the associative property under multiplication of $R(\theta)$ and the property $R(\theta)^{-1} \neq R(-\theta)$. If the point $[(u_x + v_x)/2 \quad (u_y + v_y)/2 \quad (u_z + v_z)/2 \quad 1]^{\#}$ is chosen, the angle of rotation will always be 180°.

The angle θ between **u** and **v** is given by the dot product of the unit vectors defining the direction of **u** and **v** with respect to the origin, so

$$\cos\theta = \frac{1}{u_x^2 + u_y^2 + u_z^2} [u_x \quad u_y \quad u_z] \begin{bmatrix} c_3c_2 & s_3c_1 + c_3s_2s_1 & s_3s_1 - c_3s_2c_1 \\ -s_3c_2 & c_3c_1 - s_3s_2s_1 & c_3s_1 + s_3s_2s_1 \\ s_2 & -c_2s_1 & c_2c_1 \end{bmatrix} \begin{bmatrix} u_x \\ u_y \\ u_z \end{bmatrix} \tag{2.18}$$

Notice that θ is a function of **u** as well as the Euler angles. The normalized plane of rotation $[a \quad b \quad c \quad 0]$ that contains u and v is given by

$$\begin{bmatrix} a \\ b \\ c \end{bmatrix} = \frac{1}{u_x^2 + u_y^2 + u_z^2} \begin{bmatrix} u_x \\ u_y \\ u_z \end{bmatrix} \times \begin{bmatrix} v_x \\ v_y \\ v_z \end{bmatrix} \tag{2.19}$$

Rotations about the major axes on frames [Craig 1986; Wolovich 1987] are used extensively by others, since they consider rotation in terms of the Euler angles; that is, a general rotation is performed about the three orthogonal major axes. This approach is not employed here, since rotation in planes of movement is found to be more efficient and simpler, especially when applied to robotic movement. Specifically, to form equation (2.17) requires $16\tau_1 + 4\tau_2 + 6\tau_3$, where τ_1 is the arithmetic time for multiplication or division, τ_2 is the time for addition or subtraction, and τ_3 is the time for a trigonometric function evaluation. On the other hand, employing equation (2.3) to rotate a vector that is then added to a point requires $3\tau_1 + 3\tau_2 + 2\tau_3$. Often we can make the assumption that computer time is a multiple of the arithmetic times; the relation between the arithmetic times is $\tau_1 = 3\tau_2 = \tau_3/3$, so the ratio of computer times under the rotation in planes approach versus the Euler angle approach is 15 : 58. Homogeneous equations used with the transformation of frames have the clean look of simplicity not verified by computational efficiency.

Rotation by $\pi/2$ about the x-, y- and z-axes reduces equations (2.15) and (2.16) to

$$R_x\left(\frac{\pi}{2}\right) = \begin{bmatrix} 1 & 0 & 0 & 0 \\ 0 & 0 & -1 & 0 \\ 0 & 1 & 0 & 0 \\ 0 & 0 & 0 & 1 \end{bmatrix} \begin{bmatrix} w_x \\ w_y \\ w_z \\ 1 \end{bmatrix},$$

$$R_y\left(\frac{\pi}{2}\right) = \begin{bmatrix} 0 & 0 & -1 & 0 \\ 0 & 1 & 0 & 0 \\ 1 & 0 & 0 & 0 \\ 0 & 0 & 0 & 1 \end{bmatrix} \begin{bmatrix} w_x \\ w_y \\ w_z \\ 1 \end{bmatrix},$$

$$\text{and} \quad R_z\left(\frac{\pi}{2}\right) = \begin{bmatrix} 0 & -1 & 0 & 0 \\ 1 & 0 & 0 & 0 \\ 0 & 0 & 1 & \\ 0 & 0 & 0 & 1 \end{bmatrix} \begin{bmatrix} w_x \\ w_y \\ w_z \\ 1 \end{bmatrix}$$

respectively. Thus rotation about the x-axis followed by the y-axis produces

$$\begin{bmatrix} 0 & 0 & 0 & 0 \\ 0 & 1 & 0 & 0 \\ 1 & 0 & 0 & 0 \\ 0 & 0 & 0 & 1 \end{bmatrix} \begin{bmatrix} 1 & 0 & 0 & 0 \\ 0 & 0 & -1 & 0 \\ 0 & 1 & 0 & 0 \\ 0 & 0 & 0 & 1 \end{bmatrix} = \begin{bmatrix} 0 & -1 & 0 & 0 \\ 0 & 0 & -1 & 0 \\ 1 & 0 & 0 & 0 \\ 0 & 0 & 0 & 1 \end{bmatrix}$$

and rotation about the y-axis followed by the x-axis produces

$$\begin{bmatrix} 1 & 0 & 0 & 0 \\ 0 & 0 & -1 & 0 \\ 0 & 1 & 0 & 0 \\ 0 & 0 & 0 & 1 \end{bmatrix} \begin{bmatrix} 0 & 0 & -1 & 0 \\ 0 & 1 & 0 & 0 \\ 1 & 0 & 0 & 0 \\ 0 & 0 & 0 & 1 \end{bmatrix} = \begin{bmatrix} 0 & 0 & -1 & 0 \\ -1 & 0 & 0 & 0 \\ 1 & 0 & 0 & 0 \\ 0 & 0 & 0 & 1 \end{bmatrix}$$

Rotation matrices in general do not commute, so the order of rotation must be maintained.

2.1.4 Rotation of Planes

A plane $\mathbf{p} = [a \quad b \quad c \quad d]$ can be rotated about an axis $\mathbf{u} + \alpha(\mathbf{v} - \mathbf{u})$ by θ to produce a new plane $\mathbf{q} = [e \quad f \quad g \quad h]$. There are two cases to consider. In the first case the axis is on the plane, so a set of θ values produces a pencil of planes about the axis. To find $\mathbf{q}$ we rotate vector $[a \quad b \quad c]^{\#}$ by θ in the plane partially specified by $[v_x - u_x \quad v_y - u_y \quad v_z - u_z]$; that is, from equation (2.3),

$$\begin{bmatrix} e \\ f \\ g \end{bmatrix} = \begin{bmatrix} \cos\theta & -(v_z - u_z)\sin\theta & (v_y - u_y)\sin\theta \\ (v_z - u_z)\sin\theta & \cos\theta & -(v_x - u_x)\sin\theta \\ -(v_y - u_y)\sin\theta & (v_x - u_x)\sin\theta & \cos\theta \end{bmatrix} \begin{bmatrix} a \\ b \\ c \end{bmatrix} \tag{2.20}$$

and

$$h = -(u_x e + u_y f + u_z g). \tag{2.21}$$

The other case is when the axis and plane can intersect and the plane is rotated about the point of intersection. We calculate the point of intersection $\mathbf{u}'$ using equation (1.16) as

$$\mathbf{u}' = \mathbf{u} - \frac{\mathbf{pu}}{\mathbf{p}(\mathbf{v} - \mathbf{u})}(\mathbf{v} - \mathbf{u}) \tag{2.22}$$

Rotating $\mathbf{w} = \mathbf{u}' + [a \quad b \quad c \quad 0]^\#$ by θ about the axis (as discussed in Section 2.1.3) produces a new point $\mathbf{w}'$, and $\mathbf{w}' - \mathbf{w}$ produces the vector partially defining the new plane $\mathbf{q}$. $\mathbf{q}$ is completed by noting that $\mathbf{u}'$ is on $\mathbf{q}$.

2.1.5 Scaling

Scaling by scalar factor m changes u into $\mathbf{v} = m\mathbf{u} = \mathbf{u}m$, since m is a scalar. The transformation can be written in homogeneous form as

$$G = \begin{bmatrix} m & 0 & 0 & 0 \\ 0 & m & 0 & 0 \\ 0 & 0 & m & 0 \\ 0 & 0 & 0 & 1 \end{bmatrix} \tag{2.23}$$

which has determinant $|\mathbf{G}| = m^3$ and inverse

$$\mathbf{G}^{-1} = \begin{bmatrix} 1/m & 0 & 0 & 0 \\ 0 & 1/m & 0 & 0 \\ 0 & 0 & 1/m & 0 \\ 0 & 0 & 0 & 1 \end{bmatrix} \tag{2.24}$$

The more general scaling matrix

$$\begin{bmatrix} m_1 & 0 & 0 & 0 \\ 0 & m_2 & 0 & 0 \\ 0 & 0 & m_3 & 0 \\ 0 & 0 & 0 & 1 \end{bmatrix} \tag{2.25}$$

has no meaning to us here. It is a matrix that changes the relative position of a set of points. If that set of points defines the corner points of an object, the shape of the object changes, and we do not permit this.

2.1.6 Imaging

The imaging of a point occurs when light passes through a thin convex lens, as shown in Figure 2.2. Light from point **u** is imaged to point **v**, and center point **w** of the lens is in line with these two points. The plane of the lens is **p**. If D_1 is the distance between points **u** and **w**, and D_2 is the distance between points **w** and **v**, then

$$\frac{1}{D_2} = \frac{1}{f} - \frac{1}{D_1}, \quad \text{and so} \quad \mathbf{v} = \mathbf{w} - \frac{D_2}{D_1}(\mathbf{u} - \mathbf{w}) \tag{2.26}$$

where

$$D_1^2 = (u_x - w_x)^2 + (u_y - w_y)^2 + (u_z - w_z)^2$$

$$D_2^2 = (v_x - w_x)^2 + (v_y - w_y)^2 + (v_z - w_z)^2$$

and where f is the focal length of the lens (the point to which all axial parallel rays of light will focus). Notice that $D_1 > f$ and $D_2 > f$.

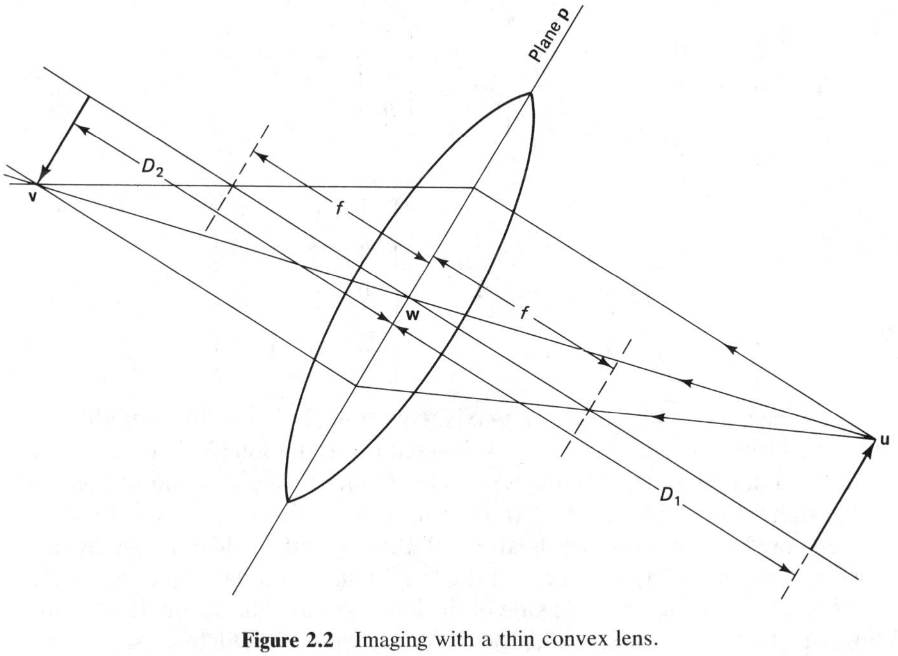

Figure 2.2 Imaging with a thin convex lens.

It would be desirable to express point $\mathbf{v}$ in terms of point $\mathbf{u}$ as in $\mathbf{v} = \mathbf{Mu}$, where $\mathbf{M}$ is some homogeneous matrix, but this is not possible except for some simple cases due to the implicit nature of equation (2.26): D_2 is a function of $\mathbf{v}$. If we assume that the axis of the lens is along the x-axis and the lens is at the origin, point $\mathbf{u}$ can be imaged to point $\mathbf{v}$ as

$$\frac{1}{v_x} = \frac{1}{f} - \frac{1}{u_x}$$

so

$$v_x = \frac{u_x f}{u_x - f}, \qquad v_y = \frac{u_y f}{u_x - f}, \quad \text{and} \quad v_z = \frac{u_z f}{u_x - f}$$

and the homogeneous transformation becomes

$$\begin{bmatrix} v_x \\ v_y \\ v_z \\ 1 \end{bmatrix} = \begin{bmatrix} \frac{f}{u_x - f} & 0 & 0 & 0 \\ 0 & \frac{f}{u_x - f} & 0 & 0 \\ 0 & 0 & \frac{f}{u_x - f} & 0 \\ 0 & 0 & 0 & 1 \end{bmatrix} \begin{bmatrix} u_x \\ u_y \\ u_z \\ 1 \end{bmatrix} \equiv \begin{bmatrix} 1 & 0 & 0 & 0 \\ 0 & 1 & 0 & 0 \\ 0 & 0 & 1 & 0 \\ 0 & 0 & 0 & \frac{u_x}{f} - 1 \end{bmatrix} \begin{bmatrix} u_x \\ u_y \\ u_z \\ 1 \end{bmatrix}$$

$$\equiv \begin{bmatrix} 1 & 0 & 0 & 0 \\ 0 & 1 & 0 & 0 \\ 0 & 0 & 1 & 0 \\ \frac{1}{f} & 0 & 0 & 1 \end{bmatrix} \begin{bmatrix} u_x \\ u_y \\ u_z \\ 1 \end{bmatrix} = \mathbf{L}_x \begin{bmatrix} u_x \\ u_y \\ u_z \\ 1 \end{bmatrix} \tag{2.27}$$

$$|\,\mathbf{L}_x\,| = 1 \quad \text{and} \quad \mathbf{L}_x^{-1} = \begin{bmatrix} 1 & 0 & 0 & 0 \\ 0 & 1 & 0 & 0 \\ 0 & 0 & 1 & 0 \\ \frac{-1}{f} & 0 & 0 & 1 \end{bmatrix} \tag{2.28}$$

If the axis of the lens is the y-axis, equation (2.27) would be modified so that the term $-1/f$ appears in the (4, 2) location, Equation (2.27) transforms a point to a unique point. An object can be defined as a collection of points in three-dimensional space. The transform will produce an image in three-dimensional space. The implication of this is that a clear image will be formed only at a film plane behind the focal point for a flat object parallel to the film plane on the opposite side of the lens; we say that the depth of field is limited. This is in contrast to the infinite depth of field that occurs in a pinhole camera (discussed in Section 2.3).

Provided that the object is small compared to its distance from the lens, the depth of field will not be a factor, and the image can be captured in two dimensions without fuzziness (as occurs in a normal camera). However, information is lost in the process because a three-dimensional object is imaged on a flat plane. This will be made evident when the pinhole camera is discussed. The nonsingular transformation $\mathbf{v} = \mathbf{Tu}$ with the special case given in equation (2.27) does not lose information.

2.2 SEQUENCES OF TRANSFORMATIONS

Recall that $\mathbf{u}$ can be a point with respect to the origin or a vector with respect to a point, so a point is a vector with respect to the origin. Consider a sequence of translations and rotations on vector $\mathbf{u}$, such as

$$\mathbf{v} = L_{xf} G_m T_{\boldsymbol{\alpha}} R_{y\theta} \mathbf{u} \tag{2.29}$$

This represents the following operations on $\mathbf{u}$:

1. Rotation of point $\mathbf{u}$ by θ within a particular plane.
2. Translation of point $R_{y\theta}\mathbf{u}$ by vector $\boldsymbol{\alpha}$.
3. Scaling of point $T_{\boldsymbol{\alpha}} R_{y\theta}\mathbf{u}$ by scalar m with respect to the origin.
4. Imaging of point $G_m T_\alpha R_{y\theta}\mathbf{u}$ along the x-axis with a lens of focal length f.

Nonsingular matrices of the form R_y, T, G, and L_x perform linear operations, and so

$$\mathbf{u} = R_{y\theta}^{-1} T_\alpha^{-1} G_m^{-1} L_{xf}^{-1} \mathbf{v} \tag{2.30}$$

Thus, L_{xf}^{-1} images point $\mathbf{v}$ along the x-axis with a lens of focal length f; G_m^{-1} scales point $L_{xf}^{-1}\mathbf{v}$ by $1/m$; T_α^{-1} translates point $G_m^{-1}L_{xf}^{-1}\mathbf{v}$ by $-\alpha$ with respect to the origin; and $R_{y\theta}^{-1}$ rotates point $T_\alpha^{-1}G_m^{-1}L_{xf}^{-1}$ by angle $-\theta$ within the plane.

Matrices L, G, T, and R do not in general commute, meaning that a sequence of operations must be performed in the given order. See the appropriate example at the end of this chapter.

The following are problems of defining sequences involving rotation and imaging in terms of homogeneous matrices:

1. Imaging matrices of the type L_{xf} requires the axis of the lens to be a major axis, whereas a more arbitrary axis is usually required. Although it is feasible to move the object so that the axis of the lens is a major axis, this mechanism is more expensive computationally and more confusing than leaving the object where it is and using equation (2.26).
2. The determination of the Euler angles to find matrices of the form $R_{y\theta}$ is inefficient and often inconvenient, particularly when considering the

movement of robots with some revolute joints, since one must resort to "frames" at intermediate joints, a mechanism shown here to be unnecessary.

2.3 PERSPECTIVE PROJECTION

A perspective view is easy to obtain with respect to any plane **p** and focal point **w**, as shown in Figure 2.3. A perspective projection is found by projecting from a point **u** on the object to focal point **w** and determining the point of intersection **v** of this line with plane **p**. The axial point $\mathbf{v}_a$ on **p** is at distance $\alpha = -(aw_x + bw_y + cw_z + d)$ from focal point **w** (according to Section 1.4.7), so

$$\mathbf{v}_a = \mathbf{w} + \alpha \begin{bmatrix} a \\ b \\ c \\ 0 \end{bmatrix}$$

The intersection of the line between **u** and **w** with plane **p**, as given by equation (1.16), is the point

$$\mathbf{v} = \mathbf{u} - \frac{\mathbf{pu}}{\mathbf{p}(\mathbf{w} - \mathbf{u})}(\mathbf{w} - \mathbf{u}) \tag{2.31}$$

where scalar $\mathbf{pu}/\mathbf{p}(\mathbf{w} - \mathbf{u})$ is a function of object point **u**. Thus, in general we cannot write **v** as a function of **u** in explicit homogeneous matrix form.

For some simple cases we can write the perspective projection in homogeneous matrix form. Assume **p** is in the *yz*-plane with focal point at $\mathbf{w} = [-x_c \quad 0 \quad 0 \quad 1]^{\#}$. Given a point on the object $[u_x \quad u_y \quad u_z \quad 1]^{\#}$, the perspec-

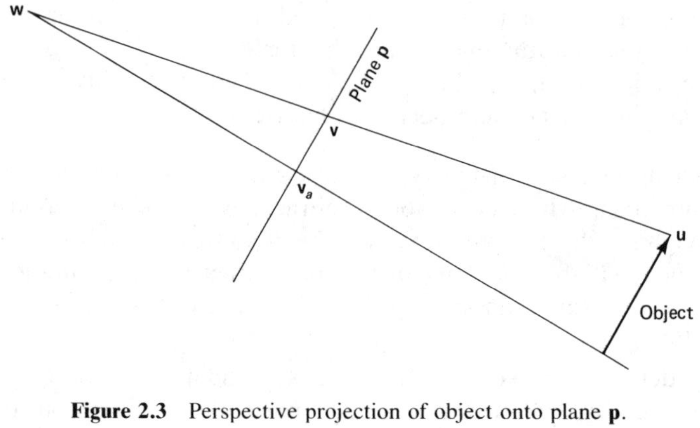

Figure 2.3 Perspective projection of object onto plane **p**.

tive point $\mathbf{v} = [v_x \quad v_y \quad v_z \quad 1]^{\#}$ is given by $v_x = 0$, $v_y = u_y/(1 + u_x/x_c)$, and $v_z = u_z/(1 + u_x/x_c)$. Thus, point $[u_x \quad u_y \quad u_z \quad 1]^{\#}$ is projected to the yz-plane as point

$$\begin{bmatrix} v_x \\ v_y \\ v_z \\ 1 \end{bmatrix} = \begin{bmatrix} 0 \\ \dfrac{u_y}{1 + u_x/x_c} \\ \dfrac{u_z}{1 + u_x/x_c} \\ 1 \end{bmatrix} = \begin{bmatrix} 0 \\ u_y \\ u_z \\ \dfrac{1 + u_x}{x_c} \end{bmatrix}$$

$$= \begin{bmatrix} 0 & 0 & 0 & 0 \\ 0 & 1 & 0 & 0 \\ 0 & 0 & 1 & 0 \\ \dfrac{1}{x_c} & 0 & 0 & 1 \end{bmatrix} \begin{bmatrix} u_x \\ u_y \\ u_z \\ 1 \end{bmatrix} = \mathbf{P} \begin{bmatrix} u_x \\ u_y \\ u_z \\ 1 \end{bmatrix} \tag{2.32}$$

Matrix $\mathbf{P}$ is singular, since the first row contains only zeros. Application of $\mathbf{P}$ to some point will lose information. In fact, any perspective projection loses information; a collection of points in Cartesian space is imaged to a collection of points in plane.

We can produce a perspective view of an object (a collection of corner points) from any point or angle. All corner points must be on the same side of the plane parallel to $\mathbf{p}$ passing through the vanishing point; otherwise the perspective projection will be unbounded. Thus, the perspective projection of an object such as a teacup should not be attempted from within the rim.

2.4 CAMERAS

A pinhole camera produces an inverted-perspective projection. Such a camera is shown in Figure 2.4. The camera is a light-tight box, apart from a single pinhole. A point on the film plane at the back of the camera will receive light from any point along the projection line shown in Figure 2.4, so the depth of view is infinite. All points in front of the pinhole and along a line passing through the pinhole strike the film plane at a unique point.

As stated previously, an ordinary lens camera will produce the same image as the pinhole camera, provided the lens aperture is very small. For the work presented here, we assume that the image is in focus, so we can use the equations governing the operation of the pinhole camera throughout. The equation governing the location of $\mathbf{v}$ is the same as obtained before, so equation (2.31) applies. Notice, however, that $\mathbf{p}$ and $\mathbf{v}$ have a different orientation in this case. As before, it is possible to express the perspective projection in explicit homogeneous representation form in some simple

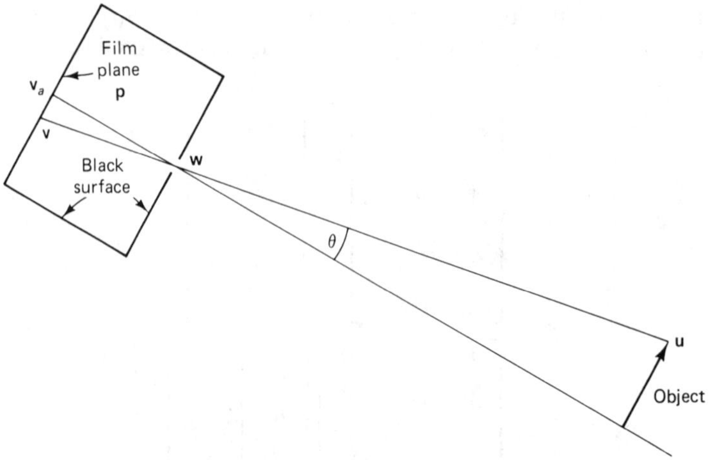

Figure 2.4 The pinhole camera.

cases. For example, suppose **w** is at the origin and the film plane is a plane parallel to the yz-plane passing through point $-x_c$; then

$$\begin{bmatrix} \bar{x} \\ \bar{y} \\ \bar{z} \\ 1 \end{bmatrix} = \begin{bmatrix} -x_c \\ -yx_c/x \\ -zx_c/x \\ 1 \end{bmatrix} \equiv \begin{bmatrix} -1 \\ -y/x \\ -z/x \\ 1/x_c \end{bmatrix} \equiv \begin{bmatrix} x \\ y \\ z \\ -x/x_c \end{bmatrix}$$

$$= \begin{bmatrix} 1 & 0 & 0 & 0 \\ 0 & 1 & 0 & 0 \\ 0 & 0 & 1 & 0 \\ -1/x_c & 0 & 0 & 0 \end{bmatrix} \begin{bmatrix} x \\ y \\ z \\ 1 \end{bmatrix} = \mathbf{P}_c \begin{bmatrix} x \\ y \\ z \\ 1 \end{bmatrix} \tag{2.33}$$

which is a singular transformation, since the last column contains only zeros.

It should be pointed out that the multiple-lens system used in practical cameras produces a noninverted image. The transformation to produce such an image, equivalent to equation (2.33), replaces $-1/x_c$ with $1/f$.

2.5 IMAGING OF OBJECTS

The imaging of an object is performed corner point by corner point using equation (2.31). In order to determine what will be on the film of the camera, the orientation of the camera must be established by a unit vector in the film plane, which then becomes the ordinate. If we consider the camera to be hand-held, the human input establishes an orientation (the eyes define a vertical), but we do not have this attribute in an artificial vision system; instead, a vector chosen in some manner will be used. The axial point on the film plane $\mathbf{p} = [a \quad b \quad c \quad d]$ is given by

$$\mathbf{v}_a = \mathbf{w} - \alpha \begin{bmatrix} a \\ b \\ c \\ 0 \end{bmatrix}$$

where $\mathbf{p}\mathbf{v}_a = 0$, so $\alpha = aw_x + bw_y + cw_z + d$.

An axial point on the object side of the camera is

$$\mathbf{w} + \alpha \begin{bmatrix} a \\ b \\ c \\ 0 \end{bmatrix}$$

and any vector added to this point can be used to establish camera orientation. Suppose we choose the Cartesian coordinate to be vertical. Then

$$\bar{\mathbf{u}} = \mathbf{w} + \alpha \begin{bmatrix} a \\ b \\ c \\ 0 \end{bmatrix} + \begin{bmatrix} 0 \\ 0 \\ 1 \\ 0 \end{bmatrix}$$

with image $\bar{\mathbf{v}} = \bar{\mathbf{u}} - \beta(\mathbf{w} - \bar{\mathbf{u}})$, where $\beta = \mathbf{p}\bar{\mathbf{u}}/\mathbf{p}(\mathbf{w} - \bar{\mathbf{u}})$ from equation (2.31). Thus $\bar{\mathbf{v}} - \mathbf{v}_a$ is a vector establishing the direction of the ordinate in the film plane; that is, the ordinate is the unit vector

$$\boldsymbol{\eta} = \frac{\bar{\mathbf{v}} - \mathbf{v}_a}{|\bar{\mathbf{v}} - \mathbf{v}_a|} \tag{2.34}$$

But unit vectors $[a \quad b \quad c]^\#$ and $\boldsymbol{\eta}$ are orthogonal and can define the abscissa $\boldsymbol{\mu}$ in the film plane. That is, suppose $[a \quad b \quad c]^\#$ is directed toward the object; then by the right-hand screw rule, the abscissa is defined by the unit vector

$$\boldsymbol{\mu} = \boldsymbol{\eta} \times \begin{bmatrix} a \\ b \\ c \end{bmatrix} \tag{2.35}$$

Thus, for some point on the film plane, which is defined with respect to the axial point on the film plane by vector $\boldsymbol{\nu}$, the projection of $\boldsymbol{\nu}$ on the abscissa is

$$\boldsymbol{\mu}\,|\boldsymbol{\nu}|\cos\theta = \boldsymbol{\mu}\,|\boldsymbol{\nu}|\frac{\boldsymbol{\nu}\cdot\boldsymbol{\mu}}{|\boldsymbol{\nu}||\boldsymbol{\mu}|} = \boldsymbol{\nu}\cdot\boldsymbol{\mu}\boldsymbol{\mu}$$

so $\boldsymbol{\nu} \cdot \boldsymbol{\mu}$ is the abscissa value. Similarly, the ordinate value is $\boldsymbol{\nu} \cdot \boldsymbol{\eta}$. Since we know what is on the film, we can use this information to control a robot manipulator; see example 7 in Section 8.9.

2.6 POINT LOCATION FROM TWO IMAGES

Animals use two eyes to give stereoscopic vision and depth perception. In synthetic vision in order to find the location of a point in three-dimensional space, we require knowledge of the camera(s) position(s) and

1. Two cameras
2. A moveable camera with which two independent images are captured
3. Structured light [Nakagawa 1985] (collimated light with energy in a limited waveband) from two independent sources

Consider two cameras looking at a single unknown point $\mathbf{u}$, as shown in Figure 2.5. If the location of the cameras is known, then pinholes $\mathbf{w}_1$ and $\mathbf{w}_2$

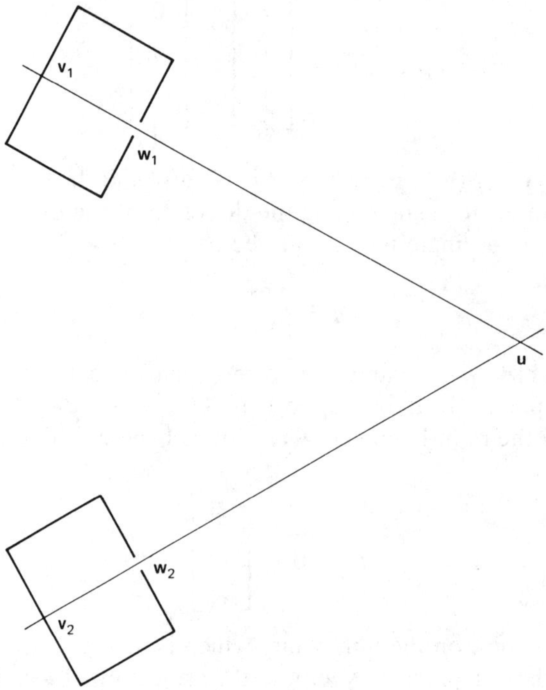

Figure 2.5 Two cameras looking at a single point.

are known. Given the image locations on the film plane, then point **u** can be determined as the intersection of the line passing through points $\mathbf{v}_1$ and $\mathbf{w}_1$ and the line passing through points $\mathbf{v}_2$ and $\mathbf{w}_2$. Use the averaging method of equation (1.4) to determine **u**.

2.7 ERRORS AND ERROR PROPAGATION DURING TRANSFORMATION AND IMAGING OPERATIONS

We have already discussed the problem of two lines meeting (they don't) in three-dimensional space and the averaging system for best estimate of the point of intersection. In this section the overall problem of errors is considered.

The width of the head of an animal is dictated more by security than facility in producing stereoscopic images. A very wide head or a head with eyes outboard on stalks is difficult to protect from damage, particularly since most animals are predator and/or prey. Unfortunately, the narrow spacing between the eyes makes depth perception difficult. It is simple to determine that an object is within reach but far more difficult to estimate the distance away of some object. Adult human eyes are about 6 cm apart. Suppose the distance of a small object at about 10 m is to be estimated. The axes of the eyes (equivalent to the axes in two pinhole cameras) will intersect at an angle of $\tan^{-1}(0.06/10) = 0.3438° = 0°20'38''$. An error of $10''$ (10 seconds of angle) in these axes will result in an estimation of distance from 9.92 m to 10.077 m, a total error of 1.57%. In most experimental setups, an error in alignment of the order of $10''$ is very difficult to achieve.

An experimental setup is not limited by the distance apart of the cameras. In order to minimize errors it is prudent to position the cameras so that the angle between the axes of the camera is of the order of 90°; the cameras would then be orthogonal. Suppose two orthogonal cameras are viewing the small object from 10 m each, and the alignment error of the cameras is $10''$, then the estimate of the distance of the object will vary by $\frac{1}{3600}$ m (a $10''$ arc over a radius of 10 m)s for an error of from 9.997 m to 10.003 m or 0.06%.

Analyses of error propagation, usually performed in the fields of Numerical Analysis or Numerical Methods [Ralston 1978], are not appropriate here since most of these analyses concern the relative magnitude of 'the pivot.' The pivot is the term that appears in the denominator in many operations. For example, when inverting a matrix of order n a total of n pivots will be calculated, and if any of these pivots are zero the inversion process fails: numeric techniques use pivotal selection to bring a non-zero pivot into the pivotal position in order for the inversion to proceed, and if this is impossible the matrix is singular, so its inverse does not exist. However, if a pivot is small according to some measure then errors are known to occur that are larger than would occur if pivotal selection is used to produce a larger pivot. Here, we do not need to invert matrices, and by proper consideration of all the operations involved in robotic kinematics can avoid any zero denominator terms and usually any denominator terms at all.

The inverse of the operation $R(\theta)$ is $R(-\theta)$, and the reverse of $G(m)$ is $G(1/m)$, where m is known to be nonzero. The determination of a plane $[a \quad b \quad c \quad d]$ from three points could involve the inverse of a 3×3 matrix, but it is numerically simpler to form two vectors from the three points and calculate the cross product to produce $[a \quad b \quad c]^{\#}$. Thus, prudent evaluation of the operations to be performed will obviate the use of pivots and so avoid almost all the conditions that produce numeric instability.

2.8 CUT-AND-PASTE OPERATIONS

Suppose a series of points $\mathbf{u}_i$, $i = 1, \ldots, n$, given in normalized plane $\mathbf{p} = [a \quad b \quad c \quad d]$ is required to have the same spatial relationship in normalized plane $\mathbf{q} = [e \quad f \quad g \quad h]$. We assume that known $\mathbf{v}_1$ in $\mathbf{q}$ corresponds to $\mathbf{u}_1$ and some orientation angle θ is known; without this or other equivalent information, we have no way of knowing the location or orientation of the set $\mathbf{v}_i$ of points. We are performing a cut-and-paste operation on the set of points $\mathbf{u}_i$—cutting them out of $\mathbf{p}$ and pasting them in the required position and orientation in $\mathbf{q}$.

The cross and dot products of the two vectors partially defining the planes produce unit vector are

$$\boldsymbol{\eta} = \begin{bmatrix} \eta_x \\ \eta_y \\ \eta_z \end{bmatrix} = \frac{\begin{bmatrix} a \\ b \\ c \end{bmatrix} \times \begin{bmatrix} e \\ f \\ g \end{bmatrix}}{\left\| \begin{bmatrix} a \\ b \\ c \end{bmatrix} \times \begin{bmatrix} e \\ f \\ g \end{bmatrix} \right\|}$$

$$= \frac{1}{\sqrt{(bg - cf)^2 + (ce - ag)^2 + (af - be)^2}} \begin{bmatrix} bg - cf \\ ce - ag \\ af - be \end{bmatrix} \tag{2.36}$$

and

$$\cos \psi = [a \quad b \quad c] \begin{bmatrix} e \\ f \\ g \end{bmatrix} = ae + bf + cg \tag{2.37}$$

Let $\zeta_i = \mathbf{u}_i - \mathbf{u}_1$, $i = 2, \ldots, n$. The protection of ζ_i on $\boldsymbol{\eta}$ produces vector $(\zeta_i \cdot \boldsymbol{\eta})\boldsymbol{\eta}$. Rotating vector $\zeta_i - (\zeta_i \cdot \boldsymbol{\eta})\boldsymbol{\eta}$ in the plane partially defined by vector $\boldsymbol{\eta}$ produces vector $\boldsymbol{\mu}_i$, i.e. using equation (2.3) we get

$$\boldsymbol{\mu}_i = \begin{bmatrix} \cos\psi & \eta_z \sin\psi & -\eta_y \sin\psi \\ -\eta_z \sin\psi & \cos\psi & \eta_x \sin\psi \\ \eta_y \sin\psi & -\eta_x \sin\psi & \cos\psi \end{bmatrix} \{\zeta_i - (\zeta_i \cdot \boldsymbol{\eta})\boldsymbol{\eta}\} \quad (2.38)$$

Vector $\boldsymbol{\mu}_i + (\zeta_i \cdot \boldsymbol{\eta})\boldsymbol{\eta}$ lies in plane $[e \quad f \quad g \quad 0]$ and rotation by angle θ about $\mathbf{v}_1$ produces

$$\mathbf{v}_i = \begin{bmatrix} \cos\theta & g\sin\theta & -f\sin\theta & 0 \\ -g\sin\theta & \cos\theta & e\sin\theta & 0 \\ f\sin\theta & -e\sin\theta & \cos\theta & 0 \\ 0 & 0 & 0 & 0 \end{bmatrix} \{\boldsymbol{\mu}_i + (\zeta_i \cdot \boldsymbol{\eta})\boldsymbol{\eta}\} + \mathbf{v}_1, \quad (2.39)$$

$$i = 2, 3, \ldots, n$$

We see in Chapter 5 that the base orientation of a robot is defined by normalized plane $\mathbf{p}_1 = [a \quad b \quad c \quad d]$ and orientation vector $\boldsymbol{\nu}$, and all subsequent planes of motion $\mathbf{p}_2, \mathbf{p}_3, \ldots,$ and joint locations $\mathbf{u}_2, \mathbf{u}_3, \ldots,$ are defined with respect to this base orientation. Typically we assume that $\mathbf{u}_1$ is the origin. Suppose the base orientation is changed to normalized $\mathbf{p}_1' = [a' \quad b' \quad c' \quad d']$ and $\boldsymbol{\nu}'$, we can determine the new planes of motion $\mathbf{p}_2', \mathbf{p}_3'$, and points $\mathbf{u}_2', \mathbf{u}_3'$ using the methodology discussed in this section.

2.9 EXAMPLES

1. Demonstrate the types of nonsingular transformation matrices capable of being written as nonimplicit homogeneous matrices that commute.

We start by noting that the nonimplicit nature of the homogeneous matrices to be considered excludes imaging except for the simple cases of imaging about a major axis. We can consider the following matrices as typical for demonstrating commutability (or lack thereof):

$$T_\alpha = \begin{bmatrix} 1 & 0 & 0 & \alpha_1 \\ 0 & 1 & 0 & \alpha_2 \\ 0 & 0 & 1 & \alpha_3 \\ 0 & 0 & 0 & 1 \end{bmatrix}, \quad T_\beta = \begin{bmatrix} 1 & 0 & 0 & \beta_1 \\ 0 & 1 & 0 & \beta_2 \\ 0 & 0 & 1 & \beta_3 \\ 0 & 0 & 0 & 1 \end{bmatrix},$$

$$G_m = \begin{bmatrix} m & 0 & 0 & 0 \\ 0 & m & 0 & 0 \\ 0 & 0 & m & 0 \\ 0 & 0 & 0 & 1 \end{bmatrix}, \qquad G_n = \begin{bmatrix} n & 0 & 0 & 0 \\ 0 & n & 0 & 0 \\ 0 & 0 & n & 0 \\ 0 & 0 & 0 & 1 \end{bmatrix}$$

$$L_x = \begin{bmatrix} 1 & 0 & 0 & 0 \\ 0 & 1 & 0 & 0 \\ 0 & 0 & 1 & 0 \\ a & 0 & 0 & 1 \end{bmatrix}, \qquad L_y = \begin{bmatrix} 1 & 0 & 0 & 0 \\ 0 & 1 & 0 & 0 \\ 0 & 0 & 1 & 0 \\ 0 & b & 0 & 1 \end{bmatrix}$$

$$R_x = \begin{bmatrix} 1 & 0 & 0 & 0 \\ 0 & \cos\theta & -\sin\theta & 0 \\ 0 & \sin\theta & \cos\theta & 0 \\ 0 & 0 & 0 & 1 \end{bmatrix}, \qquad R_y = \begin{bmatrix} \cos\phi & 0 & \sin\phi & 0 \\ 0 & 1 & 0 & 0 \\ -\sin\phi & 0 & \cos\phi & 0 \\ 0 & 0 & 0 & 1 \end{bmatrix}$$

$$T_\alpha T_\beta = \begin{bmatrix} 1 & 0 & 0 & \alpha_1 \\ 0 & 1 & 0 & \alpha_2 \\ 0 & 0 & 1 & \alpha_3 \\ 0 & 0 & 0 & 1 \end{bmatrix} \begin{bmatrix} 1 & 0 & 0 & \beta_1 \\ 0 & 1 & 0 & \beta_2 \\ 0 & 0 & 1 & \beta_3 \\ 0 & 0 & 0 & 1 \end{bmatrix} = \begin{bmatrix} 1 & 0 & 0 & \alpha_1 + \beta_1 \\ 0 & 1 & 0 & \alpha_2 + \beta_2 \\ 0 & 0 & 1 & \alpha_3 + \beta_3 \\ 0 & 0 & 0 & 1 \end{bmatrix} \equiv T_\beta T_\alpha$$

$$G_m G_n = \begin{bmatrix} m & 0 & 0 & 0 \\ 0 & m & 0 & 0 \\ 0 & 0 & m & 0 \\ 0 & 0 & 0 & 1 \end{bmatrix} \begin{bmatrix} n & 0 & 0 & 0 \\ 0 & n & 0 & 0 \\ 0 & 0 & n & 0 \\ 0 & 0 & 0 & 1 \end{bmatrix} = \begin{bmatrix} mn & 0 & 0 & 0 \\ 0 & mn & 0 & 0 \\ 0 & 0 & mn & 0 \\ 0 & 0 & 0 & 1 \end{bmatrix} \equiv G_n G_m$$

$$L_x L_y = \begin{bmatrix} 1 & 0 & 0 & 0 \\ 0 & 1 & 0 & 0 \\ 0 & 0 & 1 & 0 \\ a & 0 & 0 & 1 \end{bmatrix} \begin{bmatrix} 1 & 0 & 0 & 0 \\ 0 & 1 & 0 & 0 \\ 0 & 0 & 1 & 0 \\ 0 & b & 0 & 1 \end{bmatrix} = \begin{bmatrix} 1 & 0 & 0 & 0 \\ 0 & 1 & 0 & 0 \\ 0 & 0 & 1 & 0 \\ a & b & 0 & 1 \end{bmatrix} \equiv L_y L_x$$

$$R_x R_y = \begin{bmatrix} 1 & 0 & 0 & 0 \\ 0 & \cos\theta & -\sin\theta & 0 \\ 0 & \sin\theta & \cos\theta & 0 \\ 0 & 0 & 0 & 1 \end{bmatrix} \begin{bmatrix} \cos\phi & 0 & \sin\phi & 0 \\ 0 & 1 & 0 & 0 \\ -\sin\phi & 0 & \cos\phi & 0 \\ 0 & 0 & 0 & 1 \end{bmatrix}$$

$$= \begin{bmatrix} \cos\phi & 0 & \sin\phi & 0 \\ \sin\theta\sin\phi & \cos\theta & -\sin\theta\cos\phi & 0 \\ -\cos\theta\sin\phi & \sin\theta & \cos\theta\cos\phi & 0 \\ 0 & 0 & 0 & 1 \end{bmatrix}$$

$$R_yR_x = \begin{bmatrix} \cos\phi & 0 & \sin\phi & 0 \\ 0 & 1 & 0 & 0 \\ -\sin\phi & 0 & \cos\phi & 0 \\ 0 & 0 & 0 & 1 \end{bmatrix} \begin{bmatrix} 1 & 0 & 0 & 0 \\ 0 & \cos\theta & -\sin\theta & 0 \\ 0 & \sin\theta & \cos\theta & 0 \\ 0 & 0 & 0 & 1 \end{bmatrix}$$

$$= \begin{bmatrix} \cos\phi & \sin\phi\sin\theta & \sin\phi\cos\theta & 0 \\ 0 & \cos\theta & -\sin\theta & 0 \\ -\sin\phi & \cos\phi\sin\theta & \cos\theta\cos\phi & 0 \\ 0 & 0 & 0 & 1 \end{bmatrix} \neq R_xR_y$$

Thus we conclude that translations, scalings, and imagings about major axes commute, but rotations do not commute.

$$T_\alpha G_m = \begin{bmatrix} 1 & 0 & 0 & \alpha_1 \\ 0 & 1 & 0 & \alpha_2 \\ 0 & 0 & 1 & \alpha_3 \\ 0 & 0 & 0 & 1 \end{bmatrix} \begin{bmatrix} m & 0 & 0 & 0 \\ 0 & m & 0 & 0 \\ 0 & 0 & m & 0 \\ 0 & 0 & 0 & 1 \end{bmatrix} = \begin{bmatrix} m & 0 & 0 & \alpha_1 \\ 0 & m & 0 & \alpha_2 \\ 0 & 0 & m & \alpha_3 \\ 0 & 0 & 0 & 1 \end{bmatrix}, \quad \text{but}$$

$$G_m T_\alpha = \begin{bmatrix} m & 0 & 0 & 0 \\ 0 & m & 0 & 0 \\ 0 & 0 & m & 0 \\ 0 & 0 & 0 & 1 \end{bmatrix} \begin{bmatrix} 1 & 0 & 0 & \alpha_1 \\ 0 & 1 & 0 & \alpha_2 \\ 0 & 0 & 1 & \alpha_3 \\ 0 & 0 & 0 & 1 \end{bmatrix} = \begin{bmatrix} m & 0 & 0 & m\alpha_1 \\ 0 & m & 0 & m\alpha_2 \\ 0 & 0 & m & m\alpha_3 \\ 0 & 0 & 0 & 1 \end{bmatrix} \neq T_\alpha G_m$$

$$T_\alpha L_x = \begin{bmatrix} 1 & 0 & 0 & \alpha_1 \\ 0 & 1 & 0 & \alpha_2 \\ 0 & 0 & 1 & \alpha_3 \\ 0 & 0 & 0 & 1 \end{bmatrix} \begin{bmatrix} 1 & 0 & 0 & 0 \\ 0 & 1 & 0 & 0 \\ 0 & 0 & 1 & 0 \\ a & 0 & 0 & 1 \end{bmatrix} = \begin{bmatrix} 1 + a\alpha_1 & 0 & 0 & \alpha_1 \\ a\alpha_2 & 1 & 0 & \alpha_2 \\ a\alpha_3 & 0 & 1 & \alpha_3 \\ a & 0 & 0 & 1 \end{bmatrix}, \quad \text{but}$$

$$L_x T_\alpha = \begin{bmatrix} 1 & 0 & 0 & 0 \\ 0 & 1 & 0 & 0 \\ 0 & 0 & 1 & 0 \\ a & 0 & 0 & 1 \end{bmatrix} \begin{bmatrix} 1 & 0 & 0 & \alpha_1 \\ 0 & 1 & 0 & \alpha_2 \\ 0 & 0 & 1 & \alpha_3 \\ 0 & 0 & 0 & 1 \end{bmatrix} = \begin{bmatrix} 1 & 0 & 0 & \alpha_1 \\ 0 & 1 & 0 & \alpha_2 \\ 0 & 0 & 1 & \alpha_3 \\ a & 0 & 0 & a\alpha_1 \end{bmatrix} \neq T_\alpha L_x$$

$$G_m L_x = \begin{bmatrix} m & 0 & 0 & 0 \\ 0 & m & 0 & 0 \\ 0 & 0 & m & 0 \\ 0 & 0 & 0 & 1 \end{bmatrix} \begin{bmatrix} 1 & 0 & 0 & 0 \\ 0 & 1 & 0 & 0 \\ 0 & 0 & 1 & 0 \\ a & 0 & 0 & 1 \end{bmatrix} = \begin{bmatrix} m & 0 & 0 & 0 \\ 0 & m & 0 & 0 \\ 0 & 0 & m & 0 \\ a & 0 & 0 & 1 \end{bmatrix}, \quad \text{but}$$

$$L_x G_m = \begin{bmatrix} 1 & 0 & 0 & 0 \\ 0 & 1 & 0 & 0 \\ 0 & 0 & 1 & 0 \\ a & 0 & 0 & 1 \end{bmatrix} \begin{bmatrix} m & 0 & 0 & 0 \\ 0 & m & 0 & 0 \\ 0 & 0 & m & 0 \\ 0 & 0 & 0 & 1 \end{bmatrix} = \begin{bmatrix} m & 0 & 0 & 0 \\ 0 & m & 0 & 0 \\ 0 & 0 & m & 0 \\ am & 0 & 0 & 1 \end{bmatrix} \neq G_m L_x$$

so two different operations in the group scaling, imaging, and translation do not in general commute. It is unlikely that two of the same operation are performed together (e.g., scaling followed by scaling is redundant). The recommendation is to assume that no operations are commutable and follow strictly the order given.

---

2. A point **u** on plane $[a \quad b \quad c \quad 0]$ is distance β from the origin. Find **u** if it is as close as possible to the z-axis. If it is rotated by θ in **p** about the origin to reach point **v**, find an expression for **v** in terms of a, b, c, β, and θ. Find the horizontal distance from the pivot.

Point **u** is found as follows. A point vertically above the pivot is $[0 \quad 0 \quad 1 \quad 1]^{\#}$, and the line normal to **p** from this point is

$$\begin{bmatrix} 0 \\ 0 \\ 1 \\ 1 \end{bmatrix} + \alpha \begin{bmatrix} a \\ b \\ c \\ 0 \end{bmatrix}$$

For this line to intersect **p** we have

$$[a \quad b \quad c \quad 0]\left\{\begin{bmatrix} 0 \\ 0 \\ 1 \\ 1 \end{bmatrix} + \alpha \begin{bmatrix} a \\ b \\ c \\ 0 \end{bmatrix}\right\} = 0, \qquad \text{so } \alpha = -c$$

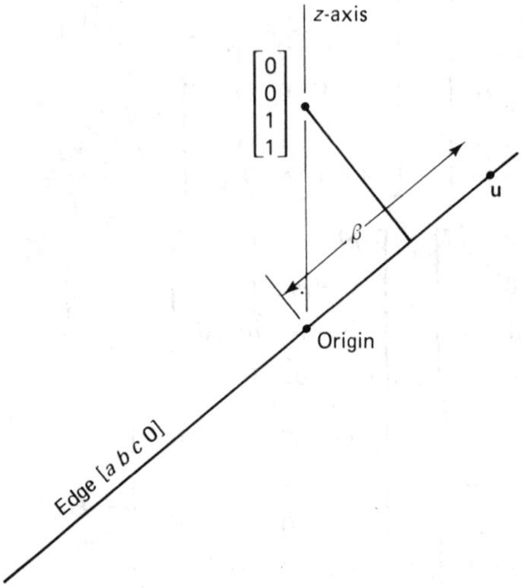

Figure 2.6 Display of a point on $[a \quad b \quad c \quad 0]$.

Therefore, a point on the plane that lies on the datum line shown in Figure 2.6 is

$$\begin{bmatrix} 0 \\ 0 \\ 1 \\ 1 \end{bmatrix} - c \begin{bmatrix} a \\ b \\ c \\ 0 \end{bmatrix} = \begin{bmatrix} -ac \\ -bc \\ 1 - c^2 \\ 1 \end{bmatrix}$$

and the distance D of this point from the origin is given by

$$D^2 = a^2c^2 + b^2c^2 + (1 - c^2)^2 = a^2c^2 + b^2c^2 + c^4 + 1 - 2c^2 = 1 - c^2$$

Thus

$$\mathbf{u} = \frac{\beta}{\sqrt{1 - c^2}} \begin{bmatrix} -ac \\ -bc \\ 1 - c^2 \\ 0 \end{bmatrix} + \begin{bmatrix} 0 \\ 0 \\ 0 \\ 1 \end{bmatrix}$$

From equation (2.3) we have the rotated location of the point as

$$\mathbf{v} = \frac{\beta}{\sqrt{1 - c^2}} \begin{bmatrix} \cos\theta & -c\sin\theta & b\sin\theta & 0 \\ c\sin\theta & \cos\theta & -a\sin\theta & 0 \\ -b\sin\theta & a\sin\theta & \cos\theta & 0 \\ 0 & 0 & 0 & 1 \end{bmatrix} \begin{bmatrix} -ac \\ -bc \\ 1 - c^2 \\ 0 \end{bmatrix} + \begin{bmatrix} 0 \\ 0 \\ 0 \\ 1 \end{bmatrix}$$

$$= \frac{\beta}{\sqrt{1 - c^2}} \begin{bmatrix} -ac\cos\theta + b\sin\theta \\ -bc\cos\theta - a\sin\theta \\ (1 - c^2)\cos\theta \\ 0 \end{bmatrix} + \begin{bmatrix} 0 \\ 0 \\ 0 \\ 1 \end{bmatrix}$$

The horizontal distance between $\mathbf{v}$ and the pivot is

$$\frac{\beta}{\sqrt{1 - c^2}} \sqrt{(-ac\cos\theta + b\sin\theta)^2 + (bc\cos\theta + a\sin\theta)^2}$$

$$= \beta\sqrt{\frac{a^2c^2\cos^2\theta + b^2\sin^2\theta + b^2c^2\cos^2\theta + a^2\sin^2\theta}{1 - c^2}}$$

$$= \beta\sqrt{\frac{c^2(1 - c^2)\cos^2\theta + (1 - c^2)\sin^2\theta}{1 - c^2}} = \beta\sqrt{c^2\cos^2\theta + \sin^2\theta}$$

---

3. Rotate vector $\boldsymbol{v}$ by $\pi/2$ about the x-axis, then by $\pi/2$ about the y-axis, and finally by $\pi/2$ about the z-axis. Perform the transformations in reverse order and compare results.

$$R_zR_yR_x = \begin{bmatrix} 0 & -1 & 0 \\ 1 & 0 & 0 \\ 0 & 0 & 1 \end{bmatrix} \begin{bmatrix} 0 & 0 & 1 \\ 0 & 1 & 0 \\ -1 & 0 & 0 \end{bmatrix} \begin{bmatrix} 1 & 0 & 0 \\ 0 & 0 & -1 \\ 0 & 1 & 0 \end{bmatrix}$$

$$= \begin{bmatrix} 0 & -1 & 0 \\ 1 & 0 & 0 \\ 0 & 0 & 1 \end{bmatrix} \begin{bmatrix} 0 & 1 & 0 \\ 0 & 0 & -1 \\ -1 & 0 & 0 \end{bmatrix} = \begin{bmatrix} 0 & 0 & 1 \\ 0 & 1 & 0 \\ -1 & 0 & 0 \end{bmatrix}$$

$$R_xR_yR_z = \begin{bmatrix} 1 & 0 & 0 \\ 0 & 0 & -1 \\ 0 & 1 & 0 \end{bmatrix}\begin{bmatrix} 0 & 0 & 1 \\ 0 & 1 & 0 \\ -1 & 0 & 0 \end{bmatrix}\begin{bmatrix} 0 & -1 & 0 \\ 1 & 0 & 0 \\ 0 & 0 & 1 \end{bmatrix}$$

$$= \begin{bmatrix} 1 & 0 & 0 \\ 0 & 0 & -1 \\ 0 & 1 & 0 \end{bmatrix}\begin{bmatrix} 0 & 0 & 1 \\ 1 & 0 & 0 \\ 0 & 1 & 0 \end{bmatrix} = \begin{bmatrix} 0 & 0 & 1 \\ 0 & -1 & 0 \\ 1 & 0 & 0 \end{bmatrix}$$

$$R_zR_yR_x\boldsymbol{\nu} = \begin{bmatrix} 0 & 0 & 1 \\ 0 & 1 & 0 \\ -1 & 0 & 0 \end{bmatrix}\begin{bmatrix} \nu_x \\ \nu_y \\ \nu_z \end{bmatrix} = \begin{bmatrix} \nu_z \\ \nu_y \\ -\nu_x \end{bmatrix}$$

$$R_xR_yR_z\boldsymbol{\nu} = \begin{bmatrix} 0 & 0 & 1 \\ 0 & -1 & 0 \\ 1 & 0 & 0 \end{bmatrix}\begin{bmatrix} \nu_x \\ \nu_y \\ \nu_z \end{bmatrix} = \begin{bmatrix} \nu_z \\ -\nu_y \\ \nu_x \end{bmatrix}$$

The difference between the two results indicates that rotations are not commutative.

---

4. Find the rotation matrix that transforms the first to the second unit vector:

$$\frac{1}{\sqrt{3}}\begin{bmatrix} 1 \\ 1 \\ 1 \end{bmatrix} \quad \text{and} \quad \frac{1}{\sqrt{6}}\begin{bmatrix} 2 \\ 1 \\ 1 \end{bmatrix}$$

The plane of rotation $[a \quad b \quad c \quad d]$ is given in part by

$$\begin{bmatrix} 1 \\ 1 \\ 1 \end{bmatrix} \times \begin{bmatrix} 2 \\ 1 \\ 1 \end{bmatrix} = \begin{bmatrix} 0 \\ 1 \\ -1 \end{bmatrix} \equiv \frac{1}{\sqrt{2}}\begin{bmatrix} 0 \\ 1 \\ -1 \end{bmatrix}$$

The angle between the vectors is given by the dot product

$$\cos\theta = \frac{1}{\sqrt{18}}[1 \quad 1 \quad 1]\begin{bmatrix} 2 \\ 1 \\ 1 \end{bmatrix} = \frac{4}{\sqrt{18}} = 0.9428$$

and $\sin\theta = 0.3333$. Thus, the rotation matrix is

$$R(\theta) = \begin{bmatrix} 0.9428 & 0.2357 & -0.2357 \\ -0.2357 & 0.9428 & 0 \\ 0.2357 & 0 & 0.9428 \end{bmatrix}$$

---

5. Show by example that the mapping of $\boldsymbol{\nu}$ into $\boldsymbol{\eta}$ by linear operator (matrix) A is not unique.

With matrix $A = \begin{bmatrix} 7 & -8 \\ 6 & -1 \end{bmatrix}$ and vector $\boldsymbol{\nu} = \begin{bmatrix} 5 \\ 3 \end{bmatrix}$, $\boldsymbol{\eta} = \begin{bmatrix} 7 & -8 \\ 6 & -1 \end{bmatrix}\begin{bmatrix} 5 \\ 3 \end{bmatrix} = \begin{bmatrix} 11 \\ 27 \end{bmatrix}$.

$$A^{-1} = \begin{bmatrix} 7 & -8 \\ 6 & -1 \end{bmatrix}^{-1} = \frac{1}{41}\begin{bmatrix} -1 & 8 \\ -6 & 7 \end{bmatrix} \quad \text{and} \quad \frac{1}{41}\begin{bmatrix} -1 & 8 \\ -6 & 7 \end{bmatrix}\begin{bmatrix} 11 \\ 27 \end{bmatrix} = \begin{bmatrix} 5 \\ 3 \end{bmatrix}$$

However, matrix $B = \begin{bmatrix} -2 & 1 \\ 1.5 & -0.5 \end{bmatrix}$ also solves the system, as in

$$\begin{bmatrix} -2 & 1 \\ 1.5 & -0.5 \end{bmatrix}\begin{bmatrix} 11 \\ 27 \end{bmatrix} = \begin{bmatrix} 5 \\ 3 \end{bmatrix}, \quad \text{and} \quad \frac{1}{41}\begin{bmatrix} -1 & 8 \\ -6 & 7 \end{bmatrix} \neq \begin{bmatrix} -2 & 1 \\ 1.5 & -0.5 \end{bmatrix}$$

---

6. Given point $\mathbf{u} = [1 \quad 1 \quad 1 \quad 1]^{\#}$, which is rotated about the x-, y- and z-axes, respectively, by $-30°$. Determine the solid angle of rotation and the plane of rotation that must pass through the origin. Use the matrix of equation (2.8) and verify that the resulting solid angle rotation matrix is valid.

Writing the matrices of equations (2.15) and (2.16) in condensed form and noting that $\sin 30 = \frac{1}{2}$ and $\cos 30 = \sqrt{3}/2$ gives

$$R_{z\theta}R_{y\theta}R_{x\theta}\begin{bmatrix} 1 \\ 1 \\ 1 \end{bmatrix} = \frac{1}{8}\begin{bmatrix} \sqrt{3} & 1 & 0 \\ -1 & \sqrt{3} & 0 \\ 0 & 0 & 2 \end{bmatrix}\begin{bmatrix} \sqrt{3} & 0 & -1 \\ 0 & 2 & 0 \\ 1 & 0 & \sqrt{3} \end{bmatrix}\begin{bmatrix} 2 & 0 & 0 \\ 0 & \sqrt{3} & 1 \\ 0 & -1 & \sqrt{3} \end{bmatrix}\begin{bmatrix} 1 \\ 1 \\ 1 \end{bmatrix}$$

$$= \frac{1}{8}\begin{bmatrix} \sqrt{3} & 1 & 0 \\ -1 & \sqrt{3} & 0 \\ 0 & 0 & 2 \end{bmatrix}\begin{bmatrix} \sqrt{3} & 0 & -1 \\ 0 & 2 & 0 \\ 1 & 0 & \sqrt{3} \end{bmatrix}\begin{bmatrix} 2 \\ \sqrt{3}+1 \\ \sqrt{3}-1 \end{bmatrix}$$

$$= \frac{1}{8}\begin{bmatrix} \sqrt{3} & 1 & 0 \\ -1 & \sqrt{3} & 0 \\ 0 & 0 & 2 \end{bmatrix}\begin{bmatrix} \sqrt{3}+1 \\ 2\sqrt{3}+2 \\ 5-\sqrt{3} \end{bmatrix}$$

$$= \frac{1}{8}\begin{bmatrix} 5+3\sqrt{3} \\ 5+\sqrt{3} \\ 10-2\sqrt{3} \end{bmatrix} = \begin{bmatrix} 1.2745 \\ 0.8415 \\ 0.8169 \end{bmatrix}$$

The solid angle of rotation with respect to the origin is

$$\cos\theta = \frac{1}{3}\begin{bmatrix} 1 \\ 1 \\ 1 \end{bmatrix} \cdot \begin{bmatrix} 1.2745 \\ 0.8415 \\ 0.8169 \end{bmatrix} = 0.9776$$

so $\theta = 12.14°$. The plane of rotation is given in part by

$$\begin{bmatrix} a \\ b \\ c \end{bmatrix} = \begin{bmatrix} 1.2745 \\ 0.8415 \\ 0.8169 \end{bmatrix} \times \begin{bmatrix} 1 \\ 1 \\ 1 \end{bmatrix} = \begin{bmatrix} 0.0246 \\ -0.4576 \\ 0.4276 \end{bmatrix}$$

or in normalized form as

$$\begin{bmatrix} a \\ b \\ c \end{bmatrix} = \pm\begin{bmatrix} 0.0392 \\ -0.7301 \\ 0.6822 \end{bmatrix}$$

Since the plane passes through the origin, $d = 0$. Using equation (2.8) we get

$$\mathbf{w} = \begin{bmatrix} 0.9776 & 0.1435 & 0.1535 & 0 \\ -0.1435 & 0.9776 & 0.0082 & 0 \\ -0.1535 & -0.0082 & 0.9776 & 0 \\ 0 & 0 & 0 & 1 \end{bmatrix} \begin{bmatrix} 1 \\ 1 \\ 1 \\ 1 \end{bmatrix} = \begin{bmatrix} 1.2746 \\ 0.8423 \\ 0.8159 \\ 1 \end{bmatrix}$$

which is the same as before (within rounding error). Notice the sensitivity to rounding error.

---

7. Given point $[1 \quad 1 \quad 0 \quad 1]^{\#}$, which is to be rotated in plane $\mathbf{p} = [-1 \quad 0 \quad 0 \quad 1]$ about a point on the x-axis by integer angles $10i°$. A pinhole camera sits with pinhole at the origin and film plane $\mathbf{q} = [1 \quad 0 \quad 0 \quad 1]$. Draw the image produced on $\mathbf{q}$. Comment on the spacing of the points and the shape produced. Repeat the problem with the pinhole of the camera at $[0 \quad 0 \quad -0.5 \quad 1]^{\#}$. Again comment on the shape produced.

The point of rotation $\mathbf{v}$ on the x-axis is given by $\mathbf{pv} = 0$ or $[-1 \quad 0 \quad 0 \quad 1]$ $[x \quad 0 \quad 0 \quad 1]^{\#} = 0$, so $-1x + 1 = 0$, or $x = 1$. Thus $\mathbf{v} = [1 \quad 0 \quad 0 \quad 1]^{\#}$. Using equation (2.6), which requires the plane of rotation to be normalized to $\mathbf{p} = [-1/\sqrt{2} \quad 0 \quad 1/\sqrt{2} \quad 1/\sqrt{2}]$, the rotated points are

$$\mathbf{u}_i = \begin{bmatrix} 1 \\ 0 \\ 0 \\ 1 \end{bmatrix} + \begin{bmatrix} \cos 10i & \frac{1}{\sqrt{2}} \sin 10i & 0 & 0 \\ -\frac{1}{\sqrt{2}} \sin 10i & \cos 10i & -\frac{1}{\sqrt{2}} \sin 10i & 0 \\ 0 & \frac{1}{\sqrt{2}} \sin 10i & \cos 10i & 0 \\ 0 & 0 & 0 & 1 \end{bmatrix} \begin{bmatrix} 0 \\ 1 \\ 0 \\ 0 \end{bmatrix}$$

$$= \begin{bmatrix} 1 + \frac{1}{\sqrt{2}} \sin 10i \\ \cos 10i \\ \frac{1}{\sqrt{2}} \sin 10i \\ 1 \end{bmatrix} = \begin{bmatrix} x_i \\ y_i \\ z_i \\ 1 \end{bmatrix}$$

Since when $i = 36$ the angle of rotation is $360° \equiv 0°$, this places an upper limit on i. Further, the points are symmetric about the xz-plane for $-9 \leq i \leq 9$ and $27 \geq i \geq 9$, so we need to solve for $-9 \leq i \leq 9$. From equation (2.31), the image points with pinhole $\mathbf{w} = [0 \quad 0 \quad 0 \quad 1]^{\#}$ are given by

$$\mathbf{v}_i = \mathbf{u}_i - \frac{\mathbf{q}\mathbf{u}_i}{\mathbf{q}(\mathbf{w} - \mathbf{u}_i)}(\mathbf{w} - \mathbf{u}_i) = \begin{bmatrix} x_i \\ y_i \\ z_i \\ 1 \end{bmatrix} - \frac{[1 \quad 0 \quad 0 \quad 1]\begin{bmatrix} x_i \\ y_i \\ z_i \\ 1 \end{bmatrix}}{[1 \quad 0 \quad 0 \quad 1]\begin{bmatrix} x_i \\ y_i \\ z_i \\ 0 \end{bmatrix}}\begin{bmatrix} x_i \\ y_i \\ z_i \\ 0 \end{bmatrix}$$

$$= \begin{bmatrix} x_i \\ y_i \\ z_i \\ 1 \end{bmatrix} - \frac{x_i + 1}{x_i}\begin{bmatrix} x_i \\ y_i \\ z_i \\ 0 \end{bmatrix} = \begin{bmatrix} -1 \\ y_i\left\{1 - \dfrac{x_i + 1}{x_i}\right\} \\ z_i\left\{1 - \dfrac{x_i + 1}{x_i}\right\} \\ 1 \end{bmatrix} = \begin{bmatrix} -1 \\ \cos 10i\left\{1 - \dfrac{2\sqrt{2} + \sin 10i}{\sqrt{2} + \sin 10i}\right\} \\ \dfrac{\sin 10i}{\sqrt{2}}\left\{1 - \dfrac{2\sqrt{2} + \sin 10i}{\sqrt{2} + \sin 10i}\right\} \\ 1 \end{bmatrix}$$

Thus we can construct the following table:

i	$\sin 10i$	$\cos 10i$	$1 - \dfrac{2\sqrt{2} + \sin 10i}{\sqrt{2} + \sin 10i}$	y_i	z_i	$\dfrac{\sqrt{2} + 0.5 \sin 10i}{\sqrt{2} + \sin 10i}$	z_i'
−9	−1	0	−3.415	0	2.415	2.207	4.622
−8	−0.9848	0.1736	−3.295	−0.572	2.295		
−7	−0.9397	0.3420	−2.982	−1.020	1.981		
−6	−0.8660	0.5000	−2.581	−1.291	1.580	1.800	3.380
−5	−0.7660	0.6428	−2.183	−1.403	1.182		
−4	−0.6428	0.7660	−1.834	−1.405	0.834		
−3	−0.5000	0.8660	−1.548	−1.341	0.547	1.273	1.820
−2	−0.3420	0.9397	−1.319	−1.239	0.319		
−1	−0.1736	0.9848	−1.140	−1.123	0.140		
0	0	1	−1	−1	0	1	1
1	0.1736	0.9848	−0.891	−0.877	−0.109		
2	0.3420	0.9397	−0.805	−0.756	−0.195		
3	0.5000	0.8660	−0.739	−0.640	−0.261	0.869	0.608
4	0.6428	0.7660	−0.688	−0.527	−0.313		
5	0.7660	0.6428	−0.649	−0.417	−0.352		
6	0.8660	0.5000	−0.620	−0.320	−0.380	0.810	0.430
7	0.9397	0.3420	−0.601	−0.206	−0.399		
8	0.9848	0.1736	−0.590	−0.102	−0.411		
9	1	0	−0.586	0	−0.414	0.793	0.379

The $\mathbf{u}_i$ lie on a circle in $\mathbf{p}$. The image points $\mathbf{v}_i$ form a circle centered at $y = 0$, $z = 1$ of radius 2. However, the spacing of the points is by no means equal, as shown in Figure 2.7. The imaging of a circle to a circle occurs only when the plane containing the original circle and the film plane are parallel.

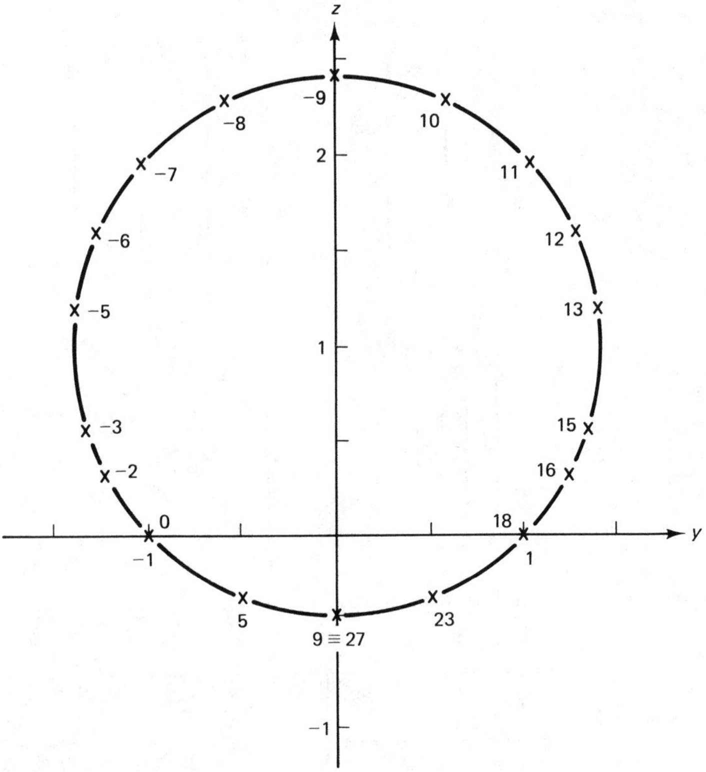

Figure 2.7

Repeating the problem with the pinhole $\mathbf{w} = [0 \quad 0 \quad -0.5 \quad 1]^{\#}$, we have

$$\mathbf{v}_i = \mathbf{u}_i - \frac{\mathbf{q}\mathbf{u}_i}{\mathbf{q}(\mathbf{w}-\mathbf{u}_i)}(\mathbf{w}-\mathbf{u}_i) = \begin{bmatrix} x_i \\ y_i \\ z_i \\ 1 \end{bmatrix} - \frac{[1 \quad 0 \quad 0 \quad 1]\begin{bmatrix} x_i \\ y_i \\ z_i \\ 1 \end{bmatrix}}{[1 \quad 0 \quad 0 \quad 1]\begin{bmatrix} x_i \\ y_i \\ z_i - 0.5 \\ 0 \end{bmatrix}} \begin{bmatrix} x_i \\ y_i \\ z_i - 0.5 \\ 0 \end{bmatrix}$$

$$= \begin{bmatrix} x_i \\ y_i \\ z_i \\ 1 \end{bmatrix} - \frac{x_i + 1}{x_i}\begin{bmatrix} x_i \\ y_i \\ z_i - 0.5 \\ 0 \end{bmatrix} = \begin{bmatrix} -1 \\ y_i\left\{1 - \dfrac{x_i + 1}{x_i}\right\} \\ z_i\left\{1 - \dfrac{x_i + 1}{x_i}\right\} + 0.5\,\dfrac{x_i + 1}{x_i} \\ 1 \end{bmatrix}$$

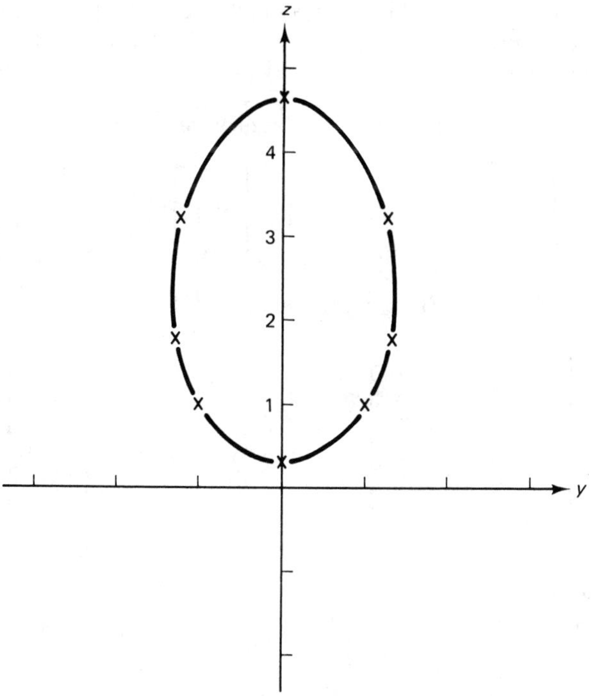

Figure 2.8

$$= \begin{bmatrix} -1 \\ \cos 10i \left\{1 - \dfrac{2\sqrt{2} + \sin 10i}{\sqrt{2} + \sin 10i}\right\} \\ \dfrac{\sin 10i}{\sqrt{2}} \left\{1 - \dfrac{2\sqrt{2} + \sin 10i}{\sqrt{2} + \sin 10i}\right\} + \dfrac{\sqrt{2} + 0.5 \sin 10i}{\sqrt{2} + \sin 10i} \\ 1 \end{bmatrix}$$

and so construct the last two columns in the table. The shape in **q** is now elliptical, as shown in Figure 2.8. It can be shown[1] that the image of any ellipse is an ellipse.

---

8. Given the three points

$$\begin{bmatrix} 1 \\ 1 \\ 0 \\ 1 \end{bmatrix}, \quad \begin{bmatrix} 2 \\ 1 \\ 0 \\ 1 \end{bmatrix}, \quad \text{and} \quad \begin{bmatrix} 1.5 \\ 2 \\ 0 \\ 1 \end{bmatrix}$$

[1] The author is indebted to Charles Dunning, president of Computer Mart, Waltham, Massachusetts, for showing this and providing an elegant proof.

Find the location of these three points in plane $\mathbf{q} = [1 \quad 1 \quad 1 \quad 1]$ when the first point given corresponds to point $[0 \quad 0 \quad -1 \quad 1]^\#$ in plane $\mathbf{q}$ and the cut-and-paste operation is performed without orientation; $\theta = 0$ in equation (2.31).

The plane $[a \quad b \quad c \quad d]$ within which the original three points lie is given by

$$\begin{bmatrix} a \\ b \\ c \end{bmatrix} = \begin{bmatrix} 1-2 \\ 1-1 \\ 0-0 \end{bmatrix} \times \begin{bmatrix} 1-1.5 \\ 1-2 \\ 0-0 \end{bmatrix} = \begin{bmatrix} -1 \\ 0 \\ 0 \end{bmatrix} \times \begin{bmatrix} -0.5 \\ -1 \\ 0 \end{bmatrix} = \begin{bmatrix} 0 \\ 0 \\ 1 \end{bmatrix}$$

From equation (2.28), and with $\mathbf{q}$ normalized to $[1 \quad 1 \quad 1 \quad 1]/\sqrt{3}$,

$$\boldsymbol{\eta} = \begin{bmatrix} 0 \\ 0 \\ 1 \end{bmatrix} \times \begin{bmatrix} \frac{1}{\sqrt{3}} \\ \frac{1}{\sqrt{3}} \\ \frac{1}{\sqrt{3}} \end{bmatrix} = \begin{bmatrix} \frac{-1}{\sqrt{3}} \\ \frac{-1}{\sqrt{3}} \\ 0 \end{bmatrix} \quad \text{and} \quad \cos\psi = [0\ 0\ 1] \begin{bmatrix} \frac{1}{\sqrt{3}} \\ \frac{1}{\sqrt{3}} \\ \frac{1}{\sqrt{3}} \end{bmatrix} = \frac{1}{\sqrt{3}}$$

so $\sin\psi = 0.8165$

$$\boldsymbol{\nu}_1 = \begin{bmatrix} 0 \\ 0 \\ 1 \end{bmatrix}, \quad \boldsymbol{\nu}_2 = \begin{bmatrix} 1 \\ 0 \\ 0 \end{bmatrix}, \quad \text{and} \quad \boldsymbol{\nu}_3 = \begin{bmatrix} 0.5 \\ 1 \\ 0 \end{bmatrix}$$

so

$$\boldsymbol{\mu}_2 = \begin{bmatrix} 0.5774 & 0 & -0.5774 \cdot 0.8165 \\ 0 & 0.5774 & 0.5774 \cdot 0.8165 \\ 0.5774 \cdot 0.8165 & -0.5774 \cdot 0.8165 & 0.5774 \end{bmatrix} \begin{bmatrix} 1 \\ 0 \\ 0 \end{bmatrix}$$

$$= \begin{bmatrix} 0.5774 & 0 & -0.4714 \\ 0 & 0.5774 & 0.4714 \\ 0.4714 & -0.4714 & 0.5774 \end{bmatrix} \begin{bmatrix} 1 \\ 0 \\ 0 \end{bmatrix} = \begin{bmatrix} 0.5774 \\ 0 \\ 0.4714 \end{bmatrix}$$

$$\boldsymbol{\mu}_3 = \begin{bmatrix} 0.5774 & 0 & -0.4714 \\ 0 & 0.5774 & 0.4714 \\ 0.4714 & -0.4714 & 0.5774 \end{bmatrix} \begin{bmatrix} 0.5 \\ 1 \\ 0 \end{bmatrix} = \begin{bmatrix} 0.2887 \\ 0.5774 \\ -0.2357 \end{bmatrix}$$

Thus

$$\mathbf{v}_2 = \begin{bmatrix} 0 + 0.5774 \\ 0 + 0 \\ -1 + 0.4714 \\ 1 \end{bmatrix} \quad \text{and} \quad \mathbf{v}_3 = \begin{bmatrix} 0 + 0.2887 \\ 0 + 0.5774 \\ -1 - 0.2357 \\ 1 \end{bmatrix}$$

9. A pinhole camera has pinhole at point $\mathbf{w} = \begin{bmatrix} -1 \\ 0 \\ 0 \\ 1 \end{bmatrix}$ and film plane $\mathbf{q} =$ $[1 \quad 0 \quad 0 \quad 2]$.

(a) Determine the range of values for α so that point $\mathbf{u} = \begin{bmatrix} \alpha \\ \alpha \\ 2\alpha \\ 1 \end{bmatrix}$ can be imaged by the camera.

(b) Determine the abscissa and ordinate values of the image of $\mathbf{u}$ for $\alpha = 0$, $\alpha = 1$, and $\alpha = 2$, assuming the camera is oriented so that a Cartesian vertical produces an image that is parallel to the ordinate. Comment on the difference in distance between the first pair and second pair of image points.

(a) From equation (2.31) the image point is given by

$$\mathbf{v} = \begin{bmatrix} \alpha \\ \alpha \\ 2\alpha \\ 1 \end{bmatrix} - \beta \begin{bmatrix} -1 - \alpha \\ -\alpha \\ -2\alpha \\ 0 \end{bmatrix}, \qquad \text{where } \beta = \frac{\mathbf{qu}}{\mathbf{q(w - u)}} = \frac{\alpha + 2}{-1 - \alpha}$$

When $\alpha = -1$, $\beta \to \infty$, and the object point cannot be imaged. The range of values that permit imaging is determined as follows. From equation (1.17),

$$\frac{\mathbf{pw}}{\mathbf{pw} - d} = \frac{-1 + 2}{-1} = -1$$

so the plane is outside $\mathbf{w}$ with respect to the origin. Thus, $\mathbf{u}$ must be on the origin side of $\mathbf{w}$, which requires $-1 < \alpha < \infty$.

(b)

$$\mathbf{v} = \begin{bmatrix} \alpha \\ \alpha \\ 2\alpha \\ 1 \end{bmatrix} - \begin{bmatrix} a + 2 \\ \dfrac{\alpha(\alpha + 2)}{\alpha + 1} \\ \dfrac{2\alpha(\alpha + 2)}{\alpha + 1} \\ 0 \end{bmatrix} = \begin{bmatrix} -2 \\ \dfrac{-\alpha}{\alpha + 1} \\ \dfrac{-2\alpha}{\alpha + 1} \\ 1 \end{bmatrix}$$

At $\alpha = 0$, $\mathbf{v} = \begin{bmatrix} -2 \\ 0 \\ 0 \\ 1 \end{bmatrix}$. At $\alpha = 1$, $\mathbf{v} = \begin{bmatrix} -2 \\ -\frac{1}{2} \\ -1 \\ 1 \end{bmatrix}$. At $\alpha = 2$, $\mathbf{v} = \begin{bmatrix} -2 \\ -\frac{2}{3} \\ -\frac{4}{3} \\ 1 \end{bmatrix}$.

An axial point on the object side of the camera is

$$\mathbf{w} + \begin{bmatrix} 1 \\ 0 \\ 0 \\ 0 \end{bmatrix} = \begin{bmatrix} 0 \\ 0 \\ 0 \\ 1 \end{bmatrix}$$

From equation (2.31) this axial point is imaged as

$$\mathbf{v} = \begin{bmatrix} 0 \\ 0 \\ 0 \\ 1 \end{bmatrix} - \frac{2}{-1} \begin{bmatrix} -1 \\ 0 \\ 0 \\ 0 \end{bmatrix} = \begin{bmatrix} -2 \\ 0 \\ 0 \\ 1 \end{bmatrix}$$

Adding a vertical component to the object point and imaging gives

$$\mathbf{v} = \begin{bmatrix} 0 \\ 0 \\ 1 \\ 1 \end{bmatrix} - \frac{2}{-1} \begin{bmatrix} -1 \\ 0 \\ -1 \\ 0 \end{bmatrix} = \begin{bmatrix} -2 \\ 0 \\ -1 \\ 1 \end{bmatrix}$$

so the unit ordinate vector is $\begin{bmatrix} 0 \\ 0 \\ -1 \end{bmatrix}$.

The abscissa is

$$\begin{bmatrix} 0 \\ 0 \\ -1 \end{bmatrix} \times \begin{bmatrix} 1 \\ 0 \\ 0 \end{bmatrix} = \begin{bmatrix} 0 \\ -1 \\ 0 \end{bmatrix}$$

The dot products of the vector $\boldsymbol{\nu}$ from the axial point on the film plane to each image point with the abscissa and ordinate give the abscissa values X and ordinate values Y; that is,

$$\alpha = 0, \qquad \boldsymbol{\nu} = \begin{bmatrix} 0 \\ 0 \\ 0 \end{bmatrix}, \qquad \text{so } X = 0,\ Y = 0$$

$$\alpha = 1, \qquad \boldsymbol{\nu} = \begin{bmatrix} 0 \\ -\frac{1}{2} \\ -1 \end{bmatrix}, \qquad \text{so } X = \tfrac{1}{2},\ Y = 1$$

$$\alpha = 2, \qquad \boldsymbol{\nu} = \begin{bmatrix} 0 \\ -\frac{2}{3} \\ -\frac{4}{3} \end{bmatrix}, \qquad \text{so } X = \tfrac{2}{3},\ Y = \tfrac{4}{3}$$

The distance between the first pair of points is $\sqrt{\frac{1}{4} + 1} = 1.118$. The distance between the second pair of points is $\sqrt{\frac{1}{36} + \frac{1}{9}} = 0.373$. The three points are in a straight line, but the change in image point location as a function of an incremental change along the object line is greatest close to the axial point (the origin in this case). Thus, the difference between the first two points will be greater than the difference between the second pair of points.

10. A circle of radius r is centered at the origin and lies on plane $\mathbf{p} = [a \quad b \quad c \quad 0]$. Define locations on the circle as a function of θ.

Suppose a point on the circle is

$$\mathbf{u} = \begin{bmatrix} x \\ y \\ z \\ 1 \end{bmatrix}$$

Then $\mathbf{pu} = ax + by + cz = 0$. Also $x^2 + y^2 + z^2 = r^2$. We have two equations and two unknowns. We arbitrarily set $z = 0$, so $ax + by = 0$ and $x^2 + y^2 = r^2$, giving

$$x = \frac{-br}{\sqrt{a^2 + b^2}} \quad \text{and} \quad y = \frac{ar}{\sqrt{a^2 + b^2}}$$

Setting

$$\boldsymbol{\nu} = \frac{r}{\sqrt{a^2 + b^2}} \begin{bmatrix} -b \\ a \\ 0 \end{bmatrix}$$

in equation (2.3),

$$\boldsymbol{\eta} = \frac{r}{\sqrt{a^2 + b^2}} \begin{bmatrix} \cos\theta & c\sin\theta & -b\sin\theta \\ -c\sin\theta & \cos\theta & a\sin\theta \\ b\sin\theta & -a\sin\theta & \cos\theta \end{bmatrix} \begin{bmatrix} -b \\ a \\ 0 \end{bmatrix}$$

$$= \frac{r}{\sqrt{a^2 + b^2}} \begin{bmatrix} -b\cos\theta + ac\sin\theta \\ a\cos\theta + bc\sin\theta \\ -(a^2 + b^2)\sin\theta \end{bmatrix}$$

---

11. A solid rectangular object has corner points

$$\begin{matrix} 1 & 2 & 3 & 4 & 5 & 6 & 7 & 8 \end{matrix}$$
$$\begin{bmatrix} 0 & 2 & 0 & 2 & 0 & 2 & 0 & 2 \\ 0 & 0 & 1 & 1 & 0 & 0 & 1 & 1 \\ 0 & 0 & 0 & 0 & 3 & 3 & 3 & 3 \\ 1 & 1 & 1 & 1 & 1 & 1 & 1 & 1 \end{bmatrix}$$

Define the object by the planes that specify its surfaces such that the vectors partially specifying the planes are pointing outward from the object. The object is viewed from a pinhole camera with hole at point $[10 \quad 10 \quad 5 \quad 1]^{\#}$. The axis of the camera intersects the origin, and the film plane has minimum distance of 1 from the pinhole. Calculate the image obtained at the film plane. Delete hidden lines so the image is a true image of a solid object.

The base of the object with corner points 1, 2, 3, and 4 is the xy-plane $[0 \quad 0 \quad 1 \quad 0]$. Corner points 1, 2, 5, and 6 are on the xz-plane $[0 \quad 1 \quad 0 \quad 0]$. Points 3, 4, 7, and 8 are on $[0 \quad 1 \quad 0 \quad -1]$. Points 1, 3, 5, and 7 are on the yz-plane $[1 \quad 0 \quad 0 \quad 0]$. Points 2, 4, 6, and 8 are on $[1 \quad 0 \quad 0 \quad -2]$. Points 5, 6, 7, and 8 are on $[0 \quad 0 \quad 1 \quad -3]$. Orienting the planes so that the vectors partially defining the planes are directed outwards from the object gives

$$S = \begin{bmatrix} 0 & 0 & -1 & 0 \\ 0 & -1 & 0 & 0 \\ 0 & 1 & 0 & -1 \\ -1 & 0 & 0 & 0 \\ 1 & 0 & 0 & -2 \\ 0 & 0 & 1 & -3 \end{bmatrix} \begin{matrix} 1 \\ 2 \\ 3 \\ 4 \\ 5 \\ 6 \end{matrix}$$

To check that the vectors partially defining the planes are oriented outward, we start by finding an interior point; such a point is $\mathbf{u} = [\frac{1}{2} \quad \frac{1}{2} \quad \frac{1}{2} \quad 1]^{\#}$. Corner point $[0 \quad 0 \quad 0 \quad 1]^{\#}$ is on plane 1, so the vector from $\mathbf{u}$ to this corner point is $\boldsymbol{\nu}_1 = [-\frac{1}{2} \quad -\frac{1}{2} \quad -\frac{1}{2}]^{\#}$, and if plane 1 is oriented correctly, the dot product of $\boldsymbol{\nu}_1$ and the vector partially defining the plane must be positive:

$$[0 \quad 0 \quad -1] \begin{bmatrix} -\frac{1}{2} \\ -\frac{1}{2} \\ -\frac{1}{2} \end{bmatrix} = \frac{1}{2}$$

We can confirm the same result for the other planes.

The average value of the corner points gives an interior point in all convex objects. A convex object can be defined as an object such that all possible lines intersect the object at zero or two points only. Concave objects present difficulties in imaging that will not be addressed here.

The pinhole is at $\mathbf{w} = [10 \quad 10 \quad 5 \quad 1]^{\#}$. The axis of the camera is defined by the line

$$\begin{bmatrix} 10 \\ 10 \\ 5 \\ 1 \end{bmatrix} + \alpha \begin{bmatrix} -10 \\ -10 \\ -5 \\ 0 \end{bmatrix}$$

where at $\alpha = 0$ the point on the line is at the pinhole and at $\alpha = 1$ the point on the line is at the origin. The distance between the pinhole and the origin is $\sqrt{10^2 + 10^2 + 5^2} = 15$. At the film plane of the camera, $\alpha = -\frac{1}{15}$, so the axial point on the film plane is given by $[10.66_6 \quad 10.66_6 \quad 5.33_3 \quad 1]^{\#}$. Thus the film plane is given by

$$[10 \quad 10 \quad 5 \quad d'] \begin{bmatrix} 10.66_6 \\ 10.66_6 \\ 5.33_3 \\ 1 \end{bmatrix} = 0, \qquad \text{so } d' = -240$$

Thus (violating the unit vector notation in order to make the arithmetic simpler) the film plane is $\mathbf{p} = [2 \quad 2 \quad 1 \quad -48]$.

If a surface of the object is to be visible from the camera, the dot product of the vector from a point on the object to the pinhole and the vector partially defining the plane coincident with the surface of the object must be positive.

For plane $\mathbf{p}_1$ and one of its corner points $\begin{bmatrix} 0 \\ 0 \\ 0 \\ 1 \end{bmatrix}$, $[0 \quad 0 \quad -1]\begin{bmatrix} 10 \\ 10 \\ 5 \end{bmatrix} = -5$,

so $\mathbf{p}_1$ cannot be seen from the camera.

For $\mathbf{p}_2$ and corner point $\begin{bmatrix} 0 \\ 0 \\ 0 \\ 1 \end{bmatrix}$, $[0 \quad -1 \quad 0]\begin{bmatrix} 10 \\ 10 \\ 5 \end{bmatrix} = -10$, so $\mathbf{p}_2$ cannot be seen.

For $\mathbf{p}_3$ and corner point $\begin{bmatrix} 0 \\ 1 \\ 0 \\ 1 \end{bmatrix}$, $[0 \quad 1 \quad 0]\begin{bmatrix} 10 \\ 10 - 1 \\ 5 \end{bmatrix} = 9$, so $\mathbf{p}_3$ is visible.

For $\mathbf{p}_4$ and corner point $\begin{bmatrix} 0 \\ 0 \\ 0 \\ 1 \end{bmatrix}$, $[-1 \quad 0 \quad 0]\begin{bmatrix} 10 \\ 10 \\ 5 \end{bmatrix} = -10$, so $\mathbf{p}_4$ cannot be seen.

For $\mathbf{p}_5$ and corner point $\begin{bmatrix} 2 \\ 0 \\ 0 \\ 1 \end{bmatrix}$, $[1 \quad 0 \quad 0]\begin{bmatrix} 10 - 2 \\ 10 \\ 5 \end{bmatrix} = 8$, so $\mathbf{p}_5$ is visible.

For $\mathbf{p}_6$ and corner point $\begin{bmatrix} 0 \\ 0 \\ 3 \\ 1 \end{bmatrix}$, $[0 \quad 0 \quad 1]\begin{bmatrix} 10 \\ 10 \\ 5 - 3 \end{bmatrix} = 2$, so $\mathbf{p}_6$ is visible.

Since $\mathbf{p}_1$, $\mathbf{p}_2$, and $\mathbf{p}_4$ meet at corner point $[0 \quad 0 \quad 0 \quad 1]^{\#}$, we conclude that this point and its associated edges are invisible on the image. Point $\mathbf{u}$ on the object is imaged to point **v** in the film plane. Using equation (2.1) on point 2,

$$\mathbf{v} = \mathbf{u} - \frac{\mathbf{pu}}{\mathbf{p}(\mathbf{w}-\mathbf{u})}(\mathbf{w}-\mathbf{u}) = \begin{bmatrix} 2 \\ 0 \\ 0 \\ 1 \end{bmatrix} - \frac{[2 \quad 2 \quad 1 \quad -48]\begin{bmatrix} 2 \\ 0 \\ 0 \\ 1 \end{bmatrix}}{[2 \quad 2 \quad 1 \quad -48]\begin{bmatrix} 10-2 \\ 10-0 \\ 5-0 \\ 0 \end{bmatrix}} \begin{bmatrix} 10-2 \\ 10-0 \\ 5-0 \\ 0 \end{bmatrix}$$

$$= \begin{bmatrix} 2 \\ 0 \\ 0 \\ 1 \end{bmatrix} + 1.0732 \begin{bmatrix} 8 \\ 10 \\ 5 \\ 0 \end{bmatrix} = \begin{bmatrix} 10.586 \\ 10.732 \\ 5.366 \\ 1 \end{bmatrix}$$

The set of visible image points is

$$\begin{array}{ccccccc} 2 & 3 & 4 & 5 & 6 & 7 & 8 \end{array}$$
$$\begin{bmatrix} 10.586 & 10.698 & 10.615 & 10.714 & 10.632 & 10.750 & 10.666 \\ 10.732 & 10.628 & 10.692 & 10.714 & 10.790 & 10.675 & 10.750 \\ 5.366 & 5.349 & 5.385 & 5.143 & 5.158 & 5.150 & 5.167 \\ 1 & 1 & 1 & 1 & 1 & 1 & 1 \end{bmatrix}$$

Suppose we set the orientation of the camera so that the image of point $[0 \quad 0 \quad 3 \quad 1]^{\#}$ is in the direction of the ordinate. Then the ordinate is

$$\begin{bmatrix} 10.714 - 10.666 \\ 10.714 - 10.666 \\ 5.143 - 5.333 \end{bmatrix} = \begin{bmatrix} 0.048 \\ 0.048 \\ -0.190 \end{bmatrix} \equiv \begin{bmatrix} 0.238 \\ 0.238 \\ -0.942 \end{bmatrix}$$

The abscissa is

$$\begin{bmatrix} 0.238 \\ 0.238 \\ -0.942 \end{bmatrix} \times \begin{bmatrix} \frac{2}{3} \\ \frac{2}{3} \\ \frac{1}{3} \end{bmatrix} = \begin{bmatrix} -0.7071 \\ 0.7071 \\ 0 \end{bmatrix}$$

EXERCISES FOR CHAPTER 2

1. If $\mathbf{v} = \mathrm{T}\mathbf{u}$, where **v** is point **u** translated by vector $\boldsymbol{\mu} = \begin{bmatrix} \mu_1 \\ \mu_2 \\ \mu_3 \end{bmatrix}$, what is T? Given point $\mathbf{u} = \begin{bmatrix} 1 \\ -1 \\ 1 \\ 1 \end{bmatrix}$ on plane $\mathbf{p} = [1 \quad 2 \quad 1 \quad 0]$, vector $\boldsymbol{\mu} = \begin{bmatrix} 0 \\ 1 \\ 1 \end{bmatrix}$. Find:

(a) Matrix **T**
(b) Point **v**
(c) Plane **q** that is parallel to **p** and passes through **v**
(d) A unit vector $\boldsymbol{\mu}$ in **p** whose projection onto the xy-plane is along the x-axis in the positive direction

---

2. Rotate vector $\begin{bmatrix} 2 \\ 2 \\ 1 \end{bmatrix}$ in plane $[1/\sqrt{2} \quad -1/\sqrt{2} \quad 0 \quad 2]$ by 30° and find the resulting vector.

---

3. Given a line $\begin{bmatrix} 1+\alpha \\ \alpha \\ -\alpha \\ 1 \end{bmatrix}$ and a point $\mathbf{u} = \begin{bmatrix} 2 \\ 2 \\ 0 \\ 1 \end{bmatrix}$

(a) Find the point **v** on the line closest to **u**.
(b) Rotate vector $\mathbf{u} - \mathbf{v}$ by $\pm 30°$ about the axis given by the line.
(c) Rotate the resulting vectors by $\pm 30°$ about the axis.

---

4. Successively rotate vector $[1 \quad 0 \quad 0]^\#$ about the major axes to reach arbitrary unit vector $[x \quad y \quad z]^\#$. Obtain the plane **p** containing the origin and vectors $[1 \quad 0 \quad 0]^\#$ and $[x \quad y \quad z]^\#$. Obtain the solid angle of rotation in **p** to rotate from $[1 \quad 0 \quad 0]^\#$ to $[x \quad y \quad z]^\#$. Compare the arithmetic times involved in both types of operation.

---

5. Suppose a point **u** is rotated by 30° with respect to the x-axis followed by $-30°$ with respect to the y-axis. Determine the solid angle of rotation with respect to the origin and the plane of rotation **p** for this solid angle for the following **u**:

(a) $\begin{bmatrix} 0 \\ 1 \\ 0 \\ 1 \end{bmatrix}$ **(b)** $\begin{bmatrix} 1 \\ 0 \\ 1 \\ 1 \end{bmatrix}$ **(c)** $\begin{bmatrix} 1 \\ 1 \\ 0 \\ 1 \end{bmatrix}$ **(d)** $\begin{bmatrix} 1 \\ 1 \\ 1 \\ 1 \end{bmatrix}$

---

6. Rotate vector $\boldsymbol{\nu} = \begin{bmatrix} 1 \\ 1 \\ 0 \end{bmatrix}$ about point $\mathbf{u} = \begin{bmatrix} 1 \\ 0 \\ 0 \\ 1 \end{bmatrix}$ by θ on $\mathbf{p} = [-1 \quad 1 \quad 1 \quad 1]$ to obtain vector $\boldsymbol{\eta}$.

(a) Find $\boldsymbol{\eta}$ in terms of θ.
(b) Determine the value(s) of θ such that the distance of $\mathbf{u} + \boldsymbol{\eta}$ from the origin is a minimum.

---

7. Rotate point $\begin{bmatrix} 1 \\ 0 \\ 1 \\ 1 \end{bmatrix}$ by 45° about the y-axis and then 45° about the z-axis. What is the angle between the old and new points measured with respect to the origin?

---

8. Vector $\begin{bmatrix} 2 \\ 2 \\ 1 \end{bmatrix}$ is produced from $\begin{bmatrix} 0 \\ 2 \\ 3 \end{bmatrix}$ by application of a scaling matrix followed by a rotation matrix. Find these matrices. Are they unique? If the scaling and rotation operations are reversed, is the resultant vector unchanged?

---

9. Determine each angle and plane of rotation required to rotate the following vectors about the origin to lie on the x-axis.

(a) $\begin{bmatrix} 1 \\ 1 \\ 1 \end{bmatrix}$ **(b)** $\begin{bmatrix} 0 \\ 1 \\ 2 \end{bmatrix}$ **(c)** $\begin{bmatrix} 0 \\ 0 \\ 3 \end{bmatrix}$

---

10. Find the homogeneous matrices defining rotations about the major axes to transform point $\begin{bmatrix} 0.57735 \\ 0.57735 \\ 0.57735 \\ 1 \end{bmatrix}$ into $\begin{bmatrix} 1 \\ 0 \\ 0 \\ 1 \end{bmatrix}$. Multiply these matrices together. If the resultant matrix defines rotation in a plane, what is the plane in which rotation occurs?

---

11. Find the homogeneous matrices defining scaling and rotations about the major axes to transform point $\begin{bmatrix} 3 \\ 2 \\ 1 \\ 1 \end{bmatrix}$ into $\begin{bmatrix} 0 \\ 0 \\ 1 \\ 1 \end{bmatrix}$.

---

12. The 30°-60°-90° triangle whose hypotenuse has length 2 lies in plane $\mathbf{p} = [-2/\sqrt{6} \quad 1/\sqrt{6} \quad 1/\sqrt{6} \quad 0]$ such that the 30° apex is at point $[1 \quad 1 \quad 1 \quad 1]^{\#}$ and the origin is on the hypotenuse. Assume the 90° corner is at positive y.

(a) Determine the location of the other corner points.

Rotate the triangle by 45° in $\mathbf{p}$ about the following points and determine the location of the corner points:

(b) $\begin{bmatrix} 0 \\ 0 \\ 0 \\ 1 \end{bmatrix}$ **(c)** $\begin{bmatrix} 1 \\ 1 \\ 1 \\ 1 \end{bmatrix}$ **(d)** $\begin{bmatrix} 0 \\ 0 \\ 2 \\ 1 \end{bmatrix}$

---

13. **(a)** A point $\begin{bmatrix} 1 \\ 2 \\ 3 \\ 1 \end{bmatrix}$ is to be rotated 30° with respect to the x-axis, the y-axis, and then the z-axis. Find the resultant point and the plane containing the old point, the new point and the origin.

(b) Repeat the problem, but this time rotate the point 30° with respect to the x-axis, the z-axis, and then the y-axis.

Compare the results obtained.

---

14. A pinhole camera has pinhole at $\mathbf{w} = \begin{bmatrix} 4 \\ 4 \\ 4 \\ 1 \end{bmatrix}$ and film plane $[1 \quad 0 \quad 0 \quad -5]$. Find the images of the following points.

(a) $\begin{bmatrix} 0 \\ 0 \\ 0 \\ 1 \end{bmatrix}$ **(b)** $\begin{bmatrix} 2 \\ 1 \\ 1 \\ 1 \end{bmatrix}$ **(c)** $\begin{bmatrix} 1 \\ 4 \\ 1 \\ 1 \end{bmatrix}$

---

15. The image of a sphere of radius 1 centered at the origin is seen by a pinhole camera as the circle defined by $x = -2$, $y = 0.1 \sin \theta$, and $z = 0.1 \cos \theta$. Determine:

(a) The film plane

(b) The pinhole

---

16. The image of three points that form an equilateral triangle with length of side unity appears as a straight line with points on the x-axis at 0, 0.2 and 0.4. The pinhole is at $[0.2 \quad 0 \quad 1 \quad 1]^{\#}$. Determine the location of the triangle.

---

17. Determine the relative positions of points $\begin{bmatrix} 1 \\ 1 \\ 1 \\ 1 \end{bmatrix}$, $\begin{bmatrix} 1 \\ 1 \\ 2 \\ 1 \end{bmatrix}$, and $\begin{bmatrix} 1 \\ 2 \\ 1 \\ 1 \end{bmatrix}$ with respect to plane $\mathbf{p} = [2 \quad 2 \quad 1 \quad -5]$ and the origin. If $\mathbf{p}$ is the perspective plane and the focal point is $\begin{bmatrix} -2 \\ -2 \\ -2 \\ 1 \end{bmatrix}$, obtain and draw the projection of the triangle composed of the three corner points given here. Assume that the Cartesian vertical is the ordinate.

---

18. Find the image taken by a pinhole camera of the object defined by the corner points

$$\begin{matrix} 1 & 2 & 3 & 4 & 5 & 6 \end{matrix}$$
$$\begin{bmatrix} 0 & 0 & 1 & 1 & 0 & 1 \\ 0 & 1 & 1 & 0 & 1 & 1 \\ 0 & 0 & 0 & 0 & 1 & 1 \\ 1 & 1 & 1 & 1 & 1 & 1 \end{bmatrix}$$

if the pinhole is at $\mathbf{w} = [-3 \quad -2 \quad -1 \quad 1]^{\#}$ and the film plane crosses the line from $\mathbf{w}$ to the origin at distance unity from $\mathbf{w}$. Assume that the film plane is orthogonal to the line joining $\mathbf{w}$ to the origin. Delete hidden lines in the image.

---

19. Three points, $\begin{bmatrix} 1 \\ 1 \\ 0 \\ 1 \end{bmatrix}$, $\begin{bmatrix} 1 \\ 2 \\ 1 \\ 1 \end{bmatrix}$, and $\begin{bmatrix} 1 \\ 0 \\ 2 \\ 1 \end{bmatrix}$ form a triangle in plane $\mathbf{p}$. This triangle is to be pasted into plane $\mathbf{q} = [\cos\theta \quad 0 \quad \sin\theta \quad 0]$ such that the first point is at the origin and the side defined by the first pair of points is vertical.

(a) Find $\mathbf{p}$.

(b) Determine the second and third corner points in $\mathbf{q}$.

---

20. Given points $\begin{bmatrix} 0 \\ 0 \\ 0 \\ 1 \end{bmatrix}$, $\begin{bmatrix} 1 \\ 0 \\ 0 \\ 1 \end{bmatrix}$, $\begin{bmatrix} 1 \\ 1 \\ 0 \\ 1 \end{bmatrix}$, and $\begin{bmatrix} 1 \\ 1 \\ 1 \\ 1 \end{bmatrix}$.

(a) Find all the planes defined by all sets of three of these points.

(b) Find the minimum distance of the unused point to each of these planes.

(c) Supposing the four points given are the corner points of a solid object, define the object by the planes that intersect its surfaces and orient the planes such that the vectors partially defining the surfaces of the object face outward from the object.

(d) Suppose a pinhole camera views the object. The pinhole is at $[-1 \quad -4 \quad -4 \quad 1]^{\#}$, the axis of the camera is the line from the pinhole to the origin, and the film plane is orthogonal to the axis at a distance unity from the pinhole. The camera is oriented such that a vertical line $[0 \quad 0 \quad z \quad 1]^{\#}$ on the object appears as a vertical line on the film. Determine the image on the film.

CHAPTER THREE

Joint Specification and Robotic Classification

In Chapter 1 the relationships between points, lines, and planes were defined in Cartesian space. In Chapter 2 the rotation of a vector about a point in a plane was discussed. We will use these relationships to classify the joint systems of robot manipulators.

A robot manipulator can be described by a simple (no loops), open, lower-pair kinematic chain [Angeles 1982] of joints and rigid links [Hartenberg and Denavit 1964; Roth 1967a; Roth 1967b] that comprise the degrees of freedom (*dof*) of the robot. Lower-pair joints are defined in the next section. For apparently historic reasons, most other workers have chosen a semigeneral notation for defining joints, the so-called Denavit-Hartenberg notation [Denavit 1955]. This notation served the field of kinematic linkages well but has problems when robot manipulators are considered. Scientific investigation often proceeds from the specific to the general, but here is a case where the opposite appears to be true: the general notation is inadequate, but the specific cases satisfy all practical applications.

Here we will define six kinematically simple joints and associated geometric symbols that are largely self-explanatory. It will be shown that these symbols are of great importance, since the type of motion of each joint is implied by the symbol, and the motion of the manipulator is apparent from the chain of such symbols. Even more important, the three-dimensional motion of the manipulator is reduced without ambiguity or inconsistency to two dimensions, so the operation of the manipulator can be shown on a sheet of paper.

3.1 KINEMATICALLY SIMPLE JOINTS

The classic work on kinematics was performed in the nineteenth century [Reuleaux 1963]. Reuleaux defined a lower kinematic pair as being characterized by a bearing surface at the joint between the two links, and so the joint has 1 *dof*, 2 *dof*, or 3 *dof*; by way of contrast, an upper-pair joint between two links is characterized by a line contact (2 *dof* or 3 *dof*) or a point contact (3 *dof*), but such joints are difficult to build and control. In fact, all common robots have joints that have 1 *dof*. Reuleaux identified 12 lower pairs as the screw, revolute, prism, cylinder, cone, hyperboloid, sphere, sector (portion of revolute), tooth, and vessal and gave textual symbols for these joints such as *R* for revolute and *S* for screw. He did not attempt a spatial or geometric symbology. The understanding of systems of joints and links improved dramatically after Hartenberg and Denavit [Hartenberg 1964] gave a fine historic perspective leading to a full analysis of the kinematics of linkages involving lower- and upper-pair kinematic joints. It is the classic paper [Denavit and Hartenberg 1955] by these authors that produced the joint/link classification that carries their names and is so popular with robotic kinematicists. Other workers have continued the analysis [Roth 1967a; Roth 1967b; Duffy 1980], where the latter citation provides lists of equations for the kinematics of revolute and sliding linkages.

Sheth and Uicker defined [Sheth and Uicker 1971] a notation for joints that included what they called revolute, prismatic, cylindrical, screw, flat, spheroid, and higher pairs (that can define the motion of gears). Hunt defined another set of lower-pair joints [Hunt 1978] but developed something conceptually more important—he associated each joint (shown in detail so the form of motion possible is apparent) with a geometric symbol. Kinoshita gave symbols for some of these joints [Kinoshita 1981], but his work was not completed or fully acknowledged by others. This is unfortunate, since he also attempts to define robot capability (see Chapter 4).

Only kinematically simple robots [Stanisic and Pennock 1985] with a single *dof* in which the axes of adjacent lower-pair joints are parallel or orthogonal are considered: the axis of a joint is the direction of the links or vectors partially defining the planes on either side of the joint. All common robot manipulators have joints that can be considered kinematically simple [Paul, Shimano, and Mayer 1981a], so it is curious that prior attempts to classify such joints have not been vigorously pursued. Further, attempting to classify joints in a more general way obviates the simple geometric relationships that occur under the restriction.

Five joints involving rotation about a bearing pin are defined; these revolute joints are the pitch, roll, yaw, crank, and cylindrical joints. In addition, we have a sliding link and define a set of four right angle bends. Except where otherwise specified, we assume that all the links are straight.

A *dof* is a revolute joint followed by a rigid link. A sliding link has 1 *dof*, so a sliding link following a revolute joint comprises 2 *dof*.

A displacement occurs when the joint has a physical size that moves the succeeding link sideways. Suppose that the link succeeding a sliding, pitch, or cylindrical joint with displacement δ moves in plane $\mathbf{p} = [a \quad b \quad c \quad d]$. Then we use displacement vector given by $\pm\delta[a \quad b \quad c]^{\#}$, where the sign depends on the side of the joint displacement; recall that we set $a^2 + b^2 + c^2 = 1$. We assume, unless otherwise specified, that there is no displacement at a joint.

The six basic types of robotic joints are shown in Figure 3.1 together with their symbolic equivalents; these can model the joints used in common commercial robots. In addition we need the right-angle bends discussed in Section 3.3. A symbolic representation should be largely self-explanatory,

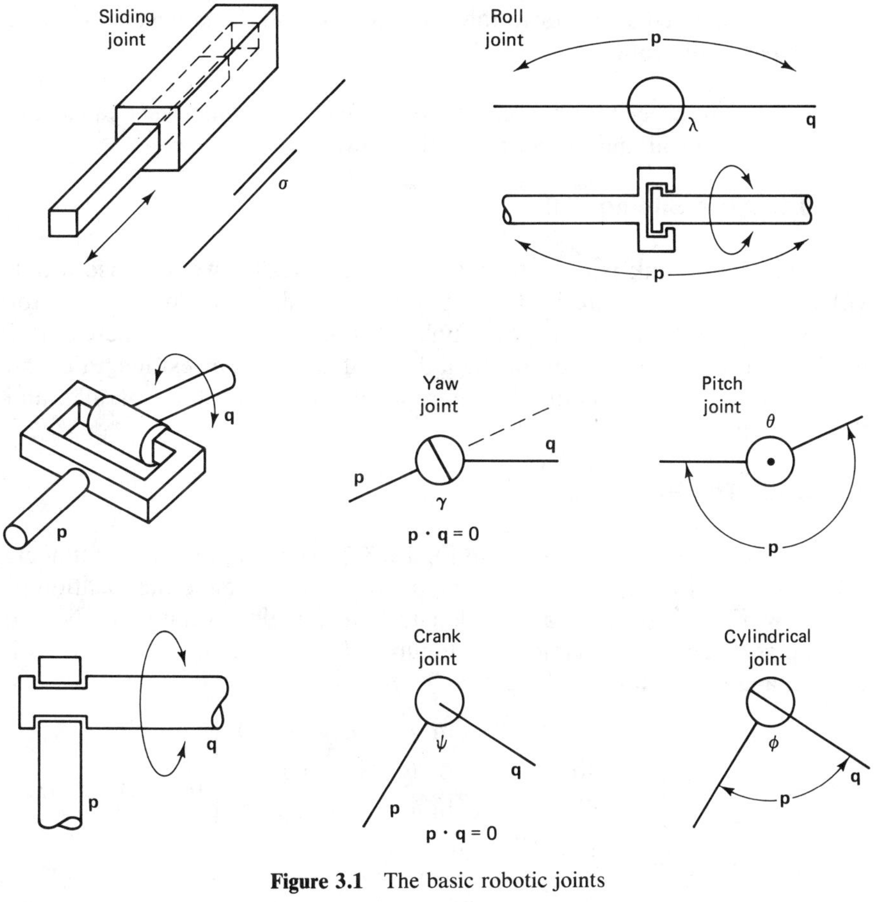

Figure 3.1 The basic robotic joints

simple, and consistent; the system used here passes all these tests and, in addition, aids in kinematic analysis (see Chapters 5 and 6). The sliding link shows two parallel lines that are close together. The pitch joint has a dot in the center, which indicates direction of the axis of the bearing surface of the joint, so the uplink (whose motion is orthogonal to the axis) lies in the plane of the paper. The roll, yaw, and cylindrical joints show the axis of the joint; the yaw joint needs the explanation of Section 3.1.4. The axis of the crank joint is the same as for the pitch joint, but the uplink (see the following definition) is connected to the joint, as is the roll joint. Full explanations are given in succeeding sections.

We define the movement of a link following a joint of a robot in terms of the plane of movement of the link preceding the joint and the status and type of the connecting joint.

Definition. When considering the links connected to a robotic joint, the *uplink* is defined as closer to the end effector, and the *downlink* is closer to the base of the robot.

The end of a joint system nearest the end effector is sometimes called the *distal end,* so distal and uplink mean the same.

3.1.1 The Sliding Link

The sliding link is easy to define mathematically: we suppose a link with a sliding joint is connected from point $\mathbf{u}$ and is directed along unit vector $\boldsymbol{\alpha} = [\alpha_1 \quad \alpha_2 \quad \alpha_3]^\#$, so the end of the link is at point $\mathbf{v} = \mathbf{u} + \sigma\boldsymbol{\alpha}$, where $\sigma \leq \overline{\sigma}$ and $\overline{\sigma}$ is the maximum length of the link. Usually the shortest length of the link is about $\overline{\sigma}/2$. The sliding link is represented by two lines parallel and close together.

3.1.2 The Pitch Joint

Consider the system shown in Figure 3.2, where a pitch joint connects links l_1 and l_2. Link l_1 moves in plane $\mathbf{p}$. We will determine the location of the end $\mathbf{w}$ of link l_2 in terms $\mathbf{u}$, $\mathbf{v}$, $\mathbf{p}$, length l_1, and angle of rotation θ. Notice that the geometry is the same as in Figure 2.1, so using equation (2.6) with some changes in symbols, we get

$$\mathbf{w} = v + \frac{l_2}{l_1}\begin{bmatrix} \cos\theta & -c\sin\theta & b\sin\theta & 0 \\ c\sin\theta & \cos\theta & -a\sin\theta & 0 \\ -b\sin\theta & a\sin\theta & \cos\theta & 0 \\ 0 & 0 & 0 & 1 \end{bmatrix}(v - u) \tag{3.1}$$

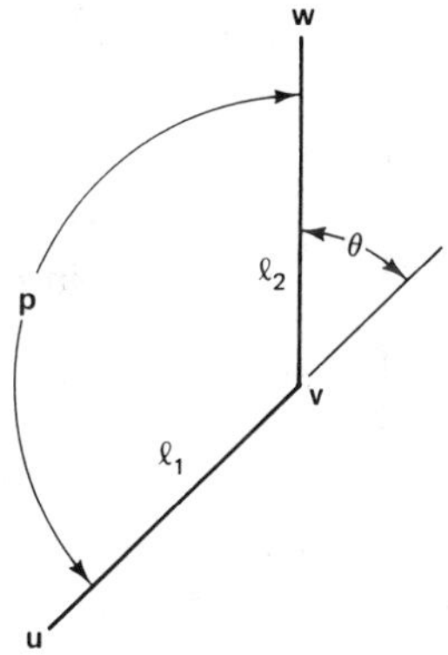

Figure 3.2 Geometry of the pitch joint

The convention adopted for positive θ is counterclockwise when viewing the link; recall we always require $[a \quad b \quad c]^\#$ to be directed out of the paper with rotation defined by the right-hand screw rule with respect to $[a \quad b \quad c]^\#$ and $\mathbf{v} - \mathbf{u}$.

Suppose $\mathbf{u} = [0 \quad 0 \quad 0 \quad 1]^\#$ and plane $\mathbf{p}$ is the yz-plane; then $\mathbf{p} = [1 \quad 0 \quad 0 \quad 0]$ and l_1 is along the z-axis. Thus, $\mathbf{v} = [0 \quad 0 \quad l_1 \quad 1]^\#$ and

$$\mathbf{w} = \begin{bmatrix} 0 \\ 0 \\ l_1 \\ 1 \end{bmatrix} + \frac{l_2}{l_1} \begin{bmatrix} \cos\theta & 0 & 0 & 0 \\ 0 & \cos\theta & -\sin\theta & 0 \\ 0 & \sin\theta & \cos\theta & 0 \\ 0 & 0 & 0 & 1 \end{bmatrix} \begin{bmatrix} 0 \\ 0 \\ l_1 \\ 0 \end{bmatrix} = \begin{bmatrix} 0 \\ -l_2\sin\theta \\ l_1 + l_2\cos\theta \\ 1 \end{bmatrix} \tag{3.2}$$

The prior example can describe a robot with two links. The first link has one end $\mathbf{u}_1$ at the origin and is oriented vertically, so $\mathbf{u}_2 = [0 \quad 0 \quad l_1 \quad 1]^\#$, and the motion of the second link is in the yz-plane, so $\mathbf{u}_3 = [0 \quad -l_2\sin\theta_1 \quad l_1 + l_2\cos\theta_1 \quad 1]^\#$. θ_1 is the angle between links l_1 and l_2. Suppose another pitch joint connects a third link l_3 to link l_2 such that the angle between l_3 and l_2 (according to the convention shown in Figure 3.2) is θ_2. All three links lie in plane $\mathbf{p} = [1 \quad 0 \quad 0 \quad 0]$ and

$$\mathbf{u}_3 = \begin{bmatrix} 0 \\ -l_2\sin\theta_1 \\ l_1 + l_2\cos\theta_1 \\ 1 \end{bmatrix} \tag{3.3}$$

so

$$\mathbf{u}_4 = \begin{bmatrix} 0 \\ -l_2\sin\theta_1 \\ l_1 + l_2\cos\theta_1 \\ 1 \end{bmatrix} + \begin{bmatrix} \cos\theta_2 & 0 & 0 & 0 \\ 0 & \cos\theta_2 & -\sin\theta_2 & 0 \\ 0 & \sin\theta_2 & \cos\theta_2 & 0 \\ 0 & 0 & 0 & 1 \end{bmatrix} \begin{bmatrix} 0 \\ -l_3\sin\theta_1 \\ l_3\cos\theta_1 \\ 0 \end{bmatrix}$$

$$= \begin{bmatrix} 0 \\ -(l_2 + l_3\cos\theta_2)\sin\theta_1 - l_3\cos\theta_1\sin\theta_2 \\ -l_3\sin\theta_1\sin\theta_2 + l_1 + (l_2 + l_3\cos\theta_2)\cos\theta_1 \\ 1 \end{bmatrix}$$

$$= \begin{bmatrix} 0 \\ -l_2\sin\theta_1 - l_3\sin(\theta_1 + \theta_2) \\ l_1 + l_2\cos\theta_1 + l_3\cos(\theta_1 + \theta_2) \\ 1 \end{bmatrix} \tag{3.4}$$

We can show for the three-pitch joint case that the end effector will be at

$$\mathbf{u}_5 = \begin{bmatrix} 0 \\ -l_2\sin\theta_1 - l_3\sin(\theta_1 + \theta_2) - l_4\sin(\theta_1 + \theta_2 + \theta_3) \\ l_1 + l_2\cos\theta_1 + l_3\cos(\theta_1 + \theta_2) + l_4\cos(\theta_1 + \theta_2 + \theta_3) \\ 1 \end{bmatrix} \tag{3.5}$$

There may be valid reasons for having a chain of three pitch joints. It is unlikely that a chain of four or more pitch joints can be justified.

3.1.3 The Roll Joint

The roll joint is shown in Figure 3.3; we use λ to designate the angle of the joint. The symbology **w**, **v**, **u** means that we view these three points in a line where **w** is on top. For this joint we have

$$\mathbf{w} = \mathbf{v} + \frac{l_2}{l_1}(\mathbf{v} - \mathbf{u}) \tag{3.6}$$

The vector partially specifying **q** is found by rotating vector $[a \quad b \quad c]^{\#}$ in the plane defined in part by the unit vector $(\mathbf{u} - \mathbf{v})/l_1$, so using equation (2.3)

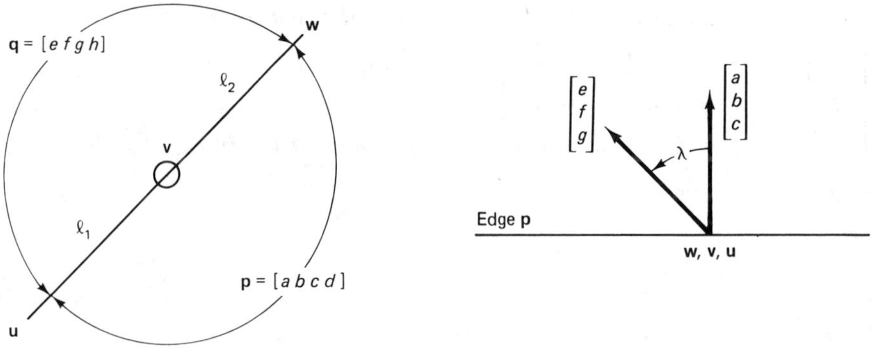

Figure 3.3 Geometry of the roll joint

we get

$$\begin{bmatrix} e \\ f \\ g \end{bmatrix} = \begin{bmatrix} \cos\lambda & \frac{(u_z - v_z)}{l_1}\sin\lambda & \frac{-(u_y - v_y)}{l_1}\sin\lambda \\ \frac{-(u_z - v_z)}{l_1}\sin\lambda & \cos\lambda & \frac{(u_x - v_x)}{l_1}\sin\lambda \\ \frac{(u_y - v_y)}{l_1}\sin\lambda & \frac{-(u_x - v_x)}{l_1}\sin\lambda & \cos\lambda \end{bmatrix} \begin{bmatrix} a \\ b \\ c \end{bmatrix} \tag{3.7}$$

3.1.4 The Yaw Joint

The yaw joint is shown symbolically in Figure 3.4(a), with its geometry in Figure 3.4(b). Link l_2 moves in $\mathbf{q} = [e \quad f \quad g \quad h]$, and $\mathbf{u}$ is also in $\mathbf{q}$. $[e \quad f \quad g]^\#$ is orthogonal to both $[a \quad b \quad c]^\#$ and $(\mathbf{u} - \mathbf{v})$ and so is defined by

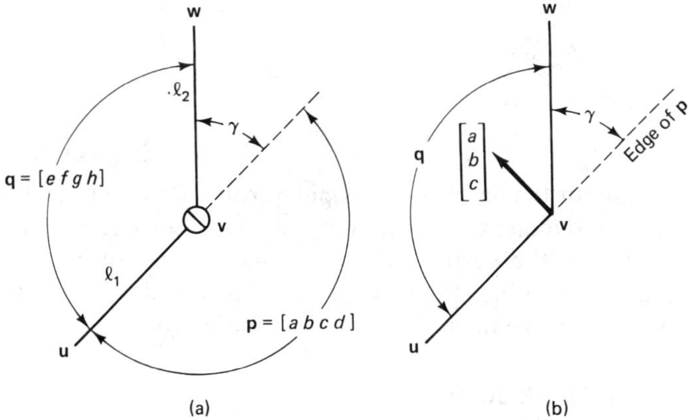

Figure 3.4 Geometry of the yaw joint

the cross product of $[a \quad b \quad c]^{\#}$ and $(\mathbf{u} - \mathbf{v})/l_1$; that is,

$$\begin{bmatrix} e \\ f \\ g \end{bmatrix} = \frac{1}{l_1}\begin{bmatrix} a \\ b \\ c \end{bmatrix} \times \begin{bmatrix} v_x - u_x \\ v_y - u_y \\ v_z - u_z \end{bmatrix} = \frac{-1}{l_1}\begin{bmatrix} b(u_z - v_z) - c(u_y - v_y) \\ c(u_x - v_x) - a(u_z - v_z) \\ a(u_y - v_y) - b(u_x - v_x) \end{bmatrix} \tag{3.8}$$

so from equation (2.8)

$$\mathbf{w} = \mathbf{v} + \frac{l_2}{l_1^2}$$

$$\begin{bmatrix} l_1\cos\tau & \{a(u_y - v_y) - b(u_x - v_x)\}\sin\tau & -\{c(u_x - v_x) - a(u_z - v_z)\}\sin\tau & 0 \\ -\{a(u_y - v_y) - b(u_x - v_x)\}\sin\tau & l_1\cos\tau & \{b(u_z - v_z) - c(u_y - v_y)\}\sin\tau & 0 \\ \{c(u_x - v_x) - a(u_z - v_z)\}\sin\tau & -\{b(u_z - v_z) - c(u_y - v_y)\}\sin\tau & l_1\cos\tau & 0 \\ 0 & 0 & 0 & 1 \end{bmatrix}$$

$$(\mathbf{v} - \mathbf{u})$$

$$= \mathbf{v} + \frac{l_2}{l_1^2}\begin{bmatrix} (v_x - u_x)l_1\cos\tau - (a\{(v_y - u_y)^2 + (v_z - u_z)^2\} - (v_x - u_x)\{b(v_y - u_y) + c(v_z - u_z)\})\sin\tau \\ (v_y - u_y)l_1\cos\tau - (b\{(v_x - u_x)^2 + (v_z - u_z)^2\} - (v_y - u_y)\{a(v_x - u_x) + c(v_z - u_z)\})\sin\tau \\ (v_z - u_z)l_1\cos\tau - (c\{(v_x - u_x)^2 + (v_y - u_y)^2\} - (v_z - u_z)\{a(v_x - u_x) + b(v_y - u_y)\})\sin\tau \\ 0 \end{bmatrix}$$

$$= \mathbf{v} + \frac{l_2}{l_1^2}\begin{bmatrix} (v_x - u_x)l_1\cos\tau - (a\{(l_1^2 - (v_x - u_x)^2\} - (v_x - u_x)\{b(v_y - u_y) + c(v_z - u_z)\})\sin\tau \\ (v_y - u_y)l_1\cos\tau - (b\{(l_1^2 - (v_y - u_y)^2\} - (v_y - u_y)\{a(v_x - u_x) + c(v_z - u_z)\})\sin\tau \\ (v_z - u_z)l_1\cos\tau - (c\{(l_1^2 - (v_z - u_z)^2\} - (v_z - u_z)\{a(v_x - u_x) + b(v_y - u_y)\})\sin\tau \\ 0 \end{bmatrix}$$

$$= \mathbf{v} + \frac{l_2}{l_1^2}\begin{bmatrix} (v_x - u_x)l_1\cos\tau - (al_1^2 - (v_x - u_x)\{a(v_x - u_x) + b(v_y - u_y) + c(v_z - u_z)\})\sin\tau \\ (v_y - u_y)l_1\cos\tau - (bl_1^2 - (v_y - u_y)\{a(v_x - u_x) + b(v_y - u_y) + c(v_z - u_z)\})\sin\tau \\ (v_z - u_z)l_1\cos\tau - (cl_1^2 - (v_z - u_z)\{a(v_x - u_x) + b(v_y - u_y) + c(v_z - u_z)\})\sin\tau \\ 0 \end{bmatrix}$$

$$= \mathbf{v} + l_2\begin{bmatrix} \frac{1}{l_1}(v_x - u_x)\cos\tau - a\sin\tau \\ \frac{1}{l_1}(v_y - u_y)\cos\tau - b\sin\tau \\ \frac{1}{l_1}(v_z - u_z)\cos\tau - c\sin\tau \\ 0 \end{bmatrix} \tag{3.9}$$

The reason for the nautical terms pitch, roll, and yaw is apparent when considering the movement of l_2 with respect to l_1. Look along the extended lower arm (l_1). **p** is the vertical plane intersecting the lower arm, and the wrist is the joint at **v**. When the hand pitches, rolls, and yaws (as does a boat at sea), this is the movement of the corresponding robotic joints.

3.1.5 The Crank Joint

The crank joint is shown symbolically in Figure 3.5(a) and geometrically in Figure 3.5(b) and is configured in such a way that the vectors par-

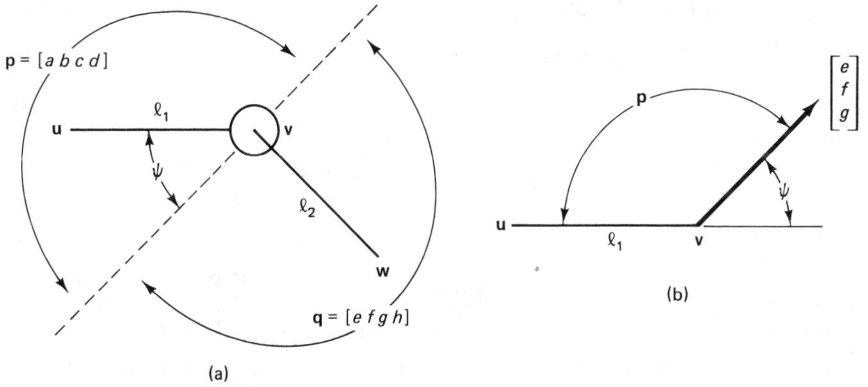

Figure 3.5 Geometry of the crank joint

tially defining **p** and link l_2 are in line and $[e \quad f \quad g]^\#$ is in **p**, so

$$\mathbf{w} = \mathbf{v} + l_2 \begin{bmatrix} a \\ b \\ c \\ 0 \end{bmatrix} \tag{3.10}$$

and

$$\begin{bmatrix} e \\ f \\ g \end{bmatrix} = \frac{1}{l_1} \begin{bmatrix} \cos\psi & -c\sin\psi & b\sin\psi \\ c\sin\psi & \cos\psi & -a\sin\psi \\ -b\sin\psi & a\sin\psi & \cos\psi \end{bmatrix} \begin{bmatrix} v_x - u_x \\ v_y - u_y \\ v_z - u_z \end{bmatrix} \tag{3.11}$$

and **q** is completed with $h = -(ev_x + fv_y + gv_z)$.

Suppose we have a chain of crank joints connected as shown in Figure 3.6, where the first crank joint is at $\mathbf{v}_1$, the first joint is at $\mathbf{v}_0$ connected to $\mathbf{v}_1$

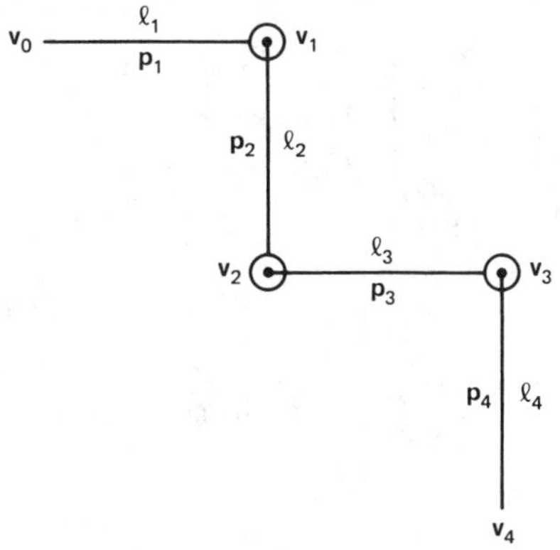

Figure 3.6 A chain of crank joints

by link l_1, and link l_i moves in $p_i = [a_i \quad b_i \quad c_i \quad d_i]$. Then from equations (3.10) and (3.11)

$$\mathbf{v}_{i+1} = \mathbf{v}_i + l_{i+1}\begin{bmatrix} a_i \\ b_i \\ c_i \\ 0 \end{bmatrix}$$

and

$$\begin{bmatrix} a_{i+1} \\ b_{i+1} \\ c_{i+1} \end{bmatrix} = \frac{1}{l_i}\begin{bmatrix} \cos\psi_i & -c_i \sin\psi_i & b_i \sin\psi_i \\ c_i \sin\psi_i & \cos\psi_i & -a_i \sin\psi_i \\ -b_i \sin\psi_i & a_i \sin\psi_i & \cos\psi_i \end{bmatrix}\begin{bmatrix} x_i - x_{i-1} \\ y_i - y_{i-1} \\ z_i - z_{i-1} \end{bmatrix} \qquad \text{for } i = 1, 2, \ldots,$$

and

$$\begin{bmatrix} a_{i+1} \\ b_{i+1} \\ c_{i+1} \end{bmatrix} = \begin{bmatrix} \cos\psi_i & -c_i \sin\psi_i & b_i \sin\psi_i \\ c_i \sin\psi_i & \cos\psi_i & -a_i \sin\psi_i \\ -b_i \sin\psi_i & a_i \sin\psi_i & \cos\psi_i \end{bmatrix}\begin{bmatrix} a_{i-1} \\ b_{i-1} \\ c_{i-1} \end{bmatrix} \qquad \text{for } i = 2, 3, \ldots .$$

Since $\mathbf{p}_3$ is independent of $\mathbf{p}_2$ and $\mathbf{p}_4$ is independent of both $\mathbf{p}_2$ and $\mathbf{p}_3$, then the vectors partially defining $\mathbf{p}_2$, $\mathbf{p}_3$, and $\mathbf{p}_4$ form a basis in Cartesian space, and any vector can be expressed as a weighted sum of these three vectors. Thus $\mathbf{p}_5$ is not independent, and a chain of four or more crank joints may be able to be replaced with a simpler equivalent. We notice that regardless of $\mathbf{p}_1$ and the direction of l_1, the angles ψ_1, ψ_2, and ψ_3 control $\mathbf{p}_4$ and the direction of l_4, so a chain of three crank joints can be used as the end effector orientation linkage system; see Section 4.3.

If the crank joint is connected the other way around (we use the symbol $\tilde{\psi}$ for this joint), we are given $\mathbf{w}$, $\mathbf{v}$, and $\mathbf{q}$ and seek $\mathbf{u}$ and $\mathbf{p}$; then from equations (3.10) and (3.11) we get

$$\begin{bmatrix} a \\ b \\ c \end{bmatrix} = -\frac{1}{l_2}\begin{bmatrix} w_x - v_x \\ w_y - v_y \\ w_z - v_z \end{bmatrix} \tag{3.12}$$

and

$$\mathbf{u} = \mathbf{v} - l_1\begin{bmatrix} \cos\tilde{\psi} & c \sin\tilde{\psi} & -b \sin\tilde{\psi} & 0 \\ -c \sin\tilde{\psi} & \cos\tilde{\psi} & a \sin\tilde{\psi} & 0 \\ b \sin\tilde{\psi} & -a \sin\tilde{\psi} & \cos\tilde{\psi} & 0 \\ 0 & 0 & 0 & 1 \end{bmatrix}\begin{bmatrix} e \\ f \\ g \\ 0 \end{bmatrix} \tag{3.13}$$

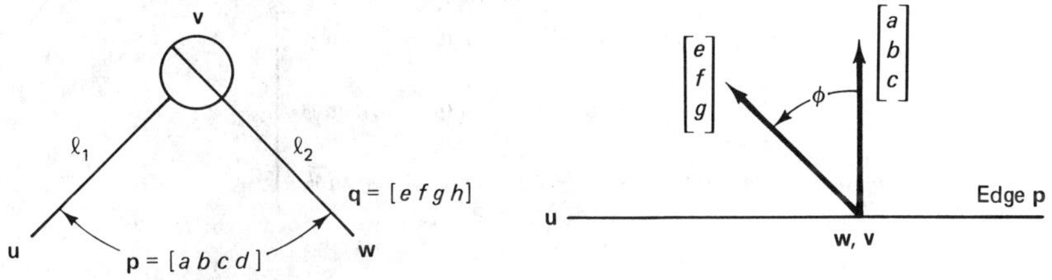

Figure 3.7 Geometry of the cylindrical joint

3.1.6 The Cylindrical Joint

The cylindrical joint is shown in Figure 3.7. Links l_1 and l_2 lie in plane **p**. The direction of **w** − **v** is found from the cross product

$$(\mathbf{v} - \mathbf{u}) \times \begin{bmatrix} a \\ b \\ c \end{bmatrix}$$

Thus

$$\mathbf{w} = \mathbf{v} - \frac{l_2}{l_1}\begin{bmatrix} b(v_z - u_z) - c(v_y - u_y) \\ c(v_x - u_x) - a(v_z - u_z) \\ a(v_y - u_y) - b(v_x - u_x) \\ 0 \end{bmatrix} \tag{3.14}$$

Recall the meaning of the symbology '**w**, **v**' in Figure 3.7(b) is that **w** and **v** are in line and **w** is on top. Here $\mathbf{q} = [e \quad f \quad g \quad h]$ is partially found from vector $[a \quad b \quad c]^{\#}$ rotated by ϕ in the plane partially defined by unit vector $(\mathbf{w} - \mathbf{v})/l_2$—that is, from equation (2.3) and equation (3.14):

$$\begin{bmatrix} e \\ f \\ g \end{bmatrix} = \frac{1}{l_1}$$

$$\begin{bmatrix} l_1\cos\phi & \{a(v_y - u_y) - b(v_x - u_x)\}\sin\phi & -\{c(v_x - u_x) - a(v_z - u_z)\}\sin\phi \\ -\{a(v_y - u_y) - b(v_x - u_x)\}\sin\phi & l_1\cos\phi & \{b(v_z - u_z) - c(v_y - u_y)\}\sin\phi \\ \{c(v_x - u_x) - a(v_z - u_z)\}\sin\phi & -\{b(v_z - u_z) - c(v_y - u_y)\}\sin\phi & l_1\cos\phi \end{bmatrix}\begin{bmatrix} a \\ b \\ c \end{bmatrix}$$

$$= \begin{bmatrix} a\cos\phi + \frac{1}{l_1}\{ab(v_y - u_y) - b^2(v_x - u_x) - c^2(v_x - u_x) + ac(v_z - u_z)\}\sin\phi \\ b\cos\phi + \frac{1}{l_1}\{-a^2(v_y - u_y) + ab(v_x - u_x) + bc(v_x - u_z) - c^2(v_y - u_y)\}sin\ \phi \\ c\cos\phi + \frac{1}{l_1}\{ac(v_x - u_x) - a^2(v_z - u_z) - b^2(v_z - u_z) + bc(v_y - u_y)\}\sin\phi \end{bmatrix}$$

$$= \begin{bmatrix} a\cos\phi + \frac{1}{l_1}\{ab(v_y - u_y) + (a^2 - 1)(v_x - u_x) + ac(v_z - u_z)\}\sin\phi \\ b\cos\phi + \frac{1}{l_1}\{(b^2 - 1)(v_y - u_y) + ab(v_x - u_x) + bc(v_x - u_z)\}\sin\phi \\ c\cos\phi + \frac{1}{l_1}\{ac(v_x - u_x) + (c^2 - 1)(v_z - u_z) + bc(v_y - u_y)\}\sin\phi \end{bmatrix}$$

$$= \begin{bmatrix} a\cos\phi - \frac{1}{l_1}(v_x - u_x)\sin\phi \\ b\cos\phi - \frac{1}{l_1}(v_y - u_y)\sin\phi \\ c\cos\phi - \frac{1}{l_1}(v_z - u_z)\sin\phi \end{bmatrix} \tag{3.15}$$

The plane is completed with $h = -\{ev_x + fv_y + gv_z\}$.

Consider a chain of cylindrical joints such as shown for crank joints in Figure 3.6. Then

$$\mathbf{v}_{i+1} = \mathbf{v}_i - \frac{l_{i+1}}{l_i}\begin{bmatrix} b_i(z_i - z_{i-1}) - c_i(y_i - y_{i-1}) \\ c_i(x_i - x_{i-1}) - a_i(z_i - z_{i-1}) \\ a_i(y_i - y_{i-1}) - b_i(x_i - x_{i-1}) \\ 0 \end{bmatrix}$$

$$\begin{bmatrix} a_{i+1} \\ b_{i+1} \\ c_{i+1} \end{bmatrix} = \begin{bmatrix} a_i\cos\phi_i - \frac{1}{l_i}(x_i - x_{i-1})\sin\phi_i \\ b_i\cos\phi_i - \frac{1}{l_i}(y_i - y_{i-1})\sin\phi_i \\ c_i\cos\phi_i - \frac{1}{l_i}(z_i - z_{i-1})\sin\phi_i \end{bmatrix}$$

As in the case of a chain of three crank joints, a chain of three cylindrical joints can be used for end effector orientation, and a chain of four or more cylindrical joints can possibly be replaced with a simpler equivalent.

Suppose the cylindrical joint is connected the other way around (and we specify this joint by $\tilde{\phi}$), so we are given **w**, **v**, and **q** and seek **u** and **p**. We find the vector partially specifying **p** from $[e \quad f \quad g]^{\#}$ rotated by $-\tilde{\phi}$ in the plane partially specified by $(\mathbf{v} - \mathbf{w})/l_2$; that is,

$$\begin{bmatrix} a \\ b \\ c \end{bmatrix} = \frac{1}{l_2}\begin{bmatrix} l_2\cos\tilde{\phi} & -(v_z - w_z)\sin\tilde{\phi} & (v_y - w_y)\sin\tilde{\phi} \\ (v_z - w_z)\sin\tilde{\phi} & l_2\cos\tilde{\phi} & -(v_x - w_x)\sin\tilde{\phi} \\ -(v_y - w_y)\sin\tilde{\phi} & (v_x - w_x)\sin\tilde{\phi} & l_2\cos\tilde{\phi} \end{bmatrix}\begin{bmatrix} e \\ f \\ g \end{bmatrix} \tag{3.16}$$

and the plane is completed with $d = -(av_x + bv_y + cv_z)$.

The cross product of the two vectors partially defining **p** and (**w** − **v**) give a vector in the same direction as vector **u** − **v**, so

$$\mathbf{u} = \mathbf{v} - \frac{l_1}{l_2}\begin{bmatrix} b(w_z - v_z) - c(w_y - v_y) \\ c(w_x - v_x) - a(w_z - v_z) \\ a(w_y - v_y) - b(w_x - v_x) \\ 0 \end{bmatrix} \tag{3.17}$$

It seems at first that joint $\tilde{\phi}$ is the same as $\tilde{\psi}$ with a 90° change in datum position, but this is not the case. Although the motion of the uplink end is the same (with a 90° change in datum), the planes of motion as defined are different.

3.2 DEFECTIVE JOINT SYSTEMS

A kinematic chain containing n joints is called *defective* if it can be replaced by a kinematic chain containing m joints, where $m < n$, such that, with the location and plane of motion of the downlink the same, the location and plane of motion of the uplink in the original chain and its equivalent are identical. The general case of a defective joint system is shown in Figure 3.8.

Right-angle bends are included in these chains and are not counted as *dof.* Chains of four pitch, four yaw, four crank (in the same orientation), and four cylindrical (in the same orientation) joints are defective. Stanisic and Pennock [Stanisic and Pennock 1985] discussed defective joint systems where two revolute joints are colinear, three revolute joints are coplanar, or two sliding joints (where each joint is preceded by a right-angle bend) are parallel. Others have discussed defective joint systems [Paul and Stevenson 1983], but since none has classified or considered defective chains of kinematically simple joints, we must start anew.

We will demonstrate some defective pairs and triads of revolute joints and show the simpler equivalents. We assume the first link moves in **p** =

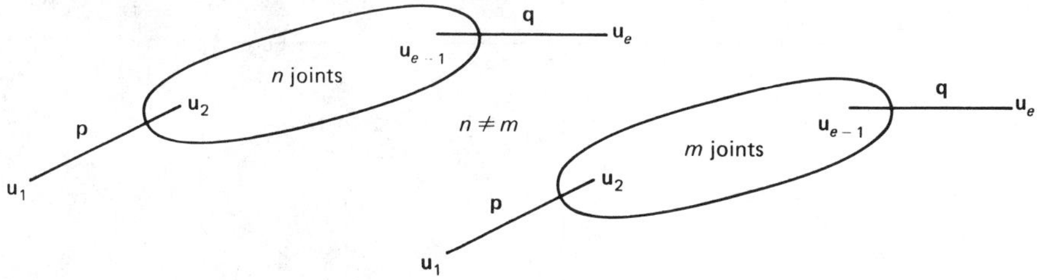

Figure 3.8 Degeneracy

$[a \quad b \quad c \quad d]$. For like pairs of joints, only the following are defective:

$$| \lambda_1 | \lambda_2 | = | \lambda_1 + \lambda_2 |$$

$$| \sigma_1 | \sigma_2 | = | \sigma_1 + \sigma_2 |$$

For dissimilar pairs, $| \phi_1 | \lambda_2 | \equiv | \phi_{1+2} |$ are defective, where the subscript 1 + 2 means that joint angles 1 and 2 are added directly. Also, $| \psi_1 | \tilde{\psi}_2 |$ and $| \phi_1 | \tilde{\phi}_2 |$ are defective, as are other combinations. We see in Chapter 4 that the capability of a robot is determined by its planes of motion, but this is valid only if the robot is not defective. Henceforth, we assume that the robot is not defective.

3.3 OTHER ROBOTIC JOINTS AND JOINT RELATIONSHIPS

The Cartesian robot (discussed in Section 3.8) requires a rigid right-angle joint between sliding links. We need to establish a convention for such right-angle joints. Consider the geometry shown in Figure 3.9, where we desig-

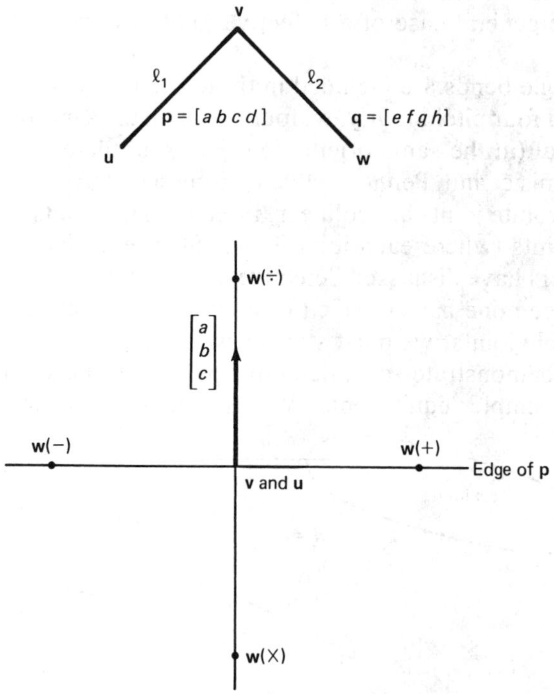

Figure 3.9 Geometry of right-angle bends

nate the four possible configurations with the symbols $+$, $-$, $\times$, and $\div$ in parentheses. This information should be placed in the crook of the joint when showing the configuration diagram. Here **q** contains the three points **u**, **v**, and **w**. From Figure 3.9 it is apparent that

$$\mathbf{w}(\div) = \mathbf{v} + l_2 \begin{bmatrix} a \\ b \\ c \\ 0 \end{bmatrix}, \qquad \mathbf{w}(\times) = \mathbf{v} - l_2 \begin{bmatrix} a \\ b \\ c \\ 0 \end{bmatrix} \tag{3.18}$$

and **q**, the plane of motion of l_2, is given by

$$\begin{bmatrix} e \\ f \\ g \end{bmatrix} = \frac{1}{l_1}(\mathbf{u} - \mathbf{v}) \quad \text{and} \quad h = -(ev_x + fv_y + gv_z) \tag{3.19}$$

for $(\times)$; the negative of this applies for $(\div)$.

For the joints $(+)$ and $(-)$, **w** is in plane **p**, so $\mathbf{q} = \mathbf{p}$ and **w** can be found by assuming that a pitch joint with $\theta = 90°$ and $-90°$, respectively, is at **v**; that is,

$$\begin{aligned} \mathbf{w}(+) &= \mathbf{v} + \frac{l_2}{l_1} \begin{bmatrix} 0 & -c & b & 0 \\ c & 0 & -a & 0 \\ -b & a & 0 & 0 \\ 0 & 0 & 0 & 1 \end{bmatrix} (\mathbf{v} - \mathbf{u}) \\ &= \mathbf{v} - \frac{l_2}{l_1} \begin{bmatrix} c(v_y - u_y) - b(v_z - u_z) \\ a(v_z - u_z) - c(v_x - u_x) \\ b(v_x - u_x) - a(v_y - u_y) \\ 0 \end{bmatrix} \end{aligned} \tag{3.20}$$

Similarly,

$$\mathbf{w}(-) = \mathbf{v} + \frac{l_2}{l_1} \begin{bmatrix} c(v_y - u_y) - b(v_z - u_z) \\ a(v_z - u_z) - c(v_x - u_x) \\ b(v_x - u_x) - a(v_y - u_y) \\ 0 \end{bmatrix} \tag{3.21}$$

The right-angle bend is usually associated with a joint, so that l_1 or l_2 as shown in Figure 3.9 will have zero length. The nonzero link can precede or succeed the right-angle bend. Thus we have three quantities to consider: the revolute joint type or the sliding link length, the type of right-angle bend, and the associated link length. These can be shown in any order, and the type of movement is dependent (in general) on this ordering. For example, the

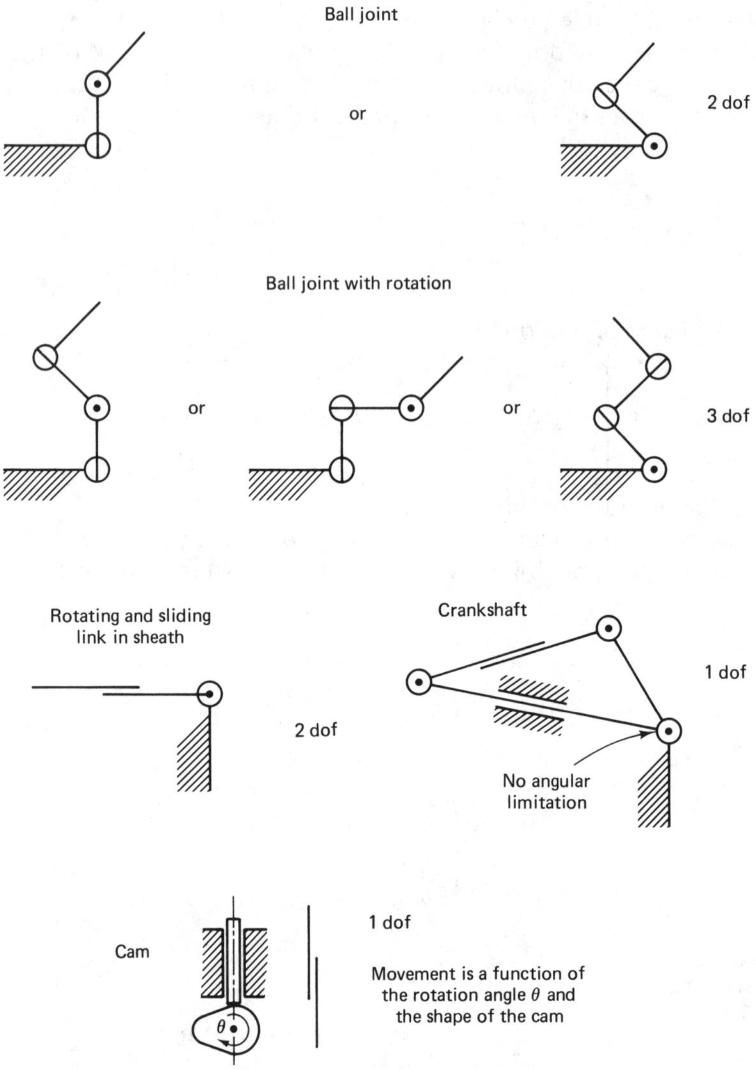

Figure 3.10 Equivalents of typical movements

specification $\theta_i(+)l_i$ means that the right-angle bend is after and tight against (link length zero followed by link length l_i) the pitch joint and link l_i succeeds the right-angle bend $(+)$. $\theta_il_i(+)$ means that the nonzero link precedes the right-angle bend. However, we will see that in practice it is always possible to associate a right-angle bend with a joint location, thus standardizing its occurrence for computer identification. This is discussed further in Section 3.7.

Any revolute joint can be modeled by any other revolute joint together with one or two right-angle bends and a change in datum. $(\times)\theta_i(\div)l_i \equiv \lambda_i l_i$ with angles equivalent. $(\times)\tilde{\phi}_i l_i \equiv \theta_i l_i$ with $\tilde{\phi}_i = -90°$ when $\theta_i = 0°$. Other equivalent systems with right-angle bends are left as an exercise.

There are other joints that are used in robotics, but these can always be expressed in terms of the joints just defined. For example, a ball joint (difficult to control and rarely used) with 2 *dof* is the combination of a cylindrical joint and pitch joint with the link between them having zero length. Movements equivalent to some typical movements are shown in Figure 3.10. The lead screw or helical joint has rotational and translational movement but only 1 *dof*. If the lead is infinite, the screw becomes a sliding joint. If the lead is zero, the joint becomes a roll joint.

3.4 DISTAL RELATIONSHIPS IN REVOLUTE JOINTS

Given a revolute joint at $\mathbf{u}_i$ whose downlink moves in plane $\mathbf{p}_j$. $\mathbf{u}_{i+1}$ and the plane of motion of $\mathbf{u}_{i+1}$ depend on the joint type, so we construct the following table, which lists the positions, angle and/or plane that determine $\mathbf{u}_{i+1}$, and the plane of motion of $\mathbf{u}_{i+1}$.

TABLE 3.1

Joint Type	$\mathbf{u}_{i+1}$	Plane of Motion of $\mathbf{u}_{i+1}$
Sliding σ	$(\mathbf{u}_{i-1}, \mathbf{u}_i, \sigma)$	$\mathbf{p}_j$
Pitch θ	$(\mathbf{u}_{i-1}, \mathbf{u}_i, \mathbf{p}_j, \theta)$	$\mathbf{p}_j$
Roll λ	$(\mathbf{u}_{i-1}, \mathbf{u}_i)$	$\mathbf{p}_{j+1}(\mathbf{u}_{i-1}, \mathbf{u}_i, \mathbf{p}_j, \lambda)$
Yaw τ	$(\mathbf{u}_{i-1}, \mathbf{u}_i, \mathbf{p}_j, \tau)$	$\mathbf{p}_{j+1}(\mathbf{p}_j, \mathbf{u}_{i-1}, \mathbf{u}_i)$
Crank ψ	$(\mathbf{u}_i, \mathbf{p}_j)$	$\mathbf{p}_{j+1}(\mathbf{p}_j, \mathbf{u}_{i-1}, \mathbf{u}_i, \psi)$
Crank $\tilde{\psi}$	$(\mathbf{u}_{i-1}, \mathbf{u}_i, \mathbf{p}_j, \tilde{\psi})$	$\mathbf{p}_{j+1}(\mathbf{u}_{i-1}, \mathbf{u}_i)$
Cylindrical ϕ	$(\mathbf{u}_{i-1}, \mathbf{u}_i, \mathbf{p}_j)$	$\mathbf{p}_{j+1}(\mathbf{u}_{i-1}, \mathbf{u}_i, \mathbf{p}_j, \phi)$
Cylindrical $\tilde{\phi}$	$(\mathbf{u}_{i-1}, \mathbf{u}_i, \mathbf{p}_{j+1})$	$\mathbf{p}_{j+1}(\mathbf{u}_{i-1}, \mathbf{u}_i, \mathbf{p}_j, \tilde{\phi})$

The set of defined revolute joints includes most revolute joint configurations. A few robots may have other joints, but these can be modeled by one of this set plus a right-angle bend. We permit the uplink of a cylindrical ϕ or crank ψ joint or the downlink of a cylindrical $\tilde{\phi}$ or crank $\tilde{\psi}$ joint to have negative length for completeness. Examples 2 and 4 demonstrate equivalent capability with different joints when using right-angle bends.

It will be shown that every common robot can be specified in terms of the joints discussed in previous sections of this chapter. It will also be shown that the symbology is simple and consistent. The mathematics of each revolute joint is different, although all that is needed to specify the fixed portion of the kinematics is a joint indicator (θ, ψ, $\tilde{\psi}$, ϕ, $\tilde{\phi}$, λ, or σ).

3.5 CURVED LINKS

Suppose the beginning of a curved link is at $\mathbf{u}$ with direction $\boldsymbol{\nu}$. A general curved link can place the other end of the curved link at $\mathbf{v}$, where the end of the link lies in direction $\boldsymbol{\eta}$. $\mathbf{v}$ and $\boldsymbol{\eta}$ can be arbitrarily chosen by the geometry of the curved link, so a total of six (three in $\mathbf{v}$ and three in $\boldsymbol{\eta}$) parameters specify the curved link. It is possible to model a curved link with fewer parameters, but then the capability of the curved link is less. For example, the Denavit-Hartenberg notation uses one parameter to model its curved link; see Section 3.6. The six parameters are equivalent to the 6 *dof* required to position the end effector of a robot at a particular location and in a particular orientation (see Chapter 4).

A manipulator with curved links is far more difficult to analyze over one with straight links and right-angle bends, yet it provides no unique functional capability. In most cases a curved link can be replaced by a straight link by reorienting the downlinks, as discussed in Section 6.1, and no further consideration is given to curved links until that section. The reasons for designing a robot with one or more curved links is not clear. The inverse kinematic analysis of a robot containing one or more curved links is about twice as costly computationwise as the same robot with the curved link(s) removed. If possible, curved links should be avoided. In this author's knowledge, all common commercial robots have straight links.

3.6 THE DENAVIT-HARTENBERG NOTATION

The Denavit-Hartenberg notation [Hartenberg and Denavit 1964; Denavit and Hartenberg 1955; Craig 1986; McCarthy 1986] is a specification for the relationship between any two revolute[1] joints and a restricted class of curved link between them. It can be used to model any joint defined in Section 3.1. One of the simplest explanations of the Denavit-Hartenberg notation is found in a work by Wolovich [Wolovich 1987].

The terminology used to define the Denavit-Hartenberg notation is changed to be more consistent with the terminology developed here, and we start by considering a revolute joint. As shown in Figure 3.11, given joint location $\mathbf{u}_i$ and two unit orthogonal vectors $\boldsymbol{\nu}_i$ and $\boldsymbol{\eta}_i$, where $\boldsymbol{\eta}_i$ is the direction followed by link l_{i-1} connected between joints at $\mathbf{u}_{i-1}$ and $\mathbf{u}_i$, which moves in plane $\mathbf{p}$, and $\boldsymbol{\nu}_i$ is the direction of the vector partially defining $\mathbf{p}$. $\boldsymbol{\nu}_{i+1}$ is a vector partially defining $\mathbf{q}$ and is found by rotating $\boldsymbol{\nu}_i$ by α_i in the plane shown. $\boldsymbol{\eta}_{i+1}$ is found by rotating $\boldsymbol{\eta}_i$ by β_i in $\mathbf{q}$. $\mathbf{u}_{i+1} = \mathbf{u}_i + l_i \boldsymbol{\eta}_{i+1}$. For

[1] It also handles a right-angle bend followed by a sliding link, a combination they call *prismatic*, but we use the simpler sliding link.

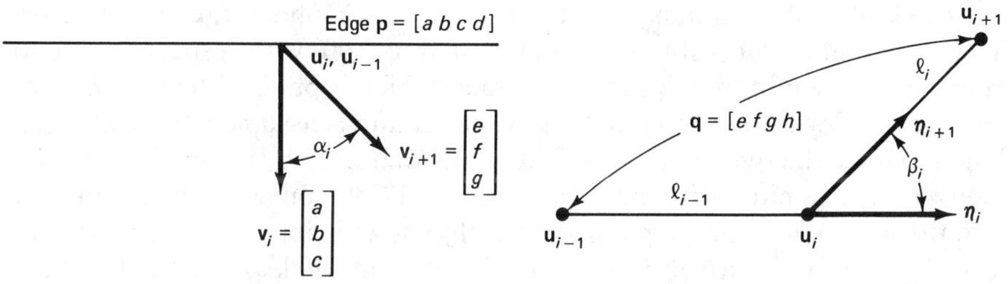

Figure 3.11 The Denavit-Hartenberg notation

the specific revolute joints considered here, we have

Pitch: $\alpha_i = 0$, so $\boldsymbol{\nu}_{i+1} = \nu_i$ and $\beta_i = \theta_i$.
Roll: $\alpha_i = \lambda_i$ and $\beta_i = 0$, so $\eta_{i+1} = \eta_i$.
Yaw: $\alpha_i = 90°$ and $\beta_i = \tau_i$.

The other revolute joints may require a right-angle bend in addition to the 2 *dof* Denavit-Hartenberg joint. For the sliding joints

(+) is $\alpha_i = 0$ and $\beta_i = -90°$.
(−) is $\alpha_i = 0$ and $\beta_i = 90°$.
(×) is $\alpha_i = -90°$ and $\beta_i = 0$.
(÷) is $\alpha_i = 90°$ and $\beta_i = 0$.

The disadvantage of Denavit-Hartenberg as far as this work is concerned is that the joint is more complex and adds no additional capability over the set of kinematically simple joints considered in Section 3.1 and the right-angle bends; also, all commonly available robots have kinematically simple joints [Paul 1981a]. Some may argue that it can handle curved links, but this argument is spurious, since its ability in this matter is one variable, whereas a general curved link requires six variables. Using kinematically simple joints gives meaning to the planes of motion, and the capability of a robot is evident from these planes alone (see Chapter 4). Denavit-Hartenberg will be used no further.

3.7 ROBOTIC SPECIFICATION

To find the end $\mathbf{u}_{i+1}$ of link l_i and the plane in which l_i moves requires length l_i, the angle of the revolute joint at $\mathbf{u}_i$, the type of revolute joint at $\mathbf{u}_i$, locations $\mathbf{u}_i$ and $\mathbf{u}_{i-1}$ for link l_{i-1}, and the plane in which l_{i-1} moves.

The Microbot training robot has 5 *dof*, not counting the end effector. The base joint is roll (with movement limited to $\pm 90°$ with respect to datum) connected by a 6-in. link l_1 to the shoulder (which is pitch with movement of $-45°$ to $+45°$ with respect to datum, where datum is assumed as for all pitch joints to have the two links in straight line), connected by a 7-in. link l_2 to the elbow (which is pitch with movement of $0°$ to $125°$), connected by a 7-in. link l_3 to the wrist with 2 *dof* (a pitch and a roll joint with approximately $-180°$ to $0°$ and $\pm 180°$, respectively), with total length $\frac{5}{8}$ in. (links l_4 and l_5) to which the end effector is connected. The schematic of the Microbot is shown in Figure 3.12.

We can define the Microbot and its position in three-dimensional space as

$$\begin{array}{cccccc} \mathbf{p}_1\overline{\mathbf{p}}_2 & & & & \overline{\mathbf{p}}_3 & \\ \left| \lambda_1 \begin{Bmatrix} +90 \\ -90 \end{Bmatrix} 6 \right. & \left| \theta_2 \begin{Bmatrix} +45 \\ -45 \end{Bmatrix} 7 \right. & \left| \theta_3 \begin{Bmatrix} +125 \\ 0 \end{Bmatrix} 7 \right. & \left| \theta_4 \begin{Bmatrix} +0 \\ -180 \end{Bmatrix} \frac{5}{16} \right. & \left| \lambda_5 \begin{Bmatrix} +180 \\ -180 \end{Bmatrix} \frac{5}{16} \right. & \Big| \\ \mathbf{u}_1 & \mathbf{u}_2 & \mathbf{u}_3 & \mathbf{u}_4 & \mathbf{u}_5 & \mathbf{u}_6 \end{array} \tag{3.22}$$

where the joint locations are at the vertical bars (so $\mathbf{u}_1$ to $\mathbf{u}_6$ can be implied), the base of the robot sits at point $\mathbf{u}_1$ in plane $\mathbf{p}_1$, and the first four links move in plane $\overline{\mathbf{p}}_2$. $\overline{\mathbf{p}}_2$ can be found, using equation (3.7), from the base orientation and position $\mathbf{u}_1$ together with λ_1. $\mathbf{u}_5$ and $\mathbf{u}_6$ (at the last vertical bar) move in plane $\overline{\mathbf{p}}_3$. Whenever a plane is given above but to the right of a vertical bar, a fold line (see Section 3.9) occurs at this joint. The meaning of the bars in $\overline{\mathbf{p}}_2$ and $\overline{\mathbf{p}}_3$ is discussed in Chapter 4. Expression (3.22) contains all the information contained in Figure 3.12, and these should be considered interchangeable.

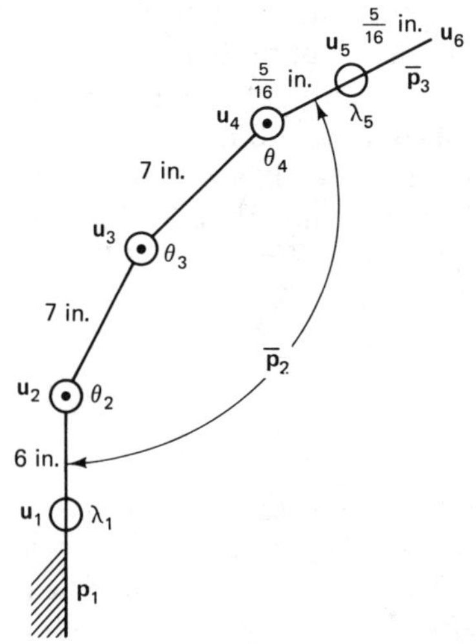

Figure 3.12 Schematic of a Microbot robot

The Microbot has linked joint movements—that is, stepping one of the motors may move more than one joint. However, this does not change the conceptual nature of the discussion here. Further, the computer control of the Microbot can be devised so that two stepper motors are driven to move one joint only.

Expression (3.22) contains redundant information. The base plane, which can be considered the prior plane of motion, is always implied and so can be deleted. The subscripts associated with the planes of motion provide no information and henceforth are deleted. The planes of motion themselves can be deduced from the central line of information. Deleting information on the planes of motion reduces expression (3.22) to

$$\left| \lambda_1 \begin{Bmatrix} +90 \\ -90 \end{Bmatrix} 6 \;\middle|\; \theta_2 \begin{Bmatrix} +45 \\ -45 \end{Bmatrix} 7 \;\middle|\; \theta_3 \begin{Bmatrix} +125 \\ 0 \end{Bmatrix} 7 \;\middle|\; \theta_4 \begin{Bmatrix} +0 \\ -180 \end{Bmatrix} \tfrac{5}{16} \;\middle|\; \lambda_5 \begin{Bmatrix} +180 \\ -180 \end{Bmatrix} \tfrac{5}{16} \right|$$

and omitting joint angle limitations further reduces the expression to

$$| \; \lambda_1 6 \; | \; \theta_2 7 \; | \; \theta_3 7 \; | \; \theta_4 \tfrac{5}{16} \; | \; \lambda_5 \tfrac{5}{16} \; |$$

Often the expression for a robot is given in terms of general link lengths l_i, so the preceding expression becomes

$$| \; \lambda_1 l_1 \; | \; \theta_2 l_2 \; | \; \theta_3 l_3 \; | \; \theta_4 l_4 \; | \; \lambda_5 l_5 \; |$$

The link lengths themselves are often implied, which gives the configuration of the robot as

$$| \; \lambda_1 \theta_2 \theta_3 \theta_4 \lambda_5 \; |$$

Since the joints are ordered, the subscripts are redundant, and the configuration can be shown in its simplest form as $| \; \lambda\theta\theta\theta\lambda \; |$.

To avoid any ambiguity, the following convention is adopted. The angles for the revolute joints are designated as follows:

θ for the pitch joint

λ for the roll joint

τ for the yaw joint

ϕ for the cylindrical joint connected in the forward orientation

$\tilde{\phi}$ for the cylindrical joint connected in the reverse orientation

$\underline{\phi}$ for the cylindrical joint with the uplink in the opposite direction than with ϕ (this is equivalent to an uplink of negative length)

$\underline{\tilde{\phi}}$ for the cylindrical joint with the uplink in the opposite direction than with $\tilde{\phi}$

ψ for the crank joint connected in the forward orientation

$\tilde{\psi}$ for the crank joint connected in the reverse orientation

$\underline{\psi}$ for the crank joint in the opposite direction than with ψ

$\underline{\tilde{\psi}}$ for the crank joint in the opposite direction than with $\tilde{\psi}$

Joint limitation information (if any) in degrees is contained in curly braces, as in

$$\left| \theta_2 \begin{Bmatrix} 80 \\ -60 \end{Bmatrix} l_2 \right| \quad \text{or} \quad \left| \tilde{\psi}_4 \begin{Bmatrix} 45 \\ -45 \end{Bmatrix} 255 \text{ mm} \right|$$

If the total swing of the joint is known but not the relationship to datum, the joint limitation information could be

$$| \ \theta_2\{140\}l_2 \ |$$

The datum positions for the joint between links l_{i-1} and l_i from which the information in the curly braces is measured are

θ	link l_i in line with link l_{i-1}
λ or ϕ	$\mathbf{q} = \mathbf{p}$
τ	link l_i in line with l_{i-1}
ψ	vector defining $\mathbf{q}$ is direction of l_{i-1}

We use English or metric measures, according to the information provided by the particular robot manufacturer, so mm stands for millimeter, m stands for meter, in. represents inches, and ft. stands for feet.

The length of the link l_i following the joint at $\mathbf{u}_i$ is after the joint limitation angles and in line with the joint type (θ, ϕ, etc.). The type of right angle turn at the end of link l_i is placed in parentheses according to the convention discussed in Section 3.3. Right-angle bends are included in the joint specifications as

$$\left| (-)\theta_2 \begin{Bmatrix} 80 \\ -60 \end{Bmatrix} l_2 \right| \quad \text{or} \quad \left| (\times)\tilde{\psi}_4 \begin{Bmatrix} 45 \\ -45 \end{Bmatrix} 255 \text{ mm} \right|$$

σ refers to a sliding link with the minimum and maximum lengths of the link given without braces (the longest length on top and the shortest length on the bottom), the information being given symbolically or numerically. There is more than one way to specify a sliding joint, and some examples are

$$\left| \theta_3 \begin{Bmatrix} 100 \\ -45 \end{Bmatrix} \begin{matrix} \bar{\sigma}_4 \\ \sigma_4 \end{matrix} \right|, \quad \left| \phi_2 \begin{matrix} 16 \text{ in.} \\ 10 \end{matrix} \right|, \quad \text{and} \quad \left| \begin{matrix} 90 \text{ mm} \\ 55 \end{matrix} \right| \left| (+) \begin{matrix} 100 \\ 65 \end{matrix} \right|$$

Notice that if the sliding link follows any revolute joint, its information is included within the bars containing that joint, but if a sliding link follows a sliding link, the symbol σ is used (as in the right-hand example just given, which can describe the first 2 *dof* of a Cartesian robot). One *dof* is a joint and a link. Notice that the number of degrees of freedom contained within the vertical lines is 1 (for a single joint) or 2 (for a revolute joint followed by a sliding joint).

The human arm can be approximately modeled by

$$\begin{array}{cccccc} \overline{\mathbf{p}}^3 & & & \overline{\mathbf{p}}^0 & \mathbf{p}^1 & \\ \mid \lambda_1 0 \mid & \theta_2 l_2 \mid & \theta_3 l_3 \mid & \theta_4 l_4 \mid & \lambda_5 l_5 \mid & \tau_6 l_6 \mid \end{array}$$

where the first 3 *dof* define the main motion of the arm to the wrist (and have the configuration of the 3 *dof* revolute robot discussed in the next section), and the last 3 *dof* define the motion of the wrist (pitch, roll, and yaw). l_2 and l_3 are large compared to l_4, l_5, and l_6. The ball joint shoulder is defined by $\lambda_1\theta_2$. The complete mechanism is often used as follows: The first 3 *dof* move the hand to the area required, and the last 3 *dof,* the built-in compliance, and the multiple *dof* in the fingers provide the fine-motor control to grasp or manipulate. In the last few years researchers have discussed a similar capability in robots [Lozano-Perez and Wesley 1984; Whitney and Edsall 1985]. Whitney and Edsall modeled the trajectories as stochastic processes in which gross parameters were entered, and sense provided fine-motor control. The method is good for noisy media (vibrations of the base, flex in links, and so on) and to take into account slop or wear in the robotic joints.

3.8 THE FIVE BASIC TYPES OF ROBOTS

The five basic robot types are revolute, cylindrical, polar, Cartesian, and SCARA, and these are shown in Figure 3.13. The revolute robot always starts with the three joints

$$\begin{array}{ccc} \overline{\mathbf{p}}_2^2 & & \\ \mid \lambda_1 l_1 \mid & \theta_2 l_2 \mid & \theta_3 l_3 \mid \end{array} \tag{3.23}$$

where the superscripts in the planes of motion are explained in Chapter 4. The subscripts refer to the number of the plane, where $\mathbf{p}_1$ is always used for the base plane; this is a useful way of specifying planes, especially when considering piggyback robots (one robot mounted on another robot). The robot is usually configured with 5 or 6 *dof.* The revolute robot is the most popular robot, and more manufacturers have this type of robot than any other type. It is used for material handling and assembly tasks of all types. It is also used for welding and paint spraying. It has a large working envelope, making it easy to interface with other equipment. Speeds up to 10 ft/s at the end effector are possible. Large models have hydraulic or pneumatic drives, whereas small and medium sized models have DC servomotor drives or even stepper motor drives. Small revolute robots have a repeatability of from ±0.002 in. to ±0.015 in., and large models have repeatability of from ±0.005 in. to ±0.2 in.

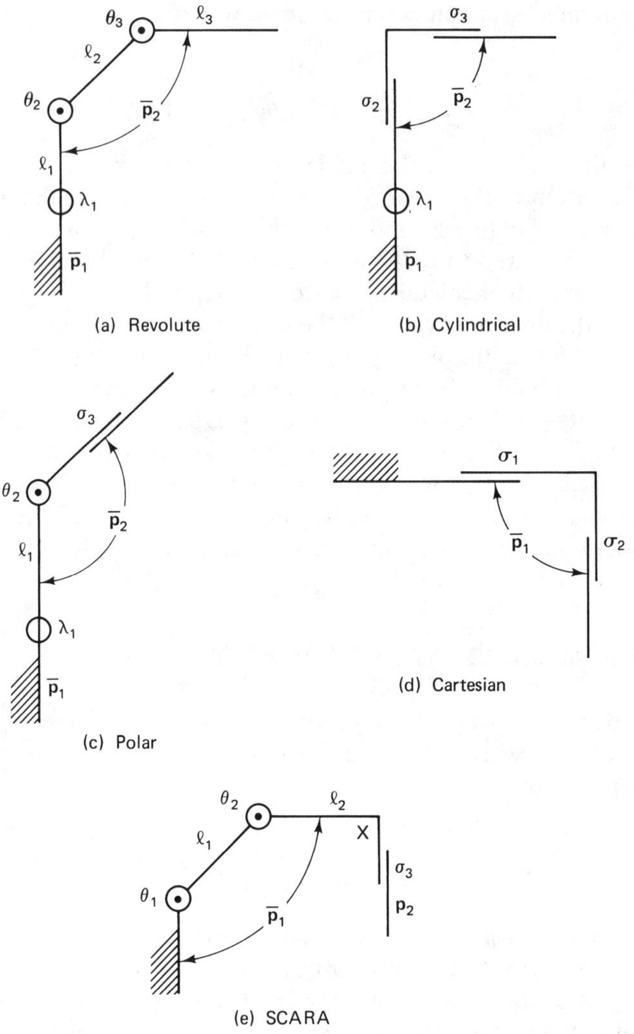

Figure 3.13 The five basic types of robots

The cylindrical robot always starts with the three joints

$$\overline{\mathbf{p}}_2^2$$
$$\left| \lambda_1 {}^{\overline{\sigma}_2}_{\sigma_2} \right| (+) {}^{\overline{\sigma}_3}_{\sigma_3} \Big| \qquad (3.24)$$

The cylindrical robot can run from small to large in size, usually with 5 or 6 *dof*. They are often used in hazardous environments. They are slow (2 to 8 ft/s) with servo drives or faster with pneumatic drives for the sliding joints on large models. The pneumatic-driven robot is sometimes called a bang-bang robot due to the mechanical settings at the ends of travel: the robot bangs against these mechanical end stops, and the end stops can be altered only manually. The trend is to use powerful and accurate (with optical encoders) servos. Small robots have accuracy of about ±0.005 in., and large robots have accuracy in the ±0.2-in. range.

The polar or spherical robot always starts with the three joints

$$\begin{array}{l} \quad\ \bar{\mathbf{p}}_2^2 \\ \big| \lambda_1 l_1 \big| \theta_2 {}^{\bar{\sigma}_3}_{\sigma_3} \big| \end{array} \tag{3.25}$$

Polar robots are often used for heavy-duty material handling and machining, usually with 5 or 6 *dof*. They are fast and accurate with electrical and/or hydraulic drives.

The Cartesian robot always starts with the two joints

$$\begin{array}{l} \quad \mathbf{p}_1^2 \\ \Big| {}^{\bar{\sigma}_1}_{\sigma_1} \Big| (+) {}^{\bar{\sigma}_2}_{\sigma_2} \Big| \end{array} \tag{3.26}$$

Cartesian robots can be configured into a gantry setup for large-scale use, such as in the automotive industry, or for small-scale use, such as an XY table. Often a third, orthogonal sliding link is used to make the robot 3 *dof* and placeable; see Section 4.3. It is rare to have this kind of robot with more than 3 *dof*. Cartesian robots are often used to transport material to and from hazardous areas. They are slow due to their dependence on chains, belts, or lead screws. They are servo- or pneumatic-driven. Larger models have accuracy from ±0.02 in. to more than ±0.2 in. The SCARA robot has limited the use of the small Cartesian robot.

The SCARA (Selective Compliant Articulative Robot for Assembly, or Selective Compliant Assembly Robot Arm) robot was developed in Japan by Professor Hiroshi Makino of Yamanashi University. It has 3 *dof* and can be described by

$$\begin{array}{l} \quad\ \mathbf{p}_1^2 \qquad\quad \mathbf{p}_2^1 \\ \big| \theta_1 l_1 \big| \theta_2 l_2 \big| (\times) \sigma_3 \big| \end{array} \tag{3.27}$$

Typically, l_1 and l_2 operate in the horizontal plane $[0 \quad 0 \quad 1 \quad -l]$, where l is the height of the base pedestal. The SCARA may be configured with a roll joint on the sliding link, and we modify the specification of this 4 *dof* robot to be

$$\begin{array}{l} \quad \mathbf{p}_1^2 \qquad\quad \mathbf{p}_2^1 \qquad\quad \bar{\mathbf{p}}_3^0 \\ \big| \theta_1 l_1 \big| \theta_2 l_2 \big| (\times) {}^{\bar{\sigma}_3}_{\sigma_3} \big| \lambda_4 0 \big| \end{array} \quad \text{or its equivalent,} \quad \begin{array}{l} \quad \mathbf{p}_1^2 \qquad\quad \bar{\mathbf{p}}_2^1 \\ \big| \theta_1 l_1 \big| \theta_2 l_2 \big| \underline{\psi}_3 {}^{\bar{\sigma}_4}_{\sigma_4} \big| \end{array}$$

It is unusual to have a SCARA-type robot with more than 4 *dof*. SCARA robots are used for high-speed, high-accuracy (±0.002-in. repeatability) small-parts assembly, such as printed circuit board populating, which require ±0.005-in. or better accuracy. Maximum speeds at the end effector are often quoted at 30 ft/s, compared to 4 ft/s for the Puma robot (which is polar). However, cycle time (the time between identical states in successive operations) is more important in determining productivity and is the parameter that should be measured, and the SCARA is usually superior in this category.

3.9 FOLD LINES

The purpose of the symbols for the joints of a robot manipulator is to present the configuration of the manipulator on a plane surface, the piece of paper. However, the robot moves in Cartesian space through one or more planes, so the planar representation must be interpreted correctly. Suppose a revolute joint at **v** connects downlink l_1 (which moves in plane **p**) with endpoints **u** and **v** to uplink l_2 (which moves in plane **q**) with endpoints **v** and **w**. We show the spatial relationship of **u**, **v**, **w**, **p**, and **q** with a fold line—the line of intersection of **p** and **q** given by the cross product of the vectors partially defining **p** and **q**.

The pitch joint produces no fold line. For the roll joint, the fold line lies along l_1 and the angle of the fold is λ, as indicated in Figure 3.14(a). To appreciate the actual planes of motion in Cartesian space, reproduce Figure 3.14(a) on a separate sheet of paper and fold it along the fold line. Keep the side marked plane **p** fixed and change the angle λ of the fold to illustrate the orientation of plane **q**.

For the yaw joint, the fold line lies along l_1 (as is evident from equation (3.7)) and the angle of the fold is 90°, as illustrated in Figure 3.14(b). We have previously indicated that the yaw joint has no orientation, so connecting it in reverse does not change its geometry. However, as shown in Figure 3.14(b), **u** and **v** are on the fold line, whereas **w** is not, so connecting the joint in reverse makes **w** and **v** lie on the fold line. The apparent inconsistency is resolved by noting that the correct location for **u** and plane **p** is obtained when using **w**, **v**, and **q**, and the location and plane of motion are the only items of interest. The reader is encouraged to verify this geometrically by constructing Figure 3.14(b) on a separate sheet and folding about the two possible fold lines to verify that **u** is unaltered.

Consider the crank joint ψ at **v**. The fold line is defined in part by vector

$$\begin{bmatrix} a \\ b \\ c \end{bmatrix} \times \begin{bmatrix} e \\ f \\ g \end{bmatrix} = \frac{1}{l_1}\begin{bmatrix} \cos\psi & -c\sin\psi & b\sin\psi \\ c\sin\psi & \cos\psi & -a\sin\psi \\ -b\sin\psi & a\sin\psi & \cos\psi \end{bmatrix}\begin{bmatrix} v_x - u_x \\ v_y - u_y \\ v_z - u_z \end{bmatrix} \times \begin{bmatrix} a \\ b \\ c \end{bmatrix}$$

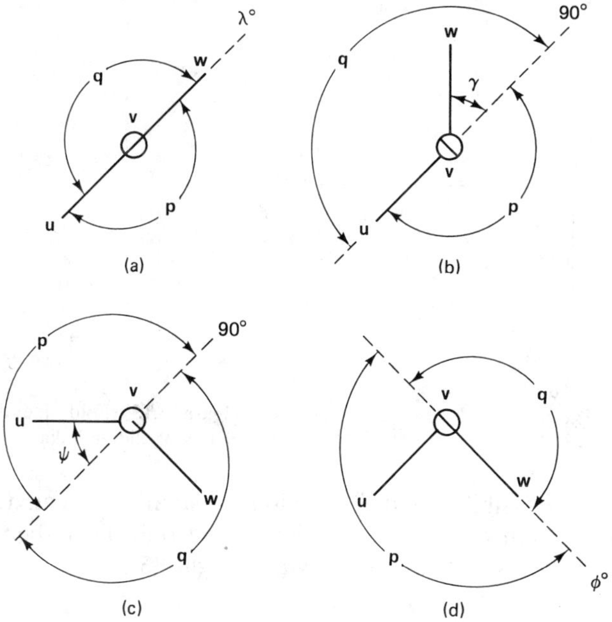

Figure 3.14 Fold lines

The interpretation of this is that the fold is always 90° and its direction is along $\mathbf{v} - \mathbf{u}$ for $\psi = 0$, swinging counterclockwise in $\mathbf{p}$ as ψ increases. This is shown in Figure 3.14(c), i.e. $\mathbf{q}$ is a function of ψ. Notice that $\mathbf{w}$ is unchanged by ψ. However, if the joint is connected the other way around (the joint is now called $\tilde{\psi}$), the uplink plane is not changed by the joint angle.

From equation 3.15, the fold line for the cylindrical joint ϕ is defined in part by

$$\begin{bmatrix} a \\ b \\ c \end{bmatrix} \times \begin{bmatrix} e \\ f \\ g \end{bmatrix} = \begin{bmatrix} a \\ b \\ c \end{bmatrix} \times \left\{ \cos\phi \begin{bmatrix} a \\ b \\ c \end{bmatrix} - \frac{\sin\phi}{l_1}(\mathbf{v} - \mathbf{u}) \right\}$$

$$= -\frac{\sin\phi}{l_1} \begin{bmatrix} a \\ b \\ c \end{bmatrix} \times (\mathbf{v} - \mathbf{u}) \equiv -\frac{\sin\phi}{l_2}(\mathbf{w} - \mathbf{v}).$$

so it lies along l_2 and the fold angle is ϕ as shown in Figure 3.14(d). Notice that $\mathbf{w}$ which lies on both $\mathbf{p}$ and $\mathbf{q}$ is uneffected by ϕ. When the joint is connected the other way around (and given the symbol $\tilde{\phi}$), the end of the uplink moves in Cartesian space with $\tilde{\phi}$ but is stationary in the uplink plane: this is evident from a study of Figure 3.14(d).

The individual planes of motion of a robot will lie between the fold lines, and a fold line will occur for every revolute joint but the pitch joint.

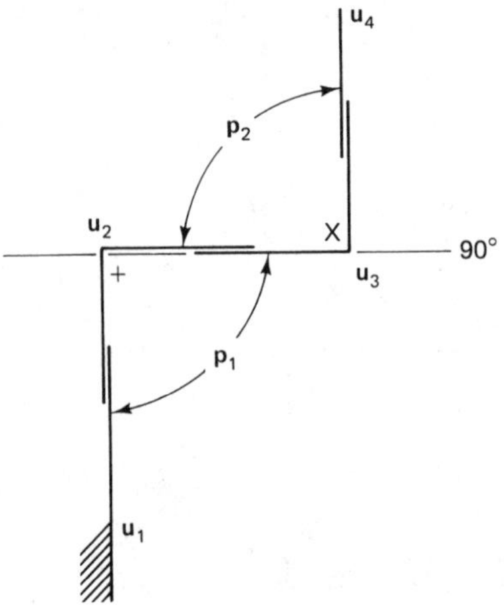

Figure 3.15 Fold line destination for chains of sliding links

We exclude the possibility of defective joint systems. We can designate the fold line for a chain of three sliding joints and right angle bends that can model a 3 *dof* Cartesian robot as shown in Figure 3.15.

3.10 EXAMPLES

1. Determine **w** and **q** for all the revolute joints as a function of the joint angle, where

$$\mathbf{u} = \begin{bmatrix} 0 \\ 0 \\ -1 \\ 1 \end{bmatrix}, \quad \mathbf{v} = \begin{bmatrix} 0 \\ 0 \\ 0 \\ 1 \end{bmatrix}, \quad \mathbf{p} = [0 \quad 1 \quad 0 \quad 0]$$

and the link lengths are unity.

We start by noting that for all revolute joints, **q** passes through **v**, which is the origin, so $h = 0$. For the pitch joint, using equation (3.1),

$$\mathbf{w} = \begin{bmatrix} 0 \\ 0 \\ 0 \\ 1 \end{bmatrix} + \begin{bmatrix} \cos\theta & 0 & \sin\theta & 0 \\ 0 & \cos\theta & 0 & 0 \\ -\sin\theta & 0 & \cos\theta & 0 \\ 0 & 0 & 0 & 1 \end{bmatrix} \begin{bmatrix} 0 \\ 0 \\ 1 \\ 0 \end{bmatrix} = \begin{bmatrix} \sin\theta \\ 0 \\ \cos\theta \\ 1 \end{bmatrix} \quad \text{and} \quad \mathbf{q} \equiv \mathbf{p}$$

For the roll joint, $\mathbf{w} = \begin{bmatrix} 0 \\ 0 \\ 1 \\ 1 \end{bmatrix}$ and

$$\begin{bmatrix} e \\ f \\ g \end{bmatrix} = \begin{bmatrix} \cos\lambda & -\sin\lambda & 0 \\ \sin\lambda & \cos\lambda & 0 \\ 0 & 0 & \cos\lambda \end{bmatrix} \begin{bmatrix} 0 \\ 1 \\ 0 \end{bmatrix} = \begin{bmatrix} -\sin\lambda \\ \cos\lambda \\ 0 \end{bmatrix}.$$

For the yaw joint, $\begin{bmatrix} e \\ f \\ g \end{bmatrix} = \begin{bmatrix} -1 \\ 0 \\ 0 \end{bmatrix}$ and

$$\mathbf{w} = \begin{bmatrix} 0 \\ 0 \\ 0 \\ 1 \end{bmatrix} + \begin{bmatrix} 0 \\ -\sin\gamma \\ \cos\gamma \\ 0 \end{bmatrix} = \begin{bmatrix} 0 \\ -\sin\gamma \\ \cos\gamma \\ 1 \end{bmatrix}.$$

For the crank joint ψ, $\mathbf{w} = \begin{bmatrix} 0 \\ 0 \\ 0 \\ 1 \end{bmatrix} - \begin{bmatrix} 0 \\ 1 \\ 0 \\ 0 \end{bmatrix} = \begin{bmatrix} 0 \\ -1 \\ 0 \\ 1 \end{bmatrix}$ and

$$\begin{bmatrix} e \\ f \\ g \end{bmatrix} = \begin{bmatrix} \cos\psi & 0 & \sin\psi \\ 0 & \cos\psi & 0 \\ -\sin\psi & 0 & \cos\psi \end{bmatrix} \begin{bmatrix} 0 \\ 0 \\ 1 \end{bmatrix} = \begin{bmatrix} \sin\psi \\ 0 \\ \cos\psi \end{bmatrix}.$$

For the crank joint $\tilde{\psi}$, $\begin{bmatrix} e \\ f \\ g \end{bmatrix} = -\begin{bmatrix} 0 \\ 0 \\ -1 \end{bmatrix} = \begin{bmatrix} 0 \\ 0 \\ 1 \end{bmatrix}$ and

$$\mathbf{w} = \begin{bmatrix} 0 \\ 0 \\ 0 \\ 1 \end{bmatrix} - \begin{bmatrix} \cos\tilde{\psi} & \sin\tilde{\psi} & 0 & 0 \\ -\sin\tilde{\psi} & \cos\tilde{\psi} & 0 & 0 \\ 0 & 0 & \cos\tilde{\psi} & 0 \\ 0 & 0 & 0 & 1 \end{bmatrix} \begin{bmatrix} 0 \\ 1 \\ 0 \\ 0 \end{bmatrix} = - \begin{bmatrix} \sin\tilde{\psi} \\ \cos\tilde{\psi} \\ 0 \\ 1 \end{bmatrix}.$$

For the cylindrical joint ϕ, $\mathbf{w} = \begin{bmatrix} 0 \\ 0 \\ 0 \\ 1 \end{bmatrix} - \begin{bmatrix} 1 \\ 0 \\ 0 \\ 0 \end{bmatrix} = \begin{bmatrix} -1 \\ 0 \\ 0 \\ 1 \end{bmatrix}$ and $\begin{bmatrix} e \\ f \\ g \end{bmatrix} = \begin{bmatrix} 0 \\ \cos\phi \\ -\sin\phi \end{bmatrix}$.

For the cylindrical joint $\tilde{\phi}$, $\begin{bmatrix} e \\ f \\ g \end{bmatrix} = \begin{bmatrix} \cos\tilde{\phi} & -\sin\tilde{\phi} & 0 \\ \sin\tilde{\phi} & \cos\tilde{\phi} & 0 \\ 0 & 0 & \cos\tilde{\phi} \end{bmatrix} \begin{bmatrix} 0 \\ 1 \\ 0 \end{bmatrix} =$

$$\begin{bmatrix} -\sin\tilde{\phi} \\ \cos\tilde{\phi} \\ 0 \end{bmatrix} \text{ and } \mathbf{w} = \begin{bmatrix} 0 \\ 0 \\ 0 \\ 1 \end{bmatrix} - \begin{bmatrix} -\cos\tilde{\phi} \\ -\sin\tilde{\phi} \\ 0 \\ 0 \end{bmatrix} = \begin{bmatrix} \cos\tilde{\phi} \\ \sin\tilde{\phi} \\ 0 \\ 1 \end{bmatrix}.$$

We summarize $\mathbf{w} = [x \quad y \quad z \quad 1]^{\#}$ and $\mathbf{q}$ for each joint type in the following table.

Joint	x	y	z	e	f	g
Pitch	$\sin\theta$	0	$\cos\theta$	0	1	0
Roll	0	0	1	$-\sin\lambda$	$\cos\lambda$	0
Yaw	0	$-\sin\gamma$	$\cos\gamma$	-1	0	0
Crank ψ	0	-1	0	$\sin\psi$	0	$\cos\psi$
Crank $\tilde{\psi}$	$\sin\tilde{\psi}$	$\cos\tilde{\psi}$	0	0	0	1
Cylindrical ϕ	-1	0	0	0	$\cos\phi$	$-\sin\phi$
Cylindrical $\tilde{\phi}$	$\cos\tilde{\phi}$	$\sin\tilde{\phi}$	0	$-\sin\tilde{\phi}$	$\cos\tilde{\phi}$	0

2. Show the true, spatial positioning of the uplink and uplink plane of motion for every defined robotic joint.

The spatial positioning is shown in Figure 3.16. The cones are used to show the vectors partially defining the planes. The fold lines are also shown where appropriate. The reader is encouraged to construct these geometries with scissors and paper.

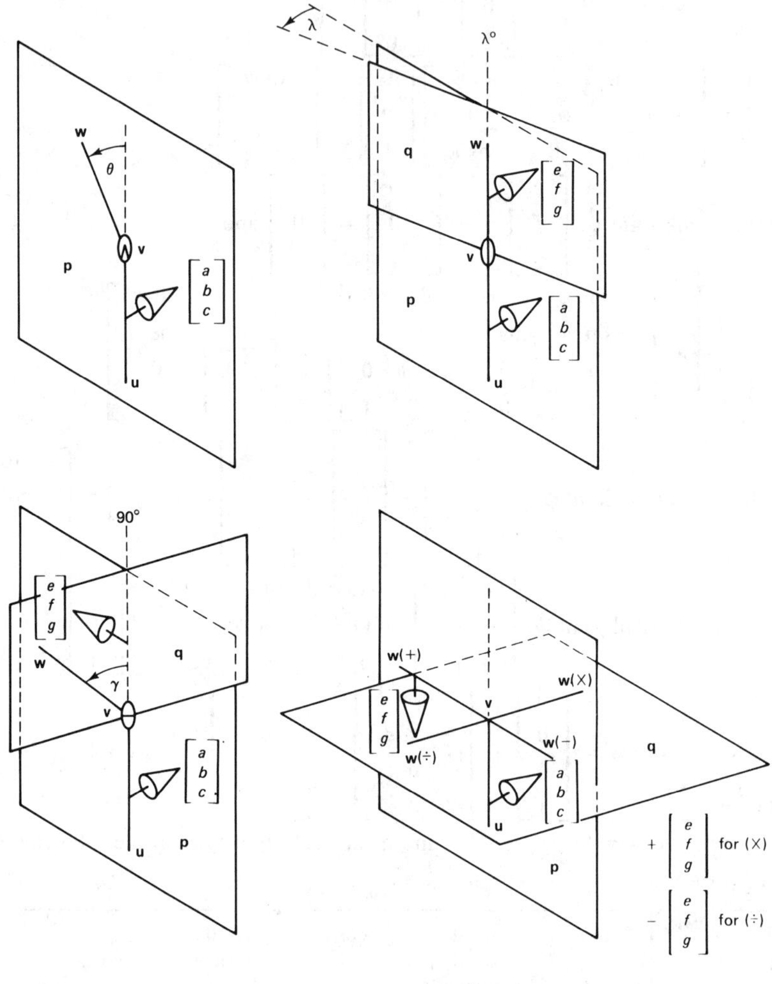

Figure 3.16 Spatial representation of joints

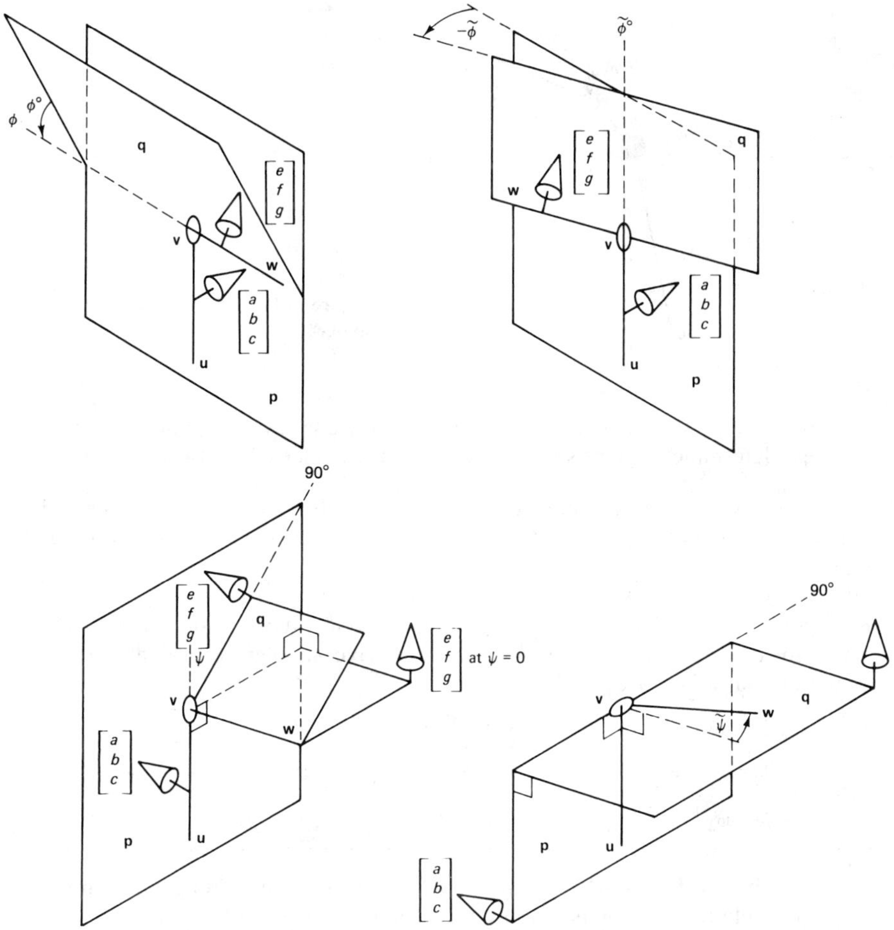

Figure 3.16 *continued*

3. Using the roll joint and one or more right-angle bends, model every other revolute joint.

$$\theta \equiv (\div)\lambda(\times) \qquad \phi \equiv (-)\lambda \qquad \tilde{\phi} \equiv \lambda(+)$$

$$\tilde{\psi} \equiv \lambda(\div) \qquad \psi \equiv (\div)\lambda \qquad \gamma \equiv (\times)\lambda(-)$$

---

4. Consider a chain of three links l_1, l_2, and l_3 connected by two revolute joints, where $l_2 = 0$, as shown in Figure 3.17. For every possible pair of revolute joints

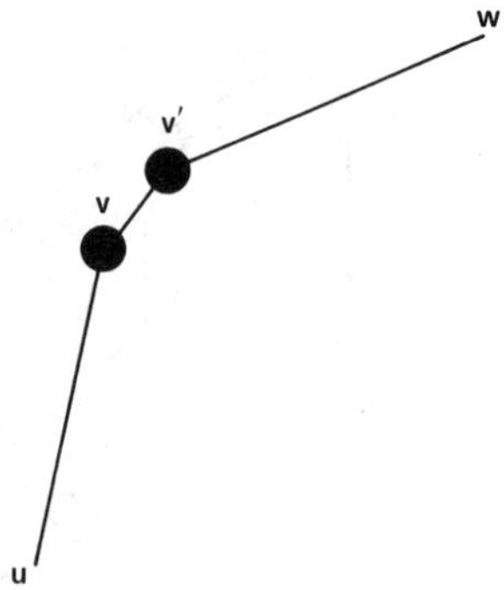

Figure 3.17 A chain of two revolute joints

determine if a simpler kinematic chain with only one revolute joint exists, and if it does, determine its joint status in terms of the angles of the two original joints.

$\mathbf{v}$ and $\mathbf{v}'$ become coincident when $l_2 = 0$. Notice that $+$, without immediate parentheses, as in $(+)$, is algebraic addition. The equivalents for pairs of the same joint are

$|\,\theta_1\,|\,\theta_2\,| \equiv |\,\theta_1+\theta_2\,|$; $|\,\lambda_1\,|\,\lambda_2\,| \equiv |\lambda_1+\lambda_2\,|$, and this is true if $l_2 \neq 0$;

$|\,\gamma_1\,|\,\gamma_2\,|$, $|\,\psi_1\,|\,\psi_2\,|$, $|\,\tilde{\psi}_1\,|\,\tilde{\psi}_2\,|$, $|\,\phi_1\,|\,\phi_2\,|$, and $|\,\tilde{\phi}_1\,|\,\tilde{\phi}_2\,|$ have no simpler equivalents. The only dissimilar revolute joint pairs that have simpler equivalents under the condition that $l_2 = 0$ are

$$|\,\phi_1\,|\,\tilde{\phi}_2\,| = |\,\gamma_{1+2}\,|, \qquad |\,\lambda_1\,|\,\tilde{\phi}_2\,| = |\,\tilde{\phi}_{1+2}\,|, \qquad |\,\psi_1\,|\,\tilde{\phi}_2\,| = |\,\theta_{1+2}\,|$$

$$|\,\phi_1\,|\,\lambda_2\,| = |\,\phi_{1+2}\,|, \qquad |\,\lambda_1\,|\,\tilde{\psi}_2\,| = |\,\tilde{\psi}_{1+2}\,|, \qquad |\,\psi_1\,|\,\lambda_2\,| = |\,\psi_{1+2}\,|$$

$$|\,\psi_1\,|\,\tilde{\psi}_2\,| = |\,\theta_{1+2}\,|$$

---

5. Model the 3 *dof* of SCARA robot $|\,\theta\theta(\times)\sigma|$ using yaw and sliding joints only. It is permissible to have joints that are immovable.

The robot would be

$$\begin{matrix}\mathbf{q}_1\,\mathbf{q}_2^1 & \mathbf{q}_3^1 & \mathbf{q}_4^1 & \mathbf{q}_5^2\end{matrix}$$

$$\left|\gamma_1 l_1\right|\gamma_2 0\left|\gamma_3 l_4\right|\gamma_4 {}^{\overline{\sigma}_5}_{\sigma_5}\Big|,$$ which is equivalent to

$$\begin{matrix}\mathbf{p}_1^2 & \mathbf{p}_2^1\end{matrix}$$

$$\left|\theta_1 l_1\right|\theta_2\left|(\times){}^{\overline{\sigma}_5}_{\sigma_5}\right|,$$

where $\mathbf{p}_1^2 = \mathbf{q}_2^1 = -\mathbf{q}_4^1$ and $\mathbf{p}_2^1 = \mathbf{q}_5^2$. Also, $\gamma_1 = \theta_1$; $\gamma_2 = 0°$; $\gamma_3 = -\theta_2$; $\gamma_4 = -90°$.

---

6. Given points $\mathbf{u} = \begin{bmatrix} 1 \\ 1 \\ 3 \\ 1 \end{bmatrix}$, $\mathbf{v} = \begin{bmatrix} 2 \\ 0 \\ 3 \\ 1 \end{bmatrix}$, and $\mathbf{p} = [3 \quad 2 \quad -2 \quad 1]$ for a link l_1 con-

nected between **u** and **v** moving in plane **p**. Verify that **u** and **v** are on **p**. Find l_1. A second link $l_2 = 1$ is attached at **v** with free end at point **w**. Find **w** in each case.
(a) A pitch joint is at **v** with $\theta = 30°$.
(b) A crank joint is at **v** with $\psi = 30°$.
(c) A cylindrical joint is at **v** with $\phi = 30°$.

$$\mathbf{pu} = [3 \quad 2 \quad -2 \quad 1]\begin{bmatrix} 1 \\ 1 \\ 3 \\ 1 \end{bmatrix} = 0 \quad \text{and} \quad \mathbf{pv} = [3 \quad 2 \quad -2 \quad 1]\begin{bmatrix} 2 \\ 0 \\ 3 \\ 1 \end{bmatrix} = 0$$

$$l_1 = \sqrt{(2-1)^2 + (0-1)^2 + (3-2)^2} = \sqrt{2} = 1.414$$

(a) In order to apply equation (3.1) the plane must be normalized to $\mathbf{p} = [3 \quad 2 \quad -2 \quad 1]/\sqrt{17} = [0.7276 \quad 0.4851 \quad -0.4851 \quad 0.2425]$. Then

$$\mathbf{w} = \begin{bmatrix} 2 \\ 0 \\ 3 \\ 1 \end{bmatrix} + \frac{1}{1.414}\begin{bmatrix} 0.866 & -0.3638 & 0.2426 & 0 \\ 0.3638 & 0.866 & 0.2426 & 0 \\ -0.2426 & -0.2426 & 0.866 & 0 \\ 0 & 0 & 0 & 1 \end{bmatrix}\begin{bmatrix} 1 \\ -1 \\ 0 \\ 0 \end{bmatrix} = \begin{bmatrix} 1.976 \\ 1.230 \\ 0 \\ 1 \end{bmatrix}$$

(b) From equation (3.10), $\mathbf{w} = \begin{bmatrix} 2 \\ 0 \\ 3 \\ 1 \end{bmatrix} - \begin{bmatrix} 0.728 \\ 0.485 \\ -0.485 \\ 0.243 \end{bmatrix} = \begin{bmatrix} 1.272 \\ -0.485 \\ 2.515 \\ 1 \end{bmatrix}$.

(c) We use equation (3.14) to get $\mathbf{w} = \begin{bmatrix} 2 \\ 0 \\ 3 \\ 1 \end{bmatrix} + \frac{1}{1.414}\begin{bmatrix} -0.485 \\ -0.485 \\ -0.728 + 0.485 \\ 0 \end{bmatrix} =$

$$\begin{bmatrix} 1.657 \\ -0.642 \\ 2.828 \\ 1 \end{bmatrix}.$$

---

7. A robot joint is at $\mathbf{v} = \begin{bmatrix} 0 \\ 0 \\ 0 \\ 1 \end{bmatrix}$, attached by link l_1 to point $\mathbf{u} = \begin{bmatrix} -2 \\ -1 \\ -1 \\ 1 \end{bmatrix}$ and by link l_2 to **w**. Link l_1 moves in plane $\mathbf{p} = [a \quad b \quad 0 \quad d]$. Find the locus of **w** if the joint is
(a) Pitch θ
(b) Crank ψ
(c) Cylindrical ϕ.

Since **p** passes through the origin at **v** then $d = 0$. **p** also passes through **u**, so $-2a - b = 0$, or $b = -2a$. Also since **p** is normalized, $a^2 + b^2 = 1$. Thus, $\mathbf{p} = [1/\sqrt{5}\ -2/\sqrt{5}\ \ 0\ \ 0]$.

(a) $l_1 = |\mathbf{v} - \mathbf{u}| = \sqrt{6}$. From equation (3.1),

$$\mathbf{w} = \begin{bmatrix} 0 \\ 0 \\ 0 \\ 1 \end{bmatrix} + \frac{l_2}{\sqrt{6}} \begin{bmatrix} \cos\theta & 0 & -\frac{2}{\sqrt{5}}\sin\theta & 0 \\ 0 & \cos\theta & \frac{-1}{\sqrt{5}}\sin\theta & 0 \\ \frac{2}{\sqrt{5}}\sin\theta & \frac{1}{\sqrt{5}}\sin\theta & \cos\theta & 0 \\ 0 & 0 & 0 & 1 \end{bmatrix} \begin{bmatrix} 2 \\ 1 \\ 1 \\ 0 \end{bmatrix}$$

$$= \frac{l_2}{\sqrt{6}} \begin{bmatrix} 2\cos\theta - \frac{2}{\sqrt{5}}\sin\theta \\ \cos\theta - \frac{1}{\sqrt{5}}\sin\theta \\ \frac{5}{\sqrt{5}}\sin\theta + \cos\theta \\ 1 \end{bmatrix}$$

(b) From equation (3.10) we have

$$\mathbf{w} = \begin{bmatrix} 0 \\ 0 \\ 0 \\ 1 \end{bmatrix} + l_2 \begin{bmatrix} \frac{1}{\sqrt{5}} \\ \frac{-2}{\sqrt{5}} \\ 0 \\ 0 \end{bmatrix}$$

(c) From equation (3.14) we have

$$\mathbf{w} = \begin{bmatrix} 0 \\ 0 \\ 0 \\ 1 \end{bmatrix} - \frac{l_2}{\sqrt{6}} \begin{bmatrix} \frac{-2}{\sqrt{5}} - 0 \\ 0 - \frac{1}{\sqrt{5}} \\ \frac{1}{\sqrt{5}} + \frac{4}{\sqrt{5}} \\ 0 \end{bmatrix} = \begin{bmatrix} \frac{2l_2}{\sqrt{30}} \\ \frac{l_2}{\sqrt{30}} \\ \frac{-5l_2}{\sqrt{30}} \\ 1 \end{bmatrix}$$

---

8. Show simpler equivalents for the defective manipulators.

(a) $|\ \lambda_1 l_1\ |\ \tilde{\phi}_2 l_2\ |\ \phi_3 l_3\ |\ \theta_4 l_4\ |$

(b) $|\ \phi_1 l_1\ |\ \psi_2 l_2\ |\ \lambda_3 l_3\ |\ \phi_4 l_4\ |\ \lambda_5 l_5\ |$

(c) $|\ \phi_1 l_1\ |\ \lambda_2 l_2\ |\ \phi_3 l_3\ |\ \phi_4 l_4\ |\ \lambda_5 l_5\ |\ \phi_6 l_6\ |$

(d) $|\ \lambda_1 l_1\ |\ \theta_2 l_2\ |\ \theta_3 l_3\ |\ \theta_4 l_4\ |\ \theta_5 l_5\ |$

(a) λ_1 is defective, so a simpler equivalent is

$$|\ \tilde{\phi}_2 l_2\ |\ \phi_3 l_3\ |\ \theta_4 l_4\ |$$

where the base position is now altered.

(b) We see that the two roll joints are defective following the crank ψ_2 and cylindrical ϕ_4, and a simpler equivalent is

$$|\ \phi_1 l_1\ |\ \psi_2 l_2 + l_3\ |\ \phi_4 l_4 + l_5\ |$$

(c) The two roll joints are defective, so a simpler equivalent is

$$|\ \phi_1 l_1 + l_2\ |\ \phi_3 l_3\ |\ \phi_4 l_4 + l_5\ |\ \phi_6 l_6\ |$$

which is a chain of four crank joints whose utility is suspect.

(d) We have a chain of four pitch joints whose use is limited to cases in which the manipulator must avoid objects in its workspace; see Section 8.8.

---

9. Determine the maximum i for the chain of joints $|\ \sigma_1(\times)\sigma_2(\times) \cdots \sigma_i(\times)\ |$ not to be defective. Assume that each pair of sliding links is separated by a right-angle bend.

We assume the location of the first two joints are at $\mathbf{u}_1$ and $\mathbf{u}_2$, where

$$\mathbf{u}_2 - \mathbf{u}_1 = \sigma_1 \begin{bmatrix} x_2 - x_1 \\ y_2 - y_1 \\ z_2 - z_1 \\ 0 \end{bmatrix}$$

moves in $\mathbf{p}_1 = [a_1 \quad b_1 \quad c_1 \quad d_1]$.

$$\mathbf{u}_3 = u_2 + \sigma_2 \begin{bmatrix} a_1 \\ b_1 \\ c_1 \\ 0 \end{bmatrix}, \qquad \phi \begin{bmatrix} a_2 \\ b_2 \\ c_2 \\ 0 \end{bmatrix} = \frac{1}{\sigma_1}(\mathbf{u}_1 - \mathbf{u}_2) = \frac{1}{\sigma_1} \begin{bmatrix} x_1 - x_2 \\ y_1 - y_2 \\ z_1 - z_2 \\ 0 \end{bmatrix}, \quad \text{and}$$

$$d_2 = -(a_2 x_2 + b_2 y_2 + c_2 z_2)$$

$$\mathbf{u}_4 = u_3 + \sigma_3 \begin{bmatrix} a_2 \\ b_2 \\ c_2 \\ 0 \end{bmatrix} = u_2 + \sigma_2 \begin{bmatrix} a_1 \\ b_1 \\ c_1 \\ 0 \end{bmatrix} + \frac{\sigma_3}{\sigma_1} \begin{bmatrix} x_1 - x_2 \\ y_1 - y_2 \\ z_1 - z_2 \\ 0 \end{bmatrix}$$

which does not provide additional motional capability over $\mathbf{u}_3$, so with $i = 3$ the system is defective. Thus, the maximum i is 2.

EXERCISES FOR CHAPTER 3

1. Determine the location of $\mathbf{u}_3$ as a function of the angle of the yaw joint at $\mathbf{u}_2 = [0 \quad 0 \quad 0 \quad 1]^{\#}$ when $\mathbf{u}_3$ moves in $\mathbf{p} = [\frac{1}{2} \quad \sqrt{3}/2 \quad 0 \quad 0]$ and where $\mathbf{u}_1 = [0 \quad 0 \quad -1 \quad 1]^{\#}$. $|\mathbf{u}_3 - \mathbf{u}_2| = 1$.

---

2. A revolute joint at $\mathbf{v} = \begin{bmatrix} 1 \\ 1 \\ 0 \\ 1 \end{bmatrix}$ connects links l_1 and l_2. The downlink end of l_1 is at $\mathbf{u} = \begin{bmatrix} 2 \\ 0 \\ 0 \\ 1 \end{bmatrix}$. l_1 moves in the plane defined by points **u, v,** and the origin. Assume $l_1 = l_2$. Find the location of the uplink end of l_2 as a function of the joint angle if the revolute joint is as given.
(a) Pitch
(b) Roll
(c) Yaw

---

3. A revolute joint is at **v** and the downlink l_1 between **u** and **v** moves in $\mathbf{p} = [a \quad b \quad c \quad d]$. Define the unit vector $\boldsymbol{\nu}$ giving the direction of link l_2 connecting **v** to **w** if the revolute joint is as follows.
(a) Pitch
(b) Yaw
(c) Crank $\tilde{\psi}$

---

4. A yaw joint is at the origin, and the downlink l_1 moves in $\mathbf{p}_1 = [0 \quad 0 \quad -1 \quad 0]$ and lies along $\boldsymbol{\nu} = \begin{bmatrix} 1 \\ 0 \\ 0 \end{bmatrix}$.
(a) Find the locations of $\mathbf{u}_2$ with the yaw joint at datum and at 30°. Assume the uplink length is l_2.
(b) Find the plane of motion of l_2.

---

5. A pitch joint at $\mathbf{v} = [x \quad y \quad 3 \quad 1]^{\#}$ has downlink of length 10 in. that moves in $\mathbf{p} = [-1 \quad 1 \quad 1 \quad 1]$ and connects $\mathbf{u} = [1 \quad 0 \quad 0 \quad 1]^{\#}$. Find **v**. If the uplink is 8 in. long, determine the location of the end of the uplink as a function of the joint angle.

---

6. If $\mathbf{u} = \begin{bmatrix} 1 \\ 1 \\ 0 \\ 1 \end{bmatrix}$, $\mathbf{v} = \begin{bmatrix} 1 \\ 0 \\ 2 \\ 1 \end{bmatrix}$, and $\mathbf{w} = \begin{bmatrix} 0 \\ 2 \\ 2 \\ 1 \end{bmatrix}$, find the plane of motion, link lengths, and joint angle if there is a pitch joint at **v**.

---

7. Repeat problem 6 for a yaw joint at **v**, finding both downlink and uplink planes of motion.

---

8. A crank joint is at the origin, the downlink lies on the positive x-axis, and the uplink lies on the positive y-axis. Find the uplink plane of motion if the joint angle is $\pi/3$.

---

9. Repeat problem 8 with a cylindrical joint at the origin.

---

10. Given a system of two yaw joints, γ_1 and γ_2, with link lengths unity. γ_1 is at $\mathbf{u}_1 = [0 \quad 0 \quad 0 \quad 1]^{\#}$. The downlink of the joint at $\mathbf{u}_1$ is vertical and its plane of motion is $\mathbf{p}_1 = [0 \quad 1 \quad 0 \quad 0]$. *Find* $\mathbf{u}_2$, $\mathbf{p}_2$, $\mathbf{u}_3$ and $\mathbf{p}_3$ when
(a) the joint angles are at datum,
(b) the joint angles are at 45°,
(c) the joint angles are at 90°.

---

11. A roll joint sits at point $\begin{bmatrix} 0 \\ 0 \\ 1 \\ 1 \end{bmatrix}$ and the end of its downlink is at $\begin{bmatrix} 0 \\ 0 \\ 0 \\ 1 \end{bmatrix}$. Its downlink moves in plane $[0.866 \quad 0.5 \quad 0 \quad 0]$. If the joint status is 45°, determine the plane of motion $[e \quad f \quad g \quad h]$ of its uplink using equation (3.7). Why is $[e \quad f \quad g]^{\#}$ a unit vector? Form the dot product of the vectors partially defining the two planes of motion and comment on the result.

---

12. Which of the following joint combinations are degenerate?

$$|\,\lambda\lambda\,|, \qquad |\,\theta\theta\,|, \qquad |\,\phi\tilde{\phi}\,|, \qquad |\,\psi\psi\,|, \qquad |\,\psi\tilde{\psi}\,|.$$

For those combinations that are degenerate, show which single joint is equivalent.

---

13. Model the pitch, crank (both orientations), and cylindrical joint (both orientations) by a right-angle bend and a roll joint; if such a model is impossible, show why.

---

14. Convert the pitch and crank ψ joints to a combination of the cylindrical joint and a right-angle bend. Verify mathematical equivalence or prove impossibility.

---

15. Sketch the symbolic diagrams similar to Figure 3.12 for the following robots:

$$|\,\phi\sigma\theta\,|, \quad |\,\theta\theta\psi\,|, \quad |\,\lambda\theta\theta\lambda\theta\,|, \quad |\,\sigma\sigma\lambda\theta\,|, \quad |\,\phi\theta\theta\phi\theta\phi\,|, \quad |\,\lambda\theta\sigma\phi\phi\phi\,|, \quad |\,\lambda\theta\theta\psi\theta\,|$$

---

16. Using yaw and right-angle bends only, model the 3 *dof* revolute robot with all datum positions matching.

---

17. Using the cylindrical joints only, model the 3 *dof* revolute robot. If more than three joints are required, these extra joints will be frozen.

---

18. Draw the schematics and fold lines for the following kinematic chains:

$$| \phi\theta\gamma\phi\theta | \qquad | \theta\theta x\theta | \qquad | \lambda\gamma\gamma | \qquad | \lambda\theta\theta\phi\phi\phi |$$

---

19. Benjamin Franklin marveled out loud on the adaptability of the human arm, and asked his audience to imagine the consequences of drinking a cup of coffee if the hand were not as well suited to the task; one might receive an earful or lapful rather than a mouthful of the brew. Suppose we model the shoulder joint as a ball joint (roll closely followed by pitch) and the elbow as a pitch joint. If you were in the audience, advise him on the minimum number and type of additional revolute joints required to perform the task.

---

20. Using the point/plane notation draw the schematics for the following robots, including the planes of motion:
   (a) GEC Little Giant $| \lambda\theta\sigma\theta |$,
   (b) GMF Robotics S-300 $| \lambda\theta\theta\lambda\theta\lambda |$,
   (c) Scrader-Bellows MotionMate MM-II $| \lambda\sigma+\sigma\lambda |$,
   (d) VSI Charlie-Screwdriver series type F $| \theta\theta\underline{\psi}\sigma |$,
   (e) Anorod 4A-1 $| \sigma+\sigma\psi\sigma |$
   (f) Eshed Robotec Scorbot-er III $| \lambda\theta\theta\theta\lambda |$.

---

21. Determine possible realizations for the revolute chains shown from an analysis of the fold lines.

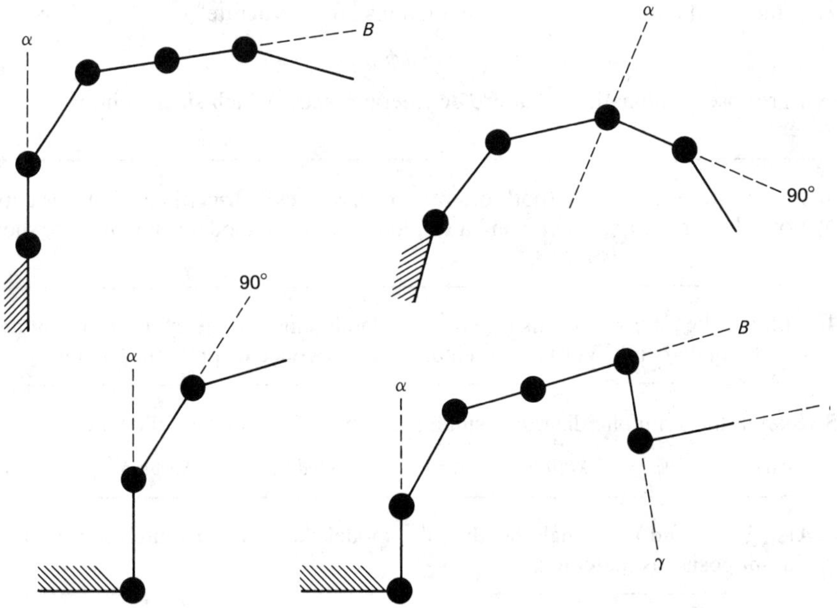

Figure 3.18 Revolute chains and their fold lines

CHAPTER FOUR

Robotic Capability

It is in this chapter that the work of Chapter 1 on planes (that may have appeared lengthy and of limited import at the time) and Chapter 3 pays off. Motion of kinematically simple manipulators is defined in terms of planes of motion. More significantly, the capability of the robot can be deduced solely from these planes [Parkin 1985c]. We assume that the kinematic chains are not defective (see Section 3.2), since a determination of the capability of a robot based on its planes of motion cannot take defective joint systems into account.

Central to the discussion is a full appreciation of fold lines and the meaning of what is meant by the planes of motion on either side of a fold line. The reader is advised to spend some time on Section 3.9 to gain a geometric feel for the line of intersection of adjacent planes of motion as a function of joint type.

4.1 PLANES OF MOTION

A plane of motion $[a \quad b \quad c \quad d]$ at the uplink side of a revolute joint is independent if a change of α degrees in the joint position results in vector $[a \quad b \quad c]$ changing direction by α degrees. From Figure 3.14 we notice that the joints λ, ψ, and ϕ produce independent planes; when the cylindrical joint is connected the other way around, we also see that $\tilde{\phi}$ produces an independent plane. Joints γ and $\tilde{\psi}$ do not produce independent planes, and the pitch joint cannot be at the end of a plane of motion (unless it is at the base or end effector). We designate independent planes with a bar, as in $\bar{\mathbf{q}}$.

The roll joint permits uplink plane $\mathbf{q}$ to be independent (and specified by $\bar{\mathbf{q}}$), but λ has no effect on location $\mathbf{w}$. If the end of $\bar{\mathbf{q}}$ is $\mathbf{w}$, $\mathbf{w}$ does not move as the joint at $\mathbf{v}$ moves; that is, $\mathbf{w}$ has 0 *dof* in $\bar{\mathbf{q}}$: we specify this plane as $\bar{\mathbf{q}}^0$. Simply put, we say λ gives $\bar{\mathbf{q}}^0$. Similarly, γ gives $\mathbf{q}^1$, ψ gives $\bar{\mathbf{q}}^0$, ϕ gives $\bar{\mathbf{q}}^0$, $\tilde{\psi}$ gives $\mathbf{q}^1$, and $\tilde{\phi}$ gives $\bar{\mathbf{q}}^0$. Notice that the sum of the bar (1 if present, 0 otherwise) and superscript define the number of *dof* in $\mathbf{q}$; in this case always 1. Also notice that as far as the planes of motion are concerned, $\lambda \equiv \psi \equiv \phi \equiv \tilde{\phi}$ and $\gamma \equiv \tilde{\psi}$.

Consider the addition of a pitch joint at $\mathbf{w}$ to the revolute joint at $\mathbf{v}$ forming plane $\mathbf{q}$. Then $\mathbf{q}$ would be specified by $\bar{\mathbf{q}}^1$ for λ, $\mathbf{q}^2$ for γ, $\bar{\mathbf{q}}^1$ for ψ, $\bar{\mathbf{q}}^1$ for ϕ, $\mathbf{q}^2$ for $\tilde{\psi}$, and $\bar{\mathbf{q}}^1$ for $\tilde{\phi}$ at $\mathbf{v}$, respectively. In this manner the type and number of *dof* in each plane of motion of a robot can be determined. The possible combinations of the planes of motion of a 3 *dof* robot are as follows:

$\bar{\mathbf{p}}^2$	$\bar{\mathbf{p}}^1\bar{\mathbf{p}}^0$	$\bar{\mathbf{p}}^0\bar{\mathbf{p}}^0\bar{\mathbf{p}}^0$	$\mathbf{p}^1\bar{\mathbf{p}}^1$	$\mathbf{p}^2\bar{\mathbf{p}}^0$	$\mathbf{p}^3$
	$\bar{\mathbf{p}}^1\mathbf{p}^1$	$\bar{\mathbf{p}}^0\bar{\mathbf{p}}^1$	$\mathbf{p}^1\bar{\mathbf{p}}^0\bar{\mathbf{p}}^0$	$\mathbf{p}^2\mathbf{p}^1$	
		$\bar{\mathbf{p}}^0\mathbf{p}^2$	$\mathbf{p}^1\bar{\mathbf{p}}^0\mathbf{p}^1$		
		$\bar{\mathbf{p}}^0\bar{\mathbf{p}}^0\mathbf{p}^1$	$\mathbf{p}^1\mathbf{p}^2$		
		$\bar{\mathbf{p}}^0\mathbf{p}^1\mathbf{p}^1$	$\mathbf{p}^1\mathbf{p}^1\bar{\mathbf{p}}^0$		
		$\bar{\mathbf{p}}^0\mathbf{p}^1\bar{\mathbf{p}}^0$	$\mathbf{p}^1\mathbf{p}^1\mathbf{p}^1$		

The five basic types of 3 *dof* robot are classified by their planes of motion as $\bar{\mathbf{p}}^2$ for the revolute, cylindrical, or polar robots, $\mathbf{p}^2\mathbf{p}^1$ or $\mathbf{p}^1\mathbf{p}^2$ for the Cartesian robot, and $\mathbf{p}^2\mathbf{p}^1$ for the SCARA robot.

Recall that as far as the planes of motion are concerned, $\lambda \equiv \psi \equiv \phi \equiv \tilde{\phi}$ and $\gamma \equiv \tilde{\psi}$. For example, replacing λ by ψ in a robot does not affect the classification of its planes of motion and so does not affect the functional capability of the robot. However, such replacement will modify the working envelope and location of all joints uplink from the modified joint. Further, the planes themselves are functions of the joint status of the robot, and switching joints modifies the numeric values of its planes.

Consider the 3 *dof* robot specified by its planes of motion $\bar{\mathbf{p}}^1\mathbf{p}^1$. The movements of the robot can be shown geometrically about its fold line as in Figure 4.1. The broken line in $\bar{\mathbf{p}}^1$ marked with 1 *dof* indicates that 1 *dof* that can alter the relative positions of the joints at the ends of $\bar{\mathbf{p}}^1$ is possible in $\bar{\mathbf{p}}^1$. Suppose the base joint is frozen; then $\mathbf{u}_4$ is capable of subscribing an arc about $\mathbf{u}_3$ in $\mathbf{p}^1$ in the first case and about $\mathbf{u}_2$ in $\bar{\mathbf{p}}^1$ in the second. Also, the single *dof* in $\mathbf{p}^1$ in the former case enables the position of $\mathbf{u}_3$ to change, and the angle of fold in the latter case enables $\bar{\mathbf{p}}^1$ to change, so $\mathbf{u}_4$ covers a volume in Cartesian space. Thus, as far as the capability of the robot is concerned, the planes are interchangeable.

This type of analysis can be conducted for any pair of planes of motion and, by extension, to any three or more planes of motion, but it should be

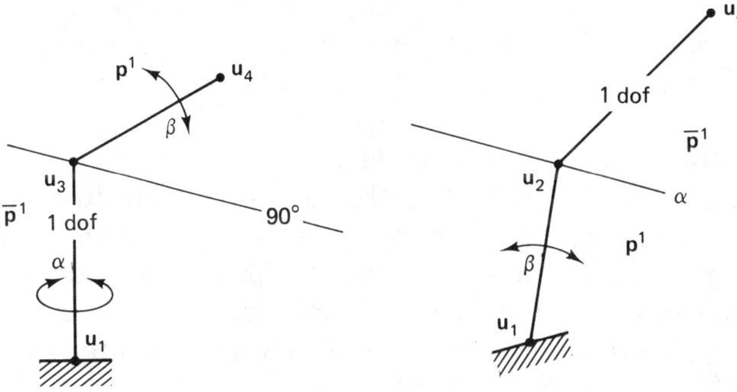

Figure 4.1 Interchange of $\mathbf{p}^1$ and $\mathbf{p}^1$

evident that interchangeability of the planes of motion does not affect the functional capability of the robot. However, there is a caveat involving the plane $\overline{\mathbf{p}}^0$. The motional capability of $\mathbf{p}^1\overline{\mathbf{p}}^0$ is not the same as $\overline{\mathbf{p}}^0\mathbf{p}^1$. However, $\overline{\mathbf{p}}^0$ can be interchanged in any chain of planes without affecting the motional capability provided it is not the end plane. Even when it is the end plane, it can often be exchanged with another plane without affecting capability depending on the motional capability required. We will see in Section 4.4 that the directable robot cannot have $\overline{\mathbf{p}}^0$ as its end plane, unlike the orientable robot.

4.2 PLACEABLE ROBOTS

Definition. A robot is said to be *placeable* if, with its end effector at some point $[x \quad y \quad z \quad 1]^{\#}$, its end effector can be placed at any point in the neighborhood (within an incremental distance) of $[x \quad y \quad z \quad 1]^{\#}$ within its working volume.

Notice that we deliberately exclude the limiting case of the robot at full stretch, since in this position the robot is poorly behaved. For example, it can behave as a stiff robot (defined later in this section). Further, in moving away from this limiting case, multiple solutions (usually two) may be possible. This will become apparent when inverse kinematic solutions are discussed in Chapter 6. The totality of neighborhood points comprise the working volume of the robot. Caution should be exercised, since it is possible for a 3 *dof* robot to have zero working volume—witness the $\overline{\mathbf{p}}^1\overline{\mathbf{p}}^0$ robot $|\ \phi\theta\lambda\ |$.

Definition. The *position space* of a robot is the set of positional and orientational requirements of the end effector plus other designated joints.

Thus, the position space of a placeable robot is the x, y, and z locations of the position of the end effector at $[x \quad y \quad z \quad 1]^{\#}$.

Definition. A *loose robot* is defined as one in which more degrees of freedom are present than are required to satisfy the position space of the robot.

Others [Hanafusa, Yoshikawa, and Nakamura 1978; Hanafusa, Yoshikawa, and Nakamura 1983; Freund 1977; Nakamura and Hanafusa 1985; Yoshikawa 1983] have employed the term *redundant* rather than loose. We find loose to be more appropriate—a loose manipulator can follow a position space that can be satisfied by another manipulator with fewer joints, and in the loose robot there is an infinite set of joint positions for every position space vector.

We will make some observations on the capabilities of the 3 *dof* robots (all 18 of them) discussed earlier. Robot $\mathbf{p}^3$ is not placeable, since it moves in one dependent plane only. A placeable robot must move in at least one independent plane or at least two dependent planes of motion. It is impossible for the plane $\bar{\mathbf{p}}^0$ to occur as the final plane in a placeable robot. We conclude that all 3 *dof* placeable robots are

$$\bar{\mathbf{p}}^2 \qquad \bar{\mathbf{p}}^1\mathbf{p}^1 \qquad \bar{\mathbf{p}}^0\bar{\mathbf{p}}^1 \qquad \bar{\mathbf{p}}^0\mathbf{p}^2 \qquad \bar{\mathbf{p}}^0\bar{\mathbf{p}}^0\mathbf{p}^1 \qquad \bar{\mathbf{p}}^0\mathbf{p}^1\mathbf{p}^1$$
$$\mathbf{p}^1\bar{\mathbf{p}}^1 \qquad \mathbf{p}^1\bar{\mathbf{p}}^0\mathbf{p}^1 \qquad \mathbf{p}^2\mathbf{p}^1 \qquad \mathbf{p}^1\mathbf{p}^2 \qquad \mathbf{p}^1\mathbf{p}^1\mathbf{p}^1$$

We see that the links of the 3 *dof* revolute $|\,\lambda\theta\theta\,|$, cylindrical $|\,\lambda\sigma+\sigma\,|$, or polar $|\,\lambda\theta\sigma\,|$ robots move in plane $\bar{\mathbf{p}}^2$, and so these robots are placeable. The 3 *dof* Cartesian robot $|\,\sigma+\sigma\times\sigma\,|$ moves in two planes $\mathbf{p}^1\mathbf{p}^2$ or $\mathbf{p}^2\mathbf{p}^1$, as indicated in Figure 3.15, and so this robot is placeable. The 3 *dof* SCARA robot moves in planes $\mathbf{p}^2\mathbf{p}^1$ and so is placeable. Notice that with a suitable realignment of the base plane, the roll, cylindrical ϕ (not $\tilde{\phi}$), and crank ψ (not $\tilde{\psi}$) joints are interchangeable at the base, so $|\,\lambda\theta\sigma\,|$ or $|\,\psi\theta\sigma\,|$ is the polar robot; see Section 5.1 for further discussion on base orientation and base joint equivalence. Further, notice that $\lambda\sigma$ is the same as $\sigma\lambda$. The only 3 *dof* robots, other than the revolute, cylindrical, and polar robots, with the characteristic of being placeable and having one plane of operation are $|\,\phi\sigma\theta\,|$ and its variants.

Robots that move such that no plane has superscript 2 or more are said to be *stiff*, sometimes requiring large changes in joint status for a moderate change in end effector position. Consider the 3 *dof* robot $|\,\phi\psi\theta\,|$. Suppose $\mathbf{p}_1 = [0 \quad 1 \quad 0 \quad -l_1]$ and l_1 is vertical, so $\mathbf{u}_2$ is at the origin. $\mathbf{u}_3$ describes a circle of radius l_2 in the xy-plane centered at the origin as ϕ_1 is scanned. With l_2 along the x-axis and l_3 vertical, $\mathbf{u}_4 = [l_2 \quad 0 \quad l_3 \quad 1]^\#$. Suppose $\mathbf{u}_4$ moves to $[l_2 \quad 0 \quad l_3 - \delta \quad 1]^\#$, where δ is small. This requires a horizontal movement in $\mathbf{u}_3$ of $\Delta\mathbf{u}_3 \cong \sqrt{2l_3\delta}$, ϕ_1 to move by $\Delta\phi_1 \cong \sqrt{2l_3\delta/l_2^2}$ rads, and ψ_2 to move by $\Delta\psi_2 \cong \sqrt{2\delta/l_3}$ rad. For $\delta = 0.01$ and $l_2 = l_3 = 1$, $\Delta\mathbf{u}_3 \cong 0.1414$, $\Delta\phi_1 \cong 0.1414$, and $\Delta\psi_2 \cong 0.1414$, so we require motion of about 14.14 times the original motion in order to follow the required path.

The rule that at least one plane of motion of the robot must have a superscript of at least 2 for that robot not to be stiff makes sense geometrically. Consider the link prior (at the downlink end) to the plane of motion under consideration to be fixed, and draw the motional capability of the position of the link at the uplink end of the plane of motion. If the plane is $\mathbf{p}^1$ or $\overline{\mathbf{p}}^1$, the uplink point follows a line trajectory as joints within the plane are moved—it has 1 *dof*. If the plane is $\mathbf{p}^2$ or $\overline{\mathbf{p}}^2$, the uplink point can cover a nonzero area as joints in the plane are moved—it has 2 *dof*. It is the ability for the end of a chain of links in a plane of motion to cover an area that makes a robot nonstiff. We say that such a robot contains a plane with area.

As a general rule we will consider as useful only robots that have one or more planes with area and so are nonstiff. This means that the list of 3 *dof* placeable, nonstiff robots is limited to $\overline{\mathbf{p}}^2$, $\overline{\mathbf{p}}^0\mathbf{p}^2$, $\mathbf{p}^1\mathbf{p}^2$, and $\mathbf{p}^2\mathbf{p}^1$.

4.3 ORIENTATION LINKAGES

The sliding link provides no orientation capability[1] and thus is excluded from the mechanisms discussed in this section. Care must be exercised when these orientation mechanisms are combined with placeable robots and described by their planes of motion when considering realizations in terms of specific joint types and introducing sliding links. In general, the sliding link should be used only in a plane with area. We will see that no orientation mechanism contains a plane with area.

Suppose a revolute joint is added to a 3 *dof* placeable robot whose end position is $\mathbf{u}_4$. If the added joint at $\mathbf{u}_4$ is roll, cylindrical ϕ, or crank ψ, the position of l_4 is unaffected by the status of the joint at $\mathbf{u}_4$, and l_4 is the axis for a pencil of planes when the joint at $\mathbf{u}_4$ is rotated; we call such a robot *pencilable*.

A 2 *dof* linkage system whose end link can be made to lie in the direction of some assigned vector regardless of the direction and plane of motion of the downlink (we call these directable mechanisms) can be described by $\overline{\mathbf{p}}^1$, $\overline{\mathbf{p}}^0\mathbf{p}^1$, or $\mathbf{p}^1\mathbf{p}^1$ but not by $\mathbf{p}^1\overline{\mathbf{p}}^0$, $\overline{\mathbf{p}}^0\overline{\mathbf{p}}^0$, or $\mathbf{p}^2$. Examples of these 2 *dof* directable mechanisms are $|\lambda\theta|$, $|\phi\gamma|$, and $|\theta\gamma|$. We can add this 2 *dof* directable mechanism to the end of any 3 *dof* placeable robot such that the end effector is at the desired position and the last link of the robot lies along any designated vector $\boldsymbol{\mu}$ and so can approach any surface in any direction—for example, orthogonally if a peg is being inserted. We call a robot with this capability *directable*.

[1] This assumes that the orientation mechanism is in isolation, but a sliding link added to a 3 *dof* placeable robot can produce a 4 *dof* robot with some orientation capabilities. An example is the sliding link added to the 3 *dof* stiff robot $\overline{\mathbf{p}}^0\overline{\mathbf{p}}^1$ to produce the 4 *dof*, nonstiff robot $\overline{\mathbf{p}}^0\overline{\mathbf{p}}^2$.

A 3 *dof* linkage system whose end link can be made to lie in the direction of some assigned vector within an assigned plane of motion regardless of the direction and plane of motion of the base link can be described by $\bar{\mathbf{p}}^1\bar{\mathbf{p}}^0$, $\bar{\mathbf{p}}^0\mathbf{p}^1\bar{\mathbf{p}}^0$, $\mathbf{p}^1\mathbf{p}^1\bar{\mathbf{p}}^0$, or $\bar{\mathbf{p}}^0\bar{\mathbf{p}}^0\bar{\mathbf{p}}^0$. Examples of these 3 *dof* orientation mechanisms are $|\,\lambda\theta\lambda\,|$, $|\,\phi\gamma\phi\,|$, $|\,\theta\gamma\lambda\,|$, and $|\,\lambda\phi\phi\,|$. If the end link of some manipulator is required to lie along a particular vector in some assigned orientation and position, then the manipulator can be any 3 *dof* placeable robot plus any 3 *dof* orientation linkage; we call such a robot *orientable*.

The planes of motion can be reordered in this 3 *dof* orientation mechanism without affecting capability. Further, the first plane of motion of the 3 *dof* orientation mechanisms described by $\mathbf{p}^1\bar{\mathbf{p}}^0\bar{\mathbf{p}}^0$ or $\mathbf{p}^1\mathbf{p}^1\bar{\mathbf{p}}^0$ can sometimes be combined with last plane of the 3 *dof* placeable robot. For example, suppose the orientation mechanism $\mathbf{p}^1\bar{\mathbf{p}}^0\bar{\mathbf{p}}^0$ starts with a pitch joint; then $\mathbf{p}^1$ is combined with the last plane of motion of the 3 *dof* placeable robot. If this 3 *dof* placeable robot is $\bar{\mathbf{p}}^2$, then the overall robot becomes $\bar{\mathbf{p}}^3\bar{\mathbf{p}}^0\bar{\mathbf{p}}^0$.

The 3 *dof* orientation mechanism $\bar{\mathbf{p}}^0\bar{\mathbf{p}}^0\bar{\mathbf{p}}^0$ is an awkward mechanism, since it is stiff with respect to orientation. Suppose the mechanism $|\,\phi\phi\phi\,|$ is used and the end effector is required to sweep an arc in a plane parallel to but distinct from the plane of motion of the base link. An incremental change in position along the arc requires large changes in the status of the cylindrical joints. Further, it will be seen in Chapter 6 that the inverse kinematic solution for robots employing this type of mechanism is difficult to obtain.

We can consider the 3 *dof* orientation mechanism in isolation as a 3 *dof* robot defined by $\bar{\mathbf{p}}^1\bar{\mathbf{p}}^0$, $\mathbf{p}^1\bar{\mathbf{p}}^0\bar{\mathbf{p}}^0$, $\mathbf{p}^1\mathbf{p}^1\bar{\mathbf{p}}^0$, or $\bar{\mathbf{p}}^0\bar{\mathbf{p}}^0\bar{\mathbf{p}}^0$ and so is seen to be stiff and unplaceable; recall that the planes can be reordered without affecting capability.

4.4 ROBOTS DEFINED BY THEIR CAPABILITIES

Definition. A placeable robot whose last link can rotate while its position is unaltered is called a *pencilable* robot.

Definition. A placeable robot is said to be *directable* if its last link can lie in any assigned direction.

Definition. A robot is said to be *orientable* if its end effector is placeable in any orientation.

Any directable robot can be made orientable by adding $\bar{\mathbf{p}}^0$.

A common commercial robot is $|\,\lambda\theta\theta\lambda\,|$, but this robot does not fit any of the preceding definitions. The 4 *dof* SCARA robot has most of the capability of this robot (if the base orientation is horizontal, the end effector can approach a horizontal surface orthogonally and rotate its end link) and in addition is usually faster and more accurate. Larger $|\lambda\theta\theta\lambda|$ robots are often used as palletizers. $|\lambda\theta\theta\lambda|$ is in the class of robots we call *drillable.*

Definition. A *drillable* robot is a 5 *dof* placeable robot whose end link can rotate about itself and whose links all lie in the first plane of motion.

Notice that the phrase "any orientation" permits rotation of the end link about itself. We usually determine robot capability in terms of the planes of motion, so the fifth joint can be ψ, ϕ, or $\tilde{\phi}$ as well as λ. The sobriquet "drillable" does not readily apply to the robot with ψ or $\tilde{\phi}_5$, so we restrict the fifth joint to be λ or ϕ.

From these definitions it is evident that a pencilable robot has at least 4 *dof,* a directable or drillable robot has at least 5 *dof,* and an orientable robot has at least 6 *dof.*

An orientable robot can be formed from any 3 *dof* placeable robot plus the 3 *dof* orientation mechanism discussed in the last section, and we can reorder the planes of motion without affecting the orientability of the robot. Consider the robot shown in Figure 4.2, which has two of its three orientation linkages near the base. Although not classified as stiff, this robot exhibits poor characteristics, since a small change in end effector orientation can necessitate large changes in joint status. Robots designed somewhat closer to the concept of a 3 *dof* placeable robot followed by a 3 *dof* orientation mechanism are more economical of joint movement for small motions of the end effector orientation.

For the ultimate in simplicity when analyzing the movements of a 6 *dof* orientable robot, we can reduce the link lengths within the orientation mechanism to zero, so the position of the end effector is the position of the 3 *dof* placeable robot. However, this scenario is unrealistic and will not be assumed.

The planes of motion for orientable robots produced from 3 *dof,* placeable robots $\bar{\mathbf{p}}^2$, $\mathbf{p}^2\mathbf{p}^1$, $\mathbf{p}^1\bar{\mathbf{p}}^1$, or $\mathbf{p}^1\mathbf{p}^1\mathbf{p}^1$ followed by 3 *dof* orientation mechanisms $\bar{\mathbf{p}}^1\bar{\mathbf{p}}^0$, $\bar{\mathbf{p}}^0\mathbf{p}^1\bar{\mathbf{p}}^0$, $\mathbf{p}^1\mathbf{p}^1\bar{\mathbf{p}}^0$, or $\bar{\mathbf{p}}^0\bar{\mathbf{p}}^0\bar{\mathbf{p}}^0$ can be reordered without altering the capabil-

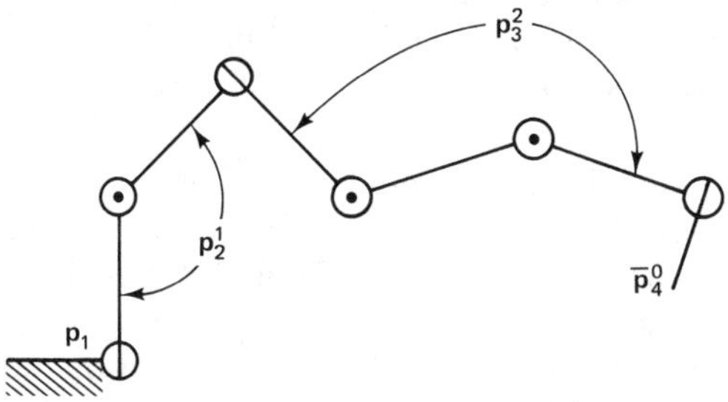

Figure 4.2 A robot produced by reordering planes

ity of the robot. Suppose the 6 *dof* orientable robot is to be nonstiff. Then the stiff 3 *dof* placeable robot $\mathbf{p}^1\bar{\mathbf{p}}^1$, $\bar{\mathbf{p}}^1\mathbf{p}^1$, or $\mathbf{p}^1\mathbf{p}^1\mathbf{p}^1$ must be combined with orientation mechanisms $\mathbf{p}^1\bar{\mathbf{p}}^0\bar{\mathbf{p}}^0$ or $\mathbf{p}^1\mathbf{p}^1\bar{\mathbf{p}}^0$ to produce the robots $\mathbf{p}^1\bar{\mathbf{p}}^2\bar{\mathbf{p}}^0\bar{\mathbf{p}}^0$, $\bar{\mathbf{p}}^1\mathbf{p}^2\bar{\mathbf{p}}^0\bar{\mathbf{p}}^0$, $\mathbf{p}^1\mathbf{p}^1\mathbf{p}^2\bar{\mathbf{p}}^0\bar{\mathbf{p}}^0$, $\mathbf{p}^1\bar{\mathbf{p}}^2\mathbf{p}^1\bar{\mathbf{p}}^0$, $\bar{\mathbf{p}}^1\mathbf{p}^2\mathbf{p}^1\bar{\mathbf{p}}^0$, or $\mathbf{p}^1\mathbf{p}^1\mathbf{p}^2\mathbf{p}^1\bar{\mathbf{p}}^0$.

Theorem 4.1. A 6 *dof* robot with two or more independent planes with area is not orientable.

The relationship between the common nonstiff robot types is shown in Table 4.1. We see that adding another *dof* to a 3 *dof* nonstiff placeable robot can make it a pencilable robot, adding another can make it a directable robot, and adding a third can make it an orientable robot.

TABLE 4.1

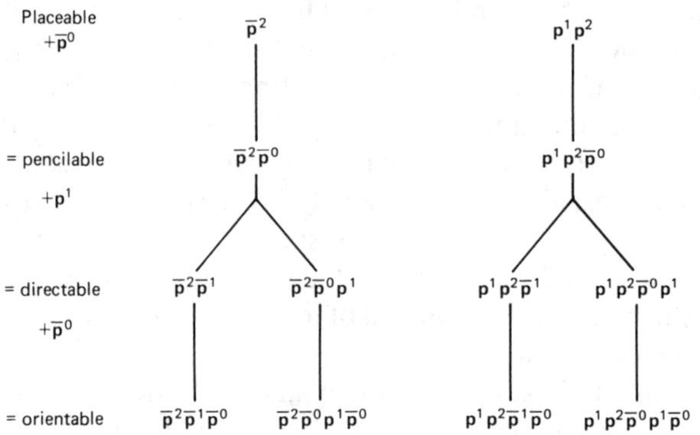

Notice that it is possible to combine the final plane of motion with a prior plane in the 3 *dof* placeable robots $\mathbf{p}^2\mathbf{p}^1$ or $\mathbf{p}^1\mathbf{p}^2$ when producing a pencilable robot; this can occur with the 4 *dof* pencilable SCARA configuration robot $|\theta\theta\times\sigma\lambda|$ with planes of motion $\mathbf{p}^2\mathbf{p}^1\bar{\mathbf{p}}^0$ which is equivalent to $|\theta\theta\underline{\psi}\sigma|$ with planes of motion $\mathbf{p}^2\bar{\mathbf{p}}^1$.

We have shown that $\mathbf{p}^1\mathbf{p}^1$ is a 2 *dof* mechanism that produces a directable robot when added to 3 *dof* placeable robots. These mechanisms can be added to placeable robots to produce nonstiff directable robots as shown in Table 4.2.

Most dexterous robots are contained in Tables 4.1 and 4.2, as are most robots with closed-form inverse kinematic solutions. Drillable robots do not fit the categorizations of Table 4.1 or Table 4.2, yet the addition of $\bar{\mathbf{p}}^0$ can produce an orientable robot; an example is $|\ \lambda\theta\theta\lambda\phi\ |$, and this robot has a closed-form inverse kinematic solution. Notice that by moving from directable to orientable robots, we exclude the orientation mechanism $\bar{\mathbf{p}}^0\bar{\mathbf{p}}^0\bar{\mathbf{p}}^0$, so the tables are not comprehensive.

TABLE 4.2

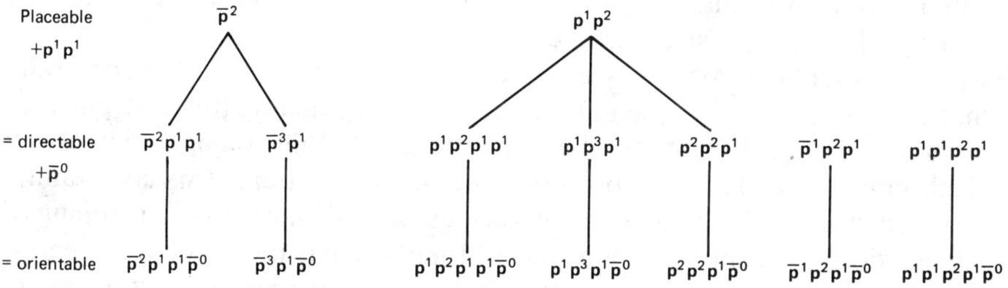

4.5 POSITION SPACE

In the preceding section we defined various types of robot. Here we present the capabilities of these robots in terms of the positional and motional capability at the end effectors; we define such capability as the position space of the robot. Given the set of joint angles of the robot (which we can call the joint space), we can find the unique position space, a procedure known as the forward kinematic solution (discussed in Chapter 5). Given the position space of the robot, we can find the joint space (which is often not unique, so we are using "space" in a loose sense), a procedure known as the inverse kinematic solution (discussed in Chapter 6).

We start with placeable robots whose position space is $\mathbf{u}_e = [x \quad y \quad z \quad 1]^\#$—that is, the end effector at $\mathbf{u}_e$ is capable of being positioned anywhere desired inside the working volume of the robot. Typically, placeable robots have 3 *dof,* and the position space imposes the three constraints x, y, and z. We require that a nonloose robot has the same number of *dof* as the position space has constraints.

A pencilable robot has $\mathbf{u}_e = [x \quad y \quad z \quad 1]^\#$ in its position space; $\mathbf{u}_e$ is unaffected by the status of the last joint of the robot. The robot has $\bar{\mathbf{p}}^0$ in its planes of motion, and freezing the joint that controls this plane still permits $\mathbf{u}_e$ to be satisfied. The total number of constraints imposed by the position space is four, three for the placeability and the fourth for end link rotation. Provided $\mathbf{u}_{e-1} = [\ddot{x} \quad \ddot{y} \quad \ddot{z} \quad 1]^\#$ can be determined, we assume the datum of the last joint is such that the final plane of motion $\bar{\mathbf{p}}^0$ passes through the base of the robot, so this datum final plane can be determined. The angle of this last joint is then chosen as the fourth entry in its position space.

A drillable robot is placeable, so it has $\mathbf{u}_e = [x \quad y \quad z \quad 1]^\#$ in its position space. The plane $\mathbf{p} = [a \quad b \quad c \quad d]$ passing through $\mathbf{u}_1$, $\mathbf{u}_2$, and $\mathbf{u}_e$ can be established, and this plane also passes through $\mathbf{u}_{e-1}$. One additional piece of information is required to determine $\mathbf{u}_{e-1}$ from the base orientation and position and $\mathbf{u}_e$. The position space is $\mathbf{u}_e$, $\mathbf{u}_{e-1}$, and the final plane of motion, a total of five independent pieces of information.

A directable robot has $\mathbf{u}_e = [x \quad y \quad z \quad 1]^\#$ and $\mathbf{u}_{e-1}$ as its position space with the constraint that $|\, \mathbf{u}_e - \mathbf{u}_{e-1} \,| = l_e$, the length of the last link. Thus, the order of the position space is five.

An orientable robot has $\mathbf{u}_e = [x \quad y \quad z \quad 1]^\#$, $\mathbf{u}_{e-1}$ (with the constraint that $|\, \mathbf{u}_e - \mathbf{u}_{e-1} \,| = l_e$, the length of the last link), and $\mathbf{p}_e$ (the last plane of motion) as its position space. Since $\mathbf{u}_e$ and $\mathbf{u}_{e-1}$ both lie on $\mathbf{p}_e$, $\mathbf{p}_e$ imposes a single constraint. Thus, the position space imposes a total of six constraints.

An inverse kinematic solution (discussed in Chapter 6) is a mapping of the position space into the joint space. Joint space is the set of joint angles, including the status of sliding links. Thus, if there is to be a direct mapping, the order of the position space must be the same as the order of the joint space. If the order of the position space is less than the order of the joint space, the robot is loose. The usual reason for using a loose robot is for obstacle avoidance, but the protocols for controlling loose robots are not fully defined; see Section 8.9.

4.6 EXAMPLES

1. Determine the logical position space of the following manipulators.

(a) $\bar{\mathbf{p}}^2\mathbf{p}^2$ **(b)** $\bar{\mathbf{p}}^2\mathbf{p}^1\mathbf{p}^1$ **(c)** $\bar{\mathbf{p}}^2\bar{\mathbf{p}}^0\mathbf{p}^1$
(d) $\mathbf{p}^2\mathbf{p}^2\bar{\mathbf{p}}^0$ **(e)** $\mathbf{p}^1\mathbf{p}^1\mathbf{p}^2\mathbf{p}^1$ **(f)** $\mathbf{p}^1\bar{\mathbf{p}}^2\mathbf{p}$
(g) $\mathbf{p}^1\bar{\mathbf{p}}^2\bar{\mathbf{p}}^1$ **(h)** $\mathbf{p}^1\bar{\mathbf{p}}^0\mathbf{p}^1\bar{\mathbf{p}}^0\mathbf{p}^1$ **(i)** $\bar{\mathbf{p}}^3\bar{\mathbf{p}}^0$
(j) $\bar{\mathbf{p}}^3\mathbf{p}^1$ **(k)** $\mathbf{p}^2\bar{\mathbf{p}}^0\bar{\mathbf{p}}^2$ **(l)** $\mathbf{p}^1\mathbf{p}^1\mathbf{p}^2$
(m) $\bar{\mathbf{p}}^2\bar{\mathbf{p}}^2\bar{\mathbf{p}}^2$ **(n)** $\mathbf{p}^2\mathbf{p}^2\mathbf{p}^2$

(a) 5 *dof,* one independent plane (which has area), so the manipulator is loose (doubly so) placeable with position vector given by the location of the end effector. The last 2 *dof* can be frozen and the manipulator becomes nonstiff placeable.

(b) 5 *dof,* three planes, where one is independent with area, so the manipulator is nonstiff directable with position vector given by the location of the ends of its last link.

(c) 5 *dof,* two independent planes (one with area), so the manipulator has the same capability as in (b).

(d) 5 *dof,* three planes (two with area), where one plane is independent, so this manipulator has the same capability as (b).

(e) 5 *dof,* four planes, one with area but no independent planes, so the manipulator is nonstiff directable.

(f) 6 *dof,* three planes, one with area and independent, so the manipulator is loose directable with the same characteristics as in (b).

(g) 6 *dof,* three planes, one with area and two independent, so the manipulator is loose directable. Provided either two independent planes are maintained (so the first *dof* can be frozen) or three planes are maintained, where at least one is independent (so any *dof* in the second or third plane can be frozen), the manipulator is still directable. Thus any one of the joints can be frozen and the manipulator becomes nonstiff directable.

(h) The manipulator is stiff and the theorems developed in this chapter do not hold for stiff robots. It is possible to extend Tables 4.1 and 4.2 to cover stiff robots, but the complexity of the problem increases by about an order of magnitude. However, as stated previously, stiff robots have very limited application and (in general) should not be considered.

(i) 5 *dof,* two independent planes, where one has area, so the manipulator is drillable.

(j) 5 *dof,* two planes, where one is independent (this plane also has area), so the manipulator is directable.

(k) 6 *dof,* three planes, where two are independent and there are two planes with area. Thus, the manipulator is not orientable, but it is loose, nonstiff directable. Further, provided we do not collapse the plane $\bar{\mathbf{p}}^0$ by freezing that joint, any other joint can be frozen to leave a nonstiff directable robot.

(l) 4 *dof,* three planes, one with area but none independent. Therefore, the manipulator is loose, nonstiff placeable, and either of the first two joints can be frozen to keep the robot nonstiff placeable.

(m) 9 *dof,* three independent planes with area, so the manipulator is loose orientable. It is difficult to determine how or where to impose other constraints (additions to the position vector) in order for the robot to be loose no longer. We can remove 2 *dof* from one plane and 1 *dof* from another by freezing the appropriate joints and the robot becomes 6 *dof,* nonstiff orientable.

(n) 6 *dof,* three planes with area, but none are independent, so the manipulator is loose placeable. We can freeze two joints in one plane (effectively removing that plane) and one joint in another plane to reduce the robot to a 3 *dof* placeable nonstiff robot.

---

2. Determine the planes of motion and capabilities of the following robots:

(a) $\mid \phi\sigma\theta\theta \mid$	**(b)** $\mid \phi\theta\phi \mid$	**(c)** $\mid \phi\sigma\phi \mid$
(d) $\mid \phi\theta\tau \mid$	**(e)** $\mid \phi\phi\phi \mid$	**(f)** $\mid \lambda\theta\theta\lambda \mid$
(g) $\mid \lambda\lambda\theta\theta \mid$	**(h)** $\mid \phi\phi\theta\theta\phi\phi \mid$	

$\bar{\mathbf{p}}^3$

(a) $|\phi_1\sigma_2|\theta_3 l_3|\theta_4 l_4|$, so the robot is loose placeable.

$\bar{\mathbf{p}}^1$ $\bar{\mathbf{p}}^0$

(b) $|\phi_1 l_1|\theta_2 l_2|\phi_3 l_3|$, so the robot is unplaceable.

$\bar{\mathbf{p}}^1$ $\bar{\mathbf{p}}^0$

(c) $|\phi_1\sigma_2|\phi_3 l_3|$, so the robot is unplaceable.

$\bar{\mathbf{p}}^1$ $\mathbf{p}^1$

(d) $|\phi_1 l_1|\theta_2 l_2|\gamma_3 l_3|$, so the robot is stiff placeable.

$\bar{\mathbf{p}}^0$ $\bar{\mathbf{p}}^0$ $\bar{\mathbf{p}}^0$

(e) $|\phi_1 l_1|\phi_2 l_2|\phi_3 l_3|$, so the robot is unplaceable.

$\bar{\mathbf{p}}^2$ $\bar{\mathbf{p}}^0$

(f) $|\lambda_1 l_1|\theta_2 l_2|\theta_3 l_3|\lambda_4 l_4|$, so the robot is pencilable.

(g) $|\overset{\bar{\mathbf{p}}^0}{\lambda_1 l_1}|\overset{\bar{\mathbf{p}}^2}{\lambda_2 l_2}|\theta_3 l_3|\theta_4 l_4|$, where the first two joints are redundant.

If we remove the first joint, we are left with a 3 *dof* nonstiff placeable robot.

(h) $|\overset{\bar{\mathbf{p}}^0}{\phi_1 l_1}|\overset{\bar{\mathbf{p}}^2}{\phi_2 l_2}|\theta_3 l_3|\theta_4 l_4|\overset{\bar{\mathbf{p}}^0}{\phi_5 l_5}|\overset{\bar{\mathbf{p}}^0}{\phi_6 l_6}|$, so the robot is nonstiff orientable.

EXERCISES FOR CHAPTER 4

1. Complete the mathematical specifications, draw the schematics, and define a logical position space for the following robots:

$$\left|\overset{\mathbf{p}_1}{\underset{\mathbf{u}_1}{\phi_1 \frac{\sigma_2}{\bar{\sigma}_2}}}\right|\theta_3 l_2\left|\theta_4 l_3\right| \qquad |\overset{\mathbf{p}_1}{\underset{\mathbf{u}_1}{\phi_1 l_1}}|\psi_2 l_2|\theta_3 l_3|$$

$$\left|\overset{\mathbf{p}_1}{\underset{\mathbf{u}_1}{\theta_1 l_1}}\right|\theta_2 \frac{\sigma_3}{\bar{\sigma}_3}\left|\phi_4 l_4\right| \qquad \left|\overset{\mathbf{p}_1\bar{\mathbf{p}}_2^2}{\underset{\mathbf{u}_1}{\phi_1 l_1}}\right| l_2 \left|\frac{\sigma_4}{\bar{\sigma}_4}\right|$$

---

2. Complete the specification of

$$|\phi_1 l_1|\theta_2 l_2|\psi_3 l_3|\theta_4 l_4|$$

Show a schematic of the robot with fold line, etc.

---

3. Determine the planes of motion and capabilities of the following robots:

(a) $|\lambda\theta\theta\lambda\theta|$ **(d)** $|\sigma{+}\sigma{\times}\sigma|$
(b) $|\theta\theta\phi|$ **(e)** $|\theta\theta\psi\theta\lambda\theta|$
(c) $|\phi\theta\sigma\lambda\theta|$ **(f)** $|\theta\theta\tilde{\psi}\theta\lambda\theta|$

---

4. Determine the positional capability of the robot

$$|\phi_1 l_1|\theta_2 l_2|\theta_3 l_3|\psi_4 l_4|\theta_5 l_5|$$

---

5. Successively add a degree of freedom to the 3 *dof* SCARA robot to make it pencilable, directable, and orientable.

---

6. Successively add a degree of freedom to the 3 *dof* robot $|\phi\sigma\gamma|$ to make it pencilable, directable, and orientable.

---

7. Give two or more possible realizations for the robots with planes of motion $\bar{\mathbf{p}}^3$, $\bar{\mathbf{p}}^2\bar{\mathbf{p}}^1$, $\mathbf{p}^2\mathbf{p}^1$, $\mathbf{p}^2\mathbf{p}^2$.

---

8. Determine the logical position spaces for the robots with planes of motion $\mathbf{p}^1\bar{\mathbf{p}}^2$, $\bar{\mathbf{p}}^3$, $\mathbf{p}^2\mathbf{p}^1$, $\mathbf{p}^2\mathbf{p}^2$, $\mathbf{p}^2\bar{\mathbf{p}}^1\mathbf{p}^1$, and $\mathbf{p}^2\bar{\mathbf{p}}^0$. If no logical position space exists, state and explain.

---

9. **(a)** A robot is composed of three pitch joints and associated links and one or more right-angle bends. Choose the minimum number of right-angle bends to produce a placeable robot.
(b) Repeat part (a) but this time use three roll joints.
Note: The configuration you produce may not be unique.

---

10. Given orientation linkage

$$\begin{array}{ccccccc} | & \phi_1 l_1 & | & \phi_2 l_2 & | & \phi_3 l_3 & | \\ \mathbf{u}_1 & & \mathbf{u}_2 & & \mathbf{u}_3 & & \mathbf{u}_4 \end{array}$$

If $\mathbf{p}_1 = [0 \quad 1 \quad 0 \quad 0]$, $\boldsymbol{\nu} = \begin{bmatrix} -1 \\ 0 \\ 0 \end{bmatrix}$, and $\mathbf{u}_1 = \begin{bmatrix} 0 \\ 0 \\ 0 \\ 1 \end{bmatrix}$, find $\mathbf{u}_4$ and the final plane of motion in each case.
(a) $\phi_1 = \phi_2 = \phi_3 = 0°$
(b) $\phi_1 = \phi_2 = \phi_3 = 30°$

---

11. A 3 *dof* placeable robot is piggybacked onto another 3 *dof* placeable robot. Is it possible for such a robot to be orientable?

---

12. Suppose we have a 3 *dof* placeable robot, a 4 *dof* pencilable robot, and a 5 *dof* directable robot. Is it possible to insert a sliding link somewhere in each kinematic chain to produce a pencilable, directable, and/or orientable robot, respectively?

---

13. Determine the planes of motion and capability of the following robots:
(a) $| \lambda\theta\theta\theta\lambda\theta |$
(b) $| \theta\theta\phi |$
(c) $| \theta\theta\psi\theta\lambda\theta |$
(d) $| \sigma{+}\sigma{\times}\sigma |$
(e) $| \phi\theta\sigma\lambda\theta |$

---

14. Determine the planes of motion, capability and logical position space (if none exists, state so) for the following manipulators:
(a) $| \theta\theta{\times}\theta\lambda |$ **(b)** $| \phi\theta\theta\theta\phi |$ **(c)** $| \lambda\theta\theta\phi\gamma |$, **(d)** $| \sigma{-}\sigma{\div}\sigma\lambda\theta |$

---

15. Determine all possible configurations for the 6 *dof* orientable, revolute robot whose joint locations are $\mathbf{u}_1$ through $\mathbf{u}_7$, where a fold line lies along the straight line connecting $\mathbf{u}_3$, $\mathbf{u}_4$, and $\mathbf{u}_5$ and where a fold line lies along the straight line connecting $\mathbf{u}_5$, $\mathbf{u}_6$ and $\mathbf{u}_7$.

---

16. Determine the planes of motion and capability of $|\ \phi\theta\theta\lambda\theta\ |$. Reverse the planes of motion and draw the schematic with the fold lines using the same joints. Discuss the capability of this manipulator from its geometry.

CHAPTER FIVE

Forward Kinematic Solutions

In order to control a robot remotely, it must be kinematically analyzed and the results of this analysis entered into the controller of the robot. Kinematic analysis starts with the determination of the position of each joint given the configuration of the robot, the base position and orientation, the angles of the revolute joints, and the lengths of the sliding links of the robot. This is called the *forward kinematic solution* (FKS) of the robot. The FKS is always possible and is always possible in a preassigned number of operations; that is, it can be determined in closed form. The solution proceeds joint by joint, using the information on the prior link positions and plane of motion to determine the plane of motion of the next joint and the next joint position [Lee 1983a; Makino 1976; McInnis and Liu 1986; Paul and Zhang 1986]. The FKS is the prerequisite for obtaining the inverse kinematic solution (shown in Chapter 6), which is entered into the robot controller and forms the basis for the remote control of the robot.

The full derivations of the kinematic equations of Section 5.2 are given in Appendix B. The reader of this appendix may be intimidated by the complexity of the equations and might ask, Where does it end? It should be recognized that the number of usable configurations is limited, and the configurations considered, possibly with minor variations, encompass most of the usable configurations. The revolute robots appear the most complex, and the most common of these robots are fully analyzed. Many commercial robots are revolute with 5 or 6 *dof*, and the complexity of the equations appears to double with each additional degree of freedom produced by a nondefective revolute joint. A sliding link, as occurs in cylindrical and polar robots, does not add complexity to the equations, so these configurations produce simpler forward kinematic equations.

It is sometimes not necessary to develop the equations for the position and orientation of the end effector. For example, suppose we have the 5 *dof* directable robot $|\ \lambda\theta\theta\lambda\theta\ |$ with expanded description

$$\begin{array}{cccccc} & \bar{\mathbf{p}}_2^2 & & \bar{\mathbf{p}}_3^1 & & \\ | & \lambda_1 l_1 \ | \ \theta_2 l_2 & | \ \theta_3 l_3 & | \ \lambda_4 l_4 & | \ \theta_5 l_5 & | \\ \mathbf{u}_1 & \mathbf{u}_2 & \mathbf{u}_3 & \mathbf{u}_4 & \mathbf{u}_5 & \mathbf{u}_6 \end{array}$$

The inverse kinematic solution requires the position of the joints from known base orientation and position and that $\mathbf{u}_5$ and $\mathbf{u}_6$ be given. The base position and orientation gives $\mathbf{u}_2$. $\mathbf{u}_5$ is on $\bar{\mathbf{p}}_2$; this plane is determined from the three points $\mathbf{u}_1$, $\mathbf{u}_2$, and $\mathbf{u}_5$. $\bar{\mathbf{p}}_3$ is determined from the three points $\mathbf{u}_4$, $\mathbf{u}_5$, and $\mathbf{u}_6$, and the inverse kinematic solution is completed. Thus, the solution is a few simple calculations on top of the solution for the robot

$$\begin{array}{ccccc} & \bar{\mathbf{p}}_2^2 & & & \\ | & \lambda_1 l_1 \ | \ \theta_2 l_2 & | \ \theta_3 \ l_3 + l_4 & & | \\ \mathbf{u}_1 & \mathbf{u}_2 & \mathbf{u}_3 & & \mathbf{u}_5 \end{array}$$

The equations developed in Appendix B are canonical. Apart from simple factorizations, they are given in their simplest possible form, and no cancellations occur. The equations are developed for the general base orientation discussed in the next section. The canonical equations guarantee that the FKS (the position of the joints and determination of the planes of motion given the base orientation and position and the status of all the joints) is obtained in the fewest number of arithmetic operations. We use these canonical equations to produce canonical inverse kinematic solutions in Chapter 6.

5.1 BASE ORIENTATION

The base orientation is given by $\mathbf{p}_1 = [a_1 \quad b_1 \quad c_1 \quad d_1]$ and unit vector $\boldsymbol{\nu} = [\nu_x \quad \nu_y \quad \nu_z]^\#$, which lies in the plane of $\mathbf{p}_1$, so $a_1\nu_x + b_1\nu_y + c_1\nu_z = 0$. The first plane of operation of the robot is $\mathbf{p}_i = [a_i \quad b_i \quad c_i \quad d_i]$, where $i = 1$ if the base joint is sliding or pitch and $i = 2$ if the first joint is roll, yaw, crank, or cylindrical. The base joint is at $\mathbf{u}_1$. We determine the first plane of motion of the robot $\mathbf{p}_2$ from $\mathbf{u}_1$, $\mathbf{p}_1$, $\boldsymbol{\nu}$, and the joint type and angle (for certain revolute joints) of the base joint. The location of point $\mathbf{u}_2$ connected to revolute joint $\mathbf{u}_1$ by link l_1 depends on $\mathbf{p}_1$, $\boldsymbol{\nu}$, the type of joint at $\mathbf{u}_1$, and the joint status. Here $\boldsymbol{\nu}$ substitutes for the $\mathbf{v} - \mathbf{u}$ used in the analysis of Section 3.1, and $\mathbf{p}_1$ is the same as $\mathbf{p}$. Notice that $\mathbf{u}_2$ is unaltered by the joint status if the joint at $\mathbf{u}_1$ is roll, crank ψ, or cylindrical ϕ.

Suppose

$$\mathbf{p}_2 = [0 \quad 1 \quad 0 \quad 0], \qquad \mathbf{u}_1 = \begin{bmatrix} 0 \\ 0 \\ 0 \\ 1 \end{bmatrix}, \quad \text{and} \quad \mathbf{u}_2 = \begin{bmatrix} 0 \\ 0 \\ l_1 \\ 1 \end{bmatrix}$$

and the revolute joint at $\mathbf{u}_1$ is in its datum position (angle of zero). Then, depending on the type of joint at $\mathbf{u}_1$, the base orientation $\mathbf{p}_1$ and $\boldsymbol{\nu}$ are given by the following:

Pitch joint: $\mathbf{p}_1 = \mathbf{p}_2 = [0 \quad 1 \quad 0 \quad 0]$ and $\boldsymbol{\nu} = \begin{bmatrix} 0 \\ 0 \\ 1 \end{bmatrix}$.

Roll joint: From equations (3.6) and (3.7), $\boldsymbol{\nu} = \begin{bmatrix} 0 \\ 0 \\ 1 \end{bmatrix}$ and $\mathbf{p}_1 = [0 \quad 1 \quad 0 \quad 0]$.

Yaw joint: From equations (3.9) and (3.8), $\boldsymbol{\nu} = \begin{bmatrix} 0 \\ 0 \\ 1 \end{bmatrix}$ and $\mathbf{p}_1 = [-1 \quad 0 \quad 0 \quad 0]$.

Crank ψ joint: From equations (3.10) and (3.11) of Chapter 3, $\boldsymbol{\nu} = \begin{bmatrix} 0 \\ -1 \\ 0 \end{bmatrix}$ and $\mathbf{p}_1 = [0 \quad 0 \quad 1 \quad 0]$.

Crank $\tilde{\psi}$ joint: From equations (3.13) and (3.12), $\boldsymbol{\nu} = \begin{bmatrix} 0 \\ -1 \\ 0 \end{bmatrix}$ and $\mathbf{p}_1 = [0 \quad 0 \quad -1 \quad 0]$.

Cylindrical ϕ joint: From equations (3.14) and (3.15), $\mathbf{p}_1 = [0 \quad -1 \quad 0 \quad 0]$, $\boldsymbol{\nu} = \begin{bmatrix} 1 \\ 0 \\ 0 \end{bmatrix}$.

Cylindrical $\tilde{\phi}$ joint: From equations (3.17) and (3.16), $\mathbf{p}_1 = [0 \quad -1 \quad 0 \quad 0]$ and $\boldsymbol{\nu} = \begin{bmatrix} 1 \\ 0 \\ 0 \end{bmatrix}$.

To simplify the analysis of each type of robot, we assume without loss of generality that the base of each robot is at the origin, so $\mathbf{u}_1 = [0 \quad 0 \quad 0 \quad 1]^{\#}$. The end effector status for other base positions is calculated by adding the required base position to each joint position. Notice that with suitable realignment of base plane $\mathbf{p}_1$ and orientation vector $\boldsymbol{\nu}$, base joints λ, ϕ, ψ are interchangeable, as are θ, τ, and $\tilde{\psi}$.

The base orientation and position is similar in form to the Plücker line coordinate system [Mason and Salisbury 1985]. The Plücker line coordinates for a screw, a twist axis or a wrench axis is a six-vector [Mason and Salisbury 1985]. Here base orientation is defined by $\mathbf{p}_1 = [a \quad b \quad c \quad d]$ (three pieces of information), $\boldsymbol{\nu} = [\nu_x \quad \nu_y \quad \nu_z]^{\#}$ (three pieces of information), and a point (another three pieces of information), for a total of nine pieces of information. Only six of these pieces of information are independent. Unit vector $\boldsymbol{\nu} = [\nu_x \quad \nu_y \quad \nu_z]^{\#}$ contains two independent pieces of information, since $\nu_x^2 + \nu_y^2 + \nu_z^2 = 1$. $\mathbf{p}_1$ contains two additional independent pieces of information, since $a^2 + b^2 + c^2 = 1$ and $a\nu_x + b\nu_y + c\nu_z = 0$. $\mathbf{p}_1\mathbf{u}_1 = 0$, so $\mathbf{u}_1$ is defined by two more pieces of information. Thus, six pieces of independent information are contained in the base orientation and position.

The traditional way of producing kinematic equations for robot manipulators is to proceed joint by joint using the Denavit-Hartenberg notation and frames [Paul 1981b; Craig 1986]. Given the position and orientation of the first joint as a frame F_0 (with one point and three mutually orthogonal unit vectors), the next frame (as determined by the joint angle, link length, and twist) is $F_1 = T_1^0 F_0$, and so $F_n = T_n^{n-1} T_{n-1}^{n-2} \cdots T_1^0 F_0 = T_n^0 F_0$, where F_i is a 4×4 homogeneous matrix giving the location and orientation of frame F_i; alternatively, F_i can be considered as a position and a rotation matrix. T_n^0 is a 4×4 homogeneous matrix whose columns are three vectors followed by a point. T_n^0 can be formed independently of F_0, but the resulting product is not irreducible. Thus terms can and usually will cancel to simplify the resultant terms in F_n. This is in contrast to the irreducible forms developed in this chapter.

Here, we develop the kinematic equations for general base orientations in terms of the n variables defining the n joints of the manipulator. The equations for joint positions and planes of motion are canonical—that is, they are in their simplest form apart from some possible simple factorizations. This canonical feature is extended to the inverse kinematic solutions developed in Chapter 6. This means that when these equations are efficiently programmed into the robot controllers, the execution time is at a minimum, a desirable feature in the real-time control of robots.

It is shown in Appendix B that the kinematic equations developed in terms of the general base orientation are considerably more complex than for some simple base orientations. It might seem that it would be better to develop the equations for a simple orientation and generalize using the cut-and-paste operations discussed in Section 2.8. However, the resulting computations are more extensive, and the method should be avoided.

5.2 FORWARD KINEMATIC SOLUTIONS

The position of the end of the link l_{i+1} following a robotic joint and the plane of motion $\mathbf{p}_{j+1}$ is determined [Parkin 1985c; Paul 1981b] by some or all of the following:

1. The position of the ends of the link l_i
2. The plane of motion $\mathbf{p}_j$ of l_i
3. The joint type (pitch, roll, yaw, cylindrical, crank, sliding)
4. The joint status (angle or length)

The base orientation is defined by $\mathbf{p}_1$ and $\boldsymbol{\nu}$, and the base is assumed to be at the origin; see Section 5.1. The basic equations presented in Section 5.2.1 for the revolute robot are derived in Appendix B. Although the most complex of the basic robot types is the revolute robot, the equations of all common revolute types are derived in canonical form.

5.2.1 The Revolute Robot

The plane of motion $\mathbf{p}_2 = [a_2 \quad b_2 \quad c_2 \quad 0]$ of the 3 *dof* revolute robot $|\lambda\theta\theta|$ with general base orientation is given by

$$\begin{bmatrix} a_2 \\ b_2 \\ c_2 \end{bmatrix} = \begin{bmatrix} a_1\cos\lambda_1 + (\nu_y c_1 - \nu_z b_1)\sin\lambda_1 \\ b_1\cos\lambda_1 + (\nu_z a_1 - \nu_x c_1)\sin\lambda_1 \\ c_1\cos\lambda_1 + (\nu_x b_1 - \nu_y a_1)\sin\lambda_1 \end{bmatrix} \tag{5.1}$$

$$\mathbf{u}_2 = \begin{bmatrix} l_1\nu_x \\ l_1\nu_y \\ l_1\nu_z \\ 1 \end{bmatrix} \tag{5.2}$$

$$\mathbf{u}_3 = \begin{bmatrix} (l_1 + l_2\cos\theta_2)\nu_x - l_2\{(\nu_y c_1 - \nu_z b_1)\cos\lambda_1 - a_1\sin\lambda_1\}\sin\theta_2 \\ (l_1 + l_2\cos\theta_2)\nu_y - l_2\{(\nu_z a_1 - \nu_x c_1)\cos\lambda_1 - b_1\sin\lambda_1\}\sin\theta_2 \\ (l_1 + l_2\cos\theta_2)\nu_z - l_2\{(\nu_x b_1 - \nu_y a_1)\cos\lambda_1 - c_1\sin\lambda_1\}\sin\theta_2 \\ 1 \end{bmatrix} \tag{5.3}$$

$$\frac{\mathbf{u}_4 - \mathbf{u}_3}{l_3} = \begin{bmatrix} \nu_x\cos(\theta_2 + \theta_3) - \{(\nu_y c_1 - \nu_z b_1)\cos\lambda_1 - a_1\sin\lambda_1\}\sin(\theta_2 + \theta_3) \\ \nu_y\cos(\theta_2 + \theta_3) - \{(\nu_z a_1 - \nu_x c_1)\cos\lambda_1 - b_1\sin\lambda_1\}\sin(\theta_2 + \theta_3) \\ \nu_z\cos(\theta_2 + \theta_3) - \{(\nu_x b_1 - \nu_y a_1)\cos\lambda_1 - c_1\sin\lambda_1\}\sin(\theta_2 + \theta_3) \\ 0 \end{bmatrix} \tag{5.4}$$

Thus, $\mathbf{u}_4$ is given by

$$\begin{bmatrix} \nu_x\{l_1 + l_2\cos\theta_2 + l_3\cos(\theta_2+\theta_3)\} - \{(\nu_y c_1 - \nu_z b_1)\cos\lambda_1 - a_1\sin\lambda_1\}\{l_2\sin\theta_2 + l_3\sin(\theta_2+\theta_3)\} \\ \nu_y\{l_1 + l_2\cos\theta_2 + l_3\cos(\theta_2+\theta_3)\} - \{(\nu_z a_1 - \nu_x c_1)\cos\lambda_1 - b_1\sin\lambda_1\}\{l_2\sin\theta_2 + l_3\sin(\theta_2+\theta_3)\} \\ \nu_z\{l_1 + l_2\cos\theta_2 + l_3\cos(\theta_2+\theta_3)\} - \{(\nu_x b_1 - \nu_y a_1)\cos\lambda_1 - c_1\sin\lambda_1\}\{l_2\sin\theta_2 + l_3\sin(\theta_2+\theta_3)\} \\ 1 \end{bmatrix} \tag{5.5}$$

Drillable robots usually start with $|\,\lambda\theta\theta\,|$, and for this configuration we can deduce

$$\mathbf{u}_5 = \begin{bmatrix} \nu_x\{l_1 + l_2\cos\theta_2 + l_3\cos(\theta_2+\theta_3) + l_4\cos(\theta_2+\theta_3+\theta_4)\} \\ -\{(\nu_y c_1 - \nu_z b_1)\cos\lambda_1 - a_1\sin\lambda_1\}\{l_2\sin\theta_2 + l_3\sin(\theta_2+\theta_3) + l_4\sin(\theta_2+\theta_3+\theta_4)\} \\ \nu_y\{l_1 + l_2\cos\theta_2 + l_3\cos(\theta_2+\theta_3) + l_4\cos(\theta_2+\theta_3+\theta_4)\} \\ -\{(\nu_z a_1 - \nu_x c_1)\cos\lambda_1 - b_1\sin\lambda_1\}\{l_2\sin\theta_2 + l_3\sin(\theta_2+\theta_3) + l_4\sin(\theta_2+\theta_3+\theta_4)\} \\ \nu_z\{l_1 + l_2\cos\theta_2 + l_3\cos(\theta_2+\theta_3) + l_4\cos(\theta_2+\theta_3+\theta_4)\} \\ -\{(\nu_x b_1 - \nu_y a_1)\cos\lambda_1 - c_1\sin\lambda_1\}\{l_2\sin\theta_2 + l_3\sin(\theta_2+\theta_3) + l_4\sin(\theta_2+\theta_3+\theta_4)\} \\ 1 \end{bmatrix} \tag{5.6}$$

For $\mathbf{p}_1 = [0 \quad 1 \quad 0 \quad 0]$ and $\boldsymbol{\nu} = \begin{bmatrix} 0 \\ 0 \\ 1 \end{bmatrix}$, equations (5.5) and (5.6) simplify to

$$\mathbf{u}_4 = \begin{bmatrix} \cos\lambda_1\{l_2\sin\theta_2 + l_3\sin(\theta_2+\theta_3)\} \\ \sin\lambda_1\{l_2\sin\theta_2 + l_3\sin(\theta_2+\theta_3)\} \\ l_1 + l_2\cos\theta_2 + l_3\cos(\theta_2+\theta_3) \\ 1 \end{bmatrix} \tag{5.7}$$

and

$$\mathbf{u}_5 = \begin{bmatrix} \cos\lambda_1\{l_2\sin\theta_2 + l_3\sin(\theta_2+\theta_3) + l_4\sin(\theta_2+\theta_3+\theta_4)\} \\ \sin\lambda_1\{l_2\sin\theta_2 + l_3\sin(\theta_2+\theta_3) + l_4\sin(\theta_2+\theta_3+\theta_4)\} \\ l_1 + l_2\cos\theta_2 + l_3\cos(\theta_2+\theta_3) + l_4\cos(\theta_2+\theta_3+\theta_4) \\ 1 \end{bmatrix} \tag{5.8}$$

Alternatively, the revolute robot can be defined with a cylindrical or crank ψ joint replacing the roll base joint. The first worked example is for the 3 *dof* revolute robot $|\,\phi\theta\theta\,|$ with a specific base orientation. We could derive $\mathbf{u}_4$ for this robot in terms of the general base orientation, a lengthy procedure. Instead, we can accomplish the same purpose by determining the base orientation for the base cylindrical joint that produces all joint positions the same as in equations (5.2), (5.3), and (5.4). Since the plane of motion is to be the same for equivalent base angles, we use $\lambda_1 \equiv \phi_1 = 0$ and see that $\bar{\mathbf{p}}_2$ is the same as in equation (5.1), which is $[a_1 \quad b_1 \quad c_1 \quad d_1]$. This $\bar{\mathbf{p}}_2$ is the same as $\mathbf{p}_1$ used with the cylindrical joint. Using equation (3.17), we see that the orientation vector is given by

$$\begin{bmatrix} \bar{\nu}_x \\ \bar{\nu}_y \\ \bar{\nu}_z \end{bmatrix} = -\begin{bmatrix} b_1\nu_z - c_1\nu_y \\ c_1\nu_x - a_1\nu_z \\ a_1\nu_y - b_1\nu_x \end{bmatrix} = \begin{bmatrix} \nu_x \\ \nu_y \\ \nu_z \end{bmatrix} \times \begin{bmatrix} a_1 \\ b_1 \\ c_1 \end{bmatrix}$$

where the two vectors are orthogonal, so

$$\begin{bmatrix} \nu_x \\ \nu_y \\ \nu_z \end{bmatrix} = -\begin{bmatrix} \bar{\nu}_x \\ \bar{\nu}_y \\ \bar{\nu}_z \end{bmatrix} \times \begin{bmatrix} a_1 \\ b_1 \\ c_1 \end{bmatrix} = -\begin{bmatrix} \bar{\nu}_y c_1 - \bar{\nu}_z b_1 \\ \bar{\nu}_z a_1 - \bar{\nu}_x c_1 \\ \bar{\nu}_x b_1 - \bar{\nu}_y a_1 \end{bmatrix}$$

Substituting this vector in equation (5.5) gives $\mathbf{u}_4$ as

$$\begin{bmatrix} -(\bar{\nu}_y c_1 - \bar{\nu}_z b_1)\{l_1 + l_2\cos\theta_2 + l_3\cos(\theta_2+\theta_3)\} - \{\bar{\nu}_x\cos\phi_1 - a_1\sin\phi_1\}\{l_2\sin\theta_2 + l_3\sin(\theta_2+\theta_3)\} \\ -(\bar{\nu}_z a_1 - \bar{\nu}_x c_1)\{l_1 + l_2\cos\theta_2 + l_3\cos(\theta_2+\theta_3)\} - \{\bar{\nu}_y\cos\phi_1 - b_1\sin\phi_1\}\{l_2\sin\theta_2 + l_3\sin(\theta_2+\theta_3)\} \\ -(\bar{\nu}_x b_1 - \bar{\nu}_y a_1)\{l_1 + l_2\cos\theta_2 + l_3\cos(\theta_2+\theta_3)\} - \{\bar{\nu}_z\cos\phi_1 - c_1\sin\phi_1\}\{l_2\sin\theta_2 + l_3\sin(\theta_2+\theta_3)\} \\ 0 \end{bmatrix} \tag{5.9}$$

A 4 *dof* pencilable robot $|\lambda\theta\theta\lambda|$ has two planes of motion $\bar{\mathbf{p}}_2^2$ and $\bar{\mathbf{p}}_3^0$. $\mathbf{u}_5$ is the same as $\mathbf{u}_4$ of equation (5.5), with l_3 replaced by $l_3 + l_4$, so

$$\mathbf{u}_5 = \begin{bmatrix} \nu_x\{l_1 + l_2\cos\theta_2 + (l_3+l_4)\cos(\theta_2+\theta_3)\} \\ \quad - \{(\nu_y c_1 - \nu_z b_1)\cos\lambda_1 - a_1\sin\lambda_1\}\{l_2\sin\theta_2 + (l_3+l_4)\sin(\theta_2+\theta_3)\} \\ \nu_y\{l_1 + l_2\cos\theta_2 + (l_3+l_4)\cos(\theta_2+\theta_3)\} \\ \quad - \{(\nu_z a_1 - \nu_x c_1)\cos\lambda_1 - b_1\sin\lambda_1\}\{l_2\sin\theta_2 + (l_3+l_4)\sin(\theta_2+\theta_3)\} \\ \nu_z\{l_1 + l_2\cos\theta_2 + (l_3+l_4)\cos(\theta_2+\theta_3)\} \\ \quad - \{(\nu_x b_1 - \nu_y a_1)\cos\lambda_1 - c_1\sin\lambda_1\}\{l_2\sin\theta_2 + (l_3+l_4)\sin(\theta_2+\theta_3)\} \\ 1 \end{bmatrix} \tag{5.10}$$

and with the simple base orientation $\mathbf{p}_1 = [0 \quad 1 \quad 0 \quad 0]$ and $\boldsymbol{\nu} = \begin{bmatrix} 0 \\ 0 \\ 1 \end{bmatrix}$

$$\mathbf{u}_5 = \begin{bmatrix} \cos\lambda_1\{l_2\sin\theta_2 + (l_3+l_4)\sin(\theta_2+\theta_3)\} \\ \sin\lambda_1\{l_2\sin\theta_2 + (l_3+l_4)\sin(\theta_2+\theta_3)\} \\ l_1 + l_2\cos\theta_2 + (l_3+l_4)\cos(\theta_2+\theta_3) \\ 1 \end{bmatrix} \tag{5.11}$$

$\bar{\mathbf{p}}_3^0$ is given in part by

$$\begin{bmatrix} a_3 \\ b_3 \\ c_3 \end{bmatrix} = \begin{bmatrix} \{a_1\cos\lambda_1 + (\nu_y c_1 - \nu_z b_1)\sin\lambda_1\}\cos\lambda_4 \\ + \{\{(b_1\nu_z - c_1\nu_y)\cos\lambda_1 + a_1\sin\lambda_1\}\cos(\theta_2+\theta_3) - \nu_x\sin(\theta_2+\theta_3)\}\sin\lambda_4 \\ \{b_1\cos\lambda_1 + (\nu_z a_1 - \nu_x c_1)\sin\lambda_1\}\cos\lambda_4 \\ + \{\{(-a_1\nu_z + c_1\nu_x)\cos\lambda_1 + b_1\sin\lambda_1\}\cos(\theta_2+\theta_3) - \nu_y\sin(\theta_2+\theta_3)\}\sin\lambda_4 \\ \{c_1\cos\lambda_1 + (\nu_x b_1 - \nu_y a_1)\sin\lambda_1\}\cos\lambda_4 \\ + \{\{(a_1\nu_y - b_1\nu_x)\cos\lambda_1 + c_1\sin\lambda_1\}\cos(\theta_2+\theta_3) - \nu_z\sin(\theta_2+\theta_3)\}\sin\lambda_4 \end{bmatrix} \tag{5.12}$$

For the specific base orientation $\mathbf{p}_1 = [0 \quad 1 \quad 0 \quad 0]$ and $\boldsymbol{\nu} = \begin{bmatrix} 0 \\ 0 \\ 1 \end{bmatrix}$, we get

$$\begin{bmatrix} a_3 \\ b_3 \\ c_3 \end{bmatrix} = \begin{bmatrix} -\sin\lambda_1\cos\lambda_4 + \cos\lambda_1\cos(\theta_2+\theta_3)\sin\lambda_4 \\ \sin\lambda_1\cos(\theta_2+\theta_3)\sin\lambda_4 + \cos\lambda_1\cos\lambda_4 \\ \sin(\theta_2+\theta_3)\sin\lambda_4 \end{bmatrix} \tag{5.13}$$

A 5 *dof* directable robot $|\lambda\theta\theta\lambda\theta|$ has two planes of motion $\bar{\mathbf{p}}_2^2$ and $\bar{\mathbf{p}}_3^1$ and the same $\mathbf{u}_5$ as in equation (5.10). $\bar{\mathbf{p}}_3^1$ is given by equation (5.12).

$$\mathbf{u}_6 = \begin{bmatrix} \nu_x\{l_1 + l_2\cos\theta_2 + (l_3+l_4)\cos(\theta_2+\theta_3) + l_5\{\cos(\theta_2+\theta_3)\cos\theta_5 - \sin(\theta_2+\theta_3)\cos\lambda_4\sin\theta_5\}\} \\ - \{(c_1\nu_y - b_1\nu_z)\cos\lambda_1 - a_1\sin\lambda_1\}\{l_2\sin\theta_2 + (l_3+l_4)\sin(\theta_2+\theta_3) \\ + l_5\{\sin(\theta_2+\theta_3)\cos\theta_5 + \cos(\theta_2+\theta_3)\cos\lambda_4\sin\theta_5\}\} \\ - l_5\{(c_1\nu_y - b_1\nu_z)\sin\lambda_1 + a_1\cos\lambda_1\}\sin\lambda_4\sin\theta_5 \\ \nu_y\{l_1 + l_2\cos\theta_2 + (l_3+l_4)\cos(\theta_2+\theta_3) + l_5\{\cos(\theta_2+\theta_3)\cos\theta_5 - \sin(\theta_2+\theta_3)\cos\lambda_4\sin\theta_5\}\} \\ - \{(a_1\nu_z - c_1\nu_x)\cos\lambda_1 - b_1\sin\lambda_1\}\{l_2\sin\theta_2 + (l_3+l_4)\sin(\theta_2+\theta_3) \\ - l_5\{\sin(\theta_2+\theta_3)\cos\theta_5 + \cos(\theta_2+\theta_3)\cos\lambda_4\sin\theta_5\}\} \\ - l_5\{(a_1\nu_z - c_1\nu_x)\sin\lambda_1 + b_1\cos\lambda_1\}\sin\lambda_4\sin\theta_5 \\ \nu_z\{l_1 + l_2\cos\theta_2 + (l_3+l_4)\cos(\theta_2+\theta_3) + l_5\{\cos(\theta_2+\theta_3)\cos\theta_5 - \sin(\theta_2+\theta_3)\cos\lambda_4\sin\theta_5\}\} \\ - \{(b_1\nu_x - a_1\nu_y)\cos\lambda_1 - c_1\sin\lambda_1\}\{l_2\sin\theta_2 + (l_3+l_4)\sin(\theta_2+\theta_3) \\ - l_5\{\sin(\theta_2+\theta_3)\cos\theta_5 + \cos(\theta_2+\theta_3)\cos\lambda_4\sin\theta_5\}\} \\ - l_5\{(b_1\nu_x - a_1\nu_y)\sin\lambda_1 + c_1\cos\lambda_1\}\sin\lambda_4\sin\theta_5 \\ 1 \end{bmatrix} \tag{5.14}$$

For the robot $|\lambda\theta\theta\phi|$, $\mathbf{u}_5$ can be calculated using equation (5.5), with θ_3 replaced by $\theta_3 + \tan^{-1}(l_4/l_3) = \theta_3'$ and l_3 replaced by $\sqrt{l_3^2 + l_4^2}$, so

$$\mathbf{u}_5 = \begin{bmatrix} \nu_x\{l_1 + l_2\cos\theta_2 + l_3\cos(\theta_2+\theta_3')\} - \{(\nu_y c_1 - \nu_z b_1)\cos\lambda_1 - a_1\sin\lambda_1\}\{l_2\sin\theta_2 + l_3\sin(\theta_2+\theta_3')\} \\ \nu_y\{l_1 + l_2\cos\theta_2 + l_3\cos(\theta_2+\theta_3')\} - \{(\nu_z a_1 - \nu_x c_1)\cos\lambda_1 - b_1\sin\lambda_1\}\{l_2\sin\theta_2 + l_3\sin(\theta_2+\theta_3')\} \\ \nu_z\{l_1 + l_2\cos\theta_2 + l_3\cos(\theta_2+\theta_3')\} - \{(\nu_x b_1 - \nu_y a_1)\cos\lambda_1 - c_1\sin\lambda_1\}\{l_2\sin\theta_2 + l_3\sin(\theta_2+\theta_3')\} \\ 1 \end{bmatrix} \tag{5.15}$$

For the specific base orientation $\mathbf{p}_1 = [0 \quad 1 \quad 0 \quad 0]$ and $\boldsymbol{\nu} = \begin{bmatrix} 0 \\ 0 \\ 1 \end{bmatrix}$, equation (5.15) becomes

$$\mathbf{u}_5 = \begin{bmatrix} -\cos\lambda_1\{l_2\sin\theta_2 - \sqrt{l_3^2 + l_4^2}\sin(\theta_2 + \theta_3')\} \\ -\sin\lambda_1\{l_2\sin\theta_2 - \sqrt{l_3^2 + l_4^2}\sin(\theta_2 + \theta_3')\} \\ l_1 + l_2\cos\theta_2 - \sqrt{l_3^2 + l_4^2}\cos(\theta_2 + \theta_3') \\ 1 \end{bmatrix} \tag{5.16}$$

5.2.2 The Cylindrical Robot

For the $|\lambda\sigma{+}\sigma|$ robot, the plane of motion $\overline{\mathbf{p}}_2$ within which all the links move is the same as for the revolute robot and is given by equation (5.1). $\mathbf{u}_2$ is the same as in equation (5.2), with l_1 replaced by σ_2. $\mathbf{u}_3$ is obtained by adding σ_3 times the cross product of vectors $[a_2 \quad b_2 \quad c_2]^{\#}$ and $\boldsymbol{\nu}$ to $\mathbf{u}_2$, so

$$\mathbf{u}_3 = \begin{bmatrix} \sigma_2\nu_x \\ \sigma_2\nu_y \\ \sigma_2\nu_z \\ 1 \end{bmatrix} - \sigma_3 \begin{bmatrix} b_2\nu_z - c_2\nu_y \\ c_2\nu_x - a_2\nu_z \\ a_2\nu_y - b_2\nu_x \\ 0 \end{bmatrix}$$

$$= \begin{bmatrix} \sigma_2\nu_x \\ \sigma_2\nu_y \\ \sigma_2\nu_z \\ 1 \end{bmatrix} - \sigma_3 \begin{bmatrix} (b_1\nu_z - c_1\nu_y)\cos\lambda_1 + (\nu_z^2 a_1 - \nu_x\nu_z c_1 - \nu_x\nu_y b_1 + \nu_y^2 a_1)\sin\lambda_1 \\ (c_1\nu_x - a_1\nu_z)\cos\lambda_1 + (\nu_x^2 b_1 - \nu_x\nu_y a_1 - \nu_y\nu_z c_1 + \nu_z^2 b_1)\sin\lambda_1 \\ (a_1\nu_y - b_1\nu_x)\cos\lambda_1 + (\nu_y^2 c_1 - \nu_y\nu_z b_1 - \nu_x\nu_z a_1 + \nu_x^2 c_1)\sin\lambda_1 \\ 0 \end{bmatrix}$$

$$= \begin{bmatrix} \sigma_2\nu_x \\ \sigma_2\nu_y \\ \sigma_2\nu_z \\ 1 \end{bmatrix} - \sigma_3 \begin{bmatrix} (b_1\nu_z - c_1\nu_y)\cos\lambda_1 + a_1\sin\lambda_1 \\ (c_1\nu_x - a_1\nu_z)\cos\lambda_1 + b_1\sin\lambda_1 \\ (a_1\nu_y - b_1\nu_x)\cos\lambda_1 + c_1\sin\lambda_1 \\ 0 \end{bmatrix} \tag{5.17}$$

For the specific base orientation $\mathbf{p}_1 = [0 \quad 1 \quad 0 \quad 0]$ and $\boldsymbol{\nu} = \begin{bmatrix} 0 \\ 0 \\ 1 \end{bmatrix}$,

$$\mathbf{u}_3 = \begin{bmatrix} -\sigma_3\cos\lambda_1 \\ -\sigma_3\sin\lambda_1 \\ \sigma_2 \\ 1 \end{bmatrix} \tag{5.18}$$

5.2.3 The Cartesian Robot

Suppose we have the 3 *dof* Cartesian robot $| \sigma + \sigma \times \sigma |$. σ_1 and σ_2 lie in $\mathbf{p}_1$ and

$$\mathbf{u}_2 = \begin{bmatrix} 0 \\ 0 \\ 0 \\ 1 \end{bmatrix} + \sigma_1 \begin{bmatrix} \nu_x \\ \nu_y \\ \nu_z \\ 0 \end{bmatrix}, \qquad \mathbf{u}_3 = \mathbf{u}_2 - \sigma_2 \begin{bmatrix} \nu_y c_1 - \nu_z b_1 \\ \nu_z a_1 - \nu_x c_1 \\ \nu_x b_1 - \nu_y a_1 \\ 0 \end{bmatrix}$$

$$= \begin{bmatrix} \sigma_1 \nu_x + \sigma_2(\nu_y c_1 - \nu_z b_1) \\ \sigma_1 \nu_y + \sigma_2(\nu_z a_1 - \nu_x c_1) \\ \sigma_1 \nu_z + \sigma_2(\nu_x b_1 - \nu_y a_1) \\ 1 \end{bmatrix}$$

and

$$\mathbf{u}_4 = \mathbf{u}_3 - \sigma_3 \begin{bmatrix} a_1 \\ b_1 \\ c_1 \\ 0 \end{bmatrix} = \begin{bmatrix} \sigma_1 \nu_x + \sigma_2(\nu_y c_1 - \nu_z b_1) - \sigma_3 a_1 \\ \sigma_1 \nu_y + \sigma_2(\nu_z a_1 - \nu_x c_1) - \sigma_3 b_1 \\ \sigma_1 \nu_z + \sigma_2(\nu_x b_1 - \nu_y a_1) - \sigma_3 c_1 \\ 1 \end{bmatrix} \tag{5.19}$$

For the specific base orientation $\mathbf{p}_1 = [0 \quad 1 \quad 0 \quad 0]$ and $\boldsymbol{\nu} = [0 \quad 0 \quad 1]^{\#}$,

$$\mathbf{u}_4 = \begin{bmatrix} -\sigma_2 \\ -\sigma_3 \\ \sigma_1 \\ 1 \end{bmatrix} \tag{5.20}$$

5.2.4 The Polar Robot

For the 3 *dof* polar robot $| \lambda \theta \sigma |$, $\bar{\mathbf{p}}_2$ is given by equation (5.1) and $\mathbf{u}_3$ is the same as the $\mathbf{u}_3$ location for the 3 *dof* revolute robot as given by equation (5.3) with l_2 replaced by σ_3, so

$$\mathbf{u}_3 = \begin{bmatrix} (l_1 + \sigma_3 \cos\theta_2)\nu_x - \sigma_3\{(\nu_y c_1 - \nu_z b_1)\cos\lambda_1 - a_1 \sin\lambda_1\}\sin\theta_2 \\ (l_1 + \sigma_3 \cos\theta_2)\nu_y - \sigma_3\{(\nu_z a_1 - \nu_x c_1)\cos\lambda_1 - b_1 \sin\lambda_1\}\sin\theta_2 \\ (l_1 + \sigma_3 \cos\theta_2)\nu_z - \sigma_3\{(\nu_x b_1 - \nu_y a_1)\cos\lambda_1 - c_1 \sin\lambda_1\}\sin\theta_2 \\ 1 \end{bmatrix} \tag{5.21}$$

For the specific base orientation $\mathbf{p}_1 = [0 \quad 1 \quad 0 \quad 0]$ and $\boldsymbol{\nu} = \begin{bmatrix} 0 \\ 0 \\ 1 \end{bmatrix}$,

$$\mathbf{u}_3 = \begin{bmatrix} \sigma_3 \cos \lambda_1 \sin \theta_2 \\ \sigma_3 \sin \lambda_1 \sin \theta_2 \\ l_1 + \sigma_3 \cos \theta_2 \\ 1 \end{bmatrix} \tag{5.22}$$

A popular configuration for a polar robot (for instance, the Unimation Puma 2000) is $| \lambda\theta\sigma\theta\lambda |$. The kinematic equations are the same as the 4 *dof* $| \lambda\theta\theta\lambda |$, with l_2 replaced with σ_3. Thus, from equation (5.10)

$$\mathbf{u}_5 = \begin{bmatrix} \nu_x\{l_1 + \sigma_3 \cos \theta_2 + (l_4 + l_5)\cos(\theta_2 + \theta_4)\} \\ - \{(\nu_y c_1 - \nu_z b_1)\cos \lambda_1 - a_1 \sin \lambda_1\}\{\sigma_3 \sin \theta_2 + (l_4 + l_5)\sin(\theta_2 + \theta_4)\} \\ \nu_y\{l_1 + \sigma_3 \cos \theta_2 + (l_4 + l_5)\cos(\theta_2 + \theta_4)\} \\ - \{(\nu_z a_1 - \nu_x c_1)\cos \lambda_1 - b_1 \sin \lambda_1\}\{\sigma_3 \sin \theta_2 + (l_4 + l_5)\sin(\theta_2 + \theta_4)\} \\ \nu_z\{l_1 + \sigma_3 \cos \theta_2 + (l_4 + l_5)\cos(\theta_2 + \theta_4)\} \\ - \{(\nu_x b_1 - \nu_y a_1)\cos \lambda_1 - c_1 \sin \lambda_1\}\{\sigma_3 \sin \theta_2 + (l_4 + l_5)\sin(\theta_2 + \theta_4)\} \\ 1 \end{bmatrix}$$

and for the specific base orientation $\mathbf{p}_1 = [0 \quad 1 \quad 0 \quad 0]$ and $\boldsymbol{\nu} = \begin{bmatrix} 0 \\ 0 \\ 1 \end{bmatrix}$,

$$\mathbf{u}_5 = \begin{bmatrix} \cos \lambda_1\{\sigma_3 \sin \theta_2 + (l_4 + l_5)\sin(\theta_2 + \theta_4)\} \\ \sin \lambda_1\{\sigma_3 \sin \theta_2 + (l_4 + l_5)\sin(\theta_2 + \theta_4)\} \\ l_1 + \sigma_3 \cos \theta_2 + (l_4 + l_5)\cos(\theta_2 + \theta_4) \\ 1 \end{bmatrix} \tag{5.24}$$

5.2.5 The SCARA Robot

The 3 *dof* SCARA robot $| \theta\theta{\times}\sigma |$ can be analyzed in part from the analysis of the revolute robot. In particular, $\mathbf{u}_2$ and $\mathbf{u}_3$ are the same as $\mathbf{u}_3$ of equation (5.3) and $\mathbf{u}_4$ of equation (5.5) with $\lambda_1 = 0$, l_1 set to zero, l_2 replaced by l_1, and l_3 replaced by l_2. For the SCARA robot

$$\mathbf{u}_2 = \begin{bmatrix} l_1\{\cos \theta_1 \nu_x - (\nu_y c_1 - \nu_z b_1)\sin \theta_1\} \\ l_1\{\cos \theta_1 \nu_y - (\nu_z a_1 - \nu_x c_1)\sin \theta_1\} \\ l_1\{\cos \theta_1 \nu_z - (\nu_x b_1 - \nu_y a_1)\sin \theta_1\} \\ 1 \end{bmatrix}$$

$$\mathbf{u}_3 = \begin{bmatrix} \nu_x\{l_1\cos\theta_1 + l_2\cos(\theta_1+\theta_2)\} - (\nu_y c_1 - \nu_z b_1)\{l_1\sin\theta_1 + l_2\sin(\theta_1+\theta_2)\} \\ \nu_y\{l_1\cos\theta_1 + l_2\cos(\theta_1+\theta_2)\} - (\nu_z a_1 - \nu_x c_1)\{l_1\sin\theta_1 + l_2\sin(\theta_1+\theta_2)\} \\ \nu_z\{l_1\cos\theta_1 + l_2\cos(\theta_1+\theta_2)\} - (\nu_x b_1 - \nu_y a_1)\{l_1\sin\theta_1 + l_2\sin(\theta_1+\theta_2)\} \\ 1 \end{bmatrix} \tag{5.25}$$

From equation (3.18), we see that

$$\mathbf{u}_4 = \mathbf{u}_3 - \sigma_3 \begin{bmatrix} a_1 \\ b_1 \\ c_1 \\ 0 \end{bmatrix}$$

so

$$\mathbf{u}_4 = \begin{bmatrix} \nu_x\{l_1\cos\theta_1 + l_2\cos(\theta_1+\theta_2)\} - (\nu_y c_1 - \nu_z b_1)\{l_1\sin\theta_1 + l_2\sin(\theta_1+\theta_2)\} - \sigma_3 a_1 \\ \nu_y\{l_1\cos\theta_1 + l_2\cos(\theta_1+\theta_2)\} - (\nu_z a_1 - \nu_x c_1)\{l_1\sin\theta_1 + l_2\sin(\theta_1+\theta_2)\} - \sigma_3 b_1 \\ \nu_z\{l_1\cos\theta_1 + l_2\cos(\theta_1+\theta_2)\} - (\nu_x b_1 - \nu_y a_1)\{l_1\sin\theta_1 + l_2\sin(\theta_1+\theta_2)\} - \sigma_3 c_1 \\ 1 \end{bmatrix} \tag{5.26}$$

The usual base orientation is $\mathbf{p}_1 = [0 \quad 0 \quad -1 \quad 0]$ and $\boldsymbol{\nu} = \begin{bmatrix} 1 \\ 0 \\ 0 \end{bmatrix}$, so

$$\mathbf{u}_4 = \begin{bmatrix} \{l_1\cos\theta_1 + l_2\cos(\theta_1+\theta_2)\} \\ -\{l_1\sin\theta_1 + l_2\sin(\theta_1+\theta_2)\} \\ \sigma_3 \\ 1 \end{bmatrix} \tag{5.27}$$

Consider next the 4 *dof* of SCARA robot $|\ \theta\theta\underline{\psi}\sigma\ |$, where the underline for ψ indicates that the succeeding link has negative length; alternatively, we can consider the joint to be oriented in the other direction. From equation (3.10), we have

$$\mathbf{u}_4 = \mathbf{u}_3 + \sigma_4 \begin{bmatrix} b_1 \\ c_1 \\ a_1 \\ 0 \end{bmatrix}$$

$$= \begin{bmatrix} \nu_x\{l_1\cos\theta_1 + l_2\cos(\theta_1+\theta_2)\} - (\nu_y c_1 - \nu_z b_1)\{l_1\sin\theta_1 + l_2\sin(\theta_1+\theta_2)\} + \sigma_4 a_1 \\ \nu_y\{l_1\cos\theta_1 + l_2\cos(\theta_1+\theta_2)\} - (\nu_z a_1 - \nu_x c_1)\{l_1\sin\theta_1 + l_2\sin(\theta_1+\theta_2)\} + \sigma_4 b_1 \\ \nu_z\{l_1\cos\theta_1 + l_2\cos(\theta_1+\theta_2)\} - (\nu_x b_1 - \nu_y a_1)\{l_1\sin\theta_1 + l_2\sin(\theta_1+\theta_2)\} + \sigma_4 c_1 \\ 1 \end{bmatrix} \tag{5.28}$$

The usual base orientation is $\mathbf{p}_1 = [0 \quad 0 \quad -1 \quad 0]$ and $\boldsymbol{\nu} = \begin{bmatrix} 1 \\ 0 \\ 0 \end{bmatrix}$, so

$$\mathbf{u}_4 = \begin{bmatrix} l_1\cos\theta_1 + l_2\cos(\theta_1 + \theta_2) \\ -l_1\sin\theta_1 + l_2\sin(\theta_1 + \theta_2) \\ -\sigma_4 \\ 1 \end{bmatrix} \tag{5.29}$$

5.2.6 Unclassified Robots

The number of commercial and training robots that cannot be categorized as one of the five basic robots is small (see Appendix C). Most robots have a roll (or, with realignment of the base, cylindrical ϕ or $\tilde{\phi}$ or crank ψ) base joint, the big exception being the SCARA robot. But even the SCARA can be considered to have cylindrical base joint if the pedestal and a right-angle bend are included in the kinematic chain.

A 3 *dof* robot that has desirable characteristics (nonstiff and with a closed-form inverse kinematic solution) is the robot $|\,\lambda\sigma\theta\,|$. The position of the end effector is the same as for the $\mathbf{u}_3$ position in the revolute robot with link l_1 replaced with sliding link σ_2, so

$$\mathbf{u}_3 = \begin{bmatrix} (\sigma_2 + l_3\cos\theta_3)\nu_x - l_3\{(\nu_y c_1 - \nu_z b_1)\cos\lambda_1 - a_1\sin\lambda_1\}\sin\theta_3 \\ (\sigma_2 + l_2\cos\theta_2)\nu_y - l_3\{(\nu_z a_1 - \nu_x c_1)\cos\lambda_1 - b_1\sin\lambda_1\}\sin\theta_3 \\ (\sigma_2 + l_2\cos\theta_3)\nu_z - l_3\{(\nu_x b_1 - \nu_y a_1)\cos\lambda_1 - c_1\sin\lambda_1\}\sin\theta_3 \\ 1 \end{bmatrix} \tag{5.30}$$

The plane of motion $\bar{\mathbf{p}}_2^2$ is the same as given by equation (5.1).

The robot with planes of motion $\bar{\mathbf{p}}^0\mathbf{p}^2$ is not directly covered in the preceding kinematic analyses. Realizations of this robot are

$$\left|\begin{array}{c|c|c|} \lambda & & \\ \psi & \gamma & \theta \\ \phi & (\times)\sigma & (+)\sigma \\ \tilde{\phi} & (\div)\sigma & (-)\sigma \end{array}\right.$$

The reverse crank joint $\tilde{\psi}$ is redundant with all but the base joint $\tilde{\phi}$, and the robot with first two joints $\tilde{\phi}\tilde{\psi}$ is awkward and will be considered no further. For example, the end effector of the robot $|\,\tilde{\phi}\tilde{\psi}\sigma\,|$ is confined to the surface of a cylinder of radius l_1 with axis $\mathbf{u}_1 + \alpha\boldsymbol{\nu}$, so its working volume (see the next section) is zero. The robot types

$$\left|\begin{array}{c|c|c|} \lambda & & \\ \psi & \gamma & \theta \\ \phi & & \\ \tilde{\phi} & & \end{array}\right.$$

are variants on the revolute robot. The types

$$\left|\begin{array}{c|c|c} \lambda & & \\ \psi & & \sigma(+) \\ \phi & \gamma & \sigma(-) \\ \tilde{\phi} & & \end{array}\right|$$

are variants on the polar robot. The types

$$\left|\begin{array}{c|c|c} \lambda & & \\ \psi & \sigma(\times) & \sigma(+) \\ \phi & \sigma(\div) & \sigma(-) \\ \tilde{\phi} & & \end{array}\right|$$

are variants on the cylindrical robot. The types

$$\left|\begin{array}{c|c|c} \lambda & & \\ \psi & \sigma(\times) & \\ \phi & \sigma(\div) & \theta \\ \tilde{\phi} & & \end{array}\right|$$

are variants on the robot $|\,\lambda\sigma\theta\,|$ discussed earlier in this section.

It should be evident that the most complicated of the standard robot forms as far as kinematic analysis is concerned is the revolute robot. It has been analyzed in great detail.

5.3 WORKING ENVELOPES

The working volume of a robot is the sum total of all the locations that can be reached by the end effector [Kumar and Waldron 1981; Tsai and Soni 1983]. We call the outer surface of the working volume the *working envelope*. The working volume should be extensive enough to perform the work required and should be free of voids—places inside the working envelope that are not part of the working volume. There are no practical cases in which the working volume is in two or more separate parts, that is, where two points inside the working volume can be found such that no path can be found within the working volume to connect them.

It is possible for the robot to have a doughnut-shaped working volume with a hole, and this shape may or may not be acceptable. The determining factors are joint limitations and desired trajectories. A straight-line trajectory in our doughnut-shaped working volume is much more likely to become feasible (since it passes outside the working volume of the robot) if practical limits are placed on the angle of swing of the revolute joints. The doughnut shape is possible, since the base is assumed to rotate through 360°, but if the base is limited to a total swing of 90°, most straight-line trajectories will be confined to the working envelope. All slew trajectories (see Section 8.1) will be inside the working envelope regardless of the total swing of the base.

The effectiveness of various link-joint combinations to encompass volume in Cartesian space will now be explored for the basic 3 *dof* robots and some other robots. Others have correctly indicated that the size of the working volume and the presence of voids within this volume are important factors in choosing a robot [Tsai and Soni 1983; Gupta and Roth 1982; Lee and Yand 1983]. Lee considered several robots, including the Puma 600, the Cincinnati Milacron T_3, the Cybotech V30, and concluded that the working volume of a robot with fixed configuration is proportional to the total length of the links to the third power. This is verified here.

We define volume efficiency as the volume covered by the end effector of the robot as a function of the maximum possible volume that can be covered. Without loss of generality we assume the lengths of the links add up to unity. We will assume that the ratio of longest to shortest length of each sliding link is 2 : 1.

The robot with the maximum possible volume covered for a given total link length is the 3 *dof* robot with ball joint base and pitch middle joint with equal link lengths and no joint angle restrictions (so $\pm 180°$ travel is possible on each joint), and its working volume is the sphere of radius $r = 1$. Thus its volume is $4\pi r^3/3 = 4.1888$.

The 3 *dof* revolute robot $|\,\phi\theta\theta\,|$ with no joint restrictions or confining base has maximum volume when $l_1 = 0$ and $l_2 = l_3 = 0.5$, which produces the same working volume as before. In order to avoid voids in the working volume, we require $l_2 = l_3$. (Recall that a ball joint can be modeled by a cylindrical joint joined to a pitch joint by a link of zero length.) If the link lengths are restricted to be all the same length, the working volume would be a sphere of radius $\frac{2}{3}$ for a working volume of 1.2411, so the volume efficiency is $1.2411/4.1888 = 0.296 \equiv 29.6\%$.

The 3 *dof* cylindrical robot $|\,\phi\sigma\sigma\,|$ has as working volume the square doughnut with cross section σ_1 by σ_2, as shown in Figure 5.1. The incremental area of rotation at distance $\sigma_2 \leq x \leq 2\sigma_2$ is $2\pi\sigma_1 x$, so the working volume is given by

$$V = \int_{\sigma_2}^{2\sigma_2} 2\pi\sigma_1 x\, dx = [\pi\sigma_1 x^2]_{\sigma_2}^{2\sigma_2} = 4\pi\sigma_1\sigma_2^2 - \pi\sigma_1\sigma_2^2 = 3\pi\sigma_1\sigma_2^2$$

and with $\sigma_1 = \sigma_2 = \frac{1}{4}$, we have $V = 0.1473$, for a volume efficiency of 3.52%.

A 3 *dof* Cartesian robot $|\,\sigma(+)\sigma(\times)\sigma\,|$ has a working volume of $(l_T/3)^3 - (l_T/6)^3 = 0.03241$, for a volume efficiency of 0.774%.

The 3 *dof* polar robot $|\,\phi\theta\sigma\,|$ has a working volume between two spheres of radius $\sigma_3 = \bar{\sigma}_3/2$ and $\bar{\sigma}_3$. If the robot is designed for maximum working volume, we set $l_1 = 0$ and $\bar{\sigma}_3 = 1$, so the working volume is $4\pi/3$ −

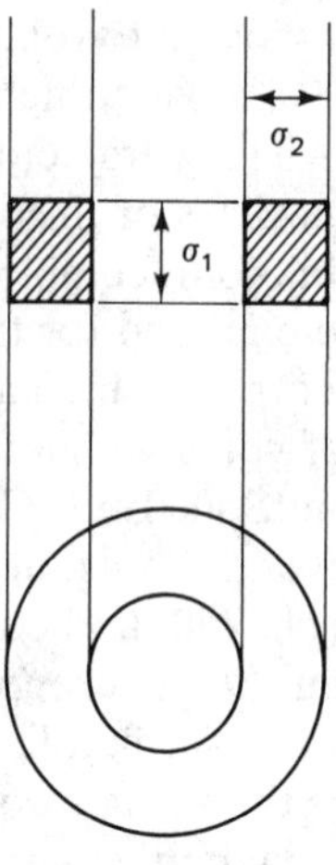

Figure 5.1 Working volume of the $|\phi\sigma{+}\sigma|$ robot

$\pi/6 = 3.665$, for a volume efficiency of 87.5%. If $l_1 = \overline{\sigma}_3$, then the working volume is 3.665/8, for a volume efficiency of 10.94%.

The 3 *dof* SCARA robot $|\theta\theta\sigma|$ with $l_1 = l_2$ has a cylindrical working volume with radius $2l_1$ and length σ_3, where the total link lengths are $2l_1 + 2\sigma_3 = 1$. Thus the volume is $V = 4\pi l_1^2\sigma_3 = 4\pi l_1^2(0.5 - l_1)$ and $dV/dl_1 = 4\pi l_1 - 12\pi l_1^2 = 0$ at a maximum or minimum, so $l_1 = \frac{1}{3}$, and $V = 0.2327$, for a volume efficiency of 5.55%.

The working volume of the robot $|\sigma(+)\sigma\tilde{\psi}|$ has a cross section shown in Figure 5.2 with height $\sigma_2/2$. The working volume of the robot is $V = \sigma_2\{\sigma_1 l_3 + \pi l_3^2\}$, and since the total link lengths are $2\sigma_1 + 2\sigma_2 + l_3$, where σ_1 and σ_2 are the shortest lengths possible for the sliding links, then

$$V = \sigma_2\left\{\frac{l_3}{2} - \sigma_2 l_3 + \left(\pi - \frac{1}{2}\right)l_3^2\right\}$$

$$\frac{\partial V}{\partial \sigma_2} = l_3\left\{\frac{1}{2} - 2\sigma_2 + \left(\pi - \frac{1}{2}\right)l_3\right\} \quad \text{and} \quad \frac{\partial V}{\partial l_3} = \sigma_2\left\{\frac{1}{2} - \sigma_2 + 2\left(\pi - \frac{1}{2}\right)l_3\right\}$$

which must be zero at a maximum or minimum, so $\sigma_2 = \frac{1}{6}$.

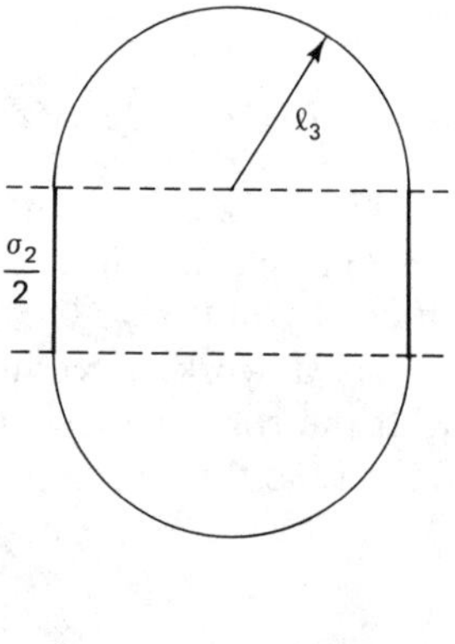

Figure 5.2 Cross section of working volume for $|\sigma{+}\sigma\tilde{\psi}|$

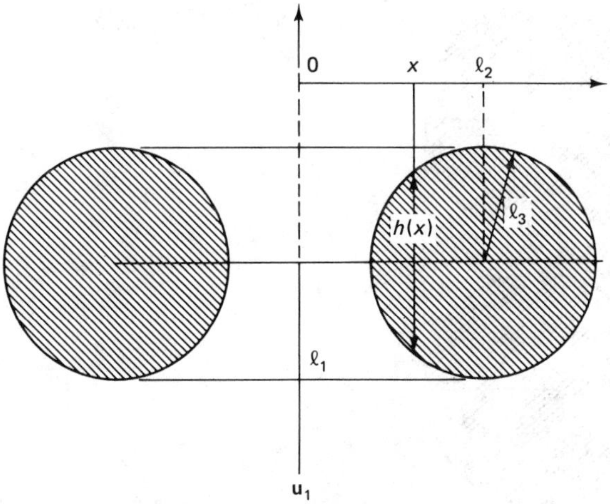

Figure 5.3 The | $\phi\phi\theta$ | robot

The working volume of the robot | $\phi\phi\theta$ | is a doughnut shape with (for $l_3 < l_2$) or without (for $l_3 > l_2$) the hole, where the center of gravity (the center of the hole, if it exists) is at $\mathbf{u}_2$ and the radius of the annular ring is l_3 at a distance l_2 from $\mathbf{u}_2$. The cross section of the doughnut shape is shown in Figure 5.3. Without any restriction from the base and with $l_1 = 0$ to maximize the working volume, the incremental height of cross section covered by the end effector is given by

$$h(x) = \begin{cases} 0, & 0 \le x \le 1 - 2l_3 \\ 2\sqrt{l_3^2 - (x - 1 + l_3)^2}, & 1 - 2l_3 \le x \le 1 \end{cases}$$

The incremental area of rotation associated with $h(x)$ is the surface area $A(x)$ of the cylinder of radius x and height $h(x)$, so

$$A(x) = 2\pi x h(x)$$

The volume of the doughnut is

$$V = \int_0^1 A(x)\,dx = \int_{1-2l_3}^1 4\pi x\sqrt{l_3^2 - (1 - l_3 - x)^2}\,dx$$

To maximize the working volume, we set $l_1 = 0$; to avoid voids in the working volume, we require $l_2 = l_3 = 0.5$, so

$$V = 4\pi \int_0^1 x\sqrt{0.25 - (0.5 - x)^2}\,dx = 4\pi \int_0^1 x\sqrt{x - x^2}\,dx$$

$$= 4\pi \left[\frac{-(x - x^2)^{3/2}}{2}\right]_0^1 + 2\pi \int_0^1 x\sqrt{x - x^2}\,dx$$

$$= 0 + \pi\,[(x - 0.5)x\sqrt{x - x^2} + 0.25\sin^{-1}(2x - 1)]_0^1 = \frac{\pi^2}{4} = 2.467$$

and the volume efficiency is 58.91%.

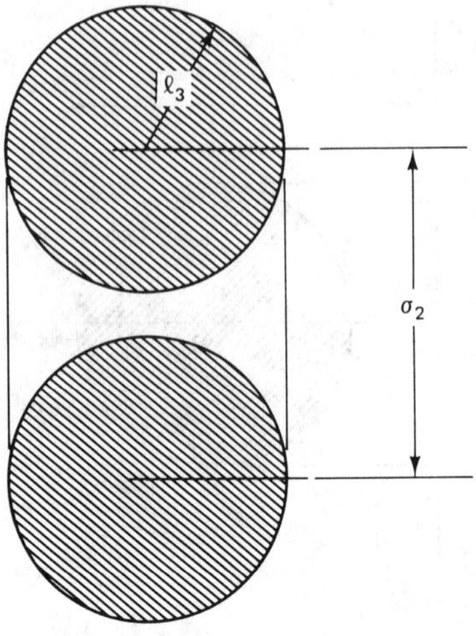

Figure 5.4 The $| \phi\sigma\theta |$ robot

A section of the working volume for the robot $| \phi\sigma\theta |$ is shown in Figure 5.4. There is a large void in the working volume that cannot be removed through choice of link lengths, so this robot should be avoided.

It should be evident that robots with one or more sliding links have smaller working volumes than robots with pitch and/or roll and/or yaw second and third joints. Robots with pitch-roll-yaw joints act somewhat like the human arm and are termed *anthropomorphic*. Anthropomorphic robots have a greater working volume than other robots. As stated previously, working volume is an important determinant in the effectiveness of a robot, but other factors may be more important. *The* most important factor is the ability of the robot to perform the task at hand.

In the preceding analysis we have assumed no limitation on the angular positions possible in the revolute joints. We saw that the revolute robot has the largest working volume, and the Cartesian robot has the smallest. With severe limitation on the angles of the revolute joints, a cross section of the working volumes for the basic robots is shown as the shaded areas in Figure 5.5. The revolute and SCARA robots have the smallest shaded areas. To investigate further, if the 3 *dof* revolute robot has pitch joints with $\pm 90°$ angles of swing, then a cross section of the working volume in $\bar{\mathbf{p}}^2$ is shown in Figure 5.6. It is evident that only with a swing approaching the maximum 360° for all the revolute joints will the working volume of a robot with revolute joints be large compared to a cylindrical or Cartesian robot.

Given the 3 *dof* orientation mechanism $| \lambda_4 l_4 | \theta_5 l_5 | \lambda_6 l_6 |$. It is not necessary for this mechanism to be without voids, since it is added to a placeable robot without voids. Suppose the 3 *dof* placeable robot is $| \lambda_1 l_1 | \theta_2 l_2 | \theta_3 l_3 |$; then the working volume of the 6 *dof* robot is $4\pi(l_2 + l_3 + l_4 + l_5 + l_6)^3/3$. However, this is not the dextrous working volume [Kumar and Waldron 1981; Craig 1986] of the robot; see Section 6.5.

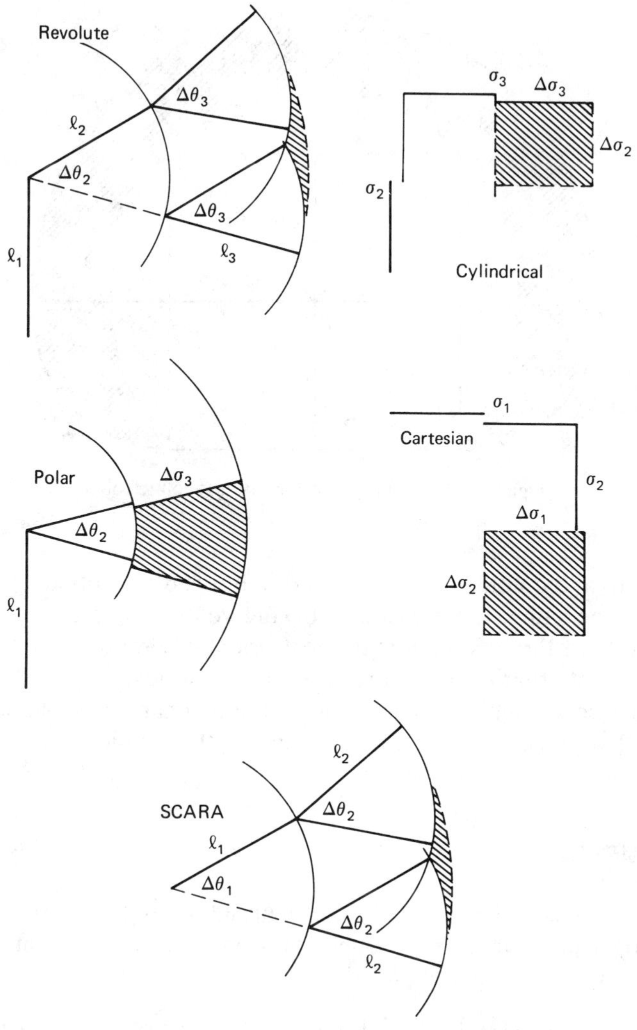

Figure 5.5 Working envelopes of basic robots

Definition. The *dextrous working volume* of a robot is the totality of points that can be reached by the end effector, where at each point the end effector can assume every possible orientation.

For the robot | $\lambda\theta\theta\lambda\theta\lambda$ | the dextrous working volume is $4\pi(l_2 + l_3 + l_4 - l_5 - l_6)^3/3$. To maximize the dextrous working volume [Lai and Lenq 1988], we must minimize lengths l_5 and l_6: this analysis holds true for every orientable robot. Suppose the robot has total link lengths of unity; then we require $l_1 = l_5 = l_6 = 0$ to maximize dextrous volume. In addition, we need

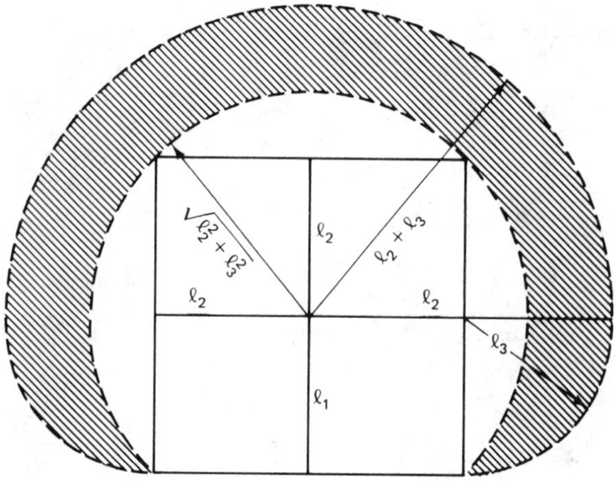

Figure 5.6 Working envelope of a revolute robot

$l_2 = l_3 + l_4$ to avoid voids, so $l_2 = 0.5$ and $l_3 + l_4 = 0.5$. Notice that if the link lengths are the same, the dextrous working volume is zero.

A study of the link lengths in most orientable robots (see Section 6.5) will demonstrate that manufacturers are aware of dextrous working volume and design accordingly. As a final note, the mechanism of the human arm results in a large dextrous working volume with no voids.

5.4 EXAMPLES

1. Assuming the base orientation $\mathbf{p}_1 = [0\ 1\ 0\ 0]$ and $\boldsymbol{\nu} = [-1\quad 0\quad 0]^{\#}$ and base joint at the origin, derive from first principles the plane of motion and joint positions for the 3 *dof* revolute robot $|\ \phi\theta\theta\ |$.

Using equations (3.14) and (3.15) of Chapter 3 we get

$$\mathbf{u}_2 = \begin{bmatrix} 0 \\ 0 \\ l_1 \\ 1 \end{bmatrix}, \quad \begin{bmatrix} a_2 \\ b_2 \\ c_2 \end{bmatrix} = \begin{bmatrix} \sin\phi_1 \\ \cos\phi_1 \\ 0 \end{bmatrix}$$

From equation (3.1),

$$\mathbf{u}_3 = \begin{bmatrix} 0 \\ 0 \\ l_1 \\ 1 \end{bmatrix} + \begin{bmatrix} \cos\theta_2 & -c_2\sin\theta_2 & b_2\sin\theta_2 & 0 \\ c_2\sin\theta_2 & \cos\theta_2 & -a_2\sin\theta_2 & 0 \\ -b_2\sin\theta_2 & a_2\sin\theta_2 & \cos\theta_2 & 0 \\ 0 & 0 & 0 & 1 \end{bmatrix} \begin{bmatrix} 0 \\ 0 \\ l_2 \\ 0 \end{bmatrix}$$

$$= \begin{bmatrix} 0 \\ 0 \\ l_1 \\ 1 \end{bmatrix} + \begin{bmatrix} \cos\theta_2 & 0 & \cos\phi_1 \sin\theta_2 & 0 \\ 0 & \cos\theta_2 & -\sin\phi_1 \sin\theta_2 & 0 \\ -\cos\phi_1 \sin\theta_2 & \sin\phi_1 \sin\theta_2 & \cos\theta_2 & 0 \\ 0 & 0 & 0 & 1 \end{bmatrix} \begin{bmatrix} 0 \\ 0 \\ l_2 \\ 0 \end{bmatrix}$$

$$= \begin{bmatrix} l_2 \cos\phi_1 \sin\theta_2 \\ -l_2 \sin\phi_1 \sin\theta_2 \\ l_1 + l_2 \cos\theta_2 \\ 1 \end{bmatrix}$$

$$\mathbf{u}_4 = \begin{bmatrix} l_2 \cos\phi_1 \sin\theta_2 \\ -l_2 \sin\phi_1 \sin\theta_2 \\ l_1 + l_2 \cos\theta_2 \\ 1 \end{bmatrix}$$

$$+ \begin{bmatrix} \cos\theta_3 & 0 & \cos\phi_1 \sin\theta_3 & 0 \\ 0 & \cos\theta_3 & -\sin\phi_1 \sin\theta_3 & 0 \\ -\cos\phi_1 \sin\theta_3 & \sin\phi_1 \sin\theta_3 & \cos\theta_3 & 0 \\ 0 & 0 & 0 & 1 \end{bmatrix} \begin{bmatrix} l_3\cos\phi_1 \sin\theta_2 \\ -l_3\sin\phi_1 \sin\theta_2 \\ l_3\cos\theta_2 \\ 0 \end{bmatrix}$$

$$= \begin{bmatrix} l_2 \cos\phi_1 \sin\theta_2 + l_3 \cos\phi_1 \sin\theta_2 \cos\theta_3 + l_3 \cos\phi_1 \cos\theta_2 \sin\theta_3 \\ -l_2 \sin\phi_1 \sin\theta_2 - l_3 \sin\phi_1 \sin\theta_2 \cos\theta_3 - l_3 \sin\phi_1 \cos\theta_2 \sin\theta_3 \\ l_1 + l_2 \cos\theta_2 - l_3 \cos^2\phi_1 \sin\theta_2 \sin\theta_3 - l_3 \sin^2\phi_1 \sin\theta_2 \sin\theta_3 + l_3 \cos\theta_2 \cos\theta_3 \\ 1 \end{bmatrix}$$

so

$$\mathbf{u}_4 = \begin{bmatrix} \cos\phi_1\{l_2 \sin\theta_2 + l_3 \sin(\theta_2 + \theta_3)\} \\ -\sin\phi_1\{l_2 \sin\theta_2 + l_3 \sin(\theta_2 + \theta_3)\} \\ l_1 + l_2 \cos\theta_2 + l_3 \cos(\theta_2 + \theta_3) \\ 1 \end{bmatrix}$$

---

2. Determine the end effector location for the five basic types of 3 *dof* robot (the Cartesian is $|\ \sigma + \sigma \times \sigma\ |$) when all revolute joints are at datum and all link lengths (including the sliding joints) have length of 1 m, and the base orientation and position are given by $\mathbf{p}_1 = [0 \quad 1 \quad 0 \quad 0]$, $\mathbf{v} = [0 \quad 0 \quad 1]^\#$, and $\mathbf{u}_1 = [0 \quad 0 \quad 0 \quad 1]^\#$.

$|\ \lambda\theta\theta\ |$: from equation (5.7), $\mathbf{u}_4 = \begin{bmatrix} 0 \\ 0 \\ 3 \\ 1 \end{bmatrix}$.

$|\ \lambda\sigma{+}\sigma\ |$: from equation (5.17), $\mathbf{u}_3 = \begin{bmatrix} 0 \\ 1 \\ 0 \\ 1 \end{bmatrix} + \begin{bmatrix} -1 \\ 0 \\ 0 \\ 0 \end{bmatrix} = \begin{bmatrix} -1 \\ 1 \\ 0 \\ 1 \end{bmatrix}$.

$| \sigma + \sigma \times \sigma |$: from section 5.2.3, $\mathbf{u}_2 = \begin{bmatrix} 0 \\ 0 \\ 1 \\ 1 \end{bmatrix}$, $\mathbf{u}_3 = \begin{bmatrix} 0+1 \\ 1+0 \\ 1+0 \\ 1 \end{bmatrix}$,

$$\mathbf{u}_4 = \begin{bmatrix} 1-0 \\ 1-0 \\ 0-1 \\ 1 \end{bmatrix} = \begin{bmatrix} -1 \\ -1 \\ 1 \\ 1 \end{bmatrix}.$$

$| \phi\theta\sigma |$: from equation (5.21), $\mathbf{u}_3 = \begin{bmatrix} 0-0 \\ 2-0 \\ 0-0 \\ 1 \end{bmatrix} = \begin{bmatrix} 0 \\ 2 \\ 0 \\ 1 \end{bmatrix}$.

$| \theta\theta \times \sigma |$: from equation (5.26), $\mathbf{u}_4 = \begin{bmatrix} 0-0-0 \\ 2-0-0 \\ 0-0-1 \\ 1 \end{bmatrix} = \begin{bmatrix} 0 \\ 2 \\ -1 \\ 1 \end{bmatrix}$.

---

3. Determine the end effector location for the five basic types of 3 *dof* robot (the Cartesian is considered to have a third orthogonal sliding joint) when all revolute joints are at datum plus $22.5i$ with sliding link lengths at $0.5 + 0.125i$, $i = 0, 1, 2, 3, 4$. The links with fixed lengths are 1 m long, and the base orientation and position is given by $\mathbf{p}_1 = [0 \ \ 0 \ \ 1 \ \ 0]$, $\mathbf{v} = [0 \ \ 1 \ \ 0]^\#$, and $\mathbf{u}_1 = [0 \ \ 0 \ \ 0 \ \ 1]^\#$. Determine the distance of the end effector from its base and the change in this distance for the given values of joint status.

From equation (5.5), the end effector position for the 3 *dof* revolute robot is given by

$$\mathbf{u}_4 = \begin{bmatrix} -\cos \lambda_1\{l_2 \sin \theta_2 + l_3 \sin(\theta_2 + \theta_3)\} \\ l_1 + l_2 \cos \theta_2 + l_3 \cos(\theta_2 + \theta_3) \\ \sin \lambda_1\{l_2 \sin \theta_2 + l_3 \sin(\theta_2 + \theta_3)\} \\ 1 \end{bmatrix}$$

From equation (5.17), the end effector position for the 3 *dof* cylindrical robot is given by

$$\mathbf{u}_3 = \begin{bmatrix} \sigma_3 \cos \lambda_1 \\ \sigma_2 \\ -\sigma_3 \sin \lambda_1 \\ 1 \end{bmatrix}$$

From equation (5.19), the end effector position for the 3 *dof* Cartesian robot is given by

$$\mathbf{u}_4 = \begin{bmatrix} \sigma_2 \\ \sigma_1 \\ -\sigma_3 \\ 1 \end{bmatrix}$$

From equation (5.21), the end effector position for the 3 *dof* polar robot is given by

$$\mathbf{u}_4 = \begin{bmatrix} -\sigma_3 \cos \lambda_1 \sin \theta_2 \\ l_1 + \sigma_3 \cos \theta_2 \\ \sigma_3 \sin \lambda_1 \sin \theta_2 \\ 1 \end{bmatrix}$$

From equation (5.26), the end effector position for the 3 *dof* SCARA robot is given by

$$\mathbf{u}_4 = \begin{bmatrix} -l_1 \sin \theta_1 - l_2 \sin (\theta_1 + \theta_2) \\ l_1 \cos \theta_1 + l_2 \cos (\theta_1 + \theta_2) \\ -\sigma_3 \\ 1 \end{bmatrix}$$

Then inserting the joint status and presenting the results in tabular form:

θ	$\cos\theta$	$\sin\theta$	$\sin\theta + \sin 2\theta$	$\cos\theta + \cos 2\theta$	σ	Rev $\mathbf{u}_4$	Cylind $\mathbf{u}_3$	Cart $\mathbf{u}_4$	Polar $\mathbf{u}_3$	Scara $\mathbf{u}_4$
0	1	0	0	2	0.5	0 3 0	0.5 0.5 0	0.5 0.5 −0.5	0 1.5 0	0 2 −0.5
22.5	0.924	0.383	1.090	1.631	0.625	1.007 0.417 2.631	0.577 0.625 −0.239	0.625 0.625 −0.625	−0.221 1.578 0.092	−1.090 1.631 −0.625
45	0.707	0.707	1.707	0.707	0.75	1.207 1.707 1.206	0.520 0.750 −0.530	0.75 0.75 −0.75	−0.375 1.530 0.375	−1.707 0.707 −0.750
67.5	0.383	0.924	1.631	−0.324	0.875	0.624 0.676 1.507	0.335 0.875 −0.809	0.825 0.825 −0.825	−0.310 1.316 0.704	−1.631 −0.324 −0.875
90	0	1	1	−1	1	0 −1 1	0 −1 −1	1 1 −1	0 1 1	−1 −1 −1
					Dmax Dmin	3 $\sqrt{2}$	$\sqrt{2}$ 0.707	$\sqrt{3}$ $\sqrt{3}/2$	1.62 $\sqrt{2}$	2.06 $\sqrt{3}$
					ΔD	1.586	$\sqrt{2}$	$\sqrt{3}/2$	0.206	0.328

where Dmax is the maximum distance of the end effector from the base for the states chosen, Dmin is the minimum distance, and ΔD is the difference in distances.

---

4. Determine the FKS for the cylindrical robot $|\,\psi\sigma\sigma\,|$ in terms of the general base orientation. Simplify the results obtained for the specific base orientation $\mathbf{p}_1 = [0 \quad 0 \quad 1 \quad 0]$ and $\mathbf{v} = [0 \quad 1 \quad 0]^{\#}$.

$$\mathbf{u}_2 = \begin{bmatrix} \sigma_2 a_1 \\ \sigma_2 b_1 \\ \sigma_2 c_1 \\ 1 \end{bmatrix},$$

and $\mathbf{u}_3$ is obtained by adding $\mathbf{u}_2$ to σ_3 times the cross product of vectors $[a_2 \quad b_2 \quad c_2]^\#$ and $(\mathbf{u}_2 - \mathbf{u}_1)/l_1 = [a_1 \quad b_1 \quad c_1]^\#$, so

$$\mathbf{u}_3 = \begin{bmatrix} \sigma_2 a_1 \\ \sigma_2 b_1 \\ \sigma_2 c_1 \\ 1 \end{bmatrix} + \sigma_3 \begin{bmatrix} b_2 c_1 - c_2 b_1 \\ c_2 a_1 - a_2 c_1 \\ a_2 b_1 - b_2 a_1 \\ 0 \end{bmatrix}$$

For the specific base orientation $\mathbf{p}_1 = [0 \quad 0 \quad 1 \quad 0]$ and $\mathbf{v} = [0 \quad 1 \quad 0]^\#$,

$$\begin{bmatrix} a_2 \\ b_2 \\ c_2 \end{bmatrix} = \begin{bmatrix} \sin\psi_1 \\ \cos\psi_1 \\ 0 \end{bmatrix} \quad \text{and} \quad \mathbf{u}_3 = \begin{bmatrix} \sigma_3 \cos\psi_1 \\ -\sigma_3 \sin\psi_1 \\ \sigma_2 \\ 1 \end{bmatrix}$$

---

5. Given the 3 *dof* mechanism $\begin{matrix} & \bar{\mathbf{p}}_2^0 & & \bar{\mathbf{p}}_3^0 & & \bar{\mathbf{p}}_4^0 & \\ | & \phi_1^1 & | & \phi_2^1 & | & \phi_3^1 & | \\ \mathbf{u}_1 & & \mathbf{u}_2 & & \mathbf{u}_3 & & \mathbf{u}_4 \end{matrix}$

If $\mathbf{u}_1$ is the origin and the base orientation is given by $\mathbf{p}_1 = [0 \quad 1 \quad 0 \quad 0]$ and $\boldsymbol{\nu} = [0 \quad 0 \quad 1]^\#$, show the motion from datum of $\mathbf{u}_4$ and $\mathbf{p}_4$ as each joint angle increases to 90°.

In tabular form the results are as follows:

ϕ_1	ϕ_2	ϕ_3	x_4	y_4	z_4	a_4	b_4	c_4	d_4
0	0	0	0	0	−1	0	1	0	0
0	0	30	0	0	−1	0	0.866	−0.5	−0.5
0	0	60	0	0	−1	0	0.5	−0.866	−0.866
0	0	90	0	0	−1	0	0	−1	−1
0	30	0	0.134	0.5	−1	0.5	0.866	0	−0.5
0	60	0	0.5	0.866	−1	0.866	0.5	0	−0.866
0	90	0	1	1	−1	1	0	0	−1
30	0	0	0	0.5	−0.866	0	0.866	0.5	0
60	0	0	0	0.866	−0.5	0	0.5	0.866	0
90	0	0	0	1	0	0	0	1	0

---

6. The sliding links in a 4 *dof* cylindrical robot $|\,\phi\sigma{+}\sigma\theta\,|$ have a 2 : 1 maximum to minimum length. If the sum of the maximum lengths of the three links is 1 m,

design the robot so the end effector covers the maximum volume in three dimensional space. Assume that the robot is mounted on an infinite horizontal base, the first link is vertical, and there are no angle limitations. The robot must be able to reach all points within its working envelope. What is the volume covered?

In order for the end effector to reach all points within the robot's working envelope, we require $\sigma_3 = l_4$. For maximum reach—and, hence, overall working volume—we require l_4 to be as large as possible. A vertical cross section of the area encompassed by the end effector of the robot is shown in Figure 5.7. As a function of x, the height $h(x)$ of the working volume V is

$$V = \int_0^{3l_4} 4\pi x h(x)\, dx$$

$$= \int_0^{l_4} 4\pi x \left(\frac{\sigma_2}{2} + l_4 - \sqrt{l_4^2 - x^2}\right) dx + \int_{l_4}^{2l_4} 4\pi x \left(\frac{\sigma_2}{2} + l_4\right) dx$$

$$+ \int_{2l_4}^{3l_4} 4\pi x \left(\frac{\sigma_2}{2} + \sqrt{l_4^2 - (x - 2l_4)^2}\right) dx$$

$$= 4\pi \left\{\frac{3\sigma_2 l_4^2}{4} - \left[\frac{1}{3}\sqrt{(l_4^2 - x^2)^3}\right]_0^{l_4} + \frac{3l_4^2}{2}\left(\frac{\sigma_2}{2} + l_4\right)\right.$$

$$\left. + \frac{5\sigma_2 l_4^2}{4} + \int_{2l_4}^{3l_4} x\sqrt{l_4^2 - (x - 2l_4)^2}\right\} dx$$

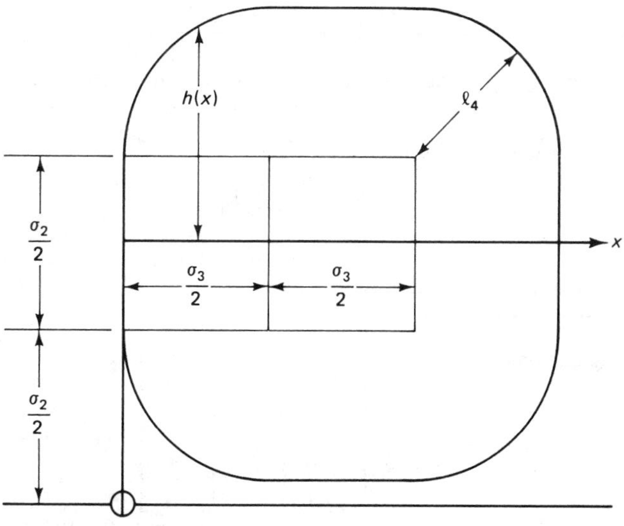

Figure 5.7 Cross section of working volume for $|\ \phi\sigma + \sigma\theta\ |$

EXERCISES FOR CHAPTER 5

1. The first two joints of a robot are at $\begin{bmatrix} 2 \\ 2 \\ 2 \\ 1 \end{bmatrix}$ and $\begin{bmatrix} 3 \\ 3 \\ 2 \\ 1 \end{bmatrix}$ and the plane of motion is $\mathbf{p} = [1/\sqrt{2} \quad -1/\sqrt{2} \quad 0 \quad 0]$. Assuming all joints are at datum, what is the base orientation $\boldsymbol{\nu}$ and $\mathbf{q}$ for every possible joint type?

---

2. The base orientation of a robot is given by $\mathbf{p}_1$ and $\boldsymbol{\nu}$. The joint locations of the n *dof* robot are $\mathbf{u}_i$, $i = 1, 2, \ldots, n + 1$. Replace the base joint with a different type of revolute joint and reorient the base such that the joint locations remain the same for the base joint at datum when the base joint is as follows.

(a) Pitch
(b) Roll
(c) Yaw
(d) Cylindrical $\tilde{\phi}$

---

3. Obtain the FKS for the robot $\underset{\mathbf{u}_1}{\overset{\mathbf{p}_1 \bar{\mathbf{p}}_2^2}{|\, \phi_1\sigma_2}} \underset{\mathbf{u}_2}{|}\, \underset{\mathbf{u}_3}{\theta_3 l_3} \,|$

---

4. Obtain the FKS for the robot $\underset{\mathbf{u}_1}{\overset{\mathbf{p}_1 \bar{\mathbf{p}}_2^2}{|\, \lambda_1 l_1}} \underset{\mathbf{u}_2}{|\, \theta_2\sigma_3} \underset{\mathbf{u}_3}{\overset{\bar{\mathbf{p}}_3^0}{|\, \phi_4 l_4}} \,|$

---

5. Obtain the FKS equations in general form for the mechanism $|\, \lambda\tau\theta \,|$.

---

6. Determine the working volume of the robot

$$\left|\lambda_1 \begin{Bmatrix} 90 \\ -90 \end{Bmatrix} 1 \,\middle|\, \theta_2 \begin{Bmatrix} 90 \\ -90 \end{Bmatrix} 1 \,\middle|\, \theta_3 \begin{Bmatrix} 90 \\ -90 \end{Bmatrix} 1 \right|$$

---

7. Repeat problem 5 with angle limitations of $\pm 45°$.

---

8. The triangular base of the manipulator $|\, \theta\phi \,|$ is bolted to the floor with screws located at corner points $\begin{bmatrix} 1 \\ -1 \\ 0 \\ 1 \end{bmatrix}$, $\begin{bmatrix} 1 \\ 1 \\ 0 \\ 1 \end{bmatrix}$, and $\begin{bmatrix} -1 \\ 1 \\ 0 \\ 1 \end{bmatrix}$. The pitch joint is at the center of the base and the first plane of motion passes through the first corner point of the base.

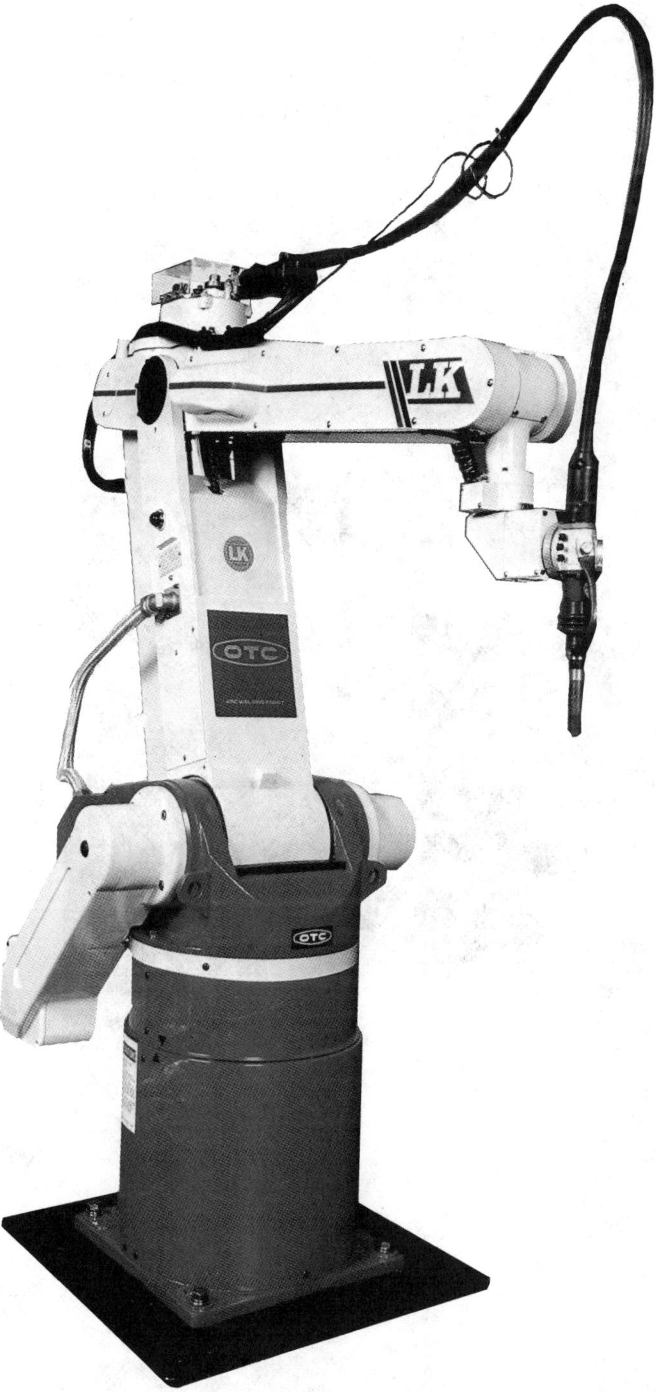

Figure 5.7a The Daihen LK

Figure 5.7b The IBM 7545

Manipulator Performance Specifications

Coordinate system	Cartesian coordinate system
Degrees of freedom	4
Maximum payload (Includes end-of-arm tooling)	5 kg (11 lb)
Repeatability	±0.05 mm (±0.002 in.)

Axis movement	Range	Max. Load	Speed mm/sec (in./sec.) Low (±5%)	Med (±5%)	High (±5%)
θ_1 axis (Arm swivel)	±120°	1 kg (2.2 lb)	690 mm/sec (27 in./sec)	1070 mm/sec (42 in./sec)	1350 mm/sec (53 in./sec)
		3 kg (6.6 lb)	690 mm/sec (27 in./sec)	1070 mm/sec (42 in./sec)	–
		5 kg (11 lb)	690 mm/sec (27 in./sec)	–	–
θ_2 axis (Arm swivel)	±125°	1 kg (2.2 lb)	690 mm/sec (27 in./sec)	1070 mm/sec (42 in./sec)	1350 mm/sec (53 in./sec)
		3 kg (6.6 lb)	690 mm/sec (27 in./sec)	1070 mm/sec (42 in./sec)	–
		5 kg (11 lb)	690 mm/sec (27 in./sec)	–	–
Roll-axis (Rotation)	±180°	5 kg (11 lb) Centered	Speed: Low 240°/sec	Med. 370°/sec	High 480°/sec
			Rotating torque = 70 kg-cm (5.1 ft-lb)		
Z-axis (Up/Down)	250 mm (9.8 in.)		Low	Med.	High
		1 kg (2.2 lb)	140 mm/sec (5.5 in./sec)	220 mm/sec (8.6 in./sec)	280 mm/sec (11.0 in./sec)
		3 kg (6.6 lb)	140 mm/sec (5.5 in./sec)	220 mm/sec (8.6 in./sec)	–
		5 kg (11 lb)	140 mm/sec (5.5 in./sec)	–	–

Lateral velocity is restricted by Z-axis position and payload.

Working area

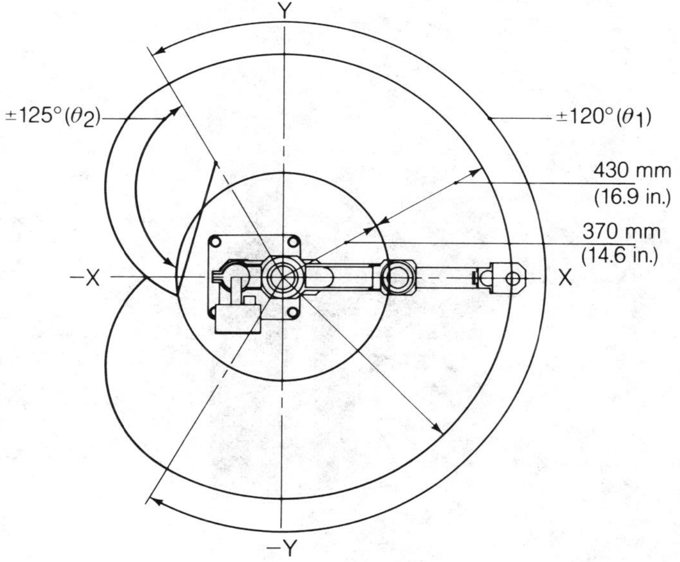

Figure 5.7b *(continued)*

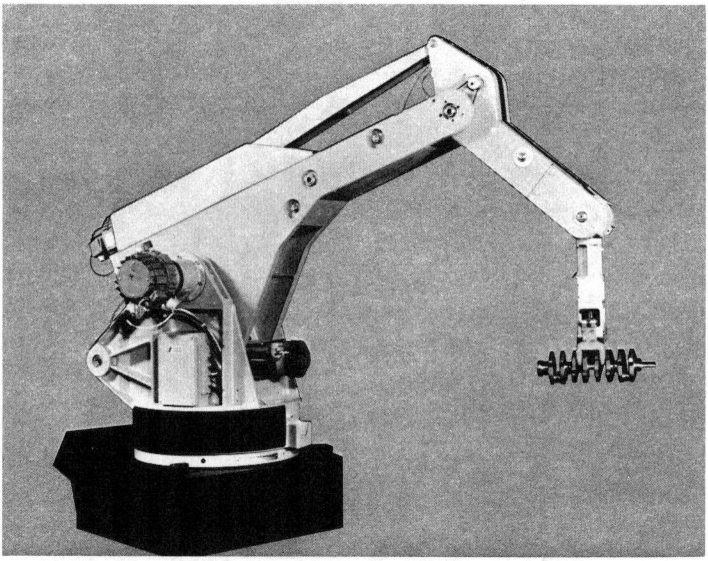

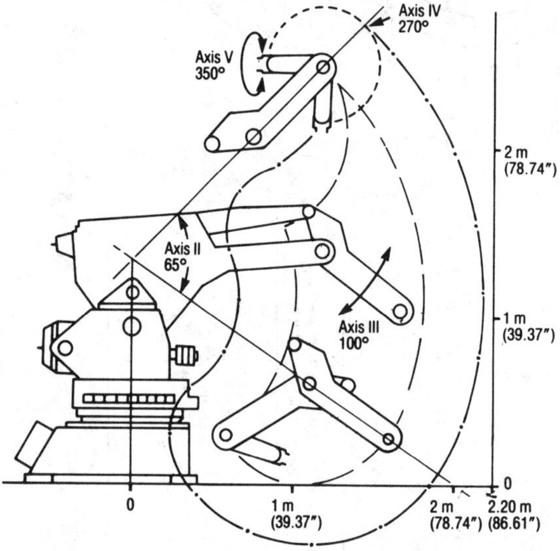

Axes		Operating area	Operating speed
Rotation	Axis I	320°	80°/sec
Vertical swivel	Axis II	65°	30°/sec
Vertical bend	Axis III	100°	50°/sec
Wrist swivel	Axis IV	270°	120°/sec
Wrist rotation	Axis V	350°	120°/sec
Transversal location unit	Axis VII	600 - 2400 (23.6 - 94.5 in.)	450 mm/sec. (17.7 in./sec)

Figure 5.7c The General Electric P6

(a) Determine the base orientation and position.
(b) Determine the joint locations for the manipulator if $l_1 = 0.5$ m and $l_2 = 0.3$ m and the joint angles are at datum.

---

9. Obtain the FKS for the mechanisms $|\,\phi\theta\theta\sigma\,|$ and $|\,\lambda\theta\theta\sigma\,|$ and compare the results.

---

10. Obtain the FKS for the mechanism $|\,\phi\theta\lambda\,|$.

---

11. Determine the position of the end effector of the Microbot Alpha II (see Appendix C) when all joints are at datum. If the base joint λ_1 is slewed from 0 to π radians, determine the location of the end effector as a function of λ_1, and the plane containing these end effector positions. Why is this plane not the same as a plane of motion of the robot?

---

12. For the Prab model 4200 (a polar robot described in Appendix C) determine each of the following.
(a) The working envelope of the robot
(b) The location of the end effector of the robot if the base orientation and position is

$$\mathbf{p}_1 = [0 \quad 1 \quad 0 \quad 0], \qquad \boldsymbol{\nu} = \begin{bmatrix} 1 \\ 0 \\ 0 \end{bmatrix}, \quad \text{and} \quad \mathbf{u}_1 = \begin{bmatrix} 0 \\ 0 \\ 0 \\ 1 \end{bmatrix}$$

and the joints are at datum with the sliding link at its maximum stretch.

---

13. Draw the schematic and identify all the joints for the robots shown in the diagrams on pages 159–162:
(a) Daihen LK
(b) IBM 7545
(c) General Electric P6

---

14. Determine the working volume of the robot

$$\left|\lambda_1 \begin{Bmatrix} 90 \\ -90 \end{Bmatrix} 1 \,\middle|\, \theta_2 \begin{Bmatrix} 90 \\ -90 \end{Bmatrix} \begin{matrix} 0.5 \\ 1 \end{matrix}\right|$$

---

15. For the manipulator $|\,\lambda_1\ 1\,|\,\theta_2\ 1\,|\,\theta_3\ 1\,|\,\lambda_4\ 0\,|\,\theta_5\ 1\,|$ the limitations on the joint angles are $\pm 90°$.

Assuming $\mathbf{p}_1 = [1 \quad 0 \quad 0 \quad 0]$, $\boldsymbol{\nu} = \begin{bmatrix} 0 \\ 0 \\ 1 \end{bmatrix}$, and $\mathbf{u}_1 = \begin{bmatrix} 0 \\ 0 \\ 0 \\ 1 \end{bmatrix}$, find the projection of the working envelope onto each plane.
(a) The xz-plane
(b) The xy-plane

CHAPTER SIX

Inverse Kinematic Solutions

The ability to control a robotic system remotely in real time hinges on the inverse kinematic solution. We saw in Chapter 5 that the forward kinematic solution can be obtained for any open kinematic chain. In this chapter we study the reverse problem—namely, given the position space of the robot, can the joint space be determined? This is known as solving the inverse kinematic problem, or obtaining the inverse kinematic solution (IKS). The IKS will not exist if

1. The order of the joint space is less than the order of the position space, or
2. The position space is chosen outside the working envelope of the robot.

If the robot is capable of satisfying the required position space, an IKS is always possible [Brady et al. 1983; Craig 1986; Ersü and Nungesser 1984; Gorla and Renaud 1984; Paul, Shimano, and Mayer 1981a; Paul 1981b; Paul and Zhang 1986; Stanisic and Pennock 1985; Sugimoto and Matsumoto 1984; Tsai and Morgan 1984; Wolovich 1987]. However, this IKS may not be obtainable algebraically, which means a closed-form solution does not exist. If a numerical, iterative solution is required, we say the IKS is in open form and call it OIKS, as opposed to the closed form, CIKS. OIKSs almost invariably preclude the effective real-time, remote control of robotic systems. Robots with OIKSs should be avoided.

Attempts to define the conditions for the existence of a CIKS for a particular robot have provided a fairly weak set of sufficient conditions [Pieper 1968; Pieper and Roth 1969]. A somewhat stronger sufficient condition developed here is summarized in Theorem 6.1. The point-plane method finds the CIKS in almost every case studied, and this includes all basic configurations with up to 6 *dof*. Thus, if it is true that all kinematically simple robots capable of satisfying their position spaces have CIKS, then the ability to satisfy the position space becomes the necessary and sufficient condition for a CIKS.

6.1 CURVED LINKS

The presence of curved links (see Section 3.5) in a robot can increase the number of arithmetic operations required in a forward kinematic solution without increasing the conceptual complexity. However, the inverse kinematic solution of a robot can be greatly complicated if curved links are present. Suppose a curved link is contained within a chain of links that defines a plane with area. No longer can that plane be drawn flat on a piece of paper, a requirement that guarantees the feasibility of a geometric solution. The problem is evident in the following example.

Consider the 3 *dof* revolute robot shown in Figure 6.1(a), which contains a simple curved link and can be specified by

$$\overbrace{\left| \phi_1 l_1 \right|}^{\bar{\mathbf{p}}_{2a}} \; \theta_2 l_{a2} \begin{bmatrix} \rho_{a2} \\ \rho_{b2} \end{bmatrix} l_{b2} \; \overbrace{\left| \theta_3 l_3 \right|}^{\bar{\mathbf{p}}_{2b}}$$

where the geometry of the curved link is shown in Figure 6.1(b). This is a simple form of curved link, which can be modeled by two straight-line segments with a rotation in two orthogonal planes, the first the plane of motion of the downlink at the connection between the straight-line segments. The line of intersection of the two planes $\bar{\mathbf{p}}_{2a}$ and $\bar{\mathbf{p}}_{2b}$ passes through the fixed joint of the curved link. Unfortunately, the IKS as presented is open. An iterative solution could proceed as follows:

1. Choose ϕ_1 to determine $\bar{\mathbf{p}}_{2a}$.
2. Choose θ_2 and so find $\bar{\mathbf{p}}_{2b}$ and $\mathbf{u}_3$.
3. Modify θ_2 to solve $|\mathbf{u}_3 - \mathbf{u}_4| - l_3 = 0$ by a standard root-finding method.
4. With θ_2 chosen as in (2), calculate $\mathbf{u}_4'$ (not the true $\mathbf{u}_4$), and form the error function $E(\phi_1) = |\mathbf{u}_4' - \mathbf{u}_4|$.
5. Use a root-finding method to find the ϕ_1 that satisfies $E(\phi_1) = 0$.

However, if the curved link in the robot just considered were straightened, the robot would become the 3 *dof* revolute robot for which the IKS is easy to

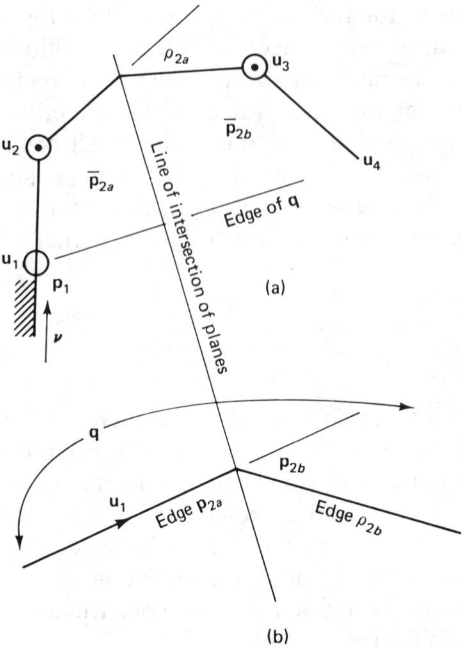

Figure 6.1 A 3 *dof* robot with a curved link

obtain. This suggests that the method of solution of robots with curved links is somehow to straighten the curved links and relocate joints and reorient links. Suppose with the robot in its datum position we perform a forward kinematic solution, starting with the base information $\mathbf{u}_1$ and $\mathbf{p}_1$, to find the position space, joint locations, and axes of the revolute joints. The axis of a revolute joint is the direction of the bearing pin of the joint, so for a pitch joint moving in plane $\mathbf{p} = [a \quad b \quad c \quad d]$, the axis is $[a \quad b \quad c]^{\#}$. We could straighten the curved link(s) with respect to the uplink plane $\bar{\mathbf{p}}_{2b}$ as shown in Figure 6.2(a). This replaces $\mathbf{u}_2$ with $\mathbf{u}_2'$. We can then replace the in-plane curved link, as shown in Figure 6.2(b), which requires a change in datum for θ_2 of α_2 and θ_3 of α_3. Unfortunately, the motion of the roll joint will now be incorrect in general; the axis of rotation given by the original location and direction of l_1 cannot be altered without affecting the motion of the robot. Thus, we have no simple way of treating a curved link. However, since all common robots have no curved links, this presents no problem. We will consider curved links no further.

6.2 IKS DETERMINED FROM THE PLANES OF MOTION

We can determine the form of the IKS from the planes of motion. Suppose the planes of motion are known. The relationships between planes are the lines of intersection of adjacent planes. If a point $\mathbf{u}_i$ on $\mathbf{p}_j$ is known, we have

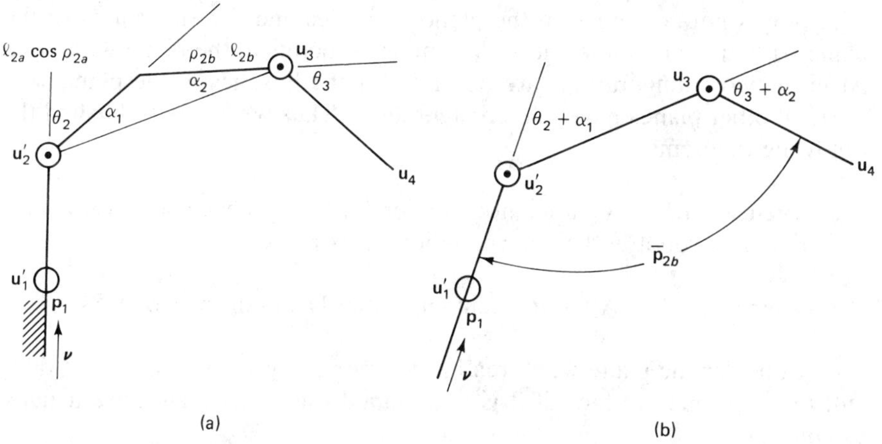

Figure 6.2 A robot reoriented to remove a curved link

the situation shown in Figure 6.3(a). Three adjacent planes of motion have the geometry shown in Figure 6.3(b). Since we know the location of the base (end effector), we have a relationship between this point and the line of intersection of the plane $\mathbf{p}_j$ and the adjacent plane. Starting from one of the known points $\mathbf{u}_i$, suppose $\mathbf{p}_j$ is $\mathbf{p}^0$ or $\mathbf{p}^1$; then the point of intersection $\mathbf{u}_{i+1}$ of the link joining $\mathbf{u}_i$ to the next joint on the line of intersection is known. We use $\mathbf{u}_{i+1}$ as known point on the next plane to find $\mathbf{u}_{i+2}$, and so on. We can start at either end to perform this analysis.

This analysis fails if $\mathbf{p}_j$ is $\mathbf{p}^2$, $\mathbf{p}^3$, and the like. However, if we are analyzing a 6 *dof*, orientable robot, from the analysis of Section 4.4 we know that at most one plane with area exists, and we can perform the analysis from both ends of the robot to determine the joint points on the lines of intersection on either side of this plane with area. The known locations of all

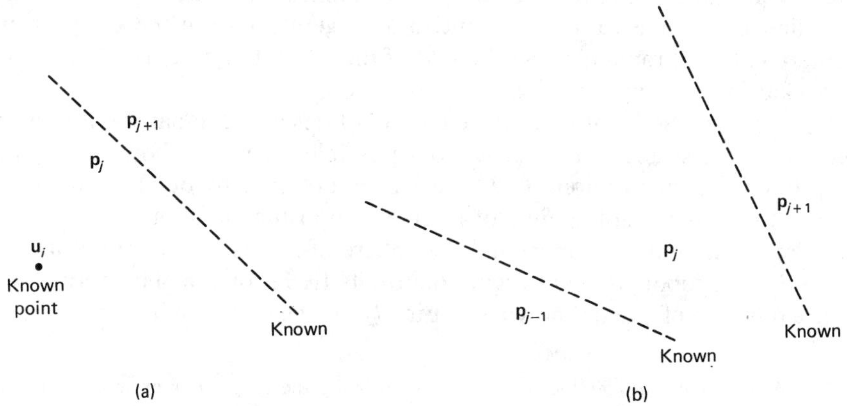

Figure 6.3 Known point and lines of intersection

joint points not contained in the plane with area and the end points of the plane with area are sufficient to determine all points in the plane with area. At most two configurations are possible for the linkages in the plane with area; all other planes have one configuration. Thus we have established the following theorems.

Theorem 6.1. A sufficient condition for the existence of a CIKS of a 6 *dof* orientable robot is the determination of its planes of motion.

Theorem 6.2. A 6 *dof* orientable robot has at most two IKSs.[1]

Consider the plane with area in the prior analysis. It is the most difficult to determine. In fact, if it is determined, the other planes are usually established.

We now consider the IKS of the complete class of 3 *dof* nonstiff, placeable robots $\bar{\mathbf{p}}_2^2$, $\bar{\mathbf{p}}_2^0\mathbf{p}_3^2$, $\mathbf{p}_1^1\mathbf{p}_2^2$ and $\mathbf{p}_1^2\mathbf{p}_2^1$. The base orientation and position give $\mathbf{p}_1$, $\boldsymbol{\nu}$, and $\mathbf{u}_1$. The status of the base joint (λ, ψ, ϕ, or $\tilde{\phi}$) or some other independent piece of information establishes the first plane of motion, $\bar{\mathbf{p}}_2^2$ or $\bar{\mathbf{p}}_2^0$. Placeability gives the end effector position $\mathbf{u}_e$. If a second plane does not exist, which occurs when the robot is specified by $\bar{\mathbf{p}}_2^2$, $\mathbf{u}_e$ provides the independent piece of information to complete the specification of $\bar{\mathbf{p}}_2^2$ and the robot $\bar{\mathbf{p}}^2$ has CIKS. For the robot $\bar{\mathbf{p}}_2^0\mathbf{p}_3^2$, the joint at known point $\mathbf{u}_2$ must be yaw or one of the right-angle bends ($\times$) or ($\div$) before or after a sliding link. In any of these cases with the given end effector position the plane $\mathbf{p}_3^2$ is established. If the first joint is yaw, crank $\tilde{\psi}$ or sliding, from equations (3.7), (3.11), or (3.19), we see that the second plane of motion $\mathbf{p}_2^2$ or $\mathbf{p}_2^1$ is readily obtained.

The reader will find Table 3.1 useful in the remainder of this section.

Consider the nonstiff, placeable robot $|\ \theta_1\tilde{\psi}_2\theta_3\ |$ with planes of motion $\mathbf{p}_1^1\mathbf{p}_2^2$ shown in Figure 6.4(a). The point of intersection of the line in $\mathbf{p}_2$ orthogonal to $\mathbf{p}_1$ and passing through $\mathbf{u}_4$ can be found from equation (1.15). The fold line is the line from this point to a tangent point on the circle in $\mathbf{p}_1$ centered at $\mathbf{u}_1$ of radius l_1, and the rest of the CIKS follows. The CIKS can be found for the robot of Figure 6.4(b).

Consider the 3 *dof* stiff, placeable robot described in part by $\mathbf{p}^1\bar{\mathbf{p}}^1$ with pitch in the first and third joints. Four possible cases are shown in Figure 6.5. For the configuration $|\ \theta\lambda\theta\ |$ the IKS is obtained by determining $\mathbf{u}_3$ as one of the two possible points of intersection of the circle of radius $l_1 + l_2$ centered at $\mathbf{u}_1$ and lying in $\mathbf{p}_1$ and the sphere of radius l_3 centered at $\mathbf{u}_4$.

For the robot $|\ \theta\psi\theta\ |$ of Figure 6.5(b), its IKS is obtained by noting that $\mathbf{u}_2$ is a function of one unknown, θ_1. Link l_2 has the direction $[a_1 \quad b_1 \quad c_1]$, so

[1] We do not consider the possibility of reversing the robot. For example, with the revolute robot, we exclude the possibility of traversing λ_1 through 180° and reversing the signs of the other revolute joints, since this is rarely possible in a practical robot.

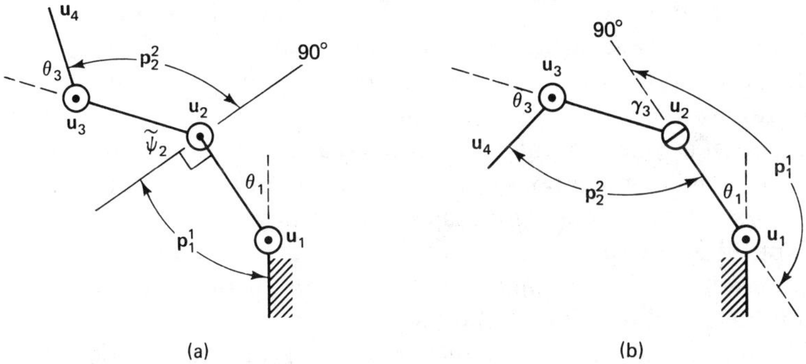

Figure 6.4 Nonstiff robots with CIKS

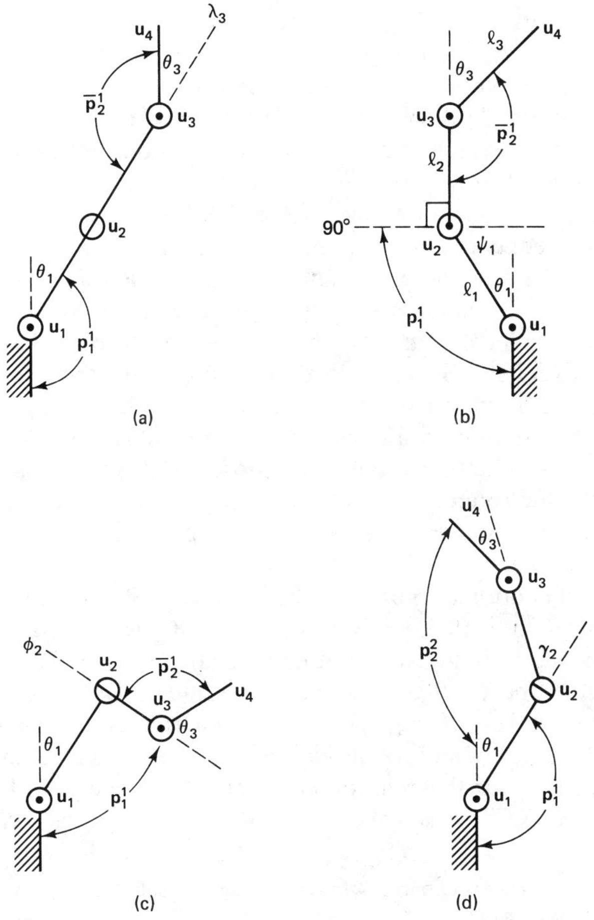

Figure 6.5 3 *dof* stiff robots

$\mathbf{u}_3$ is known in terms of θ_1. Further, $\mathbf{u}_3$ is at distance l_3 from $\mathbf{u}_4$, so θ_1 can be determined. The rest of the IKS is easily obtained.

For the robot $|\ \theta\phi\theta\ |$ of Figures 6.5(c), $\mathbf{u}_3$ is at one of the two points of intersection of the circle lying in $\mathbf{p}_1$ of radius $\sqrt{(l_1^2 + l_2^2)}$ centered at $\mathbf{u}_1$ and the surface of a sphere of radius l_3 centered at $\mathbf{u}_4$. The plane of motion $\mathbf{p}_2$ of the robot $|\ \theta\gamma\theta\ |$ of Figure 6.5(d) is established from $\mathbf{u}_1$, $\mathbf{u}_4$, and the fact that the vectors partially defining $\mathbf{p}_1$ and $\mathbf{p}_2$ are orthogonal.

Not all 3 *dof* stiff robots have CIKS that can be described in simple geometric terms. For example, the robot $|\ \theta\tilde{\phi}\theta\ |$ with planes of motion $\mathbf{p}_1^1\bar{\mathbf{p}}_2^1$ may not have a closed-form solution.

Geometric solutions are poor vehicles to program and control a robot, but the existence of a CIKS can often be established by geometric means more readily than with algebra, and any geometric solution can be transformed into an algebraic solution.[2] Sometimes, knowledge of the existence of a closed form is sufficient to accept a manipulator, with its actual solution left to a later date.

The IKSs of the basic 3 *dof* robots are shown in Figure 6.6 and yield none, one, or two solutions. The angles shown are not necessarily positive according to the standard convention. The 3 *dof* revolute robot exhibits two possible solutions. Which one is chosen depends on the context. Suppose the IKS is required for a series of position statuses, which in this case are end effector positions. Once one of the solutions is adopted, continuity locks in all subsequent solutions, and ambiguity is avoided.

Loose robots may have an infinity of IKSs. The robot $|\ \lambda\theta\theta\lambda\ |$ is loose with respect to a placeable position space but has the usual two IKSs, since λ_4 has no effect on location $\mathbf{u}_5$. On the other hand, the robot $|\ \lambda\theta\theta\theta\ |$, shown in Figure 6.7 with its IKS, is loose with regard to end effector position. For such robots a solution that is often adopted is to choose the solution that minimizes joint rotation with respect to the prior position. A plausible explanation of a manufacturer producing a loose robot is to enable the robot to control its "elbow room"—the intermediate joints and links may interfere with the work environment, and the looseness provides positional flexibility in obstacle avoidance.

A word of caution is in order. Although the methodology for obtaining a CIKS has been indicated, applying this methodology to establish the algebraic expressions in reduced (if not canonical) form is tedious. Typically, the solution for one configuration may take hours. Every robot manufacturer should provide such equations with their robots. It is imperative for the real-time control of any robot that the CIKS be available, and it is good marketing practice for the solution equations to be available for every robot. Once the CIKS is established, the equations should be written in firm-

[2] The converse may not be true—that is, an algebraic solution may not be transformed into a geometric solution.

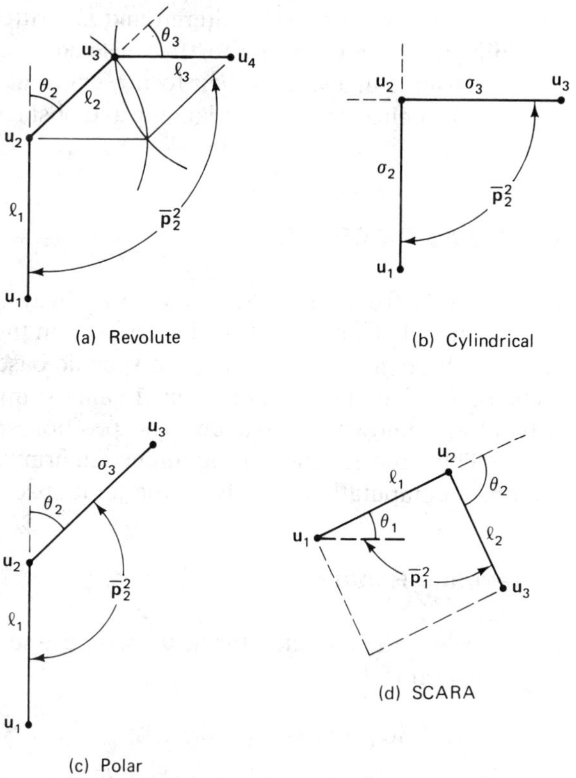

Figure 6.6 IKS for basic robots

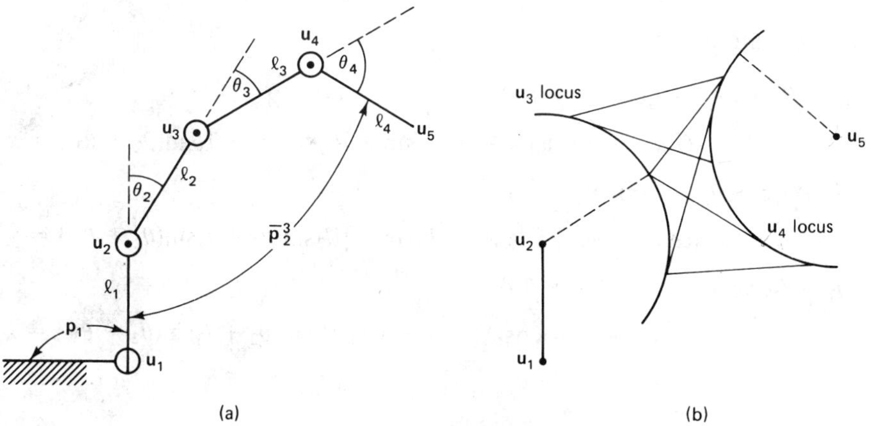

Figure 6.7 A loose, placeable robot

ware (computer software that will not be altered and is written with much care in order to minimize computational effort). These equations are transparent to the user (the user is unaware of their form), who can then concentrate on important tasks, such as trajectory planning and obstacle avoidance.

6.3 IKSs FOR THE BASIC ROBOTS

The IKS is always obtained from the FKS. The FKS is always possible, and the IKS is always obtainable from the FKS. In this section the IKSs of the basic types of robot will be given with respect to specific base orientations and positions. The base orientation and position are almost invariably fixed for the task at hand and known in advance. The position space is often obtained on-the-fly. Thus, the equations programmed in firmware should be structured to minimize computations involving the joint space variables.

6.3.1 The Revolute Robot

For the robot $|\lambda\theta\theta|$, the end effector at $\mathbf{u}_4$ is on $\bar{\mathbf{p}}_2$, so $a_2x + b_2y + c_2z = 0$ and from equation (5.1),

$$(a_1x + b_1y + c_1z)\cos\lambda_1 + \{(\nu_yc_1 - \nu_zb_1)x + (\nu_za_1 - \nu_xc_1)y + (\nu_xb_1 - \nu_ya_1)z\}\sin\lambda_1 = 0,$$

then

$$\tan\lambda_1 = \frac{-(a_1x + b_1y + c_1z)}{(\nu_yc_1 - \nu_zb_1)x + (\nu_za_1 - \nu_xc_1)y + (\nu_xb_1 - \nu_ya_1)z} \tag{6.1}$$

From equation (5.5),

$$\nu_x\{l_1 + l_2\cos\theta_2 + l_3\cos(\theta_2 + \theta_3)\} - \{(\nu_yc_1 - \nu_zb_1)\cos\lambda_1 - a_1\sin\lambda_1\}\{l_2\sin\theta_2 + l_3\sin(\theta_2 + \theta_3)\} = x$$

$$\nu_y\{l_1 + l_2\cos\theta_2 + l_3\cos(\theta_2 + \theta_3)\} - \{(\nu_za_1 - \nu_xc_1)\cos\lambda_1 - b_1\sin\lambda_1\}\{l_2\sin\theta_2 + l_3\sin(\theta_2 + \theta_3)\} = y$$

$$\nu_z\{l_1 + l_2\cos\theta_2 + l_3\cos(\theta_2 + \theta_3)\} - \{(\nu_xb_1 - \nu_ya_1)\cos\lambda_1 - c_1\sin\lambda_1\}\{l_2\sin\theta_2 + l_3\sin(\theta_2 + \theta_3)\} = z$$

or,

$$\nu_x\delta - \alpha\varepsilon = x,\ \nu_y\delta - \beta\varepsilon = y,\ \nu_z\delta - \gamma\varepsilon = z \tag{6.2}$$

where,

$$\alpha = (\nu_y c_1 - \nu_z b_1)\cos\lambda_1 - a_1 \sin\lambda_1$$

$$\beta = (\nu_z a_1 - \nu_x c_1)\cos\lambda_1 - b_1 \sin\lambda_1$$

$$\gamma = (\nu_x b_1 - \nu_y a_1)\cos\lambda_1 - c_1 \sin\lambda_1$$

$$\delta = l_1 + l_2\cos\theta_2 + l_3\cos(\theta_2 + \theta_3) \tag{6.3}$$

$$\varepsilon = l_2\sin\theta_2 + l_3\sin(\theta_2 + \theta_3) \tag{6.4}$$

and α, β and γ are known since λ_1 is established. Equations (6.2) contain the two unknowns δ and ε, and their solution depends on $\boldsymbol{\nu}$. $\boldsymbol{\nu}$ is a unit vector and at least one entry is non-zero. If $\nu_x \neq 0$, $\nu_y \neq 0$ and $\nu_z \neq 0$ we combine the first two positional equations to give

$$\delta = \frac{x\beta - y\alpha}{\nu_x\beta - \nu_y\alpha} \text{ and } \varepsilon = \frac{x\nu_y - y\nu_x}{\nu_x\beta - \nu_y\alpha} \tag{6.5}$$

If $\nu_x = 0$, $\varepsilon = x$ and δ is found from one of the other two positional equations. ε and δ are established in a similar manner if $\nu_y = 0$ or $\nu_z = 0$. Equations (6.3) and (6.4) are solved by rearranging, squaring and adding to give

$$\begin{aligned} l_3^2 &= \{\delta - l_1 - l_2\cos\theta_2\}^2 + \{\varepsilon - l_2\sin\theta_2\}^2 \\ &= (\delta - l_1)^2 + \varepsilon^2 + l_2^2 - 2l_2\{(\delta - l_1)\cos\theta_2 + \varepsilon\sin\theta_2\} \end{aligned}$$

or,

$$(\delta - l_1)\cos\theta_2 + \varepsilon\sin\theta_2 = \frac{l_2^2 - l_3^2 + (\delta - l_1)^2 + \varepsilon^2}{2l_2} = f$$

so,

$$\theta_2 = \sin^{-1}\frac{f}{\sqrt{(\delta - l_1)^2 + \varepsilon^2}} - \tan^{-1}\frac{\varepsilon}{\delta - l_1} \text{ if } \delta - l_1 \neq 0, \text{ or} \tag{6.6a}$$

$$\theta_2 = \cos^{-1}\frac{f}{\sqrt{(\delta - l_1)^2 + \varepsilon^2}} + \tan^{-1}\frac{\delta - l_1}{\varepsilon} \text{ if } \varepsilon \neq 0. \tag{6.6b}$$

From equation (6.4) we get

$$\theta_3 = \sin^{-1}\left(\frac{1}{l_3}\{-l_2\sin\theta_2 + \varepsilon\}\right) - \theta_2 \tag{6.7}$$

With the specific base orientation $\mathbf{p}_1 = [0 \;\; 1 \;\; 0 \;\; 0]$ and $\boldsymbol{\nu} = \begin{bmatrix} 0 \\ 0 \\ 1 \end{bmatrix}$, is an important special case ($\nu_x = \nu_y = 0$) not covered by the degenerate cases previously discussed and so requiring a modified solution. For this case, $\tan \lambda_1 = \frac{y}{x}$, and equations (6.6) and (6.7) provide the rest of the IKS with $\delta = z$ and $\varepsilon = x \cos \lambda_1 + y \sin \lambda_1$.

The IKS for the 4 *dof* pencilable $| \lambda\theta\theta\lambda |$ robot is obtained by replacing l_3 in the IKS for $| \lambda\theta\theta |$ with $l_3 + l_4$. λ_4 is found by calculating $\mathbf{u}_3$ and $\mathbf{u}_4$ and solving equation (5.12).

The IKS for 5 *dof* directable robot $| \lambda\theta\theta\lambda\theta |$ has $\mathbf{u}_5$ and $\mathbf{u}_6$ in the position space. We use $\mathbf{u}_5$ as in the preceding pencilable robot to find λ_1, θ_2, and θ_3. The solution is completed by calculating $\bar{\mathbf{p}}_3^1$ from $\mathbf{u}_4$, $\mathbf{u}_5$, and $\mathbf{u}_6$, thus determining λ_4 and θ_5.

The position space of the 6 *dof* orientable robot $| \lambda\theta\theta\lambda\theta\lambda |$ shown in Figure 6.8(a) is $\mathbf{u}_6$, $\mathbf{u}_7$, and $\bar{\mathbf{p}}_4^0$. $\mathbf{u}_5 = \mathbf{u}_6 + l_5(\mathbf{u}_6 - \mathbf{u}_7)/l_6$, and the solution proceeds as for the preceding directable robot to find λ_1, θ_2, θ_3, λ_4, θ_5, and $\bar{\mathbf{p}}_3^1$. The angle between $\bar{\mathbf{p}}_3^1$ and $\bar{\mathbf{p}}_4^0$ determines λ_6.

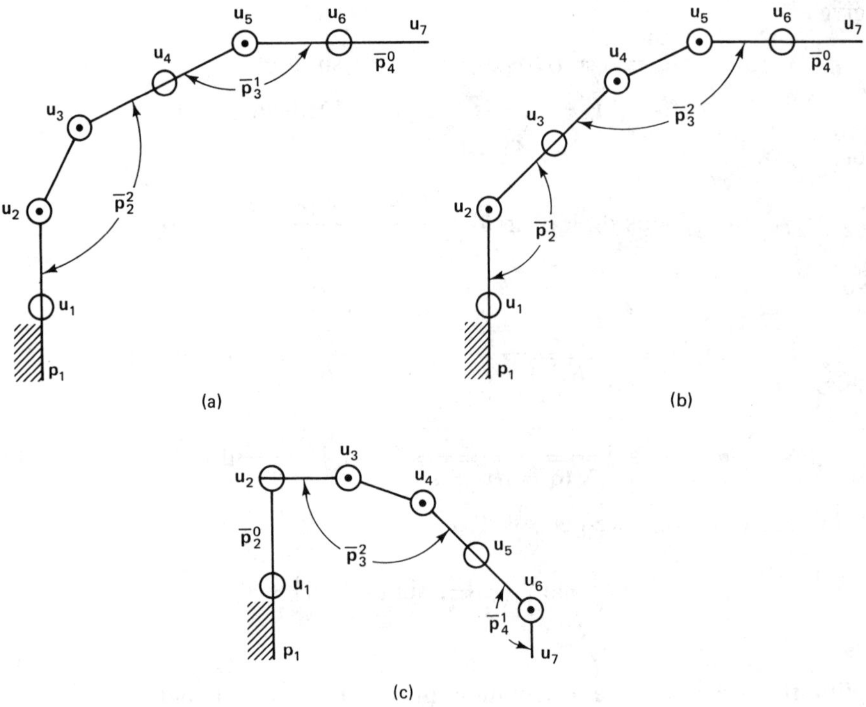

Figure 6.8 6 *dof* orientable robots

For the drillable robot $|\ \lambda\theta\theta\theta\lambda\ |$, the position of the end link (location of $\mathbf{u}_5$ and $\mathbf{u}_6$) is given. Recall that this robot is not directable, so the position must be feasible. Feasibility is guaranteed provided $\mathbf{u}_1$, $\mathbf{u}_2$, $\mathbf{u}_5$, and $\mathbf{u}_6$ all lie in the same plane. $\mathbf{u}_4$ is in line with $\mathbf{u}_6$ and $\mathbf{u}_5$ and so is known. Given $\mathbf{u}_4$, the status of the first three joints is determined as discussed earlier in this section and the CIKS is completed.

Similarly, for the robots $|\ \lambda\theta\theta\theta\phi\ |$ or $|\ \lambda\theta\theta\theta\bar{\phi}\ |$, we again use given locations $\mathbf{u}_5$ and $\mathbf{u}_6$ and the relationships developed in Section 3.1.6 to determine $\mathbf{u}_4$. For the robot $|\ \lambda\theta\theta\theta\psi\ |$ we need the joint status of ψ_5 as well as given locations $\mathbf{u}_5$ and $\mathbf{u}_6$ to determine $\mathbf{u}_4$. This robot appears to have OIKS.

The robot of Figure 6.8(b) has the same planes of motion but in a different order to the orientable robot of Figure 6.8(a). The CIKS of this robot is obtained by noting that $\mathbf{u}_2$ (which is determined by the base orientation and position), $\mathbf{u}_6$, and $\mathbf{u}_7$ (determined by the placeability and end link orientation) all lie on $\mathbf{p}_3$, so the plane with area is established and the rest of the CIKS follows.

Next consider the 6 *dof* orientable robot shown in Figure 6.8(c). This robot has the same planes of motion but in a different order to the robots of Figures 6.8(a) and 6.8(b). $\mathbf{u}_2$ (easily established), $\mathbf{u}_5$ (unknown but lying on known $\mathbf{p}_4$), and $\mathbf{u}_6$ lie on $\mathbf{p}_3$. It is possible that this robot has a CIKS, but it is not evident using the techniques advocated here. An OIKS can be devised that chooses θ_6 and so establishes $\mathbf{u}_5$ and, through it, $\mathbf{p}_3$ and $\mathbf{u}_4$. Thus $\mathbf{u}_4$ is a function of θ_6, and we modify θ_6 so that $\delta = |\ \mathbf{u}_4 - \mathbf{u}_3\ | - l_3$ is iterated to zero.

The lesson of the analysis of the robots of Figure 6.8 is that although they have the same joints (not in the same order) and the same planes of motion (not in the same order), the IKS can be radically different. We can extend this analysis by example to show that robots with the same planes of motion in the same order may not have similar IKSs. Thus, each specific robot must be analyzed individually.

6.3.2 The Cylindrical Robot

The IKS for the 3 *dof* cylindrical robot $|\ \lambda\sigma+\sigma\ |$ is derived from equation (5.17):

$$x = \sigma_2\nu_x - \sigma_3\{(b_1\nu_z - c_1\nu_y)\cos\lambda_1 + a_1\sin\lambda_1\}$$

$$y = \sigma_2\nu_y - \sigma_3\{(b_1\nu_x - a_1\nu_z)\cos\lambda_1 + b_1\sin\lambda_1\}$$

$$z = \sigma_2\nu_z - \sigma_3\{(a_1\nu_y - b_1\nu_x)\cos\lambda_1 + c_1\sin\lambda_1\}$$

so

$$\sigma_2 = \frac{z + \sigma_3\{(a_1\nu_y - b_1\nu_x)\cos\lambda_1 + c_1\sin\lambda_1\}}{\nu_z} \tag{6.8}$$

and

$$x = \frac{\nu_x}{\nu_z}(z + \sigma_3\{(a_1\nu_y - b_1\nu_x)\cos\lambda_1 + c_1\sin\lambda_1\})$$
$$- \sigma_3\{(b_1\nu_z - c_1\nu_y)\cos\lambda_1 + a_1\sin\lambda_1\}$$

$$y = \frac{\nu_y}{\nu_z}(z + \sigma_3\{(a_1\nu_y - b_1\nu_x)\cos\lambda_1 + c_1\sin\lambda_1\})$$
$$- \sigma_3\{(b_1\nu_x - a_1\nu_z)\cos\lambda_1 + b_1\sin\lambda_1\}$$

The expressions for x and y simplify to

$$x\nu_z = z\nu_x - \sigma_3\{b_1\cos\lambda_1 + (a_1\nu_z - c_1\nu_x)\sin\lambda_1\}$$
$$y\nu_z = z\nu_y + \sigma_3\{a_1\cos\lambda_1 + (c_1\nu_y - b_1\nu_z)\sin\lambda_1\}$$

so

$$\sigma_3 = \frac{z\nu_x - x\nu_z}{b_1\cos\lambda_1 + (a_1\nu_z - c_1\nu_x)\sin\lambda_1} \tag{6.9}$$

and

$$(y\nu_z - z\nu_y)\{b_1\cos\lambda_1 + (a_1\nu_z - c_1\nu_x)\sin\lambda_1\}$$
$$= (z\nu_x - x\nu_z)\{a_1\cos\lambda_1 + (c_1\nu_y - b_1\nu_z)\sin\lambda_1\}$$

giving λ_1 explicitly as

$$\lambda_1 = \tan^{-1}\left\{\frac{(y\nu_z - z\nu_y)b_1 + (x\nu_z - z\nu_x)a_1}{(z\nu_x - x\nu_z)(c_1\nu_y - b_1\nu_z) - (y\nu_z - z\nu_y)(a_1\nu_z - c_1\nu_x)}\right\}$$

$$= \tan^{-1}\left\{\frac{xa_1 + yb_1 + zc_1 - z/\nu_z}{x(b_1\nu_z - c_1\nu_y) + y(c_1\nu_x - a_1\nu_z) + z(a_1\nu_y - b_1\nu_x)}\right\} \tag{6.10}$$

The inverse kinematic solution for this robot is as follows:

1. Equation (6.10) gives λ_1.
2. Equation (6.9) gives σ_3.
3. Equation (6.8) gives σ_2.

For the simple base orientation $\mathbf{p}_1 = [0 \quad 1 \quad 0 \quad 0]$ and $\boldsymbol{\nu} = \begin{bmatrix} 0 \\ 0 \\ 1 \end{bmatrix}$, the inverse kinematic solution becomes

$$\lambda_1 = \tan^{-1}\left\{\frac{y - z}{x}\right\}, \qquad \sigma_3 = \frac{-x}{\cos\lambda_1}, \quad \text{and} \quad \sigma_2 = \frac{z}{\nu_z} \tag{6.11}$$

6.3.3 The Cartesian Robot

The IKS for the 3 *dof* Cartesian robot $|\sigma(+)\sigma(\times)\sigma|$ is found from equation (5.19) as

$$x = \sigma_1\nu_x + \sigma_2(\nu_y c_1 - \nu_z b_1) - \sigma_3 a_1$$

$$y = \sigma_1\nu_y + \sigma_2(\nu_z a_1 - \nu_x c_1) - \sigma_3 b_1$$

$$z = \sigma_1\nu_z + \sigma_2(\nu_x b_1 - \nu_y a_1) - \sigma_3 c_1$$

Writing σ_3 explicitly in the expression for z gives

$$\sigma_3 = \frac{-z + \sigma_1\nu_z + \sigma_2(\nu_x b_1 - \nu_y a_1)}{c_1} \tag{6.12}$$

Substituting for σ_3 in the expressions for σ_1 and σ_2 gives

$$xc_1 - za_1 = \sigma_1(\nu_x c_1 - \nu_z a_1) + \sigma_2\nu_y$$

and

$$yc_1 + zb_1 = \sigma_1(\nu_y c_1 - \nu_z b_1) - \sigma_2\nu_x$$

The second of these two expressions can be written

$$\sigma_2 = \frac{zb_1 - yc_1 + \sigma_1(\nu_y c_1 - \nu_z b_1)}{\nu_x} \tag{6.13}$$

and substituting for σ_2 in the first expression gives

$$\nu_x(xc_1 - za_1) - \nu_y(zb_1 - yc_1) = \sigma_1\{\nu_x^2 c_1 - \nu_x\nu_z a_1 + \nu_y^2 c_1 - \nu_x\nu_y b_1\} = \sigma_1 c_1$$

so

$$\sigma_1 = x\nu_x + y\nu_y - \frac{z(a_1\nu_x + b_1\nu_y)}{c_1} \tag{6.14}$$

For the simple base orientation $\mathbf{p}_1 = [0\quad 0\quad 1\quad 0]$, $\nu = [0\quad 1\quad 0]^\#$, and $\mathbf{u}_1 = [0\quad 0\quad 0\quad 1]^\#$,

$$\begin{bmatrix} x \\ y \\ z \\ 1 \end{bmatrix} = \begin{bmatrix} \sigma_2 \\ \sigma_1 \\ -\sigma_3 \\ 1 \end{bmatrix} \tag{6.15}$$

with evident solution.

6.3.4 The Polar Robot

The IKS for the 3 *dof* polar robot $|\lambda\theta\sigma|$ can be obtained from equation (6.1) and two of the three positional equations for the revolute robot with l_2 replaced by σ_3 and all terms involving l_3 removed:

$$\nu_x\cos\theta_2 - \alpha\sin\theta_2 = \frac{x - \nu_x l_1}{\sigma_3}$$

$$\nu_y\cos\theta_2 - \beta\sin\theta_2 = \frac{y - \nu_y l_1}{\sigma_3}$$

$$\nu_z\cos\theta_2 - \gamma\sin\theta_2 = \frac{z - \nu_z l_1}{\sigma_3}$$

From the first two equations,

$$\frac{\nu_x\cos\theta_2 - \alpha\sin\theta_2}{\nu_y\cos\theta_2 - \beta\sin\theta_2} = \frac{x - \nu_x l_1}{y - \nu_y l_1}$$

or

$$\left\{\nu_x - \frac{x - \nu_x l_1}{y - \nu_y l_1}\nu_y\right\}\cos\theta_2 - \left\{\alpha - \frac{x - \nu_x l_1}{y - \nu_y l_1}\beta\right\}\sin\theta_2 = 0$$

so

$$\theta_2 = \tan^{-1}\frac{\nu_x(y - \nu_y l_1) - \nu_y(x - \nu_x l_1)}{\alpha(y - \nu_y l_1) - \beta(x - \nu_x l_1)} \qquad (6.16)$$

Finally, provided the denominator is nonzero, the first positional equation gives

$$\sigma_3 = \frac{x - \nu_x l_1}{\nu_x\cos\theta_2 - \alpha\sin\theta_2} \qquad (6.17)$$

and the IKS is complete. If the denominator in equation (6.17) is zero, we use another one of the three original positional equations.

For the simple base orientation $\mathbf{p}_1 = [0 \quad 1 \quad 0 \quad 0]$ and $\boldsymbol{\nu} = \begin{bmatrix} 0 \\ 0 \\ 1 \end{bmatrix}$, the IKS is best obtained by direct substitution in the three positional equations:

$$\cos\lambda_1\sin\theta_2 = \frac{x}{\sigma_3}, \qquad \sin\lambda_1\sin\theta_2 = \frac{y}{\sigma_3}, \quad \text{and} \quad \cos\theta_2 = \frac{z - l_1}{\sigma_3}$$

so

$$\lambda_1 = \tan^{-1}\left\{\frac{y}{x}\right\}, \qquad \theta_2 = \tan^{-1}\frac{x}{(z - l_1)\cos\lambda_1}, \quad \text{and} \quad \sigma_3 = \frac{z - l_1}{\cos\theta_2} \qquad (6.18)$$

6.3.5 The SCARA Robot

The IKS of the SCARA robot | $\theta\theta(\times)\sigma$ | is obtained from Section 5.2.5:

$$x = \nu_x\{l_1\cos\theta_1 + l_2\cos(\theta_1 + \theta_2)\}$$
$$- (\nu_y c_1 - \nu_z b_1)\{l_1\sin\theta_1 + l_2\sin(\theta_1 + \theta_2)\} - \sigma_3 a_1$$
$$y = \nu_y\{l_1\cos\theta_1 + l_2\cos(\theta_1 + \theta_2)\}$$
$$- (\nu_z a_1 - \nu_x c_1)\{l_1\sin\theta_1 + l_2\sin(\theta_1 + \theta_2)\} - \sigma_3 b_1$$
$$z = \nu_z\{l_1\cos\theta_1 + l_2\cos(\theta_1 + \theta_2)\}$$
$$- (\nu_x b_1 - \nu_y a_1)\{l_1\sin\theta_1 + l_2\sin(\theta_1 + \theta_2)\} - \sigma_3 c_1$$

Let

$$\alpha = l_1\sin\theta_1 + l_2\sin(\theta_1 + \theta_2) \tag{6.19}$$

and

$$\beta = l_1\cos\theta_1 + l_2\cos(\theta_1 + \theta_2) \tag{6.20}$$

Then

$$x = \nu_x\beta - (\nu_y c_1 - \nu_z b_1)\alpha - \sigma_3 a_1$$
$$y = \nu_y\beta - (\nu_z a_1 - \nu_x c_1)\alpha - \sigma_3 b_1$$
$$z = \nu_z\beta - (\nu_x b_1 - \nu_y a_1)\alpha - \sigma_3 c_1$$

We eliminate σ_3 from these equations, giving

$$a_1 y - b_1 x = (a_1\nu_y - b_1\nu_x)\beta + \{b_1(\nu_y c_1 - \nu_z b_1) - a_1(\nu_z a_1 - \nu_x c_1)\}\alpha$$
$$= (a_1\nu_y - b_1\nu_x)\beta - \nu_z\alpha$$
$$a_1 z - c_1 x = (a_1\nu_z - c_1\nu_x)\beta + \{c_1(\nu_y c_1 - \nu_z b_1) - a_1(\nu_x b_1 - \nu_y a_1)\}\alpha$$
$$= (a_1\nu_z - c_1\nu_x)\beta + \nu_y\alpha$$

Eliminating α gives

$$(a_1 y - b_1 x)\nu_y + (a_1 z - c_1 x)\nu_z = \{(a_1\nu_y - b_1\nu_x)\nu_y + (a_1\nu_z - c_1\nu_x)\nu_z\}\beta = a_1\beta$$

or

$$\beta = x\nu_x + y\nu_y + z\nu_z - \frac{x}{a_1} \tag{6.21}$$

and

$$\alpha = (a_1\nu_y - b_1\nu_x)\alpha - a_1 y + b_1 x \tag{6.22}$$

$$\sigma_3 = \frac{z - \nu_z\beta - (\nu_x b_1 - \nu_y a_1)\alpha}{c_1}, \tag{6.23}$$

θ_1 and θ_2 are found by reordering, squaring, and adding equations (6.19) and (6.20) as

$$l_2^2 = (\beta - l_1\cos\theta_1)^2 + (\alpha - l_1\sin\theta_1)^2$$
$$= \alpha^2 + \beta^2 + l_1^2 - 2l_1(\alpha\sin\theta_1 + \beta\cos\theta_1)$$

so

$$\theta_1 = \sin^{-1}\frac{\alpha^2 + \beta^2 + l_1^2 - l_2^2}{2l_1\sqrt{\alpha^2 + \beta^2}} - \tan^{-1}\frac{\beta}{\alpha} \text{ if } \alpha \neq 0 \tag{6.24a}$$

or

$$\theta_1 = \cos^{-1}\frac{\alpha^2 + \beta^2 + l_1^2 - l_2^2}{2l_1\sqrt{\alpha^2 + \beta^2}} + \tan^{-1}\frac{\alpha}{\beta} \text{ if } \beta \neq 0 \tag{6.24b}$$

Substituting θ_1 in equation (6.19) or (6.20) gives θ_2, and the IKS is completed.

For the simple case $p_1 = [0 \quad 0 \quad 1 \quad 0]$ and $\nu = \begin{bmatrix} 1 \\ 0 \\ 0 \end{bmatrix}$,

$$\alpha = x,\ \beta = 0, \text{ and } \sigma_3 = -z,$$

$$\theta_1 = \sin^{-1}\frac{x^2 + l_1^2 - l_2^2}{2l_1\alpha} \tag{6.25a}$$

$$\theta_2 = \cos^{-1}\left\{\frac{x - l_1\cos\theta_1}{l_2}\right\} - \theta_1 \tag{6.25b}$$

For the 4 *dof* Scara robot $|\ \theta\theta\underline{\psi}\sigma\ |$, the IKS for position is the same as just given. $\underline{\psi}_3$ has no influence on the end effector position. Its influence is on the plane of motion $[e \quad f \quad g \quad h]$ of the end effector given by equation (3.11):

$$\begin{bmatrix} e \\ f \\ g \end{bmatrix} = \frac{1}{l_2}\begin{bmatrix} \cos\underline{\psi}_3 & -c_1\sin\underline{\psi}_3 & b_1\sin\underline{\psi}_3 \\ c_1\sin\underline{\psi}_3 & \cos\underline{\psi}_3 & -a_1\sin\underline{\psi}_3 \\ -b_1\sin\underline{\psi}_3 & a_1\sin\underline{\psi}_3 & \cos\underline{\psi}_3 \end{bmatrix}\begin{bmatrix} x_3 - x_2 \\ y_3 - y_2 \\ z_3 - z_2 \end{bmatrix}$$

$$= \begin{bmatrix} \cos\psi_3 & c_1\sin\psi_3 & -b_1\sin\psi_3 \\ -c_1\sin\psi_3 & \cos\psi_3 & a_1\sin\psi_3 \\ b_1\sin\psi_3 & -a_1\sin\psi_3 & \cos\psi_3 \end{bmatrix}$$

$$\begin{bmatrix} \nu_x\cos(\theta_1 + \theta_2) - (\nu_y c_1 - \nu_z b_1)\sin(\theta_1 + \theta_2) \\ \nu_y\cos(\theta_1 + \theta_2) - (\nu_z a_1 - \nu_x c_1)\sin(\theta_1 + \theta_2) \\ \nu_z\cos(\theta_1 + \theta_2) - (\nu_x b_1 - \nu_y a_1)\sin(\theta_1 + \theta_2) \end{bmatrix} \tag{6.26}$$

since $\psi_3 = -\underline{\psi}_3$. For the simple base orientation $\mathbf{p}_1 = [0 \quad 0 \quad -1 \quad 0]$ and $\boldsymbol{\nu} = [1 \quad 0 \quad 0]^{\#}$,

$$\begin{bmatrix} e \\ f \\ g \end{bmatrix} = \begin{bmatrix} \cos\psi_3 \cos(\theta_1 + \theta_2) + \sin\psi_3 \sin(\theta_1 - \theta_2) \\ \sin\psi_3 \cos(\theta_1 + \theta_2) - \cos\psi_3 \sin(\theta_1 - \theta_2) \\ 0 \end{bmatrix} \tag{6.27}$$

6.4 PROBLEMATICAL INVERSE KINEMATIC SOLUTIONS

Some robots are considered to have OIKS, and an iterative, open-form solution may be required [Angeles 1985; Angeles 1986; Armstrong 1979; Goldenberg, Benhabib, and Fenton 1985]. Standard philosophy is "Only in special cases may robots with six degrees of freedom be solved analytically" [Craig 1986], where the word *analytic* means *in closed form*. We have already seen that all common robot configurations have CIKS. Further study on some special types of kinematically simple robot indicates that a CIKS may require careful thought, but in no case can it be said that a CIKS is unobtainable. However, at the present time this author is not prepared to state categorically that all kinematically simple robots have CIKS.

The IKS of any *n dof* robot with kinematically simple joints whose position space has order n depends on the solution of m nonlinear, simultaneous equations, where $m \leq n$. If $m = 1$, we have a CIKS. If $m > 1$, we classify such robots as having an OIKS. However, the m equations have a solution not as computationally expensive as attempting to process such robots using the more standard Denavit-Hartenberg notations.

The 5 *dof* robot $| \lambda\theta\theta\psi\theta |$ with planes of motion $\overline{\mathbf{p}}^2\overline{\mathbf{p}}^1$ is shown in Figure 6.9. From Table 4.1 the robot is directable, and the position space is $\mathbf{u}_5$ and $\mathbf{u}_6$. With the two planes of motion $\overline{\mathbf{p}}_2 = [a_2 \quad b_2 \quad c_2 \quad 0]$ and $\overline{\mathbf{p}}_3 =$

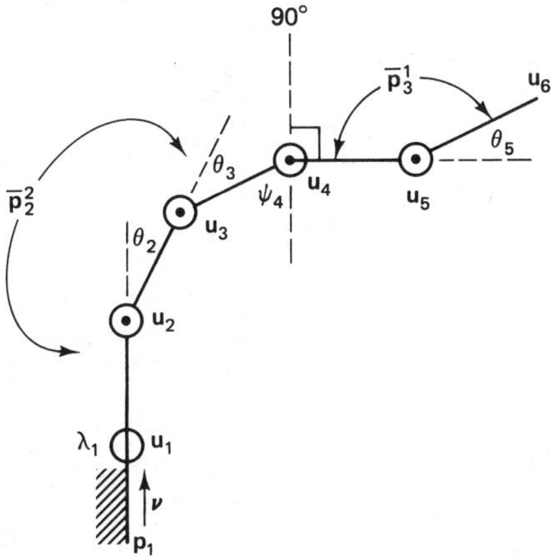

Figure 6.9 A 5 *dof* directable robot

$[a_3 \quad b_3 \quad c_3 \quad d_3]$, we see that l_4 is in the direction of vector $[a_2 \quad b_2 \quad c_2]^\#$, and so points $\mathbf{u}_1$, $\mathbf{u}_2$, and $\mathbf{u}_3$ form a triangle with $\mathbf{u}_4$ and $\mathbf{u}_5$ with a right angle at $\mathbf{u}_4$. Thus $\mathbf{u}_4$ is determined as one of the two points of intersection of three spheres:

1. Centered at known $\mathbf{u}_1$ with radius $\sqrt{|\,\mathbf{u}_5 - \mathbf{u}_1\,|^2 - l_4^2}$
2. Centered at known $\mathbf{u}_2$ with radius $\sqrt{|\,\mathbf{u}_5 - \mathbf{u}_2\,|^2 - l_4^2}$
3. Centered at known $\mathbf{u}_5$ with radius l_4

With $\mathbf{u}_4$ established, the two planes of motion can be calculated, and the remainder of CIKS is evident.

Notice that if the crank joint at $\mathbf{u}_4$ is connected in reverse, the planes of motion of the robot become $\overline{\mathbf{p}}^2\mathbf{p}^2$, and according to Table 4.1, the robot is not directable.

Most robots have simple, geometrically based IKSs [Lee and Ziegler 1983b], which can be reduced to algebraic CIKS. In fact, of the hundreds of configurations used in commercial robots, only a handful do not have geometric solutions, and it is likely that even these have CIKS.

Common 6 *dof* orientable robots that do not have geometric IKSs are the Kuka 662/100 with offset wrist, the Nachji-Fujikoshi 8601-AK and 86051-AK, and the Yaskawa Electric Co. Motoman L15 or L30, all having the configuration $|\,\lambda\theta\theta\lambda\phi\phi\,|$ shown in Figure 6.10. The popularity seems to spring from the robust mechanical construction possible in this type of orientation mechanism. From the base orientation and position we have $\mathbf{u}_1$ and

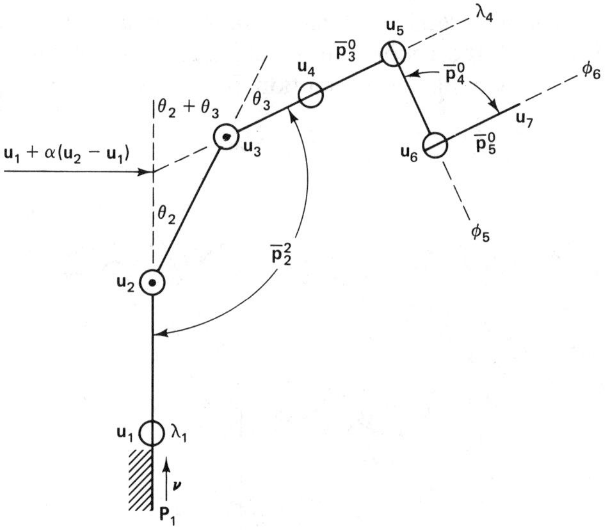

Figure 6.10 The Kuka 662/100

$\mathbf{u}_2$. From the end effector orientation and position, we have $\mathbf{u}_7$, $\mathbf{u}_6$, and $\bar{\mathbf{p}}_5$. $\mathbf{u}_5$ is on a known circle centered at $\mathbf{u}_6$, whose position is a function of ϕ_6. $\bar{\mathbf{p}}_2$ can be found from three points $\mathbf{u}_1$, $\mathbf{u}_2$, and $\mathbf{u}_5$, so $\bar{\mathbf{p}}_2$ is found as a function of ϕ_6. Link l_4 intersects line $\mathbf{u}_2 + \alpha(\mathbf{u}_1 - \mathbf{u}_2)$, and since links l_4 and l_5 are orthogonal, we have the condition $\{\mathbf{u}_2 + \alpha(\mathbf{u}_1 - \mathbf{u}_2) - \mathbf{u}_5\} \cdot \{\mathbf{u}_5 - \mathbf{u}_6\} = 0$, which determines α and, hence, $\theta_2 + \theta_3$. The line along which l_3 and l_4 lie is a known function of ϕ_6, so $\mathbf{u}_3$ is a known function of ϕ_6. Subsequently, θ_2 can be determined as a function of ϕ_6. The dot product of the vectors partially specifying $\mathbf{p}_1$ and $\mathbf{p}_2$ give cos λ_1, so λ_1 is a known function of ϕ_6. Applying the FKS gives $\mathbf{u}_5$ as a function of ϕ_6, which when equated to the circle defining $\mathbf{u}_5$ as a function of ϕ_6 establishes ϕ_6. The rest of the IKS follows. Thus, the IKS is not simple, but it is closed.

Some robots seem to have OIKS, but another problem may exist. The assumed position space may not be realizable with the robot. It is possible that some other position space can be found for such a robot and a CIKS found for this position space. It is important to determine the position space of a robot before attempting an IKS.

6.5 DEXTERITY

Consider the 3 *dof* placeable robot $\mid \theta\tilde{\phi}\theta \mid$ with planes of motion $\mathbf{p}^1\mathbf{p}^2$. This robot is not stiff, since it has a plane with area, but it is not as dextrous as a robot with an independent plane with area. What do we mean by dextrous, and can it be quantified? We can define dextrous in a placeable robot as the ability to move its end effector in any direction. More quantitatively, we can define it as the ability to move in any direction as function of the total amount of joint movement. We shall attempt to define dexterity precisely.

Kumar and Craig define a reachable workspace as the volume within which every point can be reached by a reference point on the manipulator hand [Kumar and Waldron 1981; Craig 1986]. They also define a dextrous workspace as the volume within which every point can be reached by a reference point on the manipulator hand with the hand in any desired orientation. This latter definition is not useful here, since it does not address the problem of incremental movement and stiff robots. We can produce the same definition of dextrous workspace for a 6 *dof* orientable robot by requiring joints $\mathbf{u}_4$, $\mathbf{u}_5$, $\mathbf{u}_6$, and $\mathbf{u}_7$ to be within the reachable workspace of the corresponding 3 *dof*, 4 *dof*, 5 *dof*, and 6 *dof* manipulators.

Suppose we have a placeable robot containing revolute joints only. With its end effector at a particular position $\mathbf{u}_e$, consider an incremental movement (defined in Cartesian space as a vector) of its end effector $\boldsymbol{\eta}$, where $\mid \eta \mid = \varepsilon \ll 1$. Suppose the sum total of joint movements in executing this end effector movement is $\Delta(\eta)$. Suppose we choose $\boldsymbol{\eta}$ for maximum and minimum Δ, called Δ_{max} and Δ_{min}, giving a ratio for the particular end effector position of $R(\mathbf{u}_e) = \Delta_{max}/\Delta_{min}$. The smaller $R(\mathbf{u}_e)$, the more dextrous is the robot at $\mathbf{u}_e$. However, choosing another location for the end effector will produce a different R.

Suppose every reachable position of the placeable robot is tested to find the largest ratio R_{max}, then we could quantify dexterity as R_{max}. The determination of R_{max} is very costly and is not recommended. It is likely that any attempt to quantify dexterity will be inadequate or costly. However, a few recommendations will lead to a more dexterous robot.

It has already been pointed out that stiff robots are not dexterous and are to be avoided. Also, an orientable robot should be designed with its placeable part first, since otherwise it can behave in a similar fashion to a stiff robot. In fact, the ideal dexterous, orientable robot will have its orientation linkages of zero length following the placeable portion defined by the single plane of motion $\bar{\mathbf{p}}^2$. Zero-link lengths are rarely feasible, but dexterity can be improved by correct choice of link lengths.

6.6 EXAMPLES

1. Write the canonical IKS equations for $|\lambda\theta\theta\lambda\theta\lambda|$ under a general base orientation. Assume $\nu_x \neq 0$, $\nu_y \neq 0$ and $\nu_z \neq 0$.

The position space of the robot is $\mathbf{u}_6$, $\mathbf{u}_7$ and $\bar{\mathbf{p}}_4^0$.

$$\mathbf{u}_5 = \begin{bmatrix} x_5 \\ y_5 \\ z_5 \\ 1 \end{bmatrix} = \mathbf{u}_6 + \frac{l_5}{l_6}(\mathbf{u}_6 - \mathbf{u}_7)$$

$$\tan \lambda_1 = \frac{-(a_1x_5 + b_1y_5 + c_1z_5)}{(\nu_yc_1 - \nu_zb_1)x_5 + (\nu_za_1 - \nu_xc_1)y_5 + (\nu_xb_1 - \nu_ya_1)z_5}$$

$$\delta = \frac{x_5\beta - y_5\alpha}{\nu_x\beta - \nu_y\alpha} \quad \text{and} \quad \varepsilon = \frac{x_5\nu_y - y_5\nu_x}{\nu_x\beta - \nu_y\alpha}$$

$$\frac{l_2^2 - l_3^2 + (\delta - l_1)^2 + \varepsilon^2}{2l_2} = f$$

so

$$\theta_2 = \sin^{-1}\frac{f}{\sqrt{(\delta - l_1)^2 + \varepsilon^2}} - \tan^{-1}\frac{\varepsilon}{\delta - l_1} \quad \text{if } \delta - l_1 \neq 0, \qquad \text{or}$$

$$\theta_2 = \cos^{-1}\frac{f}{\sqrt{(\delta - l_1)^2 + \varepsilon^2}} + \tan^{-1}\frac{\delta - l_1}{\varepsilon} \quad \text{if } \varepsilon \neq 0.$$

$$\theta_3 = \sin^{-1}\left\{\frac{1}{l_3}(-l_2\sin\theta_2 + \varepsilon)\right\} - \theta_2$$

$$\bar{\mathbf{p}}_3^1 = \frac{1}{l_4 l_5}(\mathbf{u}_5 - \mathbf{u}_4) \times (\mathbf{u}_6 - \mathbf{u}_5)$$

$$\lambda_4 = \cos^{-1}\bar{\mathbf{p}}_2^2 \cdot \bar{\mathbf{p}}_3^1, \qquad \theta_5 = \cos^{-1}\frac{(\mathbf{u}_5 - \mathbf{u}_4)(\mathbf{u}_6 - \mathbf{u}_5)}{\mathbf{l}_4\mathbf{l}_5}, \qquad \lambda_6 = \cos^{-1}\bar{\mathbf{p}}_3^1 \cdot \bar{\mathbf{p}}_4^0$$

---

2. Obtain the IKS for the 3 *dof* revolute robot $|\phi\theta\theta|$ with $\mathbf{p}_1 = [0 \quad 1 \quad 0 \quad 0]$, $\mathbf{v} = [1 \quad 0 \quad 0]^{\#}$, $\mathbf{u}_1 = [0 \quad 0 \quad 0 \quad 1]^{\#}$, and $\mathbf{u}_4 = [x \quad y \quad z \quad 1]^{\#}$.

Equation (5.9) gives

$$-\cos\phi_1\{l_2\sin\theta_2 + l_3\sin(\theta_2 + \theta_3)\} = x$$

$$-\sin\phi_1\{l_2\sin\theta_2 + l_3\sin(\theta_2 + \theta_3)\} = y$$

$$l_1 + l_2\cos\theta_2 + l_3\cos(\theta_2 + \theta_3) = z$$

so $\phi_1 = \tan^{-1}\{y/x\}$, $l_2\sin\theta_2 + l_3\sin(\theta_2 + \theta_3) = -y/\sin\phi_1$, and $l_2\cos\theta_2 + l_3\cos(\theta_2 + \theta_3) = z - l_1$.

Reordering and squaring both sides of the preceding equations gives

$$l_3^2\sin^2(\theta_2 + \theta_3) = \left\{l_2\sin\theta_2 + \frac{y}{\sin\phi_1}\right\}^2$$

and

$$l_3^2\cos^2(\theta_2 + \theta_3) = \{z - l_1 - l_2\cos\theta_2\}^2$$

so

$$l_3^2 = \left\{l_2\sin\theta_2 + \frac{y}{\sin\phi_1}\right\}^2 + \{z - l_1 - l_2\cos\theta_2\}^2$$

$$= l_2^2 + \frac{y^2}{\sin^2\phi_1} + (z - l_1)^2 + 2l_2\left\{\frac{y\sin\theta_2}{\sin\phi_1} - (z - l_1)\cos\theta_2\right\}$$

Hence,

$$\frac{y}{\sin\phi_1}\sin\theta_2 - (z - l_1)\cos\theta_2 = \frac{\left\{l_3^2 - l_2^2 - \dfrac{y^2}{\sin^2\phi_1} - (z - l_1)^2\right\}}{2l_2}$$

where the left-hand side can be reduced to

$$\sqrt{\frac{y^2}{\sin^2\phi_1} + (z - l_1)^2} \cdot \sin\left\{\theta_2 - \tan^{-1}\frac{(z - l_1)\sin\phi_1}{y}\right\}$$

and so θ_2 is found explicitly. Inserting ϕ_1 and θ_2 in the equation $l_2\sin\theta_2 + l_3\sin(\theta_2 + \theta_3) = y/\sin\phi_1$ establishes θ_3, and the IKS is completed.

---

3. Obtain the inverse kinematic solution for the robot $|\theta\tilde{\psi}\theta|$ of Figure 6.6(a), assuming $\mathbf{p}_1 = [0 \quad 1 \quad 0 \quad 0]$, $\mathbf{v} = [0 \quad 0 \quad 1]^{\#}$, and $\mathbf{u}_1 = [0 \quad 0 \quad 0 \quad 1]^{\#}$.

From equation (3.1) we have

$$\begin{bmatrix} x_2 \\ y_2 \\ z_2 \\ 1 \end{bmatrix} = l_1 \begin{bmatrix} \cos\theta_1 & 0 & -\sin\theta_1 & 0 \\ 0 & \cos\theta_1 & 0 & 0 \\ \sin\theta_1 & 0 & \cos\theta_1 & 0 \\ 0 & 0 & 0 & 1 \end{bmatrix} \begin{bmatrix} 0 \\ 0 \\ 1 \\ 1 \end{bmatrix} = l_1 \begin{bmatrix} -\sin\theta_1 \\ 0 \\ \cos\theta_1 \\ 1 \end{bmatrix}$$

and from equation (3.12),

$$\begin{bmatrix} e \\ f \\ g \end{bmatrix} = \frac{1}{l_1} \begin{bmatrix} x_2 - x_1 \\ y_2 - y_1 \\ z_2 - z_1 \end{bmatrix} = \begin{bmatrix} -\sin\theta_1 \\ 0 \\ \cos\theta_1 \end{bmatrix}$$

$\mathbf{p}_2^2$ passes through $\mathbf{u}_2$ and $\mathbf{u}_4$, where $\mathbf{u}_4$ is given, so $[e \quad f \quad g] \begin{bmatrix} x_4 - x_2 \\ y_4 - y_2 \\ z_4 - z_2 \end{bmatrix} = 0$, or

$$[-\sin\theta_1 \quad 0 \quad \cos\theta_1] \begin{bmatrix} x_4 + l_1 \sin\theta_1 \\ y_4 \\ z_4 - l_1 \cos\theta_1 \end{bmatrix} = 0$$

so $-x_4 \sin\theta_1 + z_4 \cos\theta_1 - l_1(\sin^2\theta_1 + \cos^2\theta_1) = z_4 \cos\theta_1 - x_4 \sin\theta_1 - l_1 = 0$. Substituting $\sin\theta_1 = 2\alpha/(1 + \alpha^2)$ and $\cos\theta_1 = (1 - \alpha^2)/(1 + \alpha^2)$, where $\alpha = \tan\theta_1/2$, then $\alpha^2(l_1 + z_4) + \alpha(2x_4) + l_1 - z_4 = 0$ and

$$\alpha = \frac{-x_4}{l_1 + z_4} \pm \frac{\sqrt{4x_4^2 - 4(l_1 + z_4)(l_1 - z_4)}}{2(l_1 + z_4)} = \frac{-x_4 \pm \sqrt{x_4^2 - l_1^2 + z_4^2}}{l_1 + z_4}.$$

Thus θ_1 and so $\mathbf{p}_2^2$ are determined. The two points $\mathbf{u}_2$ and $\mathbf{u}_4$ are on $\mathbf{p}_2^2$, so we complete the IKS by determining the position of $\mathbf{u}_3$ on $\mathbf{p}_2^2$ and thus angles ψ_2 and θ_3.

---

4. Determine the positional capabilities of the following robots and determine the IKSs with respect to these capabilities:

$$|\phi\sigma\theta\lambda| \qquad |\theta\theta\psi\sigma| \qquad |\phi\sigma\theta\tilde{\psi}| \qquad |\phi\theta\theta\phi\theta| \qquad |\phi\theta\theta\phi\theta| \qquad |\phi\theta\theta\phi\theta\tilde{\psi}|$$

$$|\phi\theta\theta\phi\theta\lambda| \qquad |\phi\theta\theta\phi\theta\psi| \qquad |\phi\theta\theta\phi\phi\tilde{\psi}|$$

$$\begin{matrix} \bar{\mathbf{p}}_2^2 & & \bar{\mathbf{p}}_3^0 & \\ |\ \phi_1\sigma_2 & |\ \theta_3 l_3 & |\ \lambda_4 l_4 & | \\ \mathbf{u}_1 & \mathbf{u}_2 & \mathbf{u}_3 & \mathbf{u}_4 \end{matrix}$$

with planes $\bar{\mathbf{p}}^2\bar{\mathbf{p}}^0$, so the robot is pencilable and $\mathbf{u}_4$ and $\bar{\mathbf{p}}_3^0$ are given. Therefore, the robot is CIKS, since $\mathbf{u}_4$ is on $\bar{\mathbf{p}}_2^2$ and this plane is established. Recall if the plane(s) with area is established, the IKS is closed. $\mathbf{u}_2$ is the point of intersection in $\bar{\mathbf{p}}^2$ where the circle of radius $l_2 + l_3$ intersects the line $\mathbf{u}_1 + \sigma_2 \boldsymbol{\nu} \times \begin{bmatrix} a \\ b \\ c \end{bmatrix}$.

$$\begin{matrix} \mathbf{p}_1^2 & & \bar{\mathbf{p}}_2^1 & \\ |\ \theta_1 l_1 & |\ \theta_2 l_2 & |\ \psi_3\sigma_4 & | \\ \mathbf{u}_1 & \mathbf{u}_2 & \mathbf{u}_3 & \mathbf{u}_4 \end{matrix}$$

with planes $\mathbf{p}^2\bar{\mathbf{p}}^1$ so this SCARA configuration is pencilable with known CIKS.

$$\begin{array}{l} \bar{\mathbf{p}}_2^2 \qquad\qquad \mathbf{p}_3^1 \\ | \phi_1\sigma_2 | \theta_3 l_3 | \tilde{\psi}_4 l_4 | \\ \mathbf{u}_1 \qquad \mathbf{u}_2 \quad \mathbf{u}_3 \quad \mathbf{u}_4 \end{array}$$

with planes $\mathbf{p}^2\bar{\mathbf{p}}^1$ so the robot is loose placeable. If $\tilde{\psi}_4$ is frozen, the robot becomes a 3 *dof* placeable robot with a displacement $\delta = -l_4$ at the end effector; refer to the introduction of Chapter 3 for the definition of δ.

$$\begin{array}{l} \bar{\mathbf{p}}_2^2 \qquad\qquad\qquad\quad \bar{\mathbf{p}}_3^0 \quad \bar{\mathbf{p}}_4^0 \\ | \phi_1 l_1 | \theta_2 l_2 | \theta_3 l_3 | \phi_4 l_4 | \phi_5 l_5 | \\ \mathbf{u}_1 \quad \mathbf{u}_2 \quad \mathbf{u}_3 \quad \mathbf{u}_4 \quad \mathbf{u}_5 \quad \mathbf{u}_6 \end{array}$$

with planes $\bar{\mathbf{p}}^2\bar{\mathbf{p}}^0\bar{\mathbf{p}}^0$, so the robot is loose pencilable. If we freeze ϕ_4 at datum, then $\mathbf{u}_6$ is on $\bar{\mathbf{p}}_2^2$ and so $\bar{\mathbf{p}}_2^2$ is established, and the rest of the IKS follows.

$$\begin{array}{l} \bar{\mathbf{p}}_2^2 \qquad\qquad\qquad\quad \bar{\mathbf{p}}_3^1 \\ | \phi_1 l_1 | \theta_2 l_2 | \theta_3 l_3 | \phi_4 l_4 | \theta_5 l_5 | \\ \mathbf{u}_1 \quad \mathbf{u}_2 \quad \mathbf{u}_3 \quad \mathbf{u}_4 \quad \mathbf{u}_5 \quad \mathbf{u}_6 \end{array}$$

with planes $\bar{\mathbf{p}}^2\bar{\mathbf{p}}^1$, so the robot is directable and $\mathbf{u}_5$ and $\mathbf{u}_6$ are given. $\mathbf{u}_5$ is on $\bar{\mathbf{p}}_2$, so $\bar{\mathbf{p}}_2$ is established and we have CIKS. The IKS proceeds as follows: given the base orientation and position, $\mathbf{u}_2$ is known. $\mathbf{u}_3$ is on $\bar{\mathbf{p}}_2$ at one of the two points of intersection of two circles, one centered at $\mathbf{u}_2$ with a radius of l_2, the other centered at $\mathbf{u}_5$ with a radius of $\sqrt{l_3^2 + l_4^2}$.

$$\begin{array}{l} \bar{\mathbf{p}}_2^2 \qquad\qquad\qquad\quad \bar{\mathbf{p}}_3^1 \qquad\quad \mathbf{p}_4^1 \\ | \phi_1 l_1 | \theta_2 l_2 | \theta_3 l_3 | \phi_4 l_4 | \theta_5 l_5 | \tilde{\psi}_6 l_6 | \\ \mathbf{u}_1 \quad \mathbf{u}_2 \quad \mathbf{u}_3 \quad \mathbf{u}_4 \quad \mathbf{u}_5 \quad \mathbf{u}_6 \quad \mathbf{u}_7 \end{array}$$

with planes $\bar{\mathbf{p}}^2\bar{\mathbf{p}}^1\mathbf{p}^1$ so the robot is loose directable. We can freeze a joint such as θ_5 or $\tilde{\psi}_6$ and so find $\bar{\mathbf{p}}_2$, since $\mathbf{u}_5$ is on this plane; the IKS proceeds as before.

$$\begin{array}{l} \bar{\mathbf{p}}_2^2 \qquad\qquad\qquad\quad \bar{\mathbf{p}}_3^1 \qquad\quad \bar{\mathbf{p}}_4^0 \\ | \phi_1 l_1 | \theta_2 l_2 | \theta_3 l_3 | \phi_4 l_4 | \theta_5 l_5 | \lambda_6 l_6 | \\ \mathbf{u}_1 \quad \mathbf{u}_2 \quad \mathbf{u}_3 \quad \mathbf{u}_4 \quad \mathbf{u}_5 \quad \mathbf{u}_6 \quad \mathbf{u}_7 \end{array}$$

with planes $\bar{\mathbf{p}}^2\bar{\mathbf{p}}^1\bar{\mathbf{p}}^0$ so the robot is orientable, and we calculate $\mathbf{u}_5$ from given $\mathbf{u}_6$ and $\mathbf{u}_7$. $\mathbf{u}_5$ lies on $\bar{\mathbf{p}}_2$, so $\bar{\mathbf{p}}_2$ is established and the CIKS is obtained as before.

$$\begin{array}{l} \bar{\mathbf{p}}_2^2 \qquad\qquad\qquad\quad \bar{\mathbf{p}}_3^1 \qquad\quad \bar{\mathbf{p}}_4^0 \\ | \phi_1 l_1 | \theta_2 l_2 | \theta_3 l_3 | \phi_4 l_4 | \theta_5 l_5 | \psi_6 l_6 | \\ \mathbf{u}_1 \quad \mathbf{u}_2 \quad \mathbf{u}_3 \quad \mathbf{u}_4 \quad \mathbf{u}_5 \quad \mathbf{u}_6 \quad \mathbf{u}_7 \end{array}$$

with planes $\bar{\mathbf{p}}^2\bar{\mathbf{p}}^1\bar{\mathbf{p}}^0$; thus, the robot is orientable and we are given $\bar{\mathbf{p}}_4$, $\mathbf{u}_6$, and $\mathbf{u}_7$. Unfortunately, as can be seen from Table 3.1, we need the status of the last joint ψ_6 in order to establish $\mathbf{u}_5$; notice that we are solving the joint from the up-link end and so must consider it as joint $\tilde{\psi}$. Thus the robot has OIKS.

$$\begin{array}{l} \bar{\mathbf{p}}_2^2 \qquad\qquad\qquad\quad \bar{\mathbf{p}}_3^0 \quad \bar{\mathbf{p}}_4^0 \quad \mathbf{p}_5^1 \\ | \phi_1 l_1 | \theta_2 l_2 | \theta_3 l_3 | \phi_4 l_4 | \phi_5 l_5 | \tilde{\psi}_6 l_6 | \\ \mathbf{u}_1 \quad \mathbf{u}_2 \quad \mathbf{u}_3 \quad \mathbf{u}_4 \quad \mathbf{u}_5 \quad \mathbf{u}_6 \quad \mathbf{u}_7 \end{array}$$

with planes $\bar{\mathbf{p}}^2\bar{\mathbf{p}}^0\bar{\mathbf{p}}^0\mathbf{p}^1$, so the robot is orientable and $\mathbf{u}_6$, $\mathbf{u}_7$, and $\mathbf{p}_5$ are given. We calculate $\mathbf{u}_5$ from given $\mathbf{u}_6$ and $\mathbf{u}_7$. $\mathbf{u}_5$ lies on $\bar{\mathbf{p}}_2$, so $\bar{\mathbf{p}}_2$ is established and the CIKS obtained as for the directable robot of part (e).

5. A 3 *dof* robot can be described by $\left| \phi_1 \;\; \begin{matrix}\bar{\sigma}_2\\ \sigma_2\end{matrix} \right| \left. \theta_3 \;\; l_3 \right|$. If the sliding joint has $\sigma_2 = 0.5$ and $\bar{\sigma}_2 = 1$ and $l_3 = 1$, sketch the working envelope of the robot. Assume no limitation on the angular rotations and that the robot is mounted on an infinite base.

Obtain the inverse kinematic solution of the robot assuming the base is at the origin. Show the joint status for the end effector at

$$\mathbf{u} = \begin{bmatrix} -0.5 \\ 0 \\ 1.5 \\ 1 \end{bmatrix} \quad \text{and} \quad \mathbf{v} = \begin{bmatrix} 0 \\ 0.5 \\ 0 \\ 1 \end{bmatrix}$$

Is a straight-line solution $\mathbf{u} + \alpha(\mathbf{v} - \mathbf{u})$, $0 \le \alpha \le 1$, possible? In particular, can the midpoint $0.5(\mathbf{u} + \mathbf{v})$ be reached? If the joint status at $\mathbf{u}$ and $\mathbf{v}$ are $\boldsymbol{\xi}_u$ and $\boldsymbol{\xi}_v$ respectively, a path $\boldsymbol{\xi}_u + \alpha(\boldsymbol{\xi}_v - \boldsymbol{\xi}_u)$ is possible (this is called a slew solution and such solutions are discussed further in section 8.1). What is the position of the end effector of this path for $\alpha = 0.5$. Compare this to the geometric midpoint.

A cross section through the working volume is shown crosshatched in Figure 6.11. Notice the extensive central void. From equation (5.3),

$$\mathbf{u}_3 = \begin{bmatrix} -\cos\phi_1 \sin\theta_2 \\ -\sin\phi_1 \sin\theta_2 \\ l_1 + l_2 \cos\theta_2 \\ 1 \end{bmatrix} = \begin{bmatrix} x \\ y \\ z \\ 1 \end{bmatrix}$$

so given x, y, and z, $\phi_1 = \tan^{-1} y/x$, $\theta_2 = -\sin^{-1}\left(\dfrac{x}{\cos\phi_1}\right)$ and $l_1 = z - l_2\cos\theta_2$ give the IKS. For $x = -0.5$, $y = 0.707$, $z = 1.5$, $\phi_1 = 125.27°$, $\theta_2 = \sin^{-1} 0.866 = 60°$, and $l_1 = 1.5 - 0.5 = 1$. For $x = 0.5$, $y = -0.5$, $z = 0$, $\phi_1 = -45°$, $\theta_2 = \sin^{-1} 0.707 = 135°$, and $l_1 = 0 + 0.707 = 0.707$.

Note: The angles were chosen to fit the situation: $\sin^{-1} 0.707 = (4n + 1)\, 90 \pm$

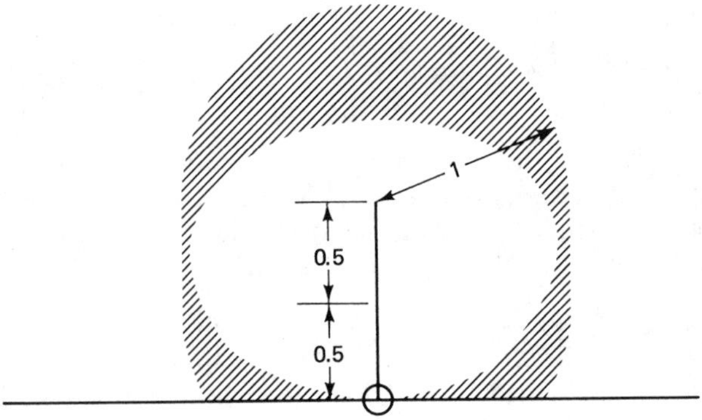

Figure 6.11 Cross section of working volume of $|\,\phi\sigma\theta\,|$

45 for $n = 0, 1, \ldots,$ and 135 fits the requirement for l_1 with $z = 0$. A straight-line solution for the end effector to move between $\mathbf{u}$ and $\mathbf{v}$ is impossible, as can be seen by comparing the midpoint of the proposed path $(\mathbf{u} + \mathbf{v})/2 = 0.5 \begin{bmatrix} 0 \\ 0.207 \\ 1.5 \\ 2 \end{bmatrix} = \begin{bmatrix} 0 \\ 0.1035 \\ 0.75 \\ 1 \end{bmatrix}$ with the shaded area.

The slew solution can be written

$$\begin{bmatrix} \phi_1(\alpha) \\ \theta_2(\alpha) \\ l_1(\alpha) \end{bmatrix} = \begin{bmatrix} 125.27 \\ 60 \\ 1 \end{bmatrix} - \alpha \begin{bmatrix} 170.27 \\ -75 \\ 0.293 \end{bmatrix}$$

for $0 \le \alpha \le 1$. The midpoint of the slew path is given by

$$\begin{bmatrix} \phi_1(0.5) \\ \theta_2(0.5) \\ l_1(0.5) \end{bmatrix} = \begin{bmatrix} 40.135 \\ 97.5 \\ 0.8535 \end{bmatrix}$$

and the location of this point is $\begin{bmatrix} -0.758 \\ -0.639 \\ 0.723 \\ 1 \end{bmatrix}$, considerably different from the midpoint of the end points.

---

6. Given a 3 *dof* polar robot $|\lambda\theta\sigma|$ such that the total link length is 1 m, design the robot such that the volume covered by the end effector at the free end of the last link is a maximum. Assume that the sliding link has a ratio of maximum length to minimum length of 2 : 1. Assume a large horizontal base, no joint angle limitations, and a vertical first link. Find the working volume of the robot.

With the base of the robot $[0 \quad 0 \quad 0 \quad 1]^\#$, the end effector of the robot must move from point $[0.8 \quad 0 \quad 0 \quad 1]^\#$ to point $[0.4 \quad 0.4 \quad 0.4 \quad 1]^\#$ in a straight line. Define the motion of the joints of the robot in order to follow this line.

There is only one solution to the problem of the design of the robot: $l_1 = 0$ and $\sigma_2 = 1$. The robot has an outer envelope of a hemisphere of radius 1 m. It cannot reach within $\frac{1}{2}$ m of the origin, so the volume covered is

$$\frac{2\pi}{3}\left(1 - \frac{1}{8}\right) = 1.833 \text{ m}^3$$

Consider the plane $[0 \quad 1/\sqrt{2} \quad 1/\sqrt{2} \quad 0]$ containing the two points $[0.8 \quad 0 \quad 0 \quad 1]^\#$ and $\bar{\mathbf{p}}^2 = [0.4 \quad 0.4 \quad 0.4 \quad 1]^\#$ and passing through the origin. This plane contains the

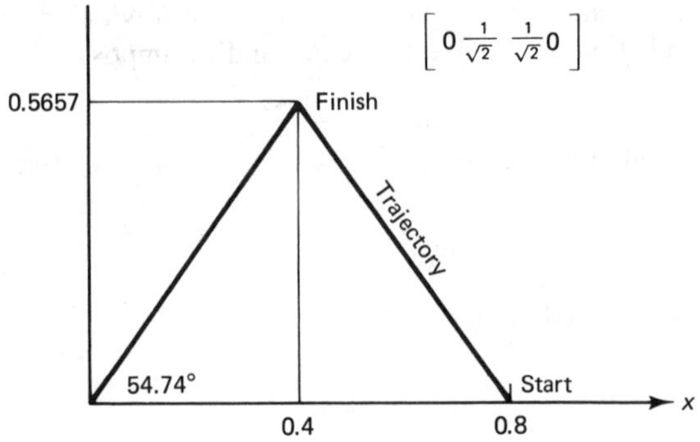

Figure 6.12

links of the robot as it moves in a straight line between the two points. The situation in this plane is shown in Figure 6.12. The revolute joint starts off with angle $\theta = 0$ and ends with $\theta = \tan^{-1} 0.2\sqrt{2}/0.4 = 35.3°$. The line in $\bar{\mathbf{p}}^2$ joining the two points in $\bar{\mathbf{p}}^2$ is given by

$$\begin{bmatrix} 0.8 + \alpha(0.4 - 0.8) \\ 0 + \alpha(0.5657 - 0) \end{bmatrix} = \begin{bmatrix} 0.8 + 0.4\alpha \\ 0.5657\alpha \end{bmatrix}$$

so to follow this line, the three joints will be $\lambda_1 = 45°$, $\theta_2 = 90 - 54.76\alpha$, and

$$\sigma_3 = 0.8\sqrt{0.75\alpha^2 - \alpha + 1}$$

where the distance of a point on the line from the origin D is given by

$$D^2 = (0.8 - 0.4\alpha)^2 + 0.5657^2\alpha^2 = 0.48\alpha^2 - 0.64\alpha + 0.64 = 0.64(0.75\alpha^2 - \alpha + 1)$$

---

7. Determine the capability of the manipulator $|\,\phi\theta\theta\lambda\theta\psi\,|$. Discuss a method of finding its IKS in terms of this capability.

The full expression for the manipulator is

$$\begin{matrix} & \bar{\mathbf{p}}_2^2 & & & \bar{\mathbf{p}}_3^1 & & & \bar{\mathbf{p}}_4^0 & \\ & |\,\phi_1 l_1 & |\,\theta_2 l_2 & |\,\theta_3 l_3 & |\,\lambda_4 l_4 & |\,\theta_5 l_5 & |\,\psi_6 l_6\,| & & \\ \mathbf{u}_1 & \mathbf{u}_2 & \mathbf{u}_3 & \mathbf{u}_4 & \mathbf{u}_5 & \mathbf{u}_6 & \mathbf{u}_7 & & \end{matrix}$$

The planes of motion are $\bar{\mathbf{p}}^2\bar{\mathbf{p}}^1\bar{\mathbf{p}}^0$, so according to Table 1 the manipulator is orientable. Thus $\mathbf{u}_7$, $\mathbf{u}_6$, and $\bar{\mathbf{p}}_4^0$ are given. Also the base orientation and position give $\mathbf{u}_1$ and $\mathbf{u}_2$. One more piece of information on $\bar{\mathbf{p}}_2^2$ will establish this plane and show that the IKS is closed. However, it is not clear that this information is readily available.

$\bar{\mathbf{p}}_3^1$ is known from $\mathbf{u}_6$, $\mathbf{u}_7$, and $\bar{\mathbf{p}}_4^0$. One possible method of solution is to choose ψ_6, which determines $\mathbf{u}_5$, and since $\mathbf{u}_5$ lies on $\bar{\mathbf{p}}_2^2$, then $\bar{\mathbf{p}}_2^2$ is established. We can then calculate the line of intersection of $\bar{\mathbf{p}}^2$ and $\bar{\mathbf{p}}^1$ and so calculate $\mathbf{u}_3$ from this line and $\mathbf{u}_5$. We calculate $\delta(\psi_6) = |\,\mathbf{u}_3 - \mathbf{u}_2\,| - l_2$, which equals zero if ψ_6 was chosen correctly. We iterate ψ_6 until $\delta = 0$.

It is not clear whether this method can, with suitable algebraic manipulations,

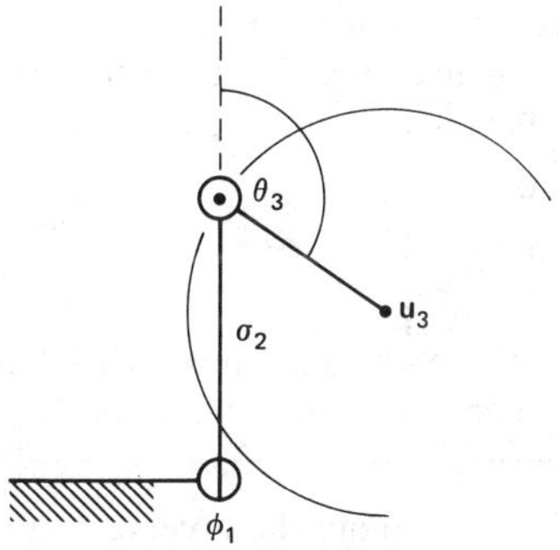

Figure 6.13 The IKS of $|\ \phi\sigma\theta\ |$

be made into a closed-form method. Only one choice (ψ_6) was made, and if δ can be written as an explicit function (albeit nonlinear) of ψ_6, the method will be closed.

---

8. Determine the plane(s) of motion of the robot $|\ \phi\sigma\theta\ |$. Is this robot placeable? If the robot is placeable, indicate geometrically the inverse kinematic solution.

There is only one plane of motion given by $\bar{\mathbf{p}}^2$, and from Section 4.2 we see that the robot is placeable. The geometric inverse kinematic solution is given in Figure 6.13.

EXERCISES FOR CHAPTER 6

1. Obtain the inverse kinematic solution for the robot

$$\begin{array}{l}\mathbf{p}_1 \\ |\ \psi_1 l_1\ |\ \theta_2 l_2\ |\ \psi_3 l_3\ |\ \theta_4 l_4\ |\ \theta_5 l_5\ |\ \psi_6 l_6\ | \\ \mathbf{u}_1 \quad \mathbf{u}_2 \quad \mathbf{u}_3 \quad \mathbf{u}_4 \quad \mathbf{u}_5 \quad \mathbf{u}_6 \quad \mathbf{u}_7\end{array}$$

---

2. For the 3 *dof* revolute robot $\begin{array}{c}\mathbf{p}_1\bar{\mathbf{p}}_2^2 \\ |\ \lambda_1\ 1\ |\ \theta_2\ 1\ |\ \theta_3\ 1\ | \\ \mathbf{u}_1 \quad \mathbf{u}_2 \quad \mathbf{u}_3 \quad \mathbf{u}_4\end{array}$ whose base is at the origin, we require its end effector to move along the line $\begin{bmatrix} 1+\alpha \\ 1 \\ 1 \\ 1 \end{bmatrix}$, $0 \le \alpha \le 2$. If $\mathbf{p}_1 = [0\ \ 1\ \ 0\ \ 0]$ and $\boldsymbol{\nu} = \begin{bmatrix} 0 \\ 0 \\ 1 \end{bmatrix}$, determine the joint status to move along the required path. Is it possible to reach every point along the path? If it is not, show what happens in the IKS equations.

---

3. A polar robot sits at the origin and its first link is vertical. $l_1 = 1$ m and the sliding link can extend from 0.5 to 1 m. Not all points along the end effector path $\begin{bmatrix} 0.5 \\ 0 \\ 0.5 + \alpha \\ 1 \end{bmatrix}$, $-0.5 \le \alpha \le 1$, can be reached. Determine the reachable section of the path, and show what happens to the IKS calculations when a nonfeasible end effector location is chosen.

---

4. Indicate how to obtain the inverse kinematic solution for the robot

$$| \theta_1 l_1 | \phi_2 l_2 | \theta_3 l_3 |$$

---

5. For the cylindrical robot $\overset{\mathbf{p}_1\bar{\mathbf{p}}_2^2}{\underset{\mathbf{u}_1\qquad\mathbf{u}_2\qquad\mathbf{u}_3}{\left| \lambda_1 {}^{1}_{0.5} \right| (\div) {}^{1}_{0.5} \Big|}}$ with $\mathbf{u}_1 = \begin{bmatrix} 0 \\ 0 \\ 0 \\ 1 \end{bmatrix}$, $\boldsymbol{\nu} = \begin{bmatrix} 0 \\ 1 \\ 0 \end{bmatrix}$, and $\mathbf{p}_1 = [0.7071 \quad 0 \quad 0.7071 \quad 0]$, find the joint status to place the end effector at $\mathbf{u}_3 = [0.4 \quad 0.4 \quad 0.4 \quad 1]^{\#}$.

---

6. Determine the logical position space and obtain the inverse kinematic solution of

$$\left| \sigma_1 {}^{1}_{2} \right| \phi_2 \; 1 \left| \theta_3 \; 1 \right|$$

with respect to this position space.

---

7. Show how to obtain the inverse kinematic solution for the robot $| \psi\theta\gamma |$.

---

8. Show how to obtain the IKS of the manipulator $| \theta_1 \; 1 \; | \; \lambda_2 \; 0 \; | \; \gamma_3 \; 1 \; |$.

---

9. If the robot $| \theta\theta\gamma |$ has a logical position space show how to obtain the IKS with respect to this position space.

---

10. Show how to obtain the IKS of the placeable robot $| \lambda\theta\gamma |$.

---

11. Show how to obtain the IKS of the placeable robot $| \lambda\gamma\gamma |$.

---

12. Does the robot $| \lambda{+}\lambda{+}\sigma |$ have a logical position space? If it does, show how to obtain the IKS.

---

13. Show how to obtain the IKS of $| \psi\gamma\theta |$. Is the IKS closed?

---

14. Show how to obtain the IKS of the robot $|\ \gamma\gamma\theta\ |$.

---

15. Determine a position space for the robot

$$|\ \phi_1 l_1\ |\ \phi_2 l_2\ |\ \theta_3 l_3\ |\ \phi_4 l_4\ |\ \theta_5 l_5\ |$$

and obtain an inverse kinematic solution for this position space.

---

16. Find the inverse kinematic solution of the 3 *dof* SCARA robot $|\theta\theta \times \sigma|$ if $\mathbf{u}_1 = [0\quad 0\quad 0\quad 1]^{\#}$, $\mathbf{p}_1 = [0\quad 0\quad 1\quad 0]$, $l_1 = 1$, $l_2 = 1$, $\sigma_3 = 0.5$, $\overline{\sigma_3} = 1$, and $\mathbf{u}_4 = [1\quad 0.5\quad 0.25\quad 1]^{\#}$.

---

17. Show how the IKS can be obtained for $|\ \tilde{\phi}\theta\theta\ |$.

---

18. It does not appear that the placeable robot $|\ \theta\gamma\gamma\ |$ has a geometric IKS. Suggest the best way to obtain an algebraic IKS. Is this solution open or closed?

---

19. Find the solutions for the mechanism $|\ \theta\theta\theta\ |$ when l_3 is parallel to l_1.

---

20. The mechanism $|\ \theta\theta\theta\ |$ is used as a sprue picker. Determine a logical position space and find the IKS with respect to this position space.

---

21. Determine a logical position space and show how to find the IKS with respect to this position space for the mechanism $|\ \sigma\theta\gamma\ |$.

---

22. Suppose the loose robot $|\ \lambda\theta\theta\theta\ |$ with unity link lengths sits at the origin with its first link vertical and its end effector required to move along the line $[\alpha\quad \alpha\quad 0\quad 1]^{\#}$, $0 \le \alpha \le 2$. Find all values of the joint status to satisfy points $\alpha = 0$, $\alpha = 0.5$, and $\alpha = 1$ along this trajectory.

---

23. Obtain the IKS for the robot

$$\begin{array}{ccccc} & \mathbf{p}_1\overline{\mathbf{p}}_2^2 & & \overline{\mathbf{p}}_3^0 & \\ |\ \lambda_1\ 0.4 & |\ \theta_2\ 0.5 & |\ \theta_3\ 0.25 & |\ \lambda_4\ 0.25 & | \\ \mathbf{u}_1 & \mathbf{u}_2 & \mathbf{u}_3 & \mathbf{u}_4 & \mathbf{u}_5 \end{array}$$

if $\mathbf{u}_1 = \begin{bmatrix} 0 \\ 0 \\ 0 \\ 1 \end{bmatrix}$, $\mathbf{u}_5 = \begin{bmatrix} 0.300 \\ 0.387 \\ 0 \\ 1 \end{bmatrix}$, $\mathbf{p}_1 = [0\quad 1\quad 0\quad 0]$, $\boldsymbol{\nu} = \begin{bmatrix} 0 \\ 0 \\ 1 \end{bmatrix}$, and $\overline{\mathbf{p}}_3^0 = [1\quad 0\quad 0\quad -0.3]$.

---

24. Suppose the robot $\begin{array}{c} \mathbf{p}_1\overline{\mathbf{p}}_2^3 \\ \left| \lambda_1\ 1 \;\middle|\; \theta_2\ 1 \;\middle|\; \theta_3\ 0 \;\middle|\; \sigma_4\ {}^{1}_{0.5} \right| \\ \mathbf{u}_1 \quad \mathbf{u}_2 \quad \mathbf{u}_3 \quad \mathbf{u}_4 \qquad \mathbf{u}_5 \end{array}$ with $\mathbf{u}_1 = \begin{bmatrix} 0 \\ 0 \\ 0 \\ 1 \end{bmatrix}$ and base orientation $\mathbf{p}_1 = [1 \quad 0 \quad 0 \quad 0]$, $\boldsymbol{\nu} = \begin{bmatrix} 0 \\ 0 \\ 1 \end{bmatrix}$ is required to pick up parts from surface $[0 \quad 0 \quad 1 \quad 0]$ when its last link is vertical. Determine the size and shape of the region on surface $[0 \quad 0 \quad 1 \quad 0]$ that can be serviced.

---

25. The robot $\begin{array}{c} \overline{\mathbf{p}}_2^3 \\ |\lambda_1\ 0 \mid \theta_2\ 1 \mid \theta_3\ 1 \mid \theta_4\ 1| \end{array}$ sits at $\mathbf{u}_1 = \begin{bmatrix} 0 \\ -0.5 \\ 0 \\ 1 \end{bmatrix}$ with $\boldsymbol{\nu} = \begin{bmatrix} 0 \\ 0 \\ 1 \end{bmatrix}$ and $\mathbf{p}_1 = [0 \quad 1 \quad 0 \quad 0]$. The robot is required to remove small parts from a conveyor belt whose center line is given by $[\alpha \quad 0 \quad 0 \quad 1]^{\#}$, where $-10 \le \alpha \le 10$. The last link of the robot, l_4, is required to be vertical when it picks up a part, and the time for prehension also requires that the end effector track along the conveyor belt a distance of 0.2 m. If the conveyor belt is 0.5 m wide, determine the working area of the conveyor belt within which the end effector must first be positioned about the part to be picked up.

---

26. The robot $\begin{array}{c} \mathbf{p}_1\overline{\mathbf{p}}_2^3 \\ \left| \theta_1\ 1 \;\middle|\; \theta_2\ 1 \;\middle|\; (\times)\sigma_3\ {}^{1}_{0.5} \right| \end{array}$ sits at $\mathbf{u}_1 = \begin{bmatrix} -1 \\ 0 \\ 1 \\ 1 \end{bmatrix}$ with $\boldsymbol{\nu} = \begin{bmatrix} 0 \\ 0 \\ 1 \end{bmatrix}$ and $\mathbf{p}_1 = [0 \quad 0 \quad 1 \quad 0]$. The robot is required to remove small parts from a conveyor belt whose center line is given by $[0 \quad \alpha \quad 0 \quad 1]^{\#}$, where $-4 \le \alpha \le 4$. If the conveyor belt is 0.4 m wide, determine the complete working area of the end effector on the conveyor belt.

---

27. The robot $|\ \lambda\theta\theta\lambda\ |$ is used as a palletizer. The robot is mounted so that the base is at the origin and the base orientation is given by $\mathbf{p}_1 = [0 \quad 1 \quad 0 \quad 0]$ and $\boldsymbol{\nu} = [0 \quad 0 \quad 1]^{\#}$. The last link of the robot is required to be vertical when palletizing vertically, and the final plane of motion must be $[0 \quad 1 \quad 0 \quad d]$. If the link lengths of the robot (except $l_5 = 0$) are 1 m, determine the motions to move a package over the square from

$$\begin{bmatrix} 0.2 \\ 0 \\ 0 \\ 1 \end{bmatrix} \text{ to } \begin{bmatrix} 0.7 \\ 0 \\ 0 \\ 1 \end{bmatrix} \text{ to } \begin{bmatrix} 0.7 \\ 0.5 \\ 0 \\ 1 \end{bmatrix} \text{ to } \begin{bmatrix} 0.2 \\ 0.5 \\ 0 \\ 1 \end{bmatrix} \text{ to } \begin{bmatrix} 0.2 \\ 0 \\ 0 \\ 1 \end{bmatrix}$$

such that the package is raised by 0.5 m while keeping the x- and y-values of the end effector essential constant.

---

28. Indicate a methodology for obtaining a numeric solution for the orientable robot $|\lambda\theta\theta\phi\phi\phi|$ by choosing a joint angle, obtaining an error in the solution based on this choice, and refining the choice.

---

29. Determine the logical position space for the robot $|\lambda\theta\theta\psi\psi\psi|$. This robot appears to have an OIKS with respect to this position space. Indicate a methodology for obtaining a numeric solution for this robot.

---

30. Obtain the IKS for the orientable robot $|\lambda\theta\theta\phi\theta\lambda|$ when the link lengths are unity.

---

31. Determine the logical position space for the robot $|\lambda\theta\theta\theta\phi|$. Indicate how to obtain the IKS with respect to this position space. If the link lengths are $l_1 = 5$, $l_2 = 10$, $l_3 = 8$, $l_4 = 6$, and $l_5 = 4$, determine the working envelope, assuming $\pm 90°$ angular swing of the revolute joints with respect to datum.

---

32. Establish a logical position space and obtain the IKS for the robot $|\lambda\theta\theta\theta\lambda\phi|$.

CHAPTER SEVEN

Parametric Description of Curves

Fundamental to the path control of robots is the mathematical description of curves. We define a curve as a finite trajectory in Cartesian space with defined endpoints. There are several ways of defining a curve, from piecewise segmentation of the curve to continuous function approximations. Continuous function approximation is usually accomplished with polynomials but may be a specification by series of orthogonal functions [Churchill 1963] (the generalized Fourier series). Contained within the family of the sets of functions orthogonal over an interval are a variety of orthogonal polynomials, but their study is beyond the scope of this work. Here, we consider the parametric description of curves by polynomials.

Curves may be as simple as a straight line (a polynomial of order 1) or so complicated that they require a high-order polynomial description. The shape of the curve is not all that is under consideration. Velocity control is an implicit part of the path-control problem, and it is shown in this chapter that parametric polynomial descriptions of curves can be devised that control the velocity.

Polynomial curve fitting is a well-established subject [Ralston and Rabinowitz 1978].

The problem of describing a curve by

$$\begin{bmatrix} x \\ y(x) \\ z(x) \\ 1 \end{bmatrix}$$

where $y(x)$ and $z(x)$ are polynomial functions of x, will be illustrated. A circle of radius r centered at the origin in the xy-plane is

$$\begin{bmatrix} x \\ \pm\sqrt{r^2 - x^2} \\ 0 \\ 1 \end{bmatrix}$$

For $-r < x < r$, no problems are encountered, but as $x \to r$, the slope $dy/dx \to \infty$. The distance between points is uneven for equal change in x: The distance between points $(0, r)$ and $(0.1r, 0.995r)$ is

$$\sqrt{0.1^2r^2 + 0.005^2r^2} = 0.10r$$

but the distance between points $(r, 0)$ and $(0.9r, 0.436r)$ is

$$\sqrt{0.1^2r^2 + 0.436^2r^2} = 0.447r$$

Further, there are two values for y for every value of x. To avoid these problems, a parametric description of the circle can be used. From the polar description of the circle $(x + jy) = r \angle \theta$, we have $x = r \cos \theta$ and $y = r \sin \theta$. The complete circle is subscribed when θ changes from ψ to $\psi + 2\pi$ radians for any ψ, and the spacing of any pair of points for the same change in θ is independent of θ. Further, no multiple-value problems are encountered.

Suppose the curve is defined parametrically such that a point on the curve is given by $[x(t) \quad y(t) \quad z(t) \quad 1]^{\#}$. Then

$$\frac{dy}{dx} = \frac{dy(t)/dt}{dx(t)/dt}$$

where both $dy(t)/dt$ and $dx(t)/dt$ are bounded, although $\dfrac{dy}{dx}$ may be unbounded.

To summarize, a nonparametric polynomial description of a closed curve (with starting point and endpoint the same) is unsatisfactory, since slopes become unbounded for at least two points, the spacing of points is uneven, and multiple solutions cause confusion. A parametric description can overcome these problems. However, a nonparametric description may be satisfactory for arcs in which the polynomial is single-valued, and a closed curve can be composed of several such arcs. Another consideration is the ordering of the points. In polynomial fitting the ordering of points is not of consideration, but in parametric descriptions the ordering of the points as well as the location of the points determines the curve. Further, it is possible in a parametric description to have two points at the same location. Henceforth, except for simple and well-defined cases, all curves will be specified parametrically.

7.1 PARABOLIC BLENDING

Parabolic blending is a technique that melds quadratic sections of curves together to produce a continuous curve with continuous first derivatives [Overhauser 1968]. Given $n + 1$ points

$$\mathbf{u}_i = \begin{bmatrix} x_i \\ y_i \\ z_i \\ 1 \end{bmatrix} = \begin{bmatrix} x(i\Delta t) \\ y(i\Delta t) \\ z(i\Delta t) \\ 1 \end{bmatrix}, \qquad i = 1, 2, \ldots, n + 1. \tag{7.1}$$

Suppose $P_i(t)$, $i = 1, 2, \ldots, n - 1$, is a quadratic that passes through the x valve of the three points $\mathbf{u}_i$, $\mathbf{u}_{i+1}$, and $\mathbf{u}_{i+2}$.

Let

$$Q_i(t) = \left\{1 - \frac{s}{\Delta t}\right\} P_i(t) + \frac{s}{\Delta t} P_{i+1}(t), \qquad 0 \le s \le \Delta t \tag{7.2}$$

For s small, $Q_i(t) \cong P_i(t)$, and for $s \to \Delta t$, $Q_i(t) \cong P_{i+1}(t)$. Thus, $Q_i(t)$ can be considered the blending of the two quadratics $P_i(t)$ and $P_{i+1}(t)$. The blending of two quadratics is shown in Figure 7.1.

A simple way of presenting the data is to construct a divided-difference table to the second divided difference. The first divided difference between t_1 and t_2 is defined as

$$\frac{\qquad}{t_1, t_2} = \frac{x_2 - x_1}{t_2 - t_1} = \frac{\qquad}{t_2, t_1} \tag{7.3}$$

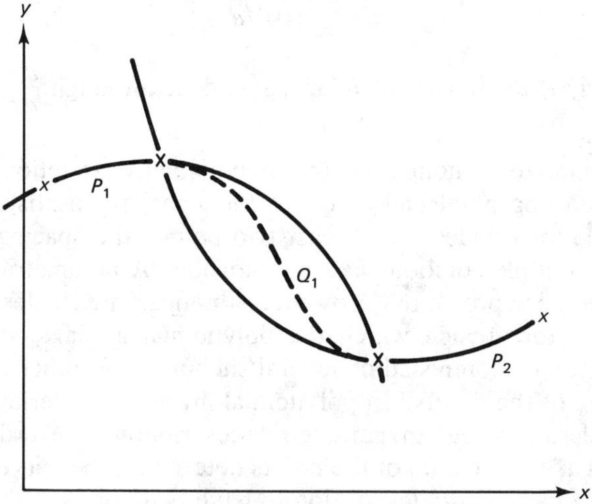

Figure 7.1 Parabolic blending

and the second divided difference between t_1, t_2, and t_3 is defined as

$$\overline{t_1, t_2, t_3} = \frac{\overline{t_3, t_2} - \overline{t_2, t_1}}{t_3 - t_1} \tag{7.4}$$

The divided difference table to the second divided difference is

$$\begin{array}{cccc}
t_1 & x_1 & & \\
 & & \overline{t_1, t_2} & \\
t_2 & x_2 & & \overline{t_1, t_2, t_3} \\
 & & \overline{t_2, t_3} & \\
t_3 & x_3 & & \overline{t_2, t_3, t_4} \\
 & & \overline{t_3, t_4} & \\
t_4 & x_4 & & \overline{t_3, t_4, t_5} \\
 & & \overline{t_4, t_5} & \\
t_5 & x_5 & & \overline{t_4, t_5, t_6} \\
 & & \overline{t_5, t_6} & \\
t_6 & x_6 & & \\
 & \vdots & &
\end{array}$$

The divided-difference table is a simple and systematic way of generating the divided differences, since we note that a divided difference is the difference of the two quantities to the immediate left divided by the time range. The paraboli for the divided difference table just given are constructed by reading a diagonal line downwards to the right as

$$\left.\begin{aligned}
P_1(t) &= x_1 + (t - t_1)\{\overline{t_1, t_2} + (t - t_2)\overline{t_1, t_2, t_3}\} \\
P_2(t) &= x_2 + (t - t_2)\{\overline{t_2, t_3} + (t - t_3)\overline{t_2, t_3, t_4}\} \\
P_3(t) &= x_3 + (t - t_3)\{\overline{t_3, t_4} + (t - t_4)\overline{t_3, t_4, t_5}\} \\
&\;\vdots \\
Q_1(t) &= \left\{1 - \frac{s}{t_2 - t_1}\right\} P_1(t) + \frac{s}{t_2 - t_1} P_2(t) \qquad \text{valid for } t_1 \le t \le t_2 \\
Q_2(t) &= \left\{1 - \frac{s}{t_3 - t_2}\right\} P_2(t) + \frac{s}{t_3 - t_2} P_3(t) \qquad \text{valid for } t_2 \le t \le t_3 \\
&\;\vdots
\end{aligned}\right\} \tag{7.5}$$

We notice that there are two running variables defining $Q_i(t)$, s, and t. However, over the range $i\Delta t \le t \le (i + 1)\Delta t$ for which $Q_i(t)$ is valid, we have $t = i\Delta t + s$. Thus we can write $Q_i(t) = Q_i(i\Delta t + s)$ as a cubic polynomial in t or a cubic polynomial in s.

Example

Perform parabolic blending on the data points (0, 0), (1, A), (2, A), (3, 0), (4, 0), (5, A), and (6, A) of (x, t).

The divided-difference table for this function is

$$
\left.\begin{array}{lllll}
0 & 0 & & & \\
 & & A & & \\
1 & A & & -\frac{A}{2} & P_1(t) = t\left\{A - (t-1)\frac{A}{2}\right\} = \frac{At}{2}\{3-t\} \\
 & & 0 & & \\
2 & A & & -\frac{A}{2} & P_2(t) = A + (t-1)\left\{0 - (t-2)\frac{A}{2}\right\} = \frac{At}{2}(3-t) \\
 & & -A & & \\
3 & 0 & & \frac{A}{2} & P_3(t) = A + (t-2)\left\{-A + (t-3)\frac{A}{2}\right\} = \frac{A}{2}\{t^2 - 7t + 12\} \\
 & & 0 & & \\
4 & 0 & & \frac{A}{2} & P_4(t) = 0 + (t-3)\left\{0 + (t-4)\frac{A}{2}\right\} = \frac{A}{2}\{t^2 - 7t + 12\} \\
 & & A & & \\
5 & A & & -\frac{A}{2} & P_5(t) = 0 + (t-4)\left\{A - (t-5)\frac{A}{2}\right\} = \frac{A}{2}\{-t^2 + 11t - 28\} \\
 & & 0 & & \\
6 & A & & &
\end{array}\right\} \quad (7.6)
$$

$$
\begin{aligned}
Q_1(t) &= (1-s)P_1(t) + sP_2(t) = (1-s)\frac{At}{2}(3-t) + \frac{sAt}{2}(3-t) \\
&= \frac{At}{2}(3-t) \qquad (7.7)
\end{aligned}
$$

is valid for $1 \le t \le 2$, with $s = t - 1$.

$$
\begin{aligned}
Q_2(t) &= (1-s)\frac{At}{2}(3-t) + s\frac{A}{2}\{t^2 - 7t + 12\} \\
&= \frac{A}{2}\{2t^3 - 11t^2 + 23t - 24\} \qquad (7.8)
\end{aligned}
$$

is valid for $2 \le t \le 3$, with $s = t - 2$.

$$
Q_3(t) = P_3(t) = \frac{A}{2}\{t^2 - 7t + 12\} \qquad (7.9)
$$

since $P_3(t) = P_4(t)$, and this is valid for $3 \le t \le 4$ with $s = t - 3$.

$$
\begin{aligned}
Q_4(t) &= (1-s)\frac{A}{2}\{t^2 - 7t + 12\} + s\frac{A}{2}\{-t^2 + 11t - 28\} \\
&= \frac{A}{2}\{-2t^3 + 19t^2 - 63t + 76\} \qquad (7.10)
\end{aligned}
$$

is valid for $4 \le t \le 5$ with $s = t - 4$.

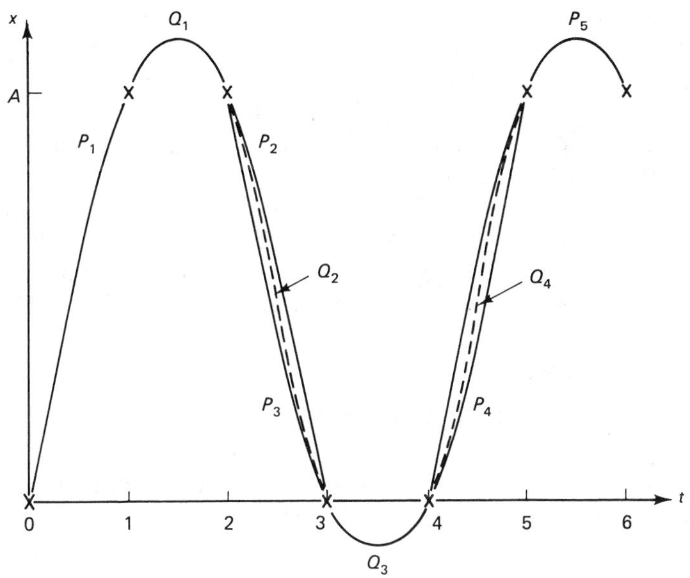

Figure 7.2

The complete P and Q paraboli are shown in Figure 7.2. Notice that for the end sections $0 \leq t \leq 1$ and $5 \leq t \leq 6$, the values of $Q(t)$ are $P_1(t)$ and $P_5(t)$, respectively.

Parabolic blending can be used to generate curves that are multivalued in a single parameter (x, y, or z) by combining sections with variables that are in different parameters. For example, suppose a curve is modeled by three parabolic sections where the first is univalued in x, and the third is univalued in y. Provided the second section is univalued in x and y, it can be described in terms of x or y, parabolic blending can be satisfactory even though the actual curve is double-valued in y at the start and double-valued in x at the end.

A similar technique to parabolic blending breaks the overall curve into sections and uses a third-order polynomial that matches the endpoints and the derivatives at the endpoints using cubic splines [Adams 1974]. The name *spline* was used to describe the method of laying out the hull of a ship—long, flexible (usually balsa wood) splines were laid out and constrained at certain points to follow a desired shape, and between the constrained points, the function behaves like what we now call a cubic spline [Denman 1971]. In the next section we discuss Newton's interpolation polynomial. This polynomial form was chosen over the Lagrange interpolating polynomial [Ralston 1978], since it is computationally more efficient to evaluate. However, the form of the Lagrange interpolating polynomial can be extended not only to pass through required points but also to pass through the endpoints with matching derivatives; the polynomial is then called a Hermite interpolating polynomial [Ralston and Rabinowitz 1978]. A Hermite interpolating polynomial between two points has four constraints (two positional, two derivatives), and so a cubic spline is produced. It must be pointed out that a prerequisite to success in all techniques is to ensure that the function is univalued over each segment.

Parabolic blending is in fact a special case of spline-fitting in which, apart from the first section, the slope of the cubic function $Q_i(t)$ is the slope of parabola $P_{i-1}(t)$ at $t = i\Delta t$. We use the four constraints of position at $t = i\Delta t$, $t = (i + 1)\Delta t$, and $t = (i + 2)\Delta t$ as well as the slope of P_{i-1} at $i\Delta t$ to form this cubic.

7.2 NEWTON'S INTERPOLATION FORMULA

The $n + 1$ data points (t_i, x_i), $i = 1, 2, \ldots, n + 1$, can be used to construct a *poly*nomial $P(t)$ of order n, and an efficient way to find $P(t)$ is by Newton's interpolation formula [Ralston and Rabinowitz, 1978] determined via the divided-difference table. The first and second divided differences were defined in Section 7.2, and the nth divided difference is

$$\overline{t_1, t_2, \ldots, t_{n+1}} = \frac{\overline{t_2, t_3, \ldots, t_{n+1}} - \overline{t_1, t_2, \ldots, t_n}}{t_{n+1} - t_1} \tag{7.11}$$

Equations (7.3), (7.4) and (7.11) can be rewritten to yield

$$\begin{aligned} x(t) &= x_1 + (t - t_1) \cdot \overline{t, t_1} \\ \overline{t, t_1} &= \overline{t_1, t_2} + (t - t_2) \cdot \overline{t, t_1, t_2} \\ \overline{t, t_1, t_2} &= \overline{t_1, t_2, t_3} + (t - t_3) \cdot \overline{t, t_1, t_2, t_3} \\ &\ \vdots \\ \overline{t, t_1, t_2, \ldots, t_n} &= \overline{t_1, t_2, \ldots, t_{n+1}} + (t - t_{n+1}) \cdot \overline{t, t_1, t_2, \ldots, t_{n+1}} \end{aligned}$$

which by successive substitution become the infinite series

$$\begin{aligned} x(t) = x_1 + (t - t_1)\{\overline{t_1, t_2} + (t - t_2)\{\overline{t_1, t_2, t_3} \\ + (t - t_3)\{\ldots\}\}\} \end{aligned} \tag{7.12}$$

An approximation to this is the polynomial of order n

$$\begin{aligned} P(t) = x_1 + (t - t_1)\{\overline{t_1, t_2} + (t - t_2)\{\overline{t_1, t_2, t_3} \\ + \ldots (t - t_n)\overline{t_1, t_2, \ldots, t_{n+1}} \ldots\}\} \end{aligned} \tag{7.13}$$

which is the unique polynomial of order n passing through the $n + 1$ points $t_1, t_2, \ldots, t_{n+1}$. Equation (7.13) is known as *Newton's interpolation formula*. The logical way to construct the divided differences and so evaluate equation (7.13) is by the divided difference table (Table 7.1).

TABLE 7.1

t_1	x_1					
		$\overline{t_1, t_2}$				
t_2	x_2		$\overline{t_1, t_2, t_3}$			
		$\overline{t_2, t_3}$		$\overline{t_1, t_2, t_3, t_4}$		
					$\overline{t_1, t_2, t_3, t_4, t_5}$	
t_3	x_3		$\overline{t_2, t_3, t_4}$			$\overline{t_1, t_2, t_3, t_4, t_5, t_6}$
				$\overline{t_2, t_3, t_4, t_5}$		
		$\overline{t_3, t_4}$			$\overline{t_2, t_3, t_4, t_5, t_6}$	
t_4	x_4		$\overline{t_3, t_4, t_5}$			
		$\overline{t_4, t_5}$		$\overline{t_3, t_4, t_5, t_6}$		
t_5	x_5		$\overline{t_4, t_5, t_6}$			
		$\overline{t_5, t_6}$				
$\vdots$						

Although only the top diagonal is used to form $P(t)$, all divided differences are required to evaluate the final divided difference. It is possible to replace a data point without having to reconstruct the complete difference table. The terms affected by data point (t_i, x_i) are displayed with the symbol X in the following table.

t_{i-5}	x_{i-5}										...
										×	
t_{i-4}	x_{i-4}								×		...
								×		×	
t_{i-3}	x_{i-3}						×		×		...
						×		×		×	
t_{i-2}	x_{i-2}				×		×		×		...
				×		×		×		×	
t_{i-1}	x_{i-1}		×		×		×		×		...
		×		×		×		×		×	
t_i	x_i		×		×		×		×		...
		×		×		×		×		×	
t_{i+1}	x_{i+1}		×		×		×		×		...
				×		×		×		×	
t_{i+2}	x_{i+2}				×		×		×		...
						×		×		×	
t_{i+3}	x_{i+3}						×		×		...
								×		×	
t_{i+4}	x_{i+4}								×		...
										×	
t_{i+5}	x_{i+5}										...

If the data point is the first (last) on the list, it can be deleted, together with the top (bottom) diagonal. Similarly, a data point can be added to the end of the list and a new diagonal inserted. This is especially useful when using the divided difference table in an iterative solution.

To set up the divided difference table with $n + 1$ points for Newton's formula takes $n(n + 1)[\tau_1 + 2\tau_2]/2$ seconds. To evaluate this table at a particular point takes $n(\tau_1 + \tau_2)$ seconds, so the total time to evaluate an interpolated point with Newton's method is $n(n + 3)\tau_1/2 + n(n + 2)\tau_2$ seconds.

Suppose we have $n + 1$ points in Cartesian space $\mathbf{u}_i = [x_i \quad y_i \quad z_i \ 1]^{\#}$, $i = 1, \ldots, n + 1$; then we can form three divided difference tables and three polynomials in t: $P_x(t)$, $P_y(t)$, and $P_z(t)$. Notice that these polynomials will be univalued in t regardless of the shape of the space curve.

The points can be chosen to produce a desired velocity function. Differentiating equation (7.13) with respect to t produces

$$
\begin{aligned}
\frac{dP(t)}{dt} = {} & \{\overline{t_1, t_2} + (t - t_2)\{\overline{t_1, t_2, t_3} + (t - t_3)\{\overline{t_1, t_2, t_3, t_4} + (t - t_4)\{\cdot \cdot \cdot \\
& + (t - t_n)\{\overline{t_1, t_2, \ldots, t_{n+1}}\} \ldots\}\} \\
& + (t - t_1)[\{\overline{t_1, t_2, t_3} + (t - t_3)\{\overline{t_1, t_2, t_3, t_4} \\
& + \cdots + (t - t_n)\overline{t_1, t_2, \ldots, t_{n+1}} \ldots\}\} \\
& + (t - t_2)[\{\overline{t_1, t_2, t_3, t_4} \\
& + \cdots + (t - t_n)\ \overline{t_1, t_2, \ldots, t_{n+1}} \ldots\}\} \\
& + \cdots + [(t - t_{n-1})\overline{t_1, t_2, \ldots, t_{n+1}}] \ldots]]
\end{aligned}
\tag{7.14}
$$

The shape of the curve $P(t)$ is not strongly influenced by the location of t_i, $i = 1, \ldots, n + 1$. Study of equation (7.14) will show that $dP(t)/dt$ is a strong function of the points t_i and so could be chosen to produce the desired velocity characteristics. However, the choice of these points is not easily determined from equation (7.13). Instead, we demonstrate velocity control with a simple example.

Suppose we require a straight-line trajectory in Cartesian space; then

$$
\left.\frac{dP_x(t)}{dt}\right|_{t=t_i} = K_1 \left.\frac{dP_y(t)}{dt}\right|_{t=t_i} = K_2 \left.\frac{dP_z(t)}{dt}\right|_{t=t_i} \qquad \text{for all } 0 \leq t_i \leq 1
$$

If $x = 0$ at $t = 0$ and $x = 1$ at $t = 1$, then the divided difference table with equally spaced points becomes

0	0			
		1		
$\frac{1}{3}$	$\frac{1}{3}$		0	
		1		0
$\frac{2}{3}$	$\frac{2}{3}$		0	
		1		
1	1			

so $P_x(t) = 0 + (t - 0)\{1 + (t - \frac{1}{3})\{0 + (t - \frac{2}{3})0\}\} = t$, and the velocity as t moves linearly from 0 to 1 is constant. Now, suppose the points for the same straight line are unequally spaced at $(0, 0)$, $(\frac{1}{3}, \alpha)$, $(\frac{2}{3}, 1 - \alpha)$, and $(1, 1)$. Then the divided difference table becomes

0	0			
		3α		
0.3333	α		$4.5 - 13.5\alpha$	
		$3 - 6\alpha$		$-9 + 27\alpha$
0.6666	$1 - \alpha$		$-4.5 + 13.5\alpha$	
		3α		
1	1			

so

$$\begin{aligned} P_x(t) &= 0 + (t - 0)\{3\alpha + (t - 0.3333)\{4.5 - 13.5\alpha - (t - 0.6666)(9 - 27\alpha)\}\} \\ &= t\{3\alpha + (t - 0.3333)\{-(9 - 27\alpha)t + 10.5 - 31.5\alpha\}\} \\ &= t\{-(9 - 27\alpha)t^2 + (13.5 - 40.5\alpha)t - (3.5 - 13.5\alpha)\} \end{aligned}$$

and $dP_x(t)/dt = -(27 - 81\alpha)t^2 + (27 - 81\alpha)t - (3.5 - 13.5\alpha)$. If we require $dP_x(t)/dt = 0$ at $t = 0$, we have $3.5 - 13.5\alpha = 0$, or $\alpha = 0.25926$. If we require $dP_x(t)/dt = 0$ at $t = 1$, we produce the same value for α. Inserting $\alpha = 0.25926$ in the derivative polynomial gives $dP_x(t)/dt = 6t(1 - t)$. A plot of this velocity function is given in Figure 7.3. This velocity characteristic is close to the characteristic desired when a manipulator executes some path in Cartesian space; this is discussed in detail in Section 8.6.

It is possible to choose the spacing of the points along a space curve so that desired velocity characteristics result while still preserving the shape of the curve. However, the process for determining the spacing is not at all simple. The preceding example discussed spacing with respect to variable x, which is equivalent to a space curve in which y and z are constants. For more general space curves, the spacing of points in Cartesian space must be considered, but the velocity characteristics are calculated with respect to t. Considerably more work is required to develop a logical system for producing parametrically based polynomials that not only follow desired trajectories but also have desirable velocity characteristics.

We end this section by showing the analog of difference calculus to differential calculus. If the data points are equally spaced with $t_{i+1} - t_i = h$,

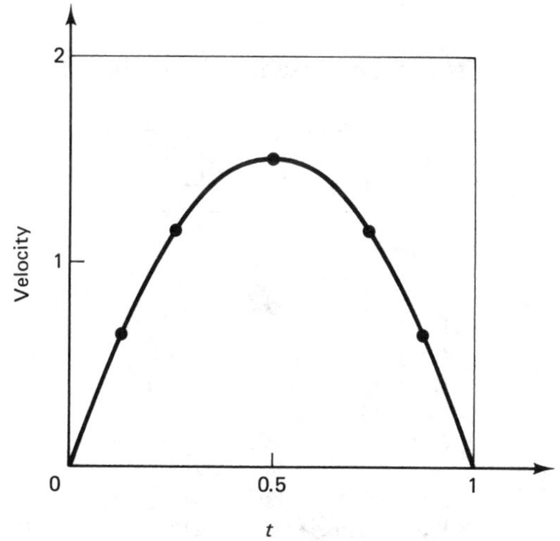

Figure 7.3 A velocity function

then

$$\overline{t_{i+1}, t_i} = \frac{x_{i+1} - x_i}{h} \tag{7.15}$$

and, as $h \to 0$,

$$\overline{t_1, t_2} = \frac{dx(t_1)}{dt}$$

and

$$\overline{t_1, t_2, \ldots, t_{n+1}} = \frac{1}{n!}\frac{d^n x(t_1)}{dt^n} \tag{7.16}$$

so the divided differences have analogues in differential calculus.

7.3 THE BEZIER FUNCTION

A parametric function with some useful properties in curve fitting is the $(n + 1)$-point Bezier function [Bezier 1971, 1972]:

$$x(t) = \sum_{i=0}^{n} x_i \begin{bmatrix} n \\ i \end{bmatrix} t^i(1 - t)^{n-i}, \qquad 0 \leq t \leq 1 \tag{7.17}$$

That is, $x(t)$ is a weighted sum of the x_i. The Bezier function is a limiting case of a B-spline [deBoor 1972; deBoor 1978]; the other limiting case is the set of straight-line segments through the ordered set of given points. The binomial coefficients are given by

$$\begin{bmatrix} n \\ i \end{bmatrix} = \frac{n!}{i!(n - i)!} \tag{7.18}$$

The binomial coefficients can be presented in a table as follows.

n	$\begin{bmatrix} n \\ 0 \end{bmatrix}$	$\begin{bmatrix} n \\ 1 \end{bmatrix}$	$\begin{bmatrix} n \\ 2 \end{bmatrix}$	$\begin{bmatrix} n \\ 3 \end{bmatrix}$	$\begin{bmatrix} n \\ 4 \end{bmatrix}$	$\begin{bmatrix} n \\ 5 \end{bmatrix}$	$\begin{bmatrix} n \\ 6 \end{bmatrix}$	$\begin{bmatrix} n \\ 7 \end{bmatrix}$	$\begin{bmatrix} n \\ 8 \end{bmatrix}$	$\begin{bmatrix} n \\ 9 \end{bmatrix}$	$\begin{bmatrix} n \\ 10 \end{bmatrix}$
0	1										
1	1	1									
2	1	2	1								
3	1	3	3	1							
4	1	4	6	4	1						
5	1	5	10	10	5	1					
6	1	6	15	20	15	6	1				
7	1	7	21	35	35	21	7	1			
8	1	8	28	56	70	56	28	8	1		
9	1	9	36	84	126	126	84	36	9	1	
10	1	10	45	120	210	252	210	120	45	10	1

The coefficients of row j can be found from the coefficients of the prior row, since coefficient (j, k) is the sum of coefficients $(j - 1, k)$ and $(j - 1, k - 1)$. As an example, the coefficient 56 in location (8, 3) is the sum of coefficients 21 and 35 in locations (7, 2) and (7, 3). Notice that $0! = 1$ and $\begin{bmatrix} n \\ n \end{bmatrix} = 1$ for all n, so at $t = 0$, $x(0) = x_0$, and at $t = 1$, $x(1) = x_n$. The derivative of $x(t)$ with respect to t is

$$\frac{d}{dt} x(t) = \sum_{i=0}^{n} x_i \begin{bmatrix} n \\ i \end{bmatrix} \{it^{i-1}(1 - t)^{n-i} - t^i(n - i)(1 - t)^{n-i-1}\} \tag{7.19}$$

At $t = 0$,

$$\frac{d}{dt} x(t) = -nx_0 + nx_1 = n(x_1 - x_0) \tag{7.20a}$$

At $t = 1$,

$$\frac{d}{dt} x(t) = n(x_n - x_{n-1}) \tag{7.20b}$$

Further, $\dfrac{dy}{dx} = \dfrac{dy/dt}{dx/dt}$, giving

$$\left.\frac{dy}{dx}\right|_{t=0} = \frac{y_1 - y_0}{x_1 - x_0} \quad \text{and} \quad \left.\frac{dy}{dx}\right|_{t=1} = \frac{y_n - y_{n-1}}{x_n - x_{n-1}}$$

Similarly,

$$\left.\frac{dz}{dx}\right|_{t=0} = \frac{z_1 - z_0}{x_1 - x_0} \quad \text{and} \quad \left.\frac{dz}{dx}\right|_{t=1} = \frac{z_n - z_{n-1}}{x_n - x_{n-1}}$$

so the slope at the each endpoint is the slope of the line joining the end pair of points.

A $t = 0.5$

$$\frac{d}{dt} x(t) = \frac{1}{2^{n-1}} \sum_{i=0}^{n} x_i \begin{bmatrix} n \\ i \end{bmatrix} (2i - n)$$

and for n even, the slope is independent of $x_{n/2}$ at $t = 0.5$.

The formulas for three, four, and five points according to equation (3.17) are as follows.

For $n = 2$,

$$x(t) = (1 - t)^2 x_0 + 2t(1 - t)x_1 + t^2 x_2 \tag{7.21}$$

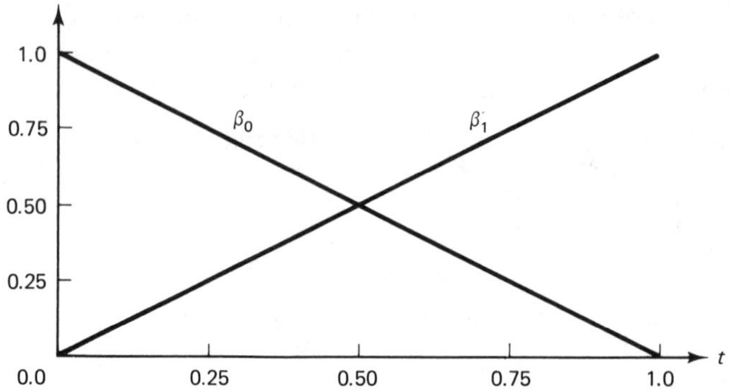

(a) $n = 1$

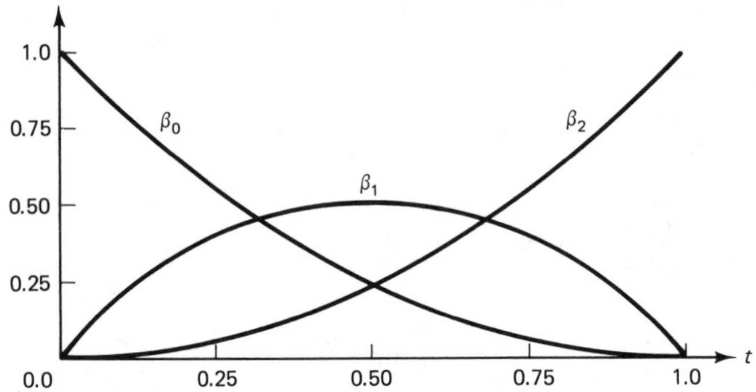

(b) $n = 2$

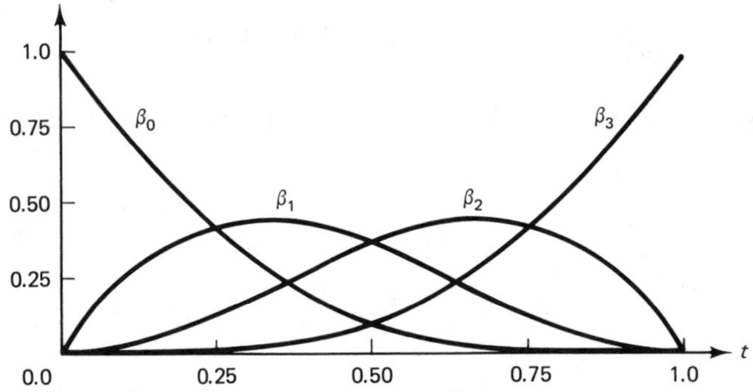

(c) $n = 3$

Figure 7.4 The Bezier blending functions

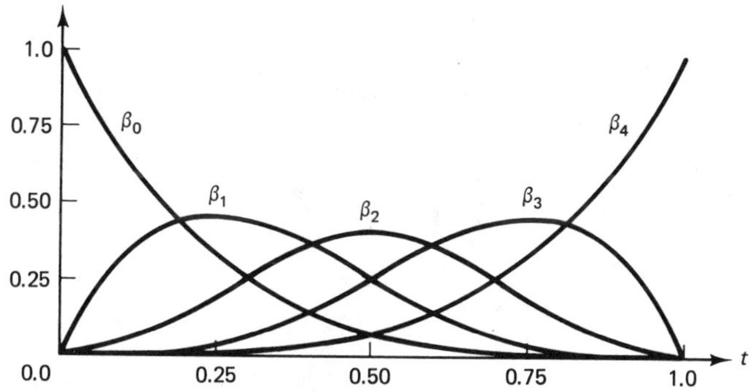

(d) $n = 4$

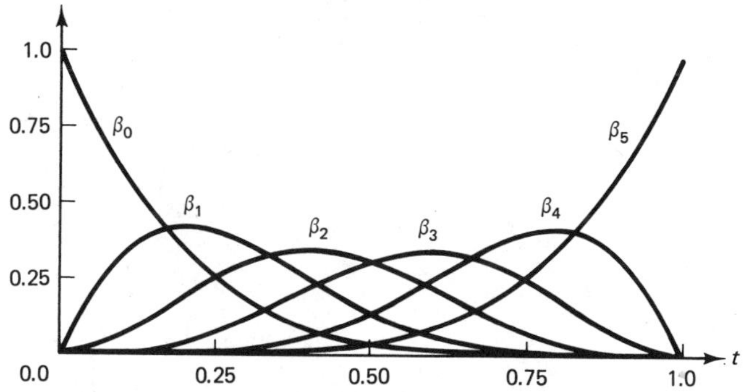

(e) $n = 5$

Figure 7.4 *Continued*

For $n = 3$,

$$x(t) = (1 - t)^3 x_0 + 3t(1 - t)^2 x_1 + 3t^2(1 - t)x_2 + t^3 x_3 \tag{7.22}$$

For $n = 4$,

$$x(t) = (1 - t)^4 x_0 + 4t(1 - t)^3 x_1 + 6t^2(1 - t)^2 x_2 + 4t^3(1 - t)x_3 + t^4 x_4 \tag{7.23}$$

It can be shown that for some t, $0 \leq t \leq 1$,

$$x(t) = b_0(t)x_0 + b_1(t)x_1 + \cdots + b_n(t)x_n \qquad \text{where } \sum_{i=0}^{n} b_i = 1 \tag{7.24}$$

The implication of this last result is that individual values $x_1, x_2, \ldots, x_{n-1}$ can be moved around to influence the shape of the Bezier function. The $b_i(t)$, $i = 0, 1, \ldots, n$, can be termed the *Bezier blending functions:* These functions are shown for $n = 0$ through 5 in Figure 7.4. A characteristic of the Bezier blending functions shows that the peaks of each function occur at $t = i/n$, $i = 0, 1, \ldots, n$; that is, a graphical plot of x versus t will require the x_i to be equally spaced at $t = i/n$. Further, the slopes of the Bezier function at the endpoints are given by the lines joining the first pair and the last pair of

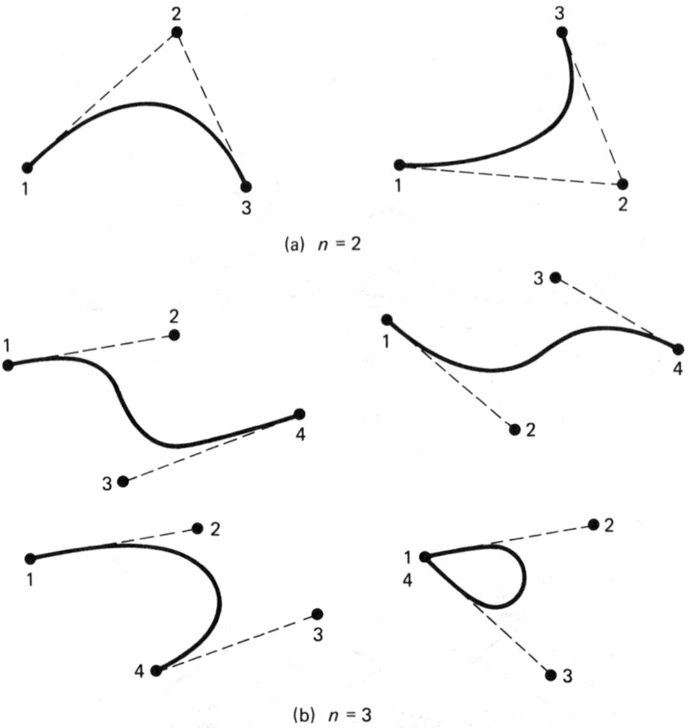

Figure 7.5 Bezier curves for three and four points

points. Thus, we can display the Bezier function with t as an independent parameter and indicate the value of the equally spaced x_i to develop geometric relationships between the function and the x_i. The x_i then shown represents the value and location of the peaks of the blending functions.

The Bezier function will not pass through points $x_1, x_2, \ldots, x_{n-1}$ except by chance. Sketches of the Bezier function under a variety of conditions for three and four points are shown in Figure 7.5; it is assumed that the four points are in a plane.

Suppose a Bezier function for $x(t)$ has points 0, α, 1, and 1. Then $x(t) = t\{3\alpha(1-t)^2 + 3t - 2t^2\}$; $x(t)$ is sketched for $\alpha = 0$, 0.5, and 1 in Figure 7.6. The slope at $t = 0$ is 3α, and the slope at $t = 1$ is zero. Next suppose we have a Bezier function with points 0, α, α, and 1. Then $x(t) = t\{3\alpha(1-t) + t^2\}$; $x(t)$ is sketched for $\alpha = 0$, 0.5, and 1 in Figure 7.7. The slope at $t = 0$ is 3α, as before, and at $t = 1$, the slope is $3 - 3\alpha$.

7.4 SIMPLE PLANE CURVES USING BEZIER FUNCTIONS

A Bezier curve of order n based on the $n + 1$ points $\mathbf{u}_i = [x_i \quad y_i \quad z_i \quad 1]^\#$, $i = 0, 1, \ldots, n$, is

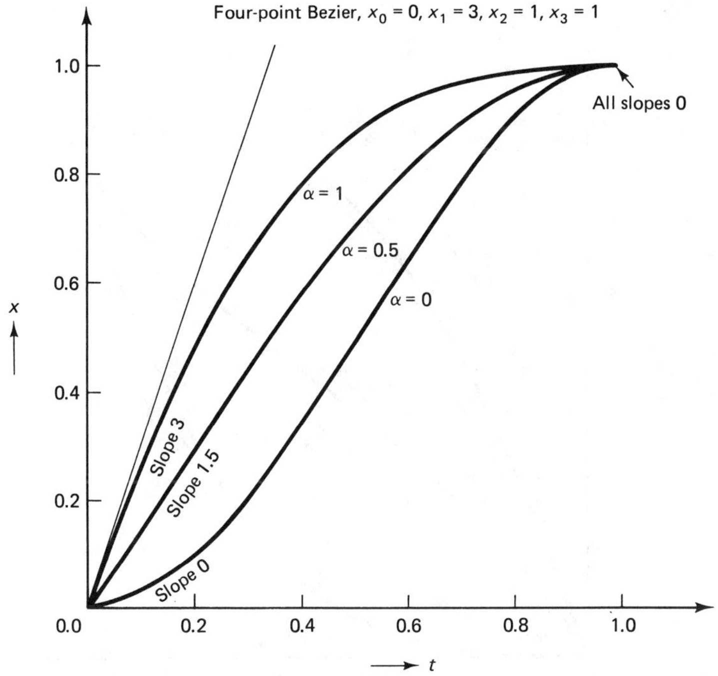

Figure 7.6 Some four-point Bezier curves

$$\mathbf{u}(t) = \begin{bmatrix} x(t) \\ y(t) \\ z(t) \\ 1 \end{bmatrix} = \begin{bmatrix} 0 \\ 0 \\ 0 \\ 1 \end{bmatrix} + \sum_{i=0}^{n} \begin{bmatrix} x_i \\ y_i \\ z_i \\ 0 \end{bmatrix} \begin{bmatrix} n \\ i \end{bmatrix} t^i(1 - t)^{n-i}, \qquad 0 \leq t \leq 1 \tag{7.25}$$

and $\mathbf{u}(0) = \mathbf{u}_0$, $\mathbf{u}(1) = \mathbf{u}_n$. Since the homogeneous representation is not necessary in the discussion that follows, we will consider points $\mathbf{u}(t)$ as 3-tuples. When we consider the interaction with the robotic workplace, the form must be converted to homogeneous form.

The Bezier curve has distinct advantages over nonparametric polynomial methods, including nonparametric spline fitting. First we note that a polynomial function in x is a single-valued function of x, whereas a Bezier curve $\mathbf{u}(t)$ can be a closed loop. Secondly, a spline fit requires calculations of gradient at the endpoints, whereas

$$\frac{d\mathbf{u}(0)}{dt} = n(\mathbf{u}_1 - \mathbf{u}_0) \quad \text{and} \quad \frac{d\mathbf{u}(1)}{dt} = n(\mathbf{u}_n - \mathbf{u}_{n-1})$$

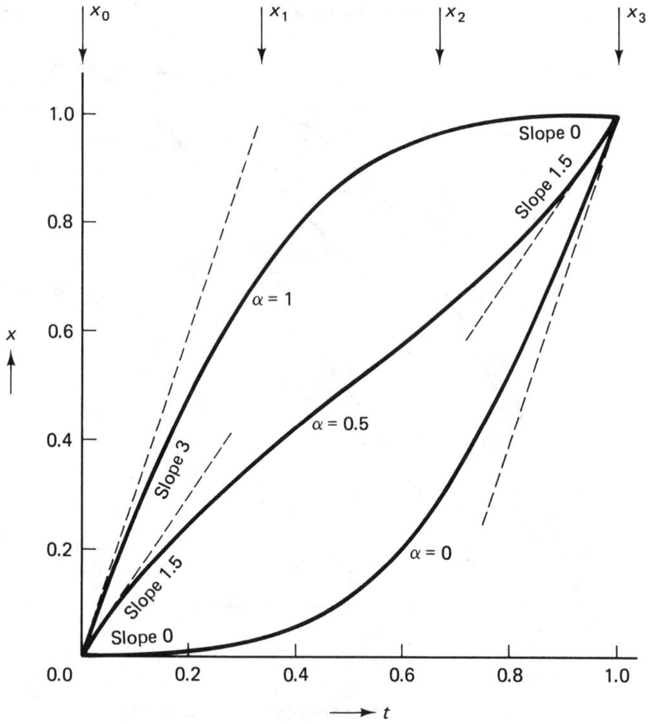

Figure 7.7 Four-point Bezier, $x_0 = 0$, $x_1 = x_2 = \alpha$, $x_3 = 1$

Further, the accuracy of the Bezier curve in following functions with continuous derivatives is surprisingly good.

The efficacy of the Bezier curve in following some desired function will be investigated. The function that is studied most is the circle, since it has a simple parametric description and is not amenable to nonparametric specification; Ellipses are modeled as simple modifications of circles. The Bezier curve should not be used when the function to be modeled has discontinuities in itself or its first derivative. If a function is discontinuous or has a discontinuous derivative at a discrete number of points, Bezier curves can be fitted between these points. That is, the type of function that should not be modeled with a single Bezier curve is $\mathbf{u}(t) = \mathbf{u}_1(t) + \mathbf{u}_2(t)$, where $\mathbf{u}_1(t) = [0 \quad 0 \quad 0 \quad 1]^\#$, $t < \bar{t}$, and $\mathbf{u}_2(t) = [1 \quad 0 \quad 0 \quad 1]^\#$, $t > \bar{t}$, for $0 \le \bar{t} \le 1$ and $0 \le t \le 1$, since

$$\mathbf{u}_1(\bar{t}) \neq \mathbf{u}_2(\bar{t}) \quad \text{or} \quad \frac{d\mathbf{u}_1(\bar{t})}{dt} \neq \frac{d\mathbf{u}_2(\bar{t})}{dt}$$

Suppose an approximation to $f(t)$, a quarter of a circle of unit radius in the xy-plane centered at the origin, is required. The Bezier curve with three points (in nonhomogeneous form)

$$\begin{bmatrix} 0 \\ 1 \\ 0 \end{bmatrix}, \quad \begin{bmatrix} 1 \\ 1 \\ 0 \end{bmatrix}, \quad \text{and} \quad \begin{bmatrix} 1 \\ 0 \\ 0 \end{bmatrix}$$

fits the endpoints and has the correct slope at the endpoints. The Bezier curve is

$$\mathbf{u}(t) = \begin{bmatrix} t(2-t) \\ 1-t^2 \\ 0 \end{bmatrix}, \quad \text{where } 0 \le t \le 1$$

The maximum error occurs at $t = 0.5$, and

$$\begin{bmatrix} x(0.5) \\ y(0.5) \\ z(0.5) \end{bmatrix} = \begin{bmatrix} 0.75 \\ 0.75 \\ 0 \end{bmatrix}, \quad \text{where } f(0.5) = \begin{bmatrix} 0.7071 \\ 0.7071 \\ 0 \end{bmatrix}$$

A Bezier curve approximation to the same quarter-circle with four points is

$$\begin{bmatrix} x(t) \\ y(t) \\ 0 \end{bmatrix} = (1-t)^3 \begin{bmatrix} 0 \\ 1 \\ 0 \end{bmatrix} + 3(1-t)^2 t \begin{bmatrix} \alpha \\ 1 \\ 0 \end{bmatrix} + 3(1-t)t^2 \begin{bmatrix} 1 \\ \alpha \\ 0 \end{bmatrix} + t^3 \begin{bmatrix} 1 \\ 0 \\ 0 \end{bmatrix} \tag{7.25}$$

and at the midpoint $t = 0.5$,

$$\begin{bmatrix} x(0.5) \\ y(0.5) \\ z(0.5) \end{bmatrix} = \begin{bmatrix} \frac{1}{2} + \frac{3\alpha}{8} \\ \frac{1}{2} + \frac{3\alpha}{8} \\ 0 \end{bmatrix}$$

Matching to the actual value of $x = y = 1/\sqrt{2}$ yields $\alpha = 0.55228$. Alternatively, if α is obtained by minimizing the maximum error (a minimax fit), we obtain $\alpha = 0.5449$. The graph of the maximum value of the error between the four-point Bezier curve and the quarter-circle as a function of α is shown in Figure 7.8; the maximum error at $\alpha = 0.5449$ is 0.00505.

Separating the variables in equation (7.26) gives

$$x(t) = 3(1-t)^2 t\alpha + 3(1-t)t^2 + t^3 = t^3(3\alpha - 2) + t^2(3 - 6\alpha) + t(3\alpha)$$

$$y(t) = (1-t)^3 + 3(1-t)^2 t + 3(1-t)t^2\alpha = t^3(2 - 3\alpha) + t^2(3\alpha - 3) + 1$$

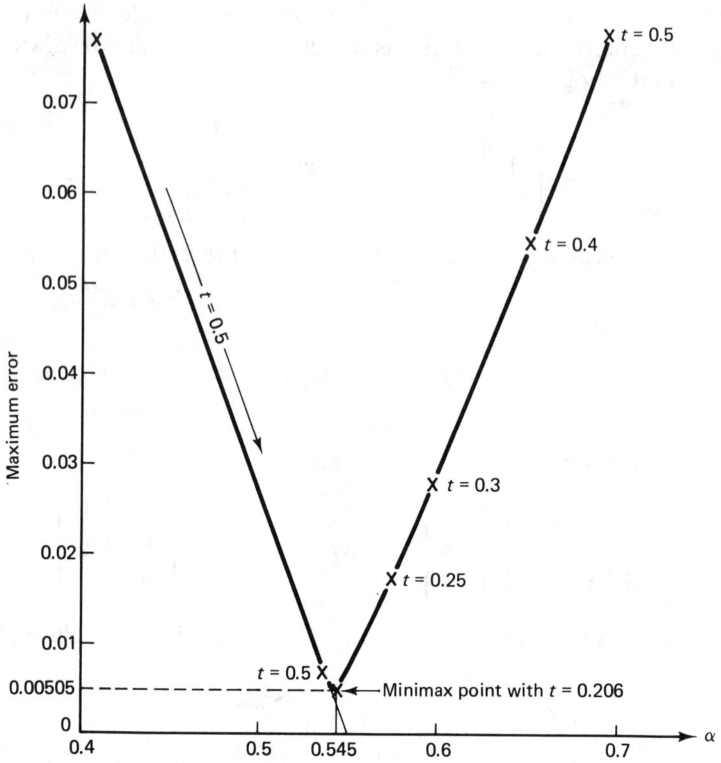

Figure 7.8 Maximum error as a function of α

Differentiating with respect to t results in

$$\frac{dx(t)}{dt} = 3t^2(3\alpha - 2) + 6t(1 - 2\alpha) + 3\alpha$$

and

$$\frac{dy(t)}{dt} = 3t^2(2 - 3\alpha) + 6t(\alpha - 1)$$

Values of these derivatives for particular values of t are given in the table.

	$t = 0$	$t = 0.5$	$t = 1$
$\frac{dx(t)}{dt}$	3α	$\frac{3}{2} - \frac{3\alpha}{4}$	0
$\frac{dy(t)}{dt}$	0	$\frac{3\alpha}{4} - \frac{3}{2}$	-3α
Velocity	3α	$\frac{3}{2}\left(\frac{\alpha^2}{2} - 2\alpha + 2\right)^{1/2}$	3α

The rate of change of position (velocity) as a function of t is given by

$$T = \sqrt{\left\{\frac{dx(t)}{dt}\right\}^2 + \left\{\frac{dy(t)}{dt}\right\}^2}$$

and these values are given in the table. For the velocity at the midpoint to equal the velocity at the ends,

$$\frac{3}{2}\left(\frac{\alpha^2}{2} - 2\alpha + 2\right)^{1/2} = 3\alpha, \quad \text{or} \quad \alpha^2 + \frac{4}{7}\alpha - \frac{4}{7} = 0$$

giving $\alpha = 0.5224$. This value is smaller than the values found for best fit by shape. Choosing the value $\alpha = 0.5523$ for best fit at the center point, the resulting velocity at the midpoint is 1.5355, somewhat smaller than the velocity at the endpoints of 1.6568.

The accuracy of a Bezier curve in modeling a semicircle is much less than when a quarter-circle is considered. A Bezier curve with four points is given by

$$\begin{bmatrix} x(t) \\ y(t) \\ z(t) \end{bmatrix} = (1-t)^3 \begin{bmatrix} 0 \\ 1 \\ 0 \end{bmatrix} + 3t(1-t)^2 \begin{bmatrix} \alpha \\ 1 \\ 0 \end{bmatrix} + 3t^2(1-t) \begin{bmatrix} \alpha \\ -1 \\ 0 \end{bmatrix} + t^3 \begin{bmatrix} 0 \\ -1 \\ 0 \end{bmatrix}$$

so

$$\begin{bmatrix} x(0.5) \\ y(0.5) \\ z(0.5) \end{bmatrix} = \begin{bmatrix} \frac{3\alpha}{4} \\ 0 \\ 0 \end{bmatrix} \quad \text{and} \quad \begin{bmatrix} x(0.25) \\ y(0.25) \\ z(0.25) \end{bmatrix} = \begin{bmatrix} \frac{9\alpha}{16} \\ \frac{11}{16} \\ 0 \end{bmatrix}$$

For best fit at $t = 0.5$, $\alpha = \frac{4}{3}$. At $t = 0.25$, a point on the circle is at $x = y = 1/\sqrt{2} = 0.7071$, but the Bezier curve has $x(0.25) = 9\alpha/16$ and $y(0.25) = 11/16 = 0.6875$. With $\alpha = \frac{4}{3}$, $x(0.25) = 0.75$. The error may not be acceptable in some applications.

A five-point Bezier curve fit to a quarter of the unit circle in the xy-plane centered at the origin is

$$\begin{bmatrix} x(t) \\ y(t) \\ z(t) \end{bmatrix} = (1-t)^4 \begin{bmatrix} 0 \\ 1 \\ 0 \end{bmatrix} + 4t(1-t)^3 \begin{bmatrix} \alpha \\ 1 \\ 0 \end{bmatrix} + 6t^2(1-t)^2 \begin{bmatrix} \beta \\ \beta \\ 0 \end{bmatrix}$$

$$+ 4t^3(1-t) \begin{bmatrix} 1 \\ \alpha \\ 0 \end{bmatrix} + t^4 \begin{bmatrix} 1 \\ 0 \\ 0 \end{bmatrix}$$

At the one-third point around the quarter circle, $x = 0.5$ and $y = 0.8660$, and the value of the Bezier curve is

$$\begin{bmatrix} x(\frac{1}{3}) \\ y(\frac{1}{3}) \\ z(\frac{1}{3}) \end{bmatrix} = \frac{1}{81}\begin{bmatrix} 9 + 32\alpha + 24\beta \\ 48 + 8\alpha + 24\beta \\ 0 \end{bmatrix}, \quad \text{so} \quad \begin{aligned} 32\alpha + 24\beta &= 31.500 \\ 8\alpha + 24\beta &= 22.146 \end{aligned}$$

Solving these two simultaneous equations gives $\alpha = 0.3898$ and $\beta = 0.7928$, and using these values at the midway point gives

$$\begin{bmatrix} x(0.5) \\ y(0.5) \\ z(0.5) \end{bmatrix} = \frac{1}{16}\begin{bmatrix} 5 + 4\alpha + 6\beta \\ 5 + 4\alpha + 6\beta \\ 0 \end{bmatrix} = \begin{bmatrix} 0.7073 \\ 0.7073 \\ 0 \end{bmatrix}$$

which is very close to the true value.

Similarly, the five-point Bezier curve fit to a semicircle in the xy-plane centered at the origin is

$$\begin{bmatrix} x(t) \\ y(t) \\ z(t) \end{bmatrix} = (1 - t)^4\begin{bmatrix} 0 \\ 1 \\ 0 \end{bmatrix} + 4t(1 - t)^3\begin{bmatrix} \alpha \\ 1 \\ 0 \end{bmatrix} + 6t^2(1 - t)^2\begin{bmatrix} \beta \\ 0 \\ 0 \end{bmatrix} + 4t^3(1 - t)\begin{bmatrix} \alpha \\ -1 \\ 0 \end{bmatrix} + t^4\begin{bmatrix} 0 \\ -1 \\ 0 \end{bmatrix}$$

Notice that the y-value is independent of both α and β and is a function of t only. At the quarter- and half-points, we have

$$\begin{bmatrix} x(0.25) \\ y(0.25) \\ z(0.25) \end{bmatrix} \begin{bmatrix} 0.4688\alpha + 0.2109\beta \\ 0.6875 \\ 0 \end{bmatrix} \quad \text{and} \quad \begin{bmatrix} x(0.5) \\ y(0.5) \\ z(0.5) \end{bmatrix} = \frac{1}{4}\begin{bmatrix} 2\alpha + 1.5\beta \\ 0 \\ 0 \end{bmatrix}$$

With $x(0.5) = 1$, then $2\alpha + 1.5\beta = 4$. With $x(0.25) = 0.7071$, $0.7071 = 0.4688\alpha + 0.2109\beta$, and so $\alpha = 0.7713$, $\beta = 1.6383$. The position of these points is shown in Figure 7.9, and although these positions appear strange, they do provide an approximation to the semicircle that could be adequate in some situations.

Suppose the five points needed to construct a Bezier curve approximation to the circle of diameter unity and center at (0, 0.5) are at

$$\begin{bmatrix} 0 & \alpha & 0 & -\alpha & 0 \\ 0 & 0 & \beta & 0 & 0 \\ 0 & 0 & 0 & 0 & 0 \end{bmatrix}$$

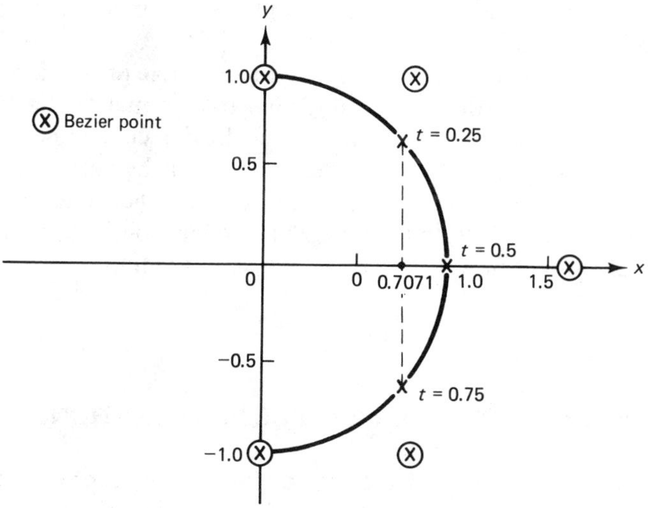

Figure 7.9

Then the Bezier curve is

$$\begin{bmatrix} x(t) \\ y(t) \\ z(t) \end{bmatrix} = 2t(1-t)\left\{ 2(1-t)^2 \begin{bmatrix} \alpha \\ 0 \\ 0 \end{bmatrix} + 3t(1-t) \begin{bmatrix} 0 \\ \beta \\ 0 \end{bmatrix} - 2t^2 \begin{bmatrix} \alpha \\ 0 \\ 0 \end{bmatrix} \right\}$$

At (0, 1), $t = 0.5$ and

$$\begin{bmatrix} x(0.5) \\ y(0.5) \\ z(0.5) \end{bmatrix} = \begin{bmatrix} 0 \\ \dfrac{3\beta}{8} \\ 0 \end{bmatrix}$$

At (0.5, 0.5), $t = 0.25$ and

$$\begin{bmatrix} x(0.25) \\ y(0.25) \\ z(0.25) \end{bmatrix} = \begin{bmatrix} \dfrac{3\alpha}{8} \\ \dfrac{27\beta}{128} \\ 0 \end{bmatrix}$$

Thus, $\alpha = \frac{4}{3}$, and there are two choices for β. For exact fit at $t = 0.5$, we require $\beta = \frac{8}{3} = 2.6667$, but this results in $y(0.25) = 0.5624$. For exact fit at $t = 0.25$, we require $\beta = \frac{64}{27} = 2.3704$. Permitting a compromise in which $\beta = (2.6667 + 2.3704)/2$ results in $y(0.5) = 0.9444$ and $y(0.25) = 0.5312$. For most robotic applications, the errors are unacceptable, so six or more points must be used to model a full circle or ellipse.

The results just given can be applied to any circle or fraction of a circle in any plane using the rotation, scaling, and transformation matrices given previously on the four or five points of the Bezier curve under consideration. Ellipses can be handled using the same results by noting that the four-point Bezier curve over a quarter of an ellipse where the center of the ellipse is at the origin and the ellipse is in the xy-plane has points $(0, b)$, $(0.55228a, b)$, $(a, 0.55228b)$ and $(a, 0)$, where the ellipse is given by

$$\frac{x^2}{a^2} + \frac{y^2}{b^2} = 1$$

7.5 SIMPLE SPACE CURVES USING BEZIER FUNCTIONS

In the last section the use of the Bezier function in approximating some simple plane curves was demonstrated. However, the power of the Bezier formulation is best demonstrated with space curves. Suppose the four points

$$\begin{bmatrix} 1 \\ 0 \\ 0 \end{bmatrix}, \quad \begin{bmatrix} 1 \\ \alpha \\ 0 \end{bmatrix}, \quad \begin{bmatrix} \alpha \\ 0 \\ 1 \end{bmatrix}, \quad \text{and} \quad \begin{bmatrix} 0 \\ 0 \\ 1 \end{bmatrix}$$

are required to approximate a curve on a unit sphere centered at the origin.

The curve becomes

$$(1-t)^3 \begin{bmatrix} 1 \\ 0 \\ 0 \end{bmatrix} + 3t(1-t)^2 \begin{bmatrix} 1 \\ \alpha \\ 0 \end{bmatrix} + 3t^2(1-t) \begin{bmatrix} \alpha \\ 0 \\ 1 \end{bmatrix} + t^3 \begin{bmatrix} 0 \\ 0 \\ 1 \end{bmatrix}$$

and at $t = \frac{1}{2}$ we have

$$\tfrac{1}{8} \begin{bmatrix} 1 \\ 0 \\ 0 \end{bmatrix} + \tfrac{3}{8} \begin{bmatrix} 1 \\ \alpha \\ 0 \end{bmatrix} + \tfrac{3}{8} \begin{bmatrix} \alpha \\ 0 \\ 1 \end{bmatrix} + \tfrac{1}{8} \begin{bmatrix} 0 \\ 0 \\ 1 \end{bmatrix} = \tfrac{1}{8} \begin{bmatrix} 4 + 3\alpha \\ 3\alpha \\ 4 \end{bmatrix} = \begin{bmatrix} x \\ y \\ z \end{bmatrix}$$

Hence, for exact fit at $t = \frac{1}{2}$, $x^2 + y^2 + z^2 = 1$, $(4 + 3\alpha)^2 + 9\alpha^2 + 16 = 64$, or $9\alpha^2 + 12\alpha - 16 = 0$, giving $\alpha = 0.8240$ or $\alpha = -2.1574$, where the negative value of α must be rejected.

At $t = \frac{1}{4}$,

$$\begin{bmatrix} x \\ y \\ z \end{bmatrix} = \tfrac{1}{64} \begin{bmatrix} 54 + 9\alpha \\ 27\alpha \\ 10 \end{bmatrix} = \begin{bmatrix} 0.9596 \\ 0.3476 \\ 0.1563 \end{bmatrix}$$

for which $\sqrt{0.9596^2 + 0.3476^2 + 0.1563^2} = 1.0325$ instead of unity.

At $t = \frac{3}{4}$,

$$\begin{bmatrix} x \\ y \\ z \end{bmatrix} = \frac{1}{64}\begin{bmatrix} 10 + 27\alpha \\ 9\alpha \\ 54 \end{bmatrix} = \begin{bmatrix} 0.5039 \\ 0.1159 \\ 0.8438 \end{bmatrix}$$

for which $\sqrt{0.5039^2 + 0.1159^2 + 0.8438^2} = 0.9896$ instead of unity.

7.6 BEZIER CURVE DETERMINATION BY POINT AND DERIVATIVE FITTING

A five-point Bezier space curve approximation to a trajectory on the unit sphere discussed previously is

$$(1-t)^4\begin{bmatrix} 1 \\ 0 \\ 0 \end{bmatrix} + 4t(1-t)^3\begin{bmatrix} 1 \\ \alpha \\ 0 \end{bmatrix} + 6t^2(1-t)^2\begin{bmatrix} \beta \\ \gamma \\ \frac{1}{2} \end{bmatrix} + 4t^3(1-t)\begin{bmatrix} \alpha \\ 0 \\ 1 \end{bmatrix} + t^4\begin{bmatrix} 0 \\ 0 \\ 1 \end{bmatrix}$$

where α, β, and γ are to be determined. The problem here is to know to match the Bezier space curve to the actual function. This is most easily performed at a series (three in this case) of points along the curve.

We consider the general case of this class of problem—that is, determining the nth-order Bezier curve that passes through $n + 1$ points or $n + 1$ constraints (points or derivatives).

7.6.1 Bezier Curve from $n + 1$ Points

Suppose the Bezier curve with points $\mathbf{f}_0, \mathbf{f}_1, \ldots, \mathbf{f}_n$ is required to pass through points $\mathbf{u}_0, \mathbf{u}_1, \ldots, \mathbf{u}_n$; $\mathbf{f}_0 = \mathbf{u}_0$, $\mathbf{f}_n = \boldsymbol{u}_n$. For the Bezier curve with $n = 2$, we match the Bezier curve to the middle point $\mathbf{u}_1$, so

$$\mathbf{u}_1 = \mathbf{f}(\tfrac{1}{2}) = \tfrac{1}{4}\mathbf{f}_0 + \tfrac{1}{2}\mathbf{f}_1 + \tfrac{1}{4}\mathbf{f}_2, \quad \text{or} \quad \mathbf{f}_1 = 2\mathbf{u}_1 - \tfrac{1}{2}(\mathbf{u}_0 + \mathbf{u}_2)$$

For $n = 3$ we match the Bezier curve at the one-third points, so

$$\mathbf{u}_1 = \mathbf{f}(\tfrac{1}{3}) = \tfrac{8}{27}\mathbf{f}_0 + \tfrac{4}{9}\mathbf{f}_1 + \tfrac{2}{9}\mathbf{f}_2 + \tfrac{1}{27}\mathbf{f}_3 \quad \text{and}$$
$$\mathbf{u}_2 = \mathbf{f}(\tfrac{2}{3}) = \tfrac{1}{27}\mathbf{f}_0 + \tfrac{2}{9}\mathbf{f}_1 + \tfrac{4}{9}\mathbf{f}_2 + \tfrac{8}{27}\mathbf{f}_3$$

or

$$2\mathbf{f}_1 + \mathbf{f}_2 = \tfrac{9}{2}\mathbf{u}_1 - \tfrac{4}{3}\mathbf{u}_0 - \tfrac{1}{6}\mathbf{u}_3$$

$$\mathbf{f}_1 + 2\mathbf{f}_2 = \tfrac{9}{2}\mathbf{u}_1 - \tfrac{1}{6}\mathbf{u}_0 - \tfrac{4}{3}\mathbf{u}_3$$

$$\begin{bmatrix} \mathbf{f}_1 \\ \mathbf{f}_2 \end{bmatrix} = \begin{bmatrix} 2 & 1 \\ 1 & 2 \end{bmatrix}^{-1} \begin{bmatrix} \frac{9}{2}\mathbf{u}_1 - \frac{4}{3}\mathbf{u}_0 - \frac{1}{6}\mathbf{u}_3 \\ \frac{9}{2}\mathbf{u}_1 - \frac{1}{6}\mathbf{u}_0 - \frac{4}{3}\mathbf{u}_3 \end{bmatrix}$$

$$= \begin{bmatrix} \frac{2}{3} & -\frac{1}{3} \\ -\frac{1}{3} & \frac{2}{3} \end{bmatrix} \begin{bmatrix} \frac{9}{2}\mathbf{u}_1 - \frac{4}{3}\mathbf{u}_0 - \frac{1}{6}\mathbf{u}_3 \\ \frac{9}{2}\mathbf{u}_1 - \frac{1}{6}\mathbf{u}_0 - \frac{4}{3}\mathbf{u}_3 \end{bmatrix}$$

For $n = 4$ we match the Bezier curve at the quarter-points, so

$$\mathbf{u}_1 = \mathbf{f}(\tfrac{1}{4}) = \tfrac{81}{256}\mathbf{f}_0 + \tfrac{27}{64}\mathbf{f}_1 + \tfrac{27}{128}\mathbf{f}_2 + \tfrac{3}{64}\mathbf{f}_3 + \tfrac{1}{256}\mathbf{f}_4$$

$$\mathbf{u}_2 = \mathbf{f}(\tfrac{1}{2}) = \tfrac{1}{16}\mathbf{f}_0 + \tfrac{1}{4}\mathbf{f}_1 + \tfrac{3}{8}\mathbf{f}_2 + \tfrac{1}{4}\mathbf{f}_3 + \tfrac{1}{16}\mathbf{f}_4$$

$$\mathbf{u}_3 = \mathbf{f}(\tfrac{3}{4}) = \tfrac{1}{256}\mathbf{f}_0 + \tfrac{3}{64}\mathbf{f}_1 + \tfrac{27}{128}\mathbf{f}_2 + \tfrac{27}{64}\mathbf{f}_3 + \tfrac{81}{256}\mathbf{f}_4$$

$$\begin{bmatrix} 18 & 9 & 2 \\ 2 & 3 & 2 \\ 2 & 9 & 18 \end{bmatrix} \begin{bmatrix} \mathbf{f}_1 \\ \mathbf{f}_2 \\ \mathbf{f}_3 \end{bmatrix} = \begin{bmatrix} \frac{128}{3}\mathbf{u}_1 - \frac{27}{2}\mathbf{u}_0 - \frac{1}{6}\mathbf{u}_4 \\ 8\mathbf{u}_2 - \frac{1}{2}\mathbf{u}_0 - \frac{1}{2}\mathbf{u}_4 \\ \frac{128}{3}\mathbf{u}_3 - \frac{1}{6}\mathbf{u}_0 - \frac{27}{2}\mathbf{u}_4 \end{bmatrix}$$

The analysis just given can be extended to higher-order curves.

7.6.2 Bezier Curve Determined from $n + 1$ Points and Derivatives

We can fit a Bezier curve to a combination of points and derivatives; we show this with examples. Suppose $\mathbf{u}(t)$ is defined with three pieces of information $\mathbf{u}_0 = \mathbf{u}(0)$, $d\mathbf{u}(0)/dt$ and $\mathbf{u}(1)$; then the Bezier curve with $n = 2$ has $\mathbf{f}_0 = \mathbf{u}_0$, $\mathbf{f}_2 = \mathbf{u}_1$, and

$$2(\mathbf{f}_1 - \mathbf{f}_0) = \frac{d\mathbf{u}(0)}{dt}$$

Given $\mathbf{u}_0$, $d\mathbf{u}(0)/dt$, $\mathbf{u}_1$, and $\mathbf{u}_2$, then the Bezier curve with $n = 3$ has $\mathbf{f}_0 = \mathbf{u}_0$, $\mathbf{f}_3 = \mathbf{u}_2$, $3(\mathbf{f}_1 - \mathbf{f}_0) = d\mathbf{u}(0)/dt$, and, at $t = \frac{2}{3}$ $\mathbf{f}_2$ is determined from

$$\tfrac{1}{27}\mathbf{f}_0 + \tfrac{2}{9}\mathbf{f}_1 + \tfrac{4}{9}\mathbf{f}_2 + \tfrac{8}{27}\mathbf{f}_3 = \mathbf{u}_1$$

Given $\mathbf{u}_0$, $d\mathbf{u}(0)/dt$, $d\mathbf{u}(1)/dt$, and $\mathbf{u}_1$, then the Bezier curve with $n = 3$ has $\mathbf{f}_0 = \mathbf{u}_0$, $\mathbf{f}_3 = \mathbf{u}_1$, $3(\mathbf{f}_1 - \mathbf{f}_0) = d\mathbf{u}(0)/dt$, and $3(\mathbf{f}_3 - \mathbf{f}_2) = d\mathbf{u}(1)/dt$. Thus, we see that points can be chosen for the Bezier curve that exhibit certain desired velocity characteristics. More is said about this when we consider velocity path control in Section 8.5.

7.7 EXAMPLES

1. A point on a unit circle can be defined parametrically as $\begin{bmatrix} \sin\theta \\ \cos\theta \\ 0 \\ 1 \end{bmatrix}$, where $0 \le \theta \le$

2π. Use four parabolic sections each over the range $\pi/2$ for θ to approximate this circle.

The set of points we use is

$$\begin{bmatrix} 0 \\ 1 \\ 0 \\ 1 \end{bmatrix}, \begin{bmatrix} 0.7071 \\ 0.7071 \\ 0 \\ 1 \end{bmatrix}, \begin{bmatrix} 1 \\ 0 \\ 0 \\ 1 \end{bmatrix}, \begin{bmatrix} 0.7071 \\ -0.7071 \\ 0 \\ 1 \end{bmatrix}, \begin{bmatrix} 0 \\ -1 \\ 0 \\ 1 \end{bmatrix},$$
$$\begin{bmatrix} -0.7071 \\ -0.7071 \\ 0 \\ 1 \end{bmatrix}, \begin{bmatrix} -1 \\ 0 \\ 0 \\ 1 \end{bmatrix}, \begin{bmatrix} -0.7071 \\ 0.7071 \\ 0 \\ 1 \end{bmatrix}, \begin{bmatrix} 0 \\ 1 \\ 0 \\ 1 \end{bmatrix}$$

which are spaced at $\pi/4$. From these points we construct divided difference tables for x and y to the second divided difference as follows:

		0.9003		
$\frac{\pi}{4}$	0.7071		−0.3358	$x(\theta) = 0 + (\theta - 0)\left\{0.9003 - \left(\theta - \frac{\pi}{4}\right) 0.3358\right\}$
		0.3729		
$\frac{\pi}{2}$	1		−0.4745	$x(\theta) = 0.7071 + \left(\theta - \frac{\pi}{4}\right)\left\{0.3729 - \left(\theta - \frac{\pi}{2}\right) 0.4745\right\}$
		−0.3729		
$\frac{3\pi}{4}$	0.7071		−0.3358	$x(\theta) = 1 + \left(\theta - \frac{\pi}{2}\right)\left\{-0.3729 - \left(\theta - \frac{3\pi}{4}\right) 0.3358\right\}$
		−0.9003		
π	0		0	$x(\theta) = 0.7071 + \left(\theta - \frac{3\pi}{4}\right)\{-0.9003 + (\theta - \pi)\, 0\}$
		−0.9003		
$\frac{5\pi}{4}$	−0.7071		0.3358	$x(\theta) = 0 + (\theta - \pi)\left\{-0.9003 + \left(\theta - \frac{5\pi}{4}\right) 0.3358\right\}$
		−0.3729		
$\frac{3\pi}{2}$	−1		0.4745	$x(\theta) = -0.7071 + \left(\theta - \frac{5\pi}{4}\right)\left\{-0.3729 + \left(\theta - \frac{3\pi}{2}\right) 0.4745\right\}$
		0.3729		
$\frac{7\pi}{4}$	−0.7071		0.3358	$x(\theta) = -1 + \left(\theta - \frac{3\pi}{2}\right)\left\{0.3729 + \left(\theta - \frac{7\pi}{4}\right) 0.3358\right\}$
		0.9003		
2π	0			

0	1			
		−0.3729		
$\frac{\pi}{4}$	0.7071		−0.3358	$y(\theta) = 1 + (\theta - 0)\left\{-0.3729 - \left(\theta - \frac{\pi}{4}\right) 0.3358\right\}$
		−0.9003		
$\frac{\pi}{2}$	0		0	$y(\theta) = 0.7071 + \left(\theta - \frac{\pi}{4}\right)\left\{-0.9003 + \left(\theta - \frac{\pi}{2}\right) 0\right\}$
		−0.9003		
$\frac{3\pi}{4}$	−0.7071		0.3358	$y(\theta) = 0 + \left(\theta - \frac{\pi}{2}\right)\left\{-0.9003 + \left(\theta - \frac{3\pi}{4}\right) 0.3358\right\}$
		−0.3729		

π	-1		0.4745	$y(\theta) = -0.7071 + \left(\theta - \frac{3\pi}{4}\right)\left\{-0.3729 + (\theta - \pi)\,0.4745\right\}$
		0.3729		
$\frac{5\pi}{4}$	-0.7071		0.3358	$y(\theta) = -1 + (\theta - \pi)\left\{0.3729 + \left(\theta - \frac{5\pi}{4}\right)0.3358\right\}$
		0.9003		
$\frac{3\pi}{2}$	0		0	$y(\theta) = -0.7071 + \left(\theta - \frac{5\pi}{4}\right)\left\{0.9003 + \left(\theta - \frac{3\pi}{2}\right)0\right\}$
		0.9003		
$\frac{7\pi}{4}$	0.7071		-0.3358	$y(\theta) = 0 + \left(\theta - \frac{3\pi}{2}\right)\left\{0.9003 - \left(\theta - \frac{7\pi}{4}\right)0.3358\right\}$
		0.3729		
2π	1			

We notice that the functions for x and y are piecewise continuous.

---

2. Determine a polynomial space curve that passes through the four points

$$\begin{bmatrix} 0 \\ 1 \\ 0 \\ 1 \end{bmatrix}, \quad \begin{bmatrix} 1 \\ 1 \\ 1 \\ 1 \end{bmatrix}, \quad \begin{bmatrix} 1 \\ 0 \\ 1 \\ 1 \end{bmatrix}, \quad \text{and} \quad \begin{bmatrix} 1 \\ -1 \\ 0 \\ 1 \end{bmatrix}$$

We construct the divided difference tables as

0	0			
		3		
$\frac{1}{3}$	1		$-\frac{9}{2}$	
		0		$\frac{9}{2}$
$\frac{2}{3}$	1		0	
		0		
1	1			

0	1			
		0		
$\frac{1}{3}$	1		$-\frac{9}{2}$	
		-3		$\frac{9}{2}$
$\frac{2}{3}$	0		0	
		-3		
1	-1			

0	0			
		3		
$\frac{1}{3}$	1		$-\frac{9}{2}$	
		0		0
$\frac{2}{3}$	1		$-\frac{9}{2}$	
		-3		
1	0			

The three polynomial functions of t are

$$x(t) = 0 + (t - 0)\{3 + (t - \tfrac{1}{3}) - \{\tfrac{9}{2} + (t - \tfrac{2}{3})\tfrac{9}{2}\}\}$$

$$= t\{3 + (t - \tfrac{1}{3})\{\tfrac{5}{3} + t\}\tfrac{9}{2}\} = t\{3 + (t^2 - 2t + \tfrac{5}{9})\tfrac{9}{2}\}$$

$$= t\left\{\frac{9t^2}{2} - 9t + \frac{11}{2}\right\}$$

$$y(t) = 1 + (t - 0)\{0 + (t - \tfrac{1}{3})\{-\tfrac{9}{2} + (t - \tfrac{2}{3})\tfrac{9}{2}\}\}$$

$$= 1 + \frac{t(t - \frac{1}{3})\{t - \frac{5}{3}\}9}{2} = 1 + \frac{t(t^2 - 2t + \frac{5}{9})9}{2}$$

$$z(t) = 0 + (t - 0)\{3 - (t - \tfrac{1}{3})\{-\tfrac{9}{2} + 0\}\} = \frac{3t(1 + 3t)}{2}$$

3. A unit sphere of radius $r = 0.5$ is centered at $[0.5 \quad 0 \quad 0 \quad 1]^\#$. Using Newton's interpolating polynomial, obtain a second-order polynomial function of x, y, and z in terms of parameter t, $0 \le t \le 1$, for the line of intersection of the surface of the sphere and the xz-plane for positive x, y, and z.

We start by identifying the sets of points as follows.

t	x	y	z
0	0	0	0
0.5	0.5	0	0.5
1	1	0	0

Thus $x \equiv t$, $y = 0$, and the divided difference table for z is

$$\begin{array}{ccccc} 0 & 0 & & & \\ & & 1 & & \\ 0.5 & 0.5 & & -2 & \\ & & -1 & & \\ 1 & 0 & & & \end{array}$$

so $z(t) = 0 + (t - 0)\{1 + (t - 0.5)\{-2\}\} = 2t(1 - t)$.

Notice that the velocity is $dz(t)/dt = 2 - 4t$, which is zero at $t = 0.5$ and a maximum at $t = 0$ and $t = 1$. This is the opposite of what is usually required for the path control of robots, where we would like the velocity at the beginning and end of the path to be zero.

---

4. Use the four points $\begin{bmatrix} 1 \\ 0 \\ 0 \\ 1 \end{bmatrix}$, $\begin{bmatrix} 1 \\ \alpha \\ 0 \\ 1 \end{bmatrix}$, $\begin{bmatrix} 0 \\ \alpha \\ 1 \\ 1 \end{bmatrix}$, and $\begin{bmatrix} 0 \\ 0 \\ 1 \\ 1 \end{bmatrix}$ in a Bezier curve to approximate a curve on the surface of a sphere of radius unity centered at the origin. Match the Bezier curve at its center point ($t = 0.5$) to the corresponding point on the sphere. Show the error between the Bezier curve and the corresponding points on the sphere at $t = 0.25$ and $t = 0.75$.

The Bezier curve (deleting the redundant fourth entry in the homogeneous representation for the points) is

$$(1 - t)^3 \begin{bmatrix} 1 \\ 0 \\ 0 \end{bmatrix} + 3(1 - t)^2 t \begin{bmatrix} 1 \\ \alpha \\ 0 \end{bmatrix} + 3(1 - t)t^2 \begin{bmatrix} 0 \\ \alpha \\ 1 \end{bmatrix} + t^3 \begin{bmatrix} 0 \\ 0 \\ 1 \end{bmatrix}$$

and at $t = \frac{1}{2}$ we have

$$\frac{1}{8}\begin{bmatrix} 1 \\ 0 \\ 0 \end{bmatrix} + \frac{3}{8}\begin{bmatrix} 1 \\ \alpha \\ 0 \end{bmatrix} + \frac{3}{8}\begin{bmatrix} 0 \\ \alpha \\ 1 \end{bmatrix} + \frac{1}{8}\begin{bmatrix} 0 \\ 0 \\ 1 \end{bmatrix} = \frac{1}{2}\begin{bmatrix} 1 \\ \dfrac{3\alpha}{2} \\ 1 \end{bmatrix}$$

so for exact fit $\frac{1}{4}\{1 + 9\alpha^2/4 + 1\} = 1$, giving $\alpha = 2\sqrt{2}/3$.

At $t = \frac{1}{4}$,

$$\begin{bmatrix} x \\ y \\ z \end{bmatrix} = \frac{1}{64}\begin{bmatrix} 54 \\ 36\alpha \\ 10 \end{bmatrix} = \begin{bmatrix} 0.84375 \\ 0.53033 \\ 0.15625 \end{bmatrix}$$

for which $\sqrt{0.84375^2 + 0.53033^2 + 0.15625^2} = 1.00875$ instead of unity. By symmetry, the error is the same at $t = \frac{3}{4}$.

---

5. Construct the best possible circle from a six-point Bezier curve and a seven-point Bezier curve. Assume the circle has unity diameter and is in the xy-plane with tangent to the circumference on the x-axis and a point on the circumference at the origin.

The six-point curve as derived from equation (7.17) is

$$\mathbf{u}(t) = (1 - t)^5\mathbf{u}_0 + 5t(1 - t)^4\mathbf{u}_1 + 10t^2(1 - t)^3\mathbf{u}_2 + 10t^3(1 - t)^2\mathbf{u}_3 + 5t^4(1 - t)\mathbf{u}_4 + t^5\mathbf{u}_5$$

With

$$\mathbf{u}_0 = \mathbf{u}_5 = \begin{bmatrix} 0 \\ 0 \\ 0 \\ 1 \end{bmatrix}, \quad \mathbf{u}_1 = \begin{bmatrix} x_1 \\ 0 \\ 0 \\ 1 \end{bmatrix} = -\mathbf{u}_4, \quad \mathbf{u}_2 = \begin{bmatrix} x_2 \\ y_2 \\ 0 \\ 1 \end{bmatrix}, \quad \text{and} \quad \mathbf{u}_3 = \begin{bmatrix} -x_2 \\ y_2 \\ 0 \\ 1 \end{bmatrix}$$

then

$$\begin{aligned} x(t) &= 5t(1 - t)^4x_1 + 10t^2(1 - t)^3x_2 - 10t^3(1 - t)^2x_2 - 5t^4(1 - t)x_1 \\ &= 5t(1 - t)(\{(1 - t)^3 - t^3\}x_1 + 2t(1 - t)(1 - 2t)x_2) \\ y(t) &= 10t^2(1 - t)^2y_2 \end{aligned}$$

At $t = 0.5$ we require $y(t) = 1$, so $\frac{10}{16}\, y_2 = 1$, or $y_2 = 1.6$.

We need two other conditions to establish x_1 and x_2. At $t = \frac{1}{6}$, $y(\frac{1}{6}) = \frac{10}{16}\,(\frac{5}{6})^2\, 1.6 = 0.3086$, at which $x = 0.46193$ for a point on the circle.

$$x(\tfrac{1}{6}) = 0.46193 = \tfrac{25}{36}\,(\{(\tfrac{5}{6})^3 - (\tfrac{1}{6})^3)x_1 + \tfrac{10}{36}\,(\tfrac{2}{3})x_2)$$

or $0.66518 = 0.57407x_1 + 0.18519x_2$.

At $t = \frac{1}{3}$, $x = 0.46193$ for a point on the circle, so

$$x(\tfrac{1}{3}) = 0.46193 = \tfrac{10}{9}\,\{(\tfrac{8}{27} - \tfrac{1}{27})x_1 + \tfrac{4}{9}\,(\tfrac{1}{3})x_2\}$$

or $0.41574 = 0.25926x_1 + 0.14815x_2$. We rewrite these two equations as

$$x_1 + 0.27840x_2 = 1.15871$$

$$x_1 + 0.57143x_2 = 1.60356$$

so $0.29303x_2 = 0.44485$, or $x_2 = 1.5181$, and $x_1 = 0.73606$.

The error at $t = 0.25$ will be determined. The correct point on the circle is $x = 0.5$ and $y = 0.25$. The actual values on the Bezier curve are

$$x(0.25) = \tfrac{15}{16}(\{(\tfrac{3}{4})^3 - \tfrac{1}{4})^3\}x_1 + \tfrac{3}{16}x_2) = \tfrac{15}{16}\,(0.40625x_1 + 0.1875x_2) = 0.54719$$

$$y(t) = \tfrac{10}{16}(\tfrac{3}{4})^2 y_2 = 0.35156y_2 = 0.5625.$$

This error is excessive, so a six-point Bezier curve designed this way is inadequate to model a circle. However, if the velocity along the curve as a function of t is permitted to vary, meaning that $x(\frac{1}{3})$ is not correct, the shape of the circle can be greatly improved. For example, if we choose $x_1 = 0.6$, $x_2 = 1.55$, and $y_2 = 1.6$, the curve shown in Figure 7.10 results.

Suppose seven points are used to model a full circle. These points would be at

$$\mathbf{u}_0 = \mathbf{u}_6 = \begin{bmatrix} 0 \\ 0 \\ 0 \\ 1 \end{bmatrix}, \quad \mathbf{u}_1 = \begin{bmatrix} x_1 \\ 0 \\ 0 \\ 1 \end{bmatrix} = -\mathbf{u}_5,$$

$$\mathbf{u}_2 = \begin{bmatrix} x_2 \\ y_2 \\ 0 \\ 1 \end{bmatrix}, \quad \mathbf{u}_3 = \begin{bmatrix} 0 \\ y_3 \\ 0 \\ 1 \end{bmatrix}, \quad \mathbf{u}_4 = \begin{bmatrix} -x_2 \\ y_2 \\ 0 \\ 1 \end{bmatrix}$$

The seven-point Bezier curve is

$$\mathbf{u}(t) = (1 - t)^6\mathbf{u}_0 + 6t(1 - t)^5\mathbf{u}_1 + 15t^2(1 - t)^4\mathbf{u}_2 + 20t^3(1 - t)^3\mathbf{u}_3 + 15t^4(1 - t)^2\mathbf{u}_4 + 6t^5(1 - t)\mathbf{u}_5 + t^6\mathbf{u}_6$$

Thus

$$\begin{aligned} x(t) &= t(1 - t)\{6(1 - t)^4x_1 + 15t(1 - t)^3x_2 - 15t^3(1 - t)x_2 - 6t^4x_1\} \\ &= 3t(1 - t)\{2\{(1 - t)^4 - t^4\}x_1 + 5t(1 - t)(1 - 2t)x_2\} \end{aligned}$$

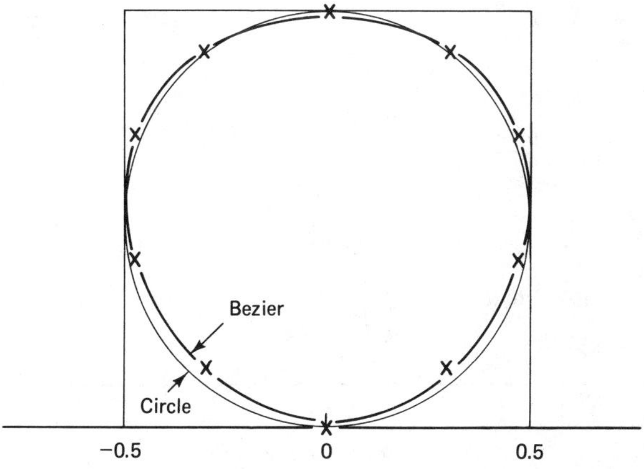

Figure 7.10 Six-point Bezier curve approximation to a circle

$$y(t) = t(1 - t)\{15t(1 - t)^3y_2 + 20t^2(1 - t)^2y_3 + 15t^3(1 - t)y_2\}$$
$$= 5t^2(1 - t)^2\{3(1 - 2t + 2t^2)y_2 + 4t(1 - t)y_3\}$$

We notice that $x(0.5) = 0$, and choose the points to fit the circle at $t = \frac{1}{6}$ and $\frac{1}{3}$.

$$x(\tfrac{1}{6}) = \tfrac{5}{12}(2\{(\tfrac{5}{6})^4 - (\tfrac{1}{6})^4\}x_1 + \tfrac{25}{108}x_2) = 0.41666\{0.96296x_1 + 0.23148x_2\}$$
$$= 0.40123x_1 + 0.09645x_2 = 0.46193$$
$$x(\tfrac{1}{3}) = \tfrac{2}{3}\{0.37037x_1 + 0.37037x_2\} = 0.24691\{x_1 + x_2\} = 0.46193$$

Rewriting these two simultaneous equations as

$$x_1 + 0.24039x_2 = 1.15128$$
$$x_1 + x_2 = 1.8708$$

with solution $0.75961x_2 = 0.71952$, or $x_2 = 0.94723$, and $x_1 = 0.92358$. (See Figure 7.11.) There is no necessity to match the y-values at the same points as chosen for x. Instead here we choose $t = \frac{1}{4}, \frac{1}{2}$, giving

$$y(\tfrac{1}{4}) = \tfrac{45}{256}\{\tfrac{15}{8}y_2 + \tfrac{3}{4}y_3\} = 0.32960y_2 + 0.13184y_3 = 0.5$$
$$y(\tfrac{1}{2}) = \tfrac{5}{16}\{\tfrac{3}{2}y_2 + y_3\} = 0.46875y_2 + 0.3125y_3 = 1.$$

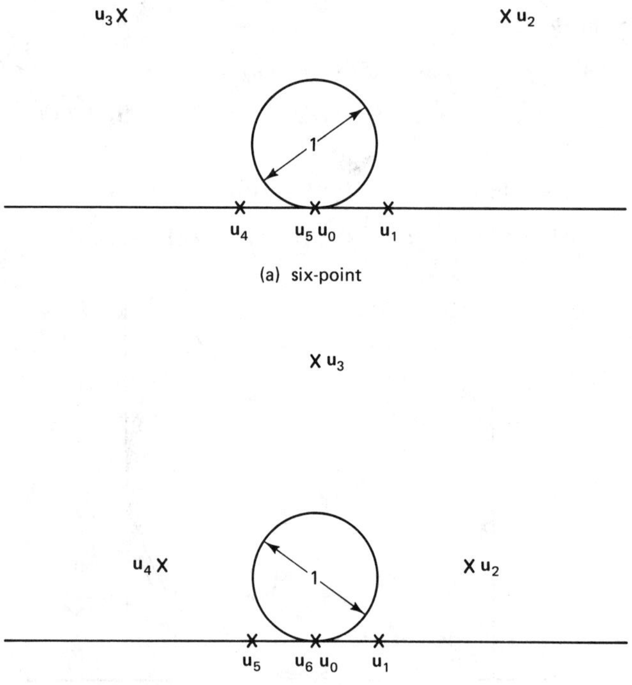

Figure 7.11 Location of points used to approximate a circle

Rewriting these two simultaneous equations,

$$y_2 + 0.4y_3 = 1.51699$$

$$y_2 + 0.66666y_3 = 2.1333$$

with solution $0.26666y_3 = 0.61634$, or $y_3 = 2.31133$, and $y_2 = 0.71111$.

As before, the shape improves if the velocity characteristics are degraded. For example, the values $x_1 = 0.5$, $x_2 = 1.25$, $y_2 = 0.6$, and $y_3 = 2.3$ would produce a shape very close to a circle.

EXERCISES FOR CHAPTER 7

1. A circle of radius 1 centered at the origin is contained in the xy-plane. Model the quarter circle for $x \geq 0$ and $y \geq 0$ by two quadratic sections and apply parabolic blending. *Hint:* Use points at angles 0°, 30°, 60°, and 90° with respect to the positive x-axis and form quadratics from the first three points and the last three points.

---

2. For $f(x) = x^3$, fit a quadratic through the three points at $x = 0$, 1 and 2 and fit a quadratic through the three points at $x = 1, 2$, and 3. Apply parabolic blending to the quadratics, and comment on the errors in the regions $0 \leq x \leq 1$, $1 \leq x \leq 2$, and $2 \leq x \leq 3$.

---

3. Model $\sin t$ over the range $0 \leq t \leq \pi/2$ by two quadratic polynomials. The first quadratic fits the function at $t = 0$, $\pi/6$, and $\pi/3$. The second quadratic fits the function at $\pi/6$, $\pi/3$, and $\pi/2$. Apply parabolic blending to the section $\pi/6 \leq t \leq \pi/3$. What is the error at $t = \pi/4$?

---

4. Using Newton's interpolating polynomial, approximate $\sin t$ over the range $0 \leq t \leq \pi/2$ using four equally spaced points. What is the error of the approximation at $t = \pi/4$?

---

5. Model the function $x(t) = \alpha t$, $0 \leq t \leq 1$, $x(t) = 2\alpha - \alpha t$, $1 \leq t \leq 2$, by a polynomial of order n with $n + 1$ equally spaced points at $t = 0$, $t = 2/n$, $t = 4/n$, . . . , $t = 2(n - 1)/n$, and $t = 2$. Use the Newton interpolating polynomial and solve for $n = 3$, 4, 5, and 6. Show the resulting polynomial approximations graphically.

---

6. Given the four data points $x(0) = 0$, $x(1) = 0$, $x(2) = 0$, and $x(T) = 1$, find the Newton interpolating polynomial that passes through all four points. Comment on the singular points that can occur for particular values of T.

---

7. Suppose the function $x(t) = 0$, $0 \leq t \leq 0.5$, $x(t) = 1$, $0.5 < t \leq 1$, is to be approximated by a Newton interpolating polynomial of order n, $n = 1, 2, 3, 4$. Assuming that the data points chosen are equally spaced, determine these polynomials.

---

8. Find the simplest polynomial function to connect the two points (0, 1) and (1, 0) such that it avoids the staircase obstacle defined by $y = 1$ for $0 \leq x < 1$ and $x = 1$, $0 \leq y < 1$. Write the function as two parametric polynomials in t, where $0 \leq t \leq 1$.

---

9. Sketch the Bezier curves that pass through the ordered sets of points shown in figure 7.12.

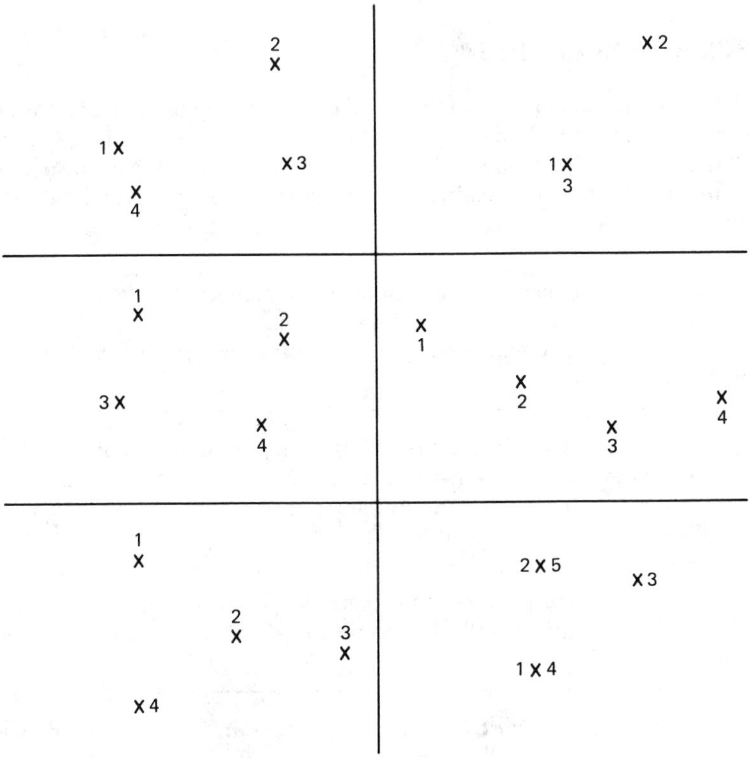

Figure 7.12 Some sets of Bezier points

---

10. Sketch the Bezier curves that pass through the ordered sets of points shown in figure 7.13.

---

11. An ellipse, whose major axis has length 2 and whose minor axis has length 2β, where $\beta < 1$, is contained in plane $\mathbf{p} = [a \quad b \quad c \quad 0]$ centered at the origin with major axis contained in the xy-plane. Model this ellipse by four sections of four-point Bezier curves. Choose the points interior to each Bezier curve for best positional fit. Draw the spatial position of the points on $\mathbf{p}$.

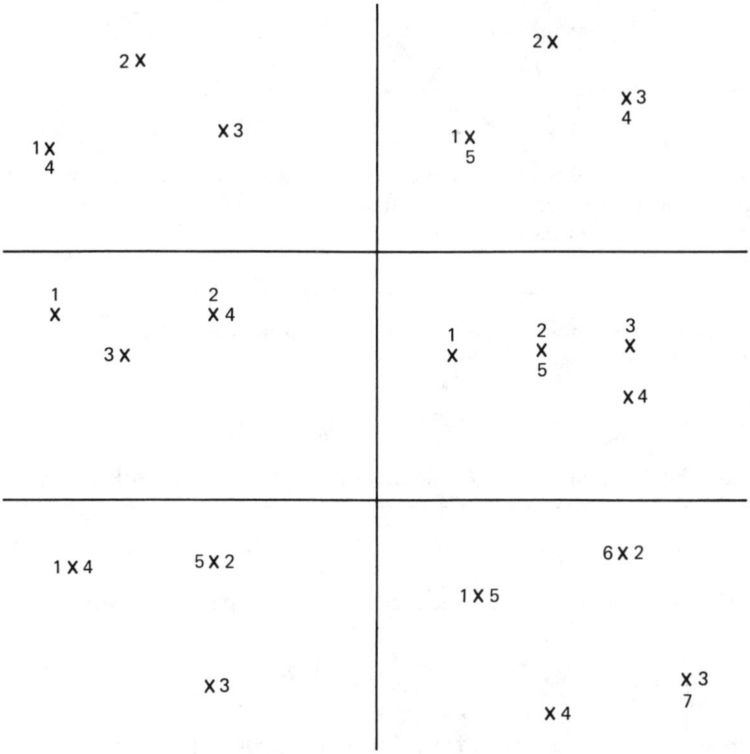

Figure 7.13 Another set of Bezier points

12. Approximate $\sin 2\pi t$ by a four-point Bezier curve over the range $0 \le t \le 1$, such that the derivatives at the ends of the curve are exact.

---

13. Approximate $\sin \pi t$ by a four-point Bezier curve over the range $0 \le t \le 1$, such that the derivatives at the ends of the curve are exact.

---

14. Approximate $\sin x$ between 0 and $\pi/2$ by a four-point Bezier curve such that the velocity at $x = \pi/2$ is zero. *Hint:* The Bezier function must run between $0 \le t \le 1$, so adjust the sine function appropriately. The Bezier curve must match the sine wave at $t = 0.5$. What is the positional error at $t = 0.25$ and $t = 0.75$?

---

15. Approximate a circle centered at the origin of radius unity lying in the xy plane by four four-point Bezier curves. The velocity at the start and end is required to be zero, and the join between adjacent Bezier curves is required to match in position and velocity. Justify the choice of Bezier points. Where do you estimate the maximum error will occur? Determine the error at this point.

---

16. $\dfrac{1 - \cos \pi t/2}{\pi t/2}$ is approximated by a five-point Bezier curve $P(t)$ between $0 \le t \le 1$ such that $P(0)$, $P(1)$, $\dfrac{dP(0)}{dt}$, $\dfrac{dP(1)}{dt}$ and $P(0.5)$ match the function. Find $P(t)$.

---

17. Find the simplest Bezier function to connect the two points (0, 1) and (1, 0) such that it avoids the staircase obstacle defined by $y = 1$ for $0 \le x < 1$ and $x = 1$, $0 \le y < 1$. Assume that the initial and final velocities are zero and that the execution time is 1 s.

---

18. Use five- and six-point Bezier functions to approximate $x(t)$, $0 \le t \le 1$, where $x(t) = 0$ and

$$\frac{d^2x}{dt^2} = \alpha, \quad 0 \le t \le 0.5, \qquad \frac{d^2x}{dt^2} = -\alpha, \quad 0.5 \le t \le 1$$

Note: The function is the ideal acceleration function given in Figure 8.4.

---

19. The end effector of a robot is required to traverse a path that starts at $\begin{bmatrix} 0 \\ 0 \\ 0 \\ 1 \end{bmatrix}$ and ends at $\begin{bmatrix} 2 \\ 0 \\ 0 \\ 1 \end{bmatrix}$. It must pass through intermediate point $\begin{bmatrix} 1 \\ 1 \\ 0 \\ 1 \end{bmatrix}$ and the velocity at the endpoints must be zero. Find the Bezier curve that satisfies these requirements.

---

20. Given function

$$x(t) = \begin{cases} 0, & 0 \le t \le 0.4 \\ 1, & 0.4 \le t \le 1 \end{cases}$$

Assume $x(0.4) = 0.4$. Determine the N-point Bezier approximating functions for $N = 4$, 5, and 6. Draw the resulting time functions.

---

21. Determine the three points in an $n = 2$ Bezier function that passes through the three points, (0, 0), (1, 0) and (1, 1) in the xz-plane.

---

22. Determine the four points in an $n = 3$ Bezier curve that passes through the four points (0, 0), (1, 0), (2, 0) and (2, 1) in the xy-plane.

---

23. Given the quadratic polynomial $x(\alpha) = 1 + \alpha + \alpha^2$. Determine the Bezier points that would produce this function over the range $0 \le \alpha \le 1$.

---

24. Given the quadratic polynomial $x(t) = 1 + \alpha + \alpha^2$. Determine the Bezier points that would produce this function over the range $-1 \leq \alpha \leq 1$.

---

25. Given the polynomial $y(t) = \alpha^3 + 1$. Determine the Bezier points that would produce this function over the range $0 \leq \alpha \leq 1$.

---

26. Given the polynomial $y(t) = \alpha^3 + \alpha + 1$. Determine the Bezier points that would produce this function over the range $0 \leq \alpha \leq 2$.

CHAPTER EIGHT

Path Control

Path control can be as simple as determining the joint status to place the end effector of the robot in two separate positions, the beginning and endpoints, and changing the joints from one status to another. It can require the end effector to avoid some region inside its working environment. It can require strict control over the complete trajectory followed by the end effector and all intermediate joints. In the most comprehensive requirement, it can require velocity control with an upper limit on acceleration along the complete curved trajectory. Although complete acceleration control is not required, there are constraints imposed by

1. The maximum torques of each actuator (joint drive motor),
2. The mechanical limitations of the joints and links, and
3. The gripping forces on the object held in the end effector.

Problems in determining position space trajectories and some suggested methods of solution are given in this chapter.

The problem of determining the joint space to produce a particular position path has been given considerable attention, for good reason [Brady et al. 1983; Freund 1977; Gupta and Roth 1982; Litvin and Casteli 1984; Luh, Walker, and Paula 1980; Luh and Lin 1981; Luh and Campbell 1982; Parkin and Hutchinson 1985a; Parkin 1985c; Paul 1979; Paul 1981b; Raibert and Horn 1978; Taylor 1979]. A paint-spraying robot must maintain a fairly constant distance from the object and keep its spray nozzle reasonably perpendicular to the surface being sprayed. Its spray nozzle is its end effector, and we have just defined a path-control problem. A seam-welding robot is required to run the weld head at a fixed distance from the seam with the wire feed rate controlled; this is a path-control problem. A SCARA robot must pick up a printed circuit board and place it in a chassis, avoiding other boards and the power supply already mounted; this is a path-control problem.

The spray gun of the paint-spray robot must not only move along a particular path, it must also move at a controlled velocity. An arc-welding robot is required not only to follow a seam but to move along the seam at a controlled rate. Here we consider position, velocity, and acceleration control [Featherstone 1983; Lozano-Perez 1981; Waldron 1982; Whitney 1969] and find Bezier function techniques preferable to other methods.

8.1 SLEW SOLUTIONS

Suppose a robot is to travel from one point to another, and the trajectory followed is unimportant. For this case a slew solution may be acceptable. A slew solution starts by determining the joint status at the beginning points and endpoints [Taylor 1979]. The joints are driven to start together and to move at constant angular (for the revolute joints) or linear (for the sliding links) speed.

Sometimes the slew rate for each joint is fixed, so some joints complete the slew before other joints. Sometimes the joint slew rate is adjusted so all joints complete their slew path at the same time. For example, for the 3 *dof* revolute robot, assume the joint status is $\lambda_1(t)$, $\theta_2(t)$, and $\theta_3(t)$, where $t = 0$ at the beginning point and $t = 1$ at the endpoint. Suppose the joint movements are required to start and stop together and the joints move at constant angular velocity. Then

$$\begin{bmatrix} \lambda_1(t) \\ \theta_2(t) \\ \theta_3(t) \end{bmatrix} = \begin{bmatrix} \lambda_1(0) \\ \theta_2(0) \\ \theta_3(0) \end{bmatrix} + t \begin{bmatrix} \lambda_1(1) - \lambda_1(0) \\ \theta_2(1) - \theta_2(0) \\ \theta_3(1) - \theta_3(0) \end{bmatrix} \tag{8.1}$$

and a slew solution is obtained. The path followed by such a solution will be straight only if the robot is the 3 *dof* Cartesian robot. We will calculate the slew motion for a series of cases involving the 3 *dof* basic robots.

8.1.1 The Revolute Robot

Suppose for the 3 *dof* revolute robot $\mathbf{u}_1$ is the origin, $\mathbf{p}_1 = [0 \quad 1 \quad 0 \quad 0]$, and $\boldsymbol{\nu} = [0 \quad 0 \quad 1]^{\#}$. If λ is fixed in its datum position, then $\bar{\mathbf{p}}_2 = [0 \quad 1 \quad 0 \quad 0]$. Suppose the links of the robot are 1 m long and the end effector is required to move from point

$$\begin{bmatrix} 1 \\ 0 \\ 1 \\ 1 \end{bmatrix} \text{ to point } \begin{bmatrix} 1 \\ 0 \\ 0 \\ 1 \end{bmatrix}$$

From Section 6.3.1 we find the joint status at the starting point as $\theta_2 = -30°$, $\theta_3 = -120°$ and that at the endpoint as $\theta_2 = -90°$, $\theta_3 = -90°$, so

$$\begin{bmatrix} \lambda_1(t) \\ \theta_2(t) \\ \theta_3(t) \end{bmatrix} = \begin{bmatrix} 0 \\ -30 \\ -120 \end{bmatrix} + t \begin{bmatrix} 0 \\ -60 \\ 30 \end{bmatrix}$$

From equation (5.7),

$$\mathbf{u}_4 = \begin{bmatrix} -\sin\theta_2(t) - \sin(\theta_2(t) + \theta_3(t)) \\ 0 \\ 1 + \cos\theta_2(t) + \cos(\theta_2(t) + \theta_3(t)) \\ 1 \end{bmatrix}$$

The path followed by the end effector is shown as the dotted line in Figure 8.1. At the parametric middle point, $t = 0.5$,

$$\mathbf{u}_4 = \begin{bmatrix} \sin 60 + \sin 165 \\ 0 \\ 1 + \cos 60 + \cos 165 \\ 1 \end{bmatrix} = \begin{bmatrix} 1.1248 \\ 0 \\ 0.5341 \\ 1 \end{bmatrix}$$

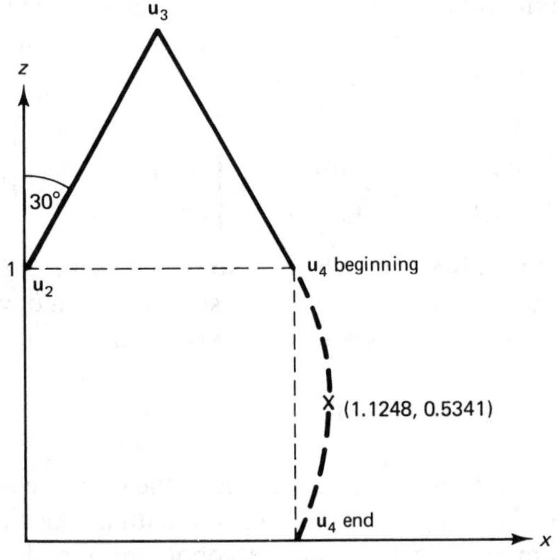

Figure 8.1 Path followed in a slew solution

Compare this to the geometric middle point [1 0 0.5 1]. We can display the slew path in a plane, since λ_1 does not change throughout this slew path. If λ_1 changes, the trajectory followed along the slew path probably cannot be displayed in a plane. However, all three variables (λ_1, θ_2, and θ_3) are single-valued functions of t.

8.1.2 The Cylindrical Robot

Suppose the sliding links of the 3 *dof* cylindrical robot with base orientation and position given by $\mathbf{p}_1 = [0 \quad 1 \quad 0 \quad 0]$, $\boldsymbol{\nu} = [0 \quad 0 \quad 1]^{\#}$, and $\mathbf{u}_1 = [0 \quad 0 \quad 0 \quad 1]^{\#}$ can vary from 0.5 m to 1 m long. A slew path is required between points $[0.5 \quad 0 \quad 0.5 \quad 1]^{\#}$ and $[0 \quad 1 \quad 1 \quad 1]^{\#}$. From Section 6.3.2, the joint status at the starting point is $\lambda_1 = 0$, $\sigma_2 = 0.5$, and $\sigma_3 = 0.5$, and the joint status at the end point is $\lambda_1 = -90°$, $\sigma_2 = 1$, and $\sigma_3 = 1$. The slew path is defined by

$$\begin{bmatrix} \lambda_1(t) \\ \sigma_2(t) \\ \sigma_3(t) \end{bmatrix} = \begin{bmatrix} 0 \\ 0.5 \\ 0.5 \end{bmatrix} + t \begin{bmatrix} -90 \\ 0.5 \\ 0.5 \end{bmatrix}$$

and the slew path can be defined in a plane in Cartesian space. In fact, any slew path for this robot can be defined in a plane.

8.1.3 The Cartesian Robot

It is a trivial task to define the slew path for the Cartesian robot, since for the end effector at $[x \quad y \quad z \quad 1]^{\#}$, $\sigma_1 = z$, $\sigma_2 = x$, and $\sigma_3 = y$, and the slew path is always a straight line in Cartesian space.

8.1.4 The Polar Robot

Consider a polar robot with base orientation and position given by $\mathbf{p}_1 = [0 \quad 1 \quad 0 \quad 0]$, $\boldsymbol{\nu} = [0 \quad 0 \quad 1]^{\#}$ and $\mathbf{u}_1 = [0 \quad 0 \quad 0 \quad 1]^{\#}$, and $l_1 = 1$, $0.7 \le \sigma_3 \le 1.4$. Suppose the robot is to traverse a slew path from $[0.5 \quad 0 \quad 1.5 \quad 1]^{\#}$ to $[0 \quad 1 \quad 1.5 \quad 1]^{\#}$; then from equation (6.17), the joint status at the start is $\lambda_1 = 0$, $\theta_2 = -45°$, and $\sigma_3 = 1/\sqrt{2}$, and at the end it is $\lambda_1 = 90°$, $\theta_2 = -63.4$, and $\sigma_3 = 1.1167$. The slew path is defined by

$$\begin{bmatrix} \lambda_1 \\ \theta_2 \\ \sigma_3 \end{bmatrix} = \begin{bmatrix} 0 \\ -45 \\ \dfrac{1}{\sqrt{2}} \end{bmatrix} + t \begin{bmatrix} 90 \\ -18.4 \\ 0.4096 \end{bmatrix}$$

In general, the slew path followed by this robot will not lie on a plane in Cartesian space.

8.1.5 The SCARA Robot

Suppose the 3 *dof* SCARA robot $|\ \theta\theta(\times)\sigma\ |$ with $\mathbf{p}_1 = [0 \quad 0 \quad -1 \quad 0]$, $\boldsymbol{\nu} = [1 \quad 0 \quad 0]^{\#}$, $\mathbf{u}_1 = [0 \quad 0 \quad 0 \quad 1]^{\#}$, $l_1 = l_2 = 1$ m, and $0.5 \le \sigma_3 \le 1$ is required to execute a slew path between points $[1 \quad 0 \quad -0.5 \quad 1]^{\#}$ and $[0.5 \quad 1 \quad -1 \quad 1]^{\#}$. From 6.3.5, the given base orientation and position gives

$$\beta = x,\ \alpha = y,\ \sigma_3 = z,\ \theta_1 = \cos^{-1}\frac{x^2 + y^2}{2\sqrt{x^2 + y^2}} + \tan^{-1}\frac{y}{x}.$$

At the starting point, $x = 1$ and $y = 0$, so $\theta_1 = \cos^{-1} 0.5 = 60°$.

$$\theta_2 = \cos^{-1}(x - \cos\theta_1) - \theta_1 = \cos^{-1}(1 - 0.5) - 60 = \pm 60 - 60 = -120.$$

N.B. the angle chosen for $\cos^{-1} 0.5$ must make the mapping from joint space to position space consistant.

At the ending point, $x = 0.5$ and $y = 1$, so

$$\theta_1 = \cos^{-1}\frac{1.25}{2\sqrt{1.25}} + \tan^{-1} 2 = 56.01 + 63.43 = 119.44°.$$

$\theta_2 = \cos^{-1}(0.5\text{-}\cos 119.44) - 119.44 = \cos^{-1}(0.9915) - 119.44 = \pm 7.47 - 119.44 = -112°$.

8.2 STRAIGHT-LINE TRAJECTORIES

It would seem that the execution of a straight-line trajectory should be a simple task for today's robots, but unfortunately this is not the case [Paul 1979]. The mapping of position space into joint space is nonlinear, so even though an exact trajectory (such as a low order polynomial) can be followed in joint space, the trajectory in position space is characterized by a polyno-mial of infinite order. Here we will discuss approximate methods for producing acceptable straight-line trajectories in position space.

Taylor's algorithm is an application of the slew path technique and permits a Cartesian space straight-line path to be followed to any desired degree of accuracy [Taylor 1979]; it produces "bounded deviation joint paths" with slew solutions between. The method can be described as follows:

1. Perform the IKS at the endpoints $\mathbf{v}_0$ and $\mathbf{v}_1$ to produce joint vectors ζ_0 and ζ_1.
2. Calculate the joint vector midpoint $\zeta_{0.5} = (\zeta_0 + \zeta_1)/2$, and perform a FKS to find $\mathbf{v}(\zeta_{0.5})$.
3. If the error $\mathbf{v}(\zeta_{0.5}) - \mathbf{v}_{0.5}$, where $\mathbf{v}_{0.5}$ is the Cartesian midpoint $(\mathbf{v}_0 + \mathbf{v}_1)/2$, is excessive, go to step 4. Otherwise, if the midpoint error is acceptable, stop the iteration.

4. Perform an IKS for $\mathbf{v}_{0.5}$ to find $\zeta(\mathbf{v}_{0.5})$, use ζ_0 and $\zeta(\mathbf{v}_{0.5})$ as the end points of one slew path, and $\zeta(\mathbf{v}_{0.5})$ and ζ_1 as the end points of another slew path. Return to step 2 using these individual slew paths.

The iteration continues until the midpoint slew error in every subpath is less than some maximum value permitted. The result is a series of slew paths, each with acceptable deviation from the required straight line path. At each intermediate point, the end points of the slew paths, the velocity will be discontinuous. Although we have not said (but will in Section 8.6) anything about the velocity on a slew path, at the intermediate points (called knot points by Taylor), the path takes an abrupt change in direction, which is undesirable. It can be argued that the method of driving and controlling the joints will smooth out the discontinuities. If the actuators (joint-drive motors) are DC servomotors, they are inductive and tend to smooth out changes, so starting the next path before the prior one is completed smoothes out the discontinuity. However, this author is uncomfortable with such a technique. A poor method of describing movement is masked by a physical phenomenon elsewhere, linking two nonlinear systems. Wherever possible, an engineer should make each part of a system self-contained, permitting the interface of a variety of different subsystems and producing robustness in design.

The solid line in Figure 8.2 shows a typical slew solution between two points. If the deviation from the required path is considered excessive, the end effector can be forced to reach one or more intermediate points, which produce the looping solutions shown as the dotted line in Figure 8.2 with discontinuous velocity at the center point (the Taylor type of solution). However, performing the IKS at the three points (two ends and the middle) and fitting a parametric quadratic polynomial for each joint angle, the solution in Cartesian space could be as shown by the dashed line in Figure 8.2, a path that has continuous velocity throughout.

Suppose we have a joint space path defined for $0 \leq \alpha \leq 1$ over m intervals, $m + 1$ equally spaced points, $(0, x_0)$, (h, x_1), $(2h, x_2)$, . . . , $(1, x_m)$,

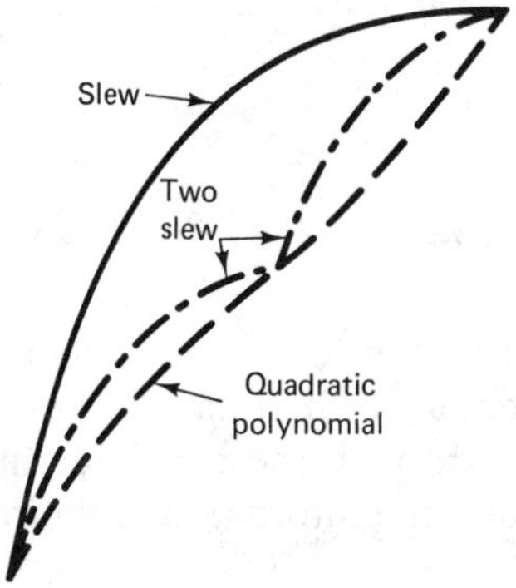

Figure 8.2 Paths

$mh = 1$, the divided difference table (see Section 7.2) is

$$
\begin{array}{lllll}
0 & x_0 & & & \\
 & & \dfrac{x_1 - x_0}{h} & & \\
h & x_1 & & \dfrac{x_2 - 2x_1 + x_0}{2h^2} & \\
 & & \dfrac{x_2 - x_1}{h} & & \dfrac{x_3 - 3x_2 + 3x_1 - x_0}{6h^3} \\
2h & x_2 & & \dfrac{x_3 - 2x_2 + x_1}{2h^2} & \qquad \dfrac{x_4 - 4x_3 + 6x_2 - 4x_1 + x_0}{24h^4} \\
 & & \dfrac{x_3 - x_2}{h} & & \dfrac{x_4 - 3x_3 + 3x_2 - x_1}{6h^3} \\
3h & x_3 & & \dfrac{x_4 - 2x_3 + x_2}{2h^2} & \\
 & & \dfrac{x_4 - x_3}{h} & & \\
4h & x_4 & & & \\
\vdots & & & &
\end{array}
$$

so the mth order polynomial in α is

$$
x(\alpha) = x_0 + \alpha \left\{ \underbrace{\frac{x_1 - x_0}{h}}_{m=1} + (\alpha - h)\left\{ \underbrace{\frac{x_2 - 2x_1 + x_0}{2h^2}}_{m=2} + (\alpha - 2h)\left\{ \underbrace{\frac{x_3 - 3x_2 + 3x_1 - x_0}{6h^3}}_{m=3} + (\alpha - 3h)\left\{ \underbrace{\frac{x_4 - 4x_3 + 6x_2 - 4x_1 + x_0}{24h^2}}_{m=4} + \ldots \right\}\right\}\right\}\right\} \tag{8.2}
$$

The vertical lines with the $m = 1$, etc. indicate the extent of the formula for that particular m, i.e. the formula to the left of that line is complete for that particular m. Execution of P(t) for $m = 2$ will produce a path such as shown by the dashed line in Figure 8.2. Example 8.3 is an application of the polynomial technique.

8.3 POLYNOMIAL TRAJECTORIES

Suppose the position space of a robot is defined as the $n + 1$ points ζ_i, $i = 0, 1, \ldots, n$, and the desired trajectory to be followed in position space is a continuous function that passes through these points. In this section we discuss techniques for determining the information required in joint space.

The starting point in all techniques is to perform the IKS on some or all of the ζ_i to produce ξ_i. Before the trajectory is completed, all points must be converted to joint space, but in some techniques trajectory execution can commence before all IKS are completed.

8.3.1 Parabolic Blending

With the jth entry of ξ designated ξ_j and the $n + 1$ values of ξ_j designated ξ_{ji}, $i = 0, 1, \ldots, n$, the divided difference table to the second order is given by

$$
\begin{array}{llll}
0 & \xi_{j0} & & \\
 & & \xi_{j1} - \xi_{j0} & \\
 & & & \dfrac{\xi_{j2} - 2\xi_{j1} + \xi_{j0}}{2} \\
1 & \xi_{j1} & & \\
 & & \xi_{j2} - \xi_{j1} & \\
 & & & \dfrac{\xi_{j3} - 2\xi_{j2} + \xi_{j1}}{2} \\
2 & \xi_{j2} & & \\
 & & \xi_{j3} - \xi_{j2} & \\
 & & & \dfrac{\xi_{j4} - 2\xi_{j3} + \xi_{j2}}{2} \\
3 & \xi_{j3} & & \\
\vdots & \vdots & \vdots & \vdots \\
 & & & \dfrac{\xi_{jn} - 2\xi_{j,n-1} + \xi_{j,n-2}}{2} \\
n-1 & \xi_{j,n-1} & & \\
 & & \xi_{jn} - \xi_{j,n-1} & \\
n & \xi_{jn} & &
\end{array}
$$

so

$$P_1(t) = \xi_{j0} + t\left\{\xi_{j1} - \xi_{j0} + \frac{(t-1)}{2}(\xi_{j2} - 2\xi_{j1} + \xi_{j0})\right\}$$

$$= \xi_{j0} + t\left(-\frac{3}{2}\xi_{j0} + 2\xi_{j1} - \frac{1}{2}\xi_{j2}\right) + t^2\left(\frac{1}{2}\xi_{j0} - \xi_{j1} + \frac{1}{2}\xi_{j2}\right) \tag{8.3a}$$

$$P_2(t) = \xi_{j1} + t\left(-\frac{3}{2}\xi_{j1} + 2\xi_{j2} - \frac{1}{2}\xi_{j3}\right) + t^2\left(\frac{1}{2}\xi_{j1} - \xi_{j2} + \frac{1}{2}\xi_{j3}\right) \tag{8.3b}$$

$$P_3(t) = \xi_{j2} + t\left(-\frac{3}{2}\xi_{j2} + 2\xi_{j3} - \frac{1}{2}\xi_{j4}\right) + t^2\left(\frac{1}{2}\xi_{j2} - \xi_{j3} + \frac{1}{2}\xi_{j4}\right) \tag{8.3c}$$

$$\vdots$$

$$Q_1(t) = (1 - s)P_1(t) + sP_2(t), \qquad 0 \le s \le 1,$$

and

$$Q_i(t) = (1 - s)P_i(t) + sP_{i+1}(t), \qquad i \le s \le i + 1,$$

or

$$Q_1(s) = (1 - s)\left\{\xi_{j0} + s\left(-\frac{3}{2}\xi_{j0} + 2\xi_{j1} - \frac{1}{2}\xi_{j2}\right) + \frac{s^2}{2}(\xi_{j0} - 2\xi_{j1} + \xi_{j2})\right\}$$
$$+ s\left\{\xi_{j1} + s\left(-\frac{3}{2}\xi_{j1} + 2\xi_{j2} - \frac{1}{2}\xi_{j3}\right) + \frac{s^2}{2}(\xi_{j1} - 2\xi_{j2} + \xi_{j3})\right\}$$

Thus,

$$Q_1(s) = \xi_{j0} + s\left(-\frac{5}{2}\xi_{j0} + 3\xi_{j1} - \frac{1}{2}\xi_{j2}\right) + s^2\left(2\xi_{j0} - \frac{9}{2}\xi_{j1} + 3\xi_{j2} - \frac{1}{2}\xi_{j3}\right)$$
$$+ s^3\left(-\frac{1}{2}\xi_{j0} + \frac{3}{2}\xi_{j1} - \frac{3}{2}\xi_{j2} + \frac{1}{2}\xi_{j3}\right), \ 0 \le s \le 1 \tag{8.4}$$

$$Q_2(s) = \xi_{j0} + s\left(-\frac{5}{2}\xi_{j0} + 3\xi_{j1} - \frac{1}{2}\xi_{j2}\right) + s^2\left(2\xi_{j0} - \frac{9}{2}\xi_{j1} + 3\xi_{j2} - \frac{1}{2}\xi_{j3}\right)$$
$$+ s^3\left(-\frac{1}{2}\xi_{j0} + \frac{3}{2}\xi_{j1} - \frac{3}{2}\xi_{j2} + \frac{1}{2}\xi_{j3}\right), \ 1 \le s \le 2 \tag{8.5}$$

8.3.2 Polynomial Fitting

A polynomial of order n in terms of running parameter t, $0 \le t \le n$, can be fitted to the $n + 1$ data points for each ξ_j. The most efficient way of constructing the polynomials is via the Newton interpolating polynomial and the divided difference table, as discussed in Section 7.2. The accuracy obtained with nth-order polynomial fitting is considerably better than with parabolic blending, particularly in all sections other than between the first and last pairs of points, since all data points influence the shape of the function at all points. The penalty for this increased accuracy is additional computational expense, which may adversely effect the chances of real time control.

If an nth-order polynomial is constructed using the Bezier curve on the ordered set of $n + 1$ data points for each ξ_j, the resulting polynomial will not pass through any intermediate points except by chance. The use of the Bezier function applied to the data points given is inappropriate. There are other uses for the Bezier function, as will be seen in later sections of this chapter.

To produce an exact straight line in Cartesian space requires a polynomial of order infinity in each of the joint variables. Also, to produce a simple trajectory in position space usually requires a polynomial of infinite order in joint space. We can approximate the function in joint space by decomposing the position space into segments in which the zero and first-order derivatives are continuous in joint space. This can be considered to be equivalent to a cubic spline [Adams 1974] or a Hermite interpolating polynomial of order 3.

The Hermite interpolating polynomial of order 3 that fits $x(t)$ and $x'(t)$ at $t = t_1$ and $t = t_2$ is given by [Ralston and Rabinowitz 1978]

$$P(t) = \frac{1}{(t_1 - t_2)^2}\left\{\left(1 + 2\frac{t - t_1}{t_2 - t_1}\right)(t - t_2)^2 x_1 + (t - t_2)^2\left(1 + 2\frac{t - t_2}{t_1 - t_2}\right)x_2 \right.$$
$$\left. + (t - t_1)(t - t_2)x_1' + (t - t_1)^2(t - t_2)x_2'\right\}$$

As an example, suppose this polynomial is used to join the two points $(0, x_1)$ and $(1, x_2)$, where the slopes are zero at both conditions. Then

$$P(t) = \{1 + 2t\}(t - 1)^2 x_1 + \{3 - 2t\}t^2 x_2$$

and $P(0) = x_1$, $P(1) = x_2$, $P'(0) = 0$, $P'(1) = 0$, so the polynomial meets the required conditions. Further, this particular polynomial is the same as the four-point Bezier function with $x_0 = x_1$ and $x_2 = x_3$. For our purposes, the Bezier function is easier to use and is more flexible.

A point in joint space maps into a point in position space. Further, polynomials that are nth-order derivative continuous in joint space are nth-order derivative continuous in position space. The mapping from one space to the other is nonlinear but well behaved.

8.4 BASIC CONSIDERATIONS IN VELOCITY CONTROL

When velocity control (position and speed of the end effector) is required [Waldron 1982; Whitney 1969], the task becomes more difficult than simple position control. The drive systems of a robot limit the acceleration permissible in a single joint, and often the velocity is also limited by frictional forces or the relationship between centrifugal forces occurring and gripping forces possible. The general problem can be classified as one of optimal velocity and acceleration control [Paul 1975; Paul 1979] or optimal path planning [Luh, Walker, and Paul 1980a].

If we suppose that the velocity and acceleration of the end effector over the required path have upper limits and that the time taken must be a minimum [Featherstone 1983; Kahn and Roth 1971; Luh and Walker 1977], then the position, speed, and acceleration of the end effector is shown in Figure 8.3 [Paul 1979]. The acceleration is constant at its maximum of $\overline{A}$ until the maximum velocity $\overline{V}$ is achieved, and this velocity is maintained until a point is reached where maximum deceleration $\overline{A}$ is applied and continued until the endpoint is reached with zero velocity at distance d from the start. If d is reduced, eventually deceleration will occur before the maximum velocity is reached, the situation shown in Figure 8.4 will occur [Paul 1979].

Applying the ideal acceleration characteristics of Figure 8.4 over distance 1 and time $0 \leq t \leq 1$, the position as a function of time is shown as the

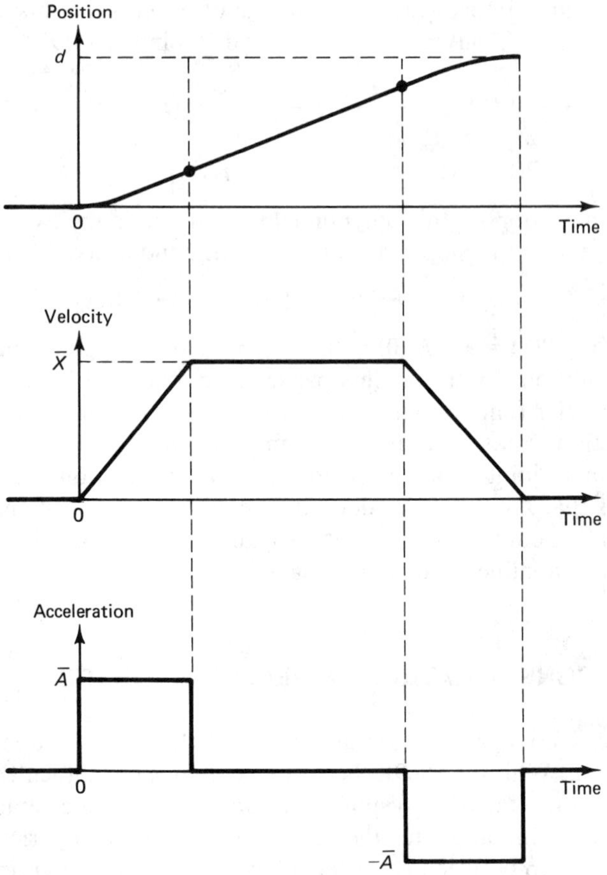

Figure 8.3 Response when velocity and acceleration are united

ideal characteristic of Figure 8.5. Compare this to $(1 - \cos \pi t)/2$ and $t^2(3 - 2t)$, also shown in Figure 8.5. The latter is based on a four-point Bezier curve discussed in Chapter 7; further discussion of the use of the Bezier curve in path control problems is given in the next section. In all three cases the velocity is zero at $t = 0$ and $t = 1$, often a necessary requirement. The ideal characteristic can be modeled by a generalized Fourier series [Churchill 1963] in terms of a set of functions, usually trigonometric or polynomial functions.

We can generalize the path-control problem for straight-line movement between points **u** and **v** such that position along the path is given by $\mathbf{u} + f(t)(\mathbf{v} - \mathbf{u})$, where $0 \leq f(t) \leq 1$, $f(t) = 0$ at $t = 0$, and $f(t) = 1$ at $t = 1$. We assume that $f(t)$ is a monotonic function of t. The path followed is a straight-

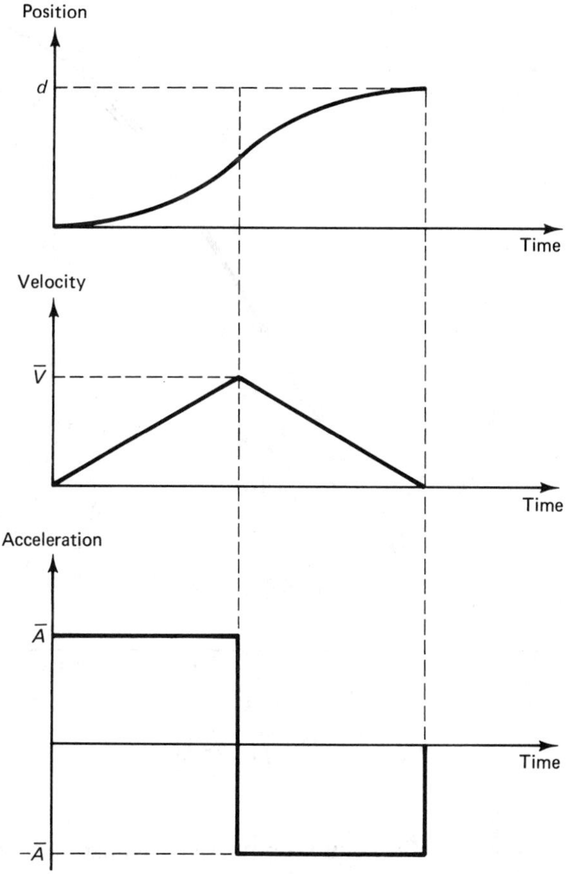

Figure 8.4 Ideal motion

line independent of $f(t)$, and the velocity along the path is dependent only on $f(t)$. In particular, if $f(t) = 1$, the velocity along the path is a constant. We will term $f(t)$ the velocity function. The motion along the path will always be toward point **v** and away from **u**.

This discussion on velocity control is solving the problem in position space, whereas the robot is driven and controlled in joint space. Translating velocity information in position space into joint space is not a simple task. Typically we have the mapping $\boldsymbol{\xi} = f(\boldsymbol{\zeta})$, where f is a nonlinear vector function. $\xi_i = f_i(\boldsymbol{\zeta})$. To map the velocities in position space into joint space requires forming the sum of the partial derivatives:

$$\xi_i = \sum_{j=1}^{n} \zeta_j \frac{\partial}{\partial \zeta_j} f_i(\boldsymbol{\zeta}), \qquad j = 1, 2, \ldots, n$$

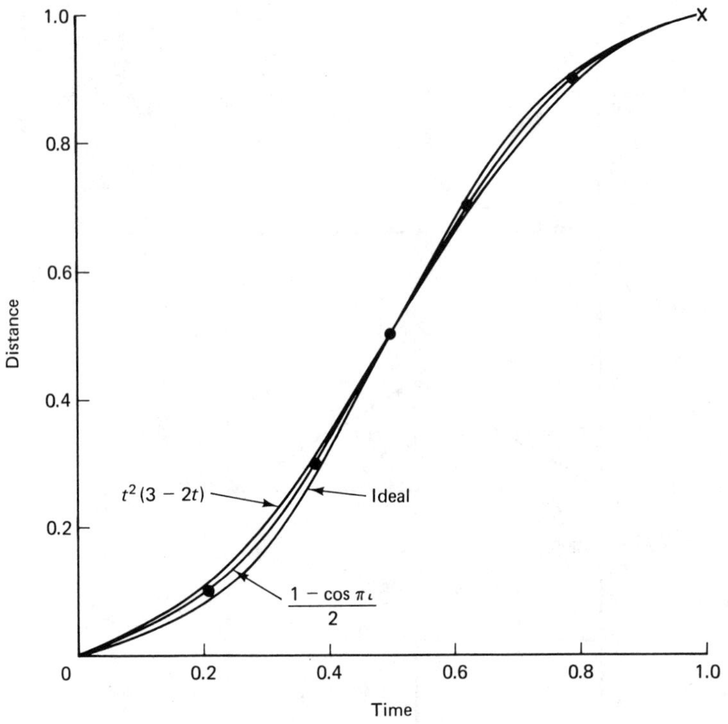

Figure 8.5 Distance profiles

For example, with the 3 *dof* revolute robot | $\lambda\theta\theta$ |, whose IKS was discussed in Section 6.3.1, the mapping of $\boldsymbol{\zeta} = [x \quad y \quad z]$ into $\boldsymbol{\xi} = [\lambda_1 \quad \theta_2 \quad \theta_3]$ is

$$\xi_1 = \tan^{-1} \frac{-a_1\zeta_1 - b_1\zeta_2 + \dfrac{1}{\nu_z}(a_1\nu_x + b_1\nu_y)\zeta_3}{(c_1\nu_y - b_1\nu_z)\zeta_1 + (a_1\nu_z - c_1\nu_x)\zeta_2 + (\nu_x b_1 - \nu_y a_1)\zeta_3} \tag{8.6}$$

$$\zeta_2 = \sin^{-1}\left\{\frac{\alpha' + \beta' + 1 - l_3/l_2}{2\sqrt{\beta'^2 + \alpha'^2}}\right\} - \tan^{-1}\frac{\alpha'}{\beta'} \tag{8.7}$$

and

$$\xi_3 = \sin^{-1}\left\{\frac{l_2(\beta' - \sin\xi_2)}{l_3}\right\} - \xi_2 \tag{8.8}$$

where

$$\beta' = \frac{y\nu_z - z\nu_y}{\{a_1\cos\xi_1 + (c_1\nu_y - b_1\nu_z)\sin\xi_1\}l_2} \tag{8.9}$$

and

$$\alpha' = \frac{z - \beta\{(\nu_x b_1 - \nu_y a_1)\cos \xi_1 - c_1 \sin \xi_1\}}{\nu_z l_2} - \frac{l_1}{l_2} \tag{8.10}$$

The calculations of the partial derivatives can be considered impractical, and we are forced to resort to approximation. The approximation we find most useful is to plan the path in position space as a polynomial function of a series of points that contain the required velocity characteristics and then to translate these points to joint space and use the same polynomial form on these joint space points. We demonstrate the technique in the next section with parabolic blending and the Bezier function.

8.5 VELOCITY PATH CONTROL

A trajectory with the required positional and velocity characteristics is planned in position space as a polynomial function of t. In this section we discuss two polynomial techniques used in joint space to approximate closely the position space trajectory.

8.5.1 Parabolic Blending

Suppose we use parabolic blending with a series of points in position space. The points are converted to joint space and parabolic blending employed with running variable t, considered to be time (or scaled real time). If the points are equally spaced, we traverse the path in position space at approximately constant speed.

Suppose we have the problem of one or more sharp changes in direction, or cusps, along the position space path. As previously discussed, parabolic blending will not accurately follow the path near these cusps. However, we can modify the technique to handle such situations. Suppose we modify the $Q_i(t)$ to be

$$Q_i(\tau) = \left\{1 - \frac{s}{d_i}\right\} P_i(\tau) + \frac{s}{d_i} P_{i+1}(\tau), \qquad 0 \le s \le d_i \tag{8.11}$$

and $\tau = d_1 + d_2 + \cdots + d_{i-1} + s$, where d_i is the distance between points $\mathbf{u}_i$ and $\mathbf{u}_{i+1}$. The shape of the series of Q functions is the same as if all d_i were unity, but by driving s linearly with time, we can modify the velocity in joint space (and so position space) with the spacing d_i. See the example at the end of this chapter.

8.5.2 Velocity Control Along a Bezier Curve

In Figure 8.5 the ideal motion characteristic (moving under maximum acceleration and deceleration at all times with no limitation on velocity) was compared to the functions $(1 - \cos \pi t)/2$ and $t^2(3 - 2t)$. The simple cubic is a four-point Bezier curve with $x_0 = 0$, $x_1 = 0$, $x_2 = 1$, and $x_3 = 1$. The closeness of this Bezier curve to the ideal characteristic is encouraging, particularly when it is noted that the ideal characteristic requires an infinite number of terms in the generalized Fourier series of orthogonal functions to model it exactly. Notice that the acceleration of the Bezier function $t^2(3 - 2t)$ is given by its second derivative as $6 - 4t$, which is a straight-line characteristic; compare this to the bang-bang acceleration of Figure 8.4, which we are attempting to duplicate. It is evident that we will need a Bezier curve with five or more points to model more accurately the bang-bang acceleration.

We ask the question, Can a path be modeled using Bezier techniques such that the velocity along that path can be controlled? Some attempts will be made to answer that question in this section.

Suppose a Bezier curve is constructed from points

$$\begin{bmatrix} x_i \\ y_i \\ z_i \\ 1 \end{bmatrix} = \begin{bmatrix} i\Delta_x \\ i\Delta_y \\ i\Delta_z \\ 1 \end{bmatrix}$$

where Δ_x, Δ_y, and Δ_z are constants and $i = 0, 1, \ldots, n$. Then

$$\begin{aligned} x(t) &= \Delta_x \sum_{i=0}^{n} i \frac{n!}{i!(n-i)!} t^i(1-t)^{n-i} \\ &= \Delta_x \Big\{ nt(1-t)^{n-1} + n(n-1)t^2(1-t)^{n-2} \\ &\quad + \frac{n(n-1)(n-2)}{2!} t^3(1-t)^{n-3} + \cdots \\ &\quad + n(n-1)t^{n-1}(1-t) + nt^n \Big\} \\ &= nt\Delta_x \Big\{\Big\{\Big\{ \cdots \Big\{\{(1-t) + (n-1)t\}(1-t) \\ &\quad + \frac{(n-1)(n-2)}{2!} t^2 \Big\}(1-t) \\ &\quad + \cdots \Big\}(1-t) + (n-1)t^{n-2}\Big\}(1-t) + t^{n-1}\Big\} = n\Delta_x t \end{aligned} \tag{8.12}$$

Notice that the velocity is $dx(t)/dt = n\Delta_x$.

If $x_i = \bar{x} + i\Delta_x$, it can be shown that $x(t) = \bar{x} + n\Delta_x t$. Since these results can be applied to $y(t)$ and $z(t)$, this means that provided the Bezier points are evenly spaced, the function $[x(t) \quad y(t) \quad z(t) \quad 1]^\#$ is a linear function of t with constant velocity over the range $0 \le t \le 1$.

Next consider the influence of the proximity of the first pair of points and the last pair of points. Suppose $x_0 = 0$, $x_1 = \alpha$, $x_2 = x_3 = 1$ and $y_0 = y_1 =$

1, $y_2 = \alpha$, and $y_3 = 0$. Then $x(t) = t\{3(1 - t)^2\alpha + 3t - 2t^2\}$ and $y(t) = (1 - t)\{1 + t - 2t^2 + 3\alpha t^2\}$, so

$$\frac{dx(t)}{dt} = 3(1 - t)\{\alpha(1 - 3t) + 2t\} \quad \text{and} \quad \frac{dy(t)}{dt} = 3t\{-2 + 2t - 3\alpha t + 2\alpha\}$$

so the actual velocity in space is

$$\sqrt{\left\{\frac{dx(t)}{dt}\right\}^2 + \left\{\frac{dy(t)}{dt}\right\}^2}$$
$$= 3\{(1 - t)^2\{\alpha(1 - 3t) + 2t\}^2 + t^2\{-2 + 2t - 3\alpha t + 2\alpha\}^2\}^{1/2} \tag{8.13}$$

At $t = 0$ and $t = 1$ the velocity is 3α, confirming prior results. At $t = 0.5$, the velocity is

$$\frac{3}{2}\left\{\left(1 - \frac{\alpha}{2}\right)^2 + \left(\frac{\alpha}{2} - 1\right)^2\right\}^{1/2} = \frac{3}{2}\left\{\frac{\alpha^2}{2} - 2\alpha + 2\right\}^{1/2} = \frac{3}{2\sqrt{2}}(\alpha - 2)$$

At $t = 0.25$ and $t = 0.75$, the velocities are

$$\frac{3}{16}\left\{34\alpha^2 - 24\alpha + 72\right\}^{1/2} \quad \text{and} \quad \frac{3}{16}\left\{34\alpha^2 - 24\alpha + 72\right\}^{1/2}$$

respectively. In tabular form, the velocities for various α and t are as follows.

t	Velocity			
	$\alpha = 0$	$\alpha = 0.5$	$\alpha = 0.55228$	$\alpha = 1$
0	0	1.5	1.6565	3
0.25	1.5910	1.5515	1.5588	1.6979
0.5	2.1213	1.5910	1.5355	1.0606
0.75	1.5910	1.5518	1.5588	1.6979
1	0	1.5	1.6568	3
Max deviation	2.1213	0.0910	0.1213	1.9394

The velocities are shown in Figure 8.6. Ideally, we want the velocity on the Bezier curve to be constant so we can control the path velocity by the velocity function alone (see Figure 7.3). The velocity of the Bezier curve approximating the quarter circle ($\alpha = 0.55228$) is almost the mirror image of the $\alpha = 0.5$ curve, indicating that the flattest velocity profile is at $\alpha \cong 0.525$. The greatest range in velocity occurs for $\alpha = 0$ and $\alpha = 1$. For $\alpha = 1$ the velocity is confined within the range 1.06063 to 3, and the maximum velocity occurs at $t = 0$ and $t = 1$. Further, as α is increased the velocity curve becomes more skewed, and the maximum velocity occurs at the endpoints rather than at $t = 0.5$. Suppose we require the velocity to be the same at

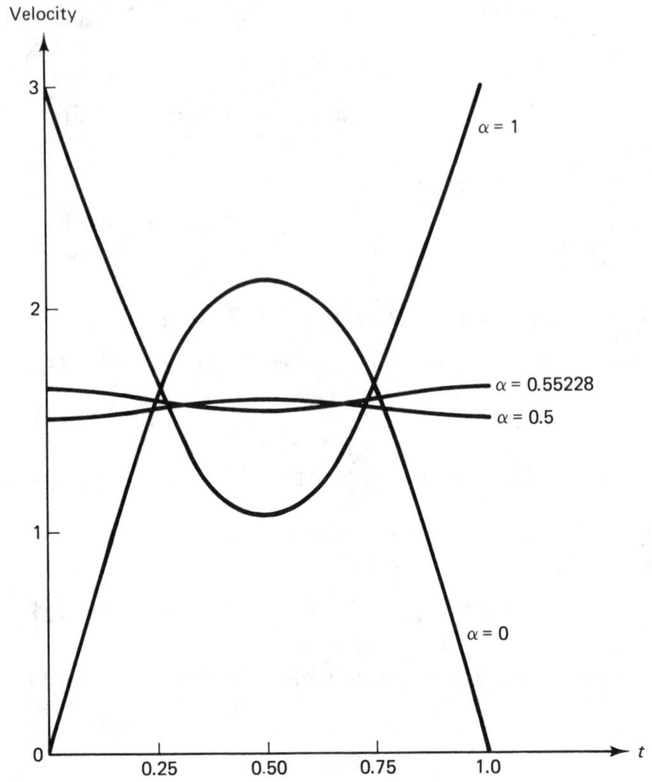

Figure 8.6 Velocities along four-point Bezier approximations to a quarter-circle

$t = 0$ and $t = 0.25$. Then $222\alpha^2 + 24\alpha - 72 = 0$, yielding $\alpha = 0.518$; for $t = 0.5$ the velocity will be 1.5719, and for $t = 0.75$ the velocity will be 1.5540.

The velocity function can be designed to take into account the velocity characteristics of the Bezier curve as well as the desired path control velocities, and the Bezier curve should be designed to provide the required shape regardless of velocity.

8.6 SPLINE TRAJECTORIES AND THE BEZIER FUNCTION

Suppose a Bezier curve is defined in position space with points ζ_i, $i = 0, 1, \ldots, m$. Performing IKSs on ζ_i produces points ξ_i in joint space. Following the derivations of Section 7.4, we note that the Bezier curve produced from these points in joint space has the following characteristics:

1. The curve passes through points $\xi_0 = \mathbf{f}(\zeta_0)$ and $\xi_m = \mathbf{f}(\xi_m)$, for which $t = 0$ and $t = 1$, respectively.

2. At $t = 0$ the slope of the curve is n times the direction given by $\xi_1 - \xi_0$, and at $t = 1$ the slope is n times the direction given by $\xi_m - \xi_{m-1}$.
3. The velocity at the end points is n times the distance between the first and last pair of points.

Since these same characteristics exist in position space and with the Bezier curves approximating short segments (so the segments are characterized by low-order polynomials in which the first two terms dominate), the distance between points ξ_i and ξ_{i+1} is proportional to the distance between points ζ_i and ζ_{i+1}. Therefore, we control the velocity over time t_i and t_{i+1} in joint space by the distance between points ζ_i and ζ_{i+1} in position space. Here we demonstrate using four-point Bezier functions.

Suppose the circle of radius 0.5 in the xz-plane centered at point $[1\ \ 0\ \ 1\ \ 1]^\#$ is to be executed in position space, starting and finishing at point $[0.5\ \ 0\ \ 1\ \ 1]^\#$ at rest, accelerating to a maximum velocity of 1 m/s, where the acceleration is limited to 2 m/s^2, by the 3 *dof* revolute robot $|\ \lambda\theta\theta\ |$ with link lengths of 1 m and base position and orientation $\mathbf{u}_1 = [0\ \ 0\ \ 0\ \ 1]^\#$, $\mathbf{p}_1 = [0\ \ 1\ \ 0\ \ 0]$, and $\boldsymbol{\nu} = [0\ \ 0\ \ 1]^\#$. We decompose the circle into four equal segments with endpoints

$$\begin{bmatrix} 0.5 \\ 0 \\ 1 \\ 1 \end{bmatrix} \text{ and } \begin{bmatrix} 1 \\ 0 \\ 1.5 \\ 1 \end{bmatrix}, \quad \begin{bmatrix} 1 \\ 0 \\ 1.5 \\ 1 \end{bmatrix} \text{ and } \begin{bmatrix} 1.5 \\ 0 \\ 1 \\ 1 \end{bmatrix},$$

$$\begin{bmatrix} 1.5 \\ 0 \\ 1 \\ 1 \end{bmatrix} \text{ and } \begin{bmatrix} 1 \\ 0 \\ 0.5 \\ 1 \end{bmatrix}, \quad \begin{bmatrix} 1 \\ 0 \\ 0.5 \\ 1 \end{bmatrix} \text{ and } \begin{bmatrix} 0.5 \\ 0 \\ 1 \\ 1 \end{bmatrix}$$

Since the slopes and velocities where segments meet must match and the starting and finishing velocities are zero, the Bezier curves are required to have points

$$\begin{bmatrix} 0.5 \\ 0 \\ 1 \\ 1 \end{bmatrix}, \quad \begin{bmatrix} 0.5 \\ 0 \\ 1 \\ 1 \end{bmatrix}, \quad \begin{bmatrix} 1-\alpha \\ 0 \\ 1.5 \\ 1 \end{bmatrix}, \text{ and } \begin{bmatrix} 1 \\ 0 \\ 1.5 \\ 1 \end{bmatrix}$$

$$\begin{bmatrix} 1 \\ 0 \\ 1.5 \\ 1 \end{bmatrix}, \quad \begin{bmatrix} 1+\alpha \\ 0 \\ 1.5 \\ 1 \end{bmatrix}, \quad \begin{bmatrix} 1.5 \\ 0 \\ 1+\beta \\ 1 \end{bmatrix}, \text{ and } \begin{bmatrix} 1.5 \\ 0 \\ 1 \\ 1 \end{bmatrix}$$

$$\begin{bmatrix} 1.5 \\ 0 \\ 1 \\ 1 \end{bmatrix}, \quad \begin{bmatrix} 1.5 \\ 0 \\ 1-\beta \\ 1 \end{bmatrix}, \quad \begin{bmatrix} 1+\gamma \\ 0 \\ 0.5 \\ 1 \end{bmatrix}, \quad \text{and} \quad \begin{bmatrix} 1 \\ 0 \\ 0.5 \\ 1 \end{bmatrix}$$

$$\begin{bmatrix} 1 \\ 0 \\ 0.5 \\ 1 \end{bmatrix}, \quad \begin{bmatrix} 1-\gamma \\ 0 \\ 0.5 \\ 1 \end{bmatrix}, \quad \begin{bmatrix} 0.5 \\ 0 \\ 1-\alpha \\ 1 \end{bmatrix}, \quad \text{and} \quad \begin{bmatrix} 0.5 \\ 0 \\ 1 \\ 1 \end{bmatrix}$$

The task now is to find α, β, and γ. The first Bezier curve is

$$\begin{bmatrix} x(t) \\ z(t) \end{bmatrix} = (1-t)^3 \begin{bmatrix} 0.5 \\ 1 \end{bmatrix} + 3(1-t)^2 t \begin{bmatrix} 0.5 \\ 1 \end{bmatrix} + 3(1-t)t^2 \begin{bmatrix} 1-\alpha \\ 1.5 \end{bmatrix} + t^3 \begin{bmatrix} 1 \\ 1.5 \end{bmatrix}$$

At the midpoint $t = 0.5$,

$$\begin{bmatrix} x(0.5) \\ z(0.5) \end{bmatrix} = 0.125 \begin{bmatrix} 6 - 3\alpha \\ 10 \end{bmatrix}$$

and the desired value is

$$\begin{bmatrix} 1 - 0.5/\sqrt{2} \\ 0.5 + 0.5/\sqrt{2} \end{bmatrix} = \begin{bmatrix} 0.6464 \\ 0.8536 \end{bmatrix}$$

This suggests a best value of $\alpha = 0.2761$. However, it was shown in Section 7.5 that the best value of α (center point matching) for the second Bezier curve when $\beta = \alpha$ is 0.5523, and the best value for the second Bezier curve to keep velocity essentially constant over the full range of t is 0.5224.

Recall that velocity planning of straight-line trajectories using Bezier functions was successful; the ideal velocity trajectories are easily and accurately followed. The lesson seems to be that the smaller the change in direction of the trajectory at the endpoint, the better. This change in direction is 0° for the straight line and 90° for the quarter-circle. Considerably more study is needed to develop an automated methodology for determining how to decompose a trajectory into segments and how to model each segment with Bezier curves.

8.7 THE STRAIGHT INSERTION PROBLEM

A common task for a manipulator is to insert a peg, screw, rivet, or electrical lead or set of leads (as in some of the computer chips, known as DIPS) through a hole [Cutkowski 1985; Lozano-Perez and Wesley 1979]. This requires the end link of the robot to move axially in the direction of the cylindrical hole. Some robots are suited for this task. For example, the

3 *dof* Cartesian robot mounted on a horizontal base can insert pegs vertically into this base. Similarly, the SCARA robot has become a standard tool for inserting DIPS in printed circuit boards. However, the 3 *dof* revolute robot cannot perform this task, regardless of the orientation of the hole. We will investigate the manipulator configurations that permit the insertion of pegs for a variety of orientations. Where the surface with the cylindrical hole is discussed, we assume that the axis of the hole is orthogonal to the plane of the surface at that point.

Directable robots can insert pegs in any position within the dextrous workspace of the robot. Listed next are simpler robots and their peg-insertion capability (we assume that the peg is in line with the end link of the robot). It is also assumed that the holes are within the dextrous workspace of each robot.

1. SCARA robots can insert pegs into a flat surface parallel to the base plane.
2. The polar robot $|\lambda\theta\sigma|$ can insert pegs in the inside surface of a sphere centered at $\mathbf{u}_2$.
3. For the Cartesian robot $|\sigma(+)\sigma(x)\sigma|$, the axis of the hole is in the vector direction given by $[a_1 \quad b_1 \quad c_1]^{\#}$, where the base plane is $\mathbf{p}_1 = [a_1 \quad b_1 \quad c_1 \quad d_1]$.
4. For the cylindrical robot $|\lambda\sigma\sigma|$ the holes are on the inside of a cylinder whose axis is the line $\mathbf{u}_2 + \alpha(\mathbf{u}_1 - \mathbf{u}_2)$.

8.8 ELBOW ROOM AND WORKPLACE INTERFERENCE

A robot is often required to operate in a cluttered workspace in which the danger of intermediate joints of the robot striking is ever present [Suh and Shin 1988; Kheradpir and Thorp 1988]. For example, consider the robot $|\lambda\theta\theta\theta|$ operating in the environment shown for one profile in Figure 8.7. To move between the two points marked ×, the end link must be carefully controlled. Suppose we move from the left to the right ×. The end link must be essentially vertical and $\mathbf{u}_4$ must move vertically downward close to the left-hand side wall until $\mathbf{u}_5$ is below the top wall. θ_4 is then decreased (the angle made more negative) until $\mathbf{u}_5$ is well clear of the top wall, and θ_3 is decreased. It is apparent that the timing sequence of joint motions is important.

When a robot is required to avoid its workplace, more degrees of freedom may be required than are necessary to satisfy the position space; that is, the robot must be loose. Recall, what we term loose in this work some call redundant [Hanafusa, Yoshikawa, and Nakamura 1978; Hanafusa, Yoshikawa, and Nakamura 1983; Yoshikawa 1983]; others have studied the problem of elbow room [Freund 1977] and collision avoidance [Lozano-

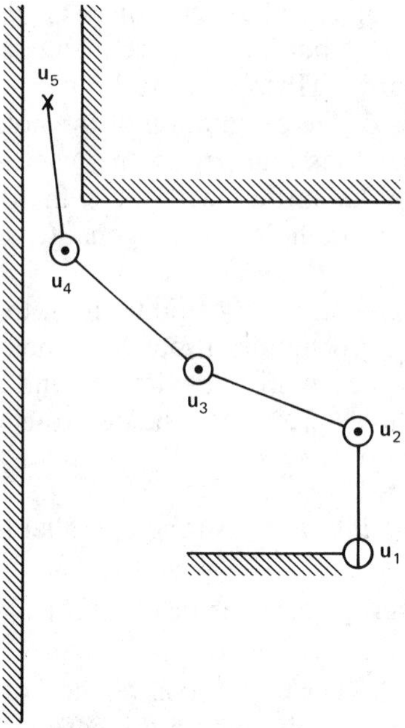

Figure 8.7

Perez 1981; Luh and Campbell 1982]. The position space usually contains m (from three to six) pieces of end effector information. It can also contain limitations on joint or link positions. If the limitations on the joint or link positions are not severe and are not encountered with frequency, possibly m *dof* are all that is required in the robot. Otherwise, more *dof* will be needed to provide elbow room, and some kind of position space must be devised to handle them.

Workplace constraint conditions usually have a binary action; that is, some of the time they do not limit movement of the robot in any way, but other times some or more of the constraint conditions are encountered [Luh and Campbell 1982]. As an example, consider the situation shown in Figure 8.8, where the end effector of the 3 *dof* robot is required to move from the first to the second marked position. A change only in θ_2 will move the robot to the required final position but will collide with the obstruction. If joint $\mathbf{u}_3$ is always more than distance l_3 from the obstruction, no interference occurs. However, this is a conservative solution, since once $\mathbf{u}_4$ is clear of the obstruction, θ_2 can be rotated clockwise and θ_3, counterclockwise. A Bezier curve between the endpoints with desirable velocity characteristics can be devised that avoids the obstacles. It can be argued that the work environment should be reconfigured so that the possibility of collisions are avoided, but this can be limiting to the tasks that we require of the robot.

The danger to the robot in striking its workplace is exacerbated, since many robots have optical encoders at their joints that are mounted outboard

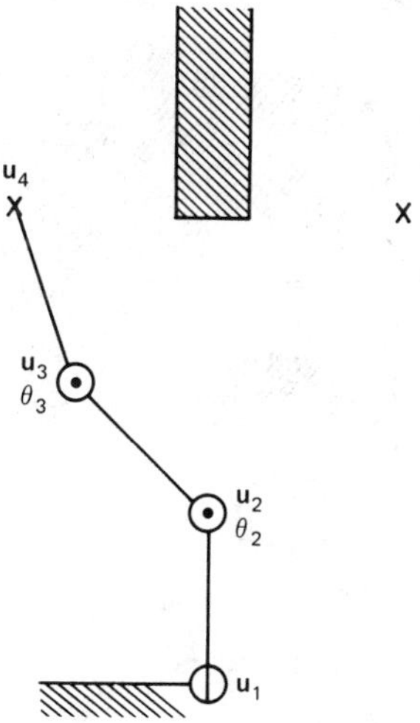

Figure 8.8 Obstacle interference

and vulnerable to damage because both of their position on the robot and their delicacy.

Suppose the path followed by the end effector of a loose robot is known, and this path avoids objects in the workplace. The task is to plan the inverse kinematic solution so that the intermediate joints avoid these objects. A primary requirement is that the robot is loose with respect to the end effector space. A question that can be asked is, even if the end effector misses the objects, is this an assurance that the intermediate joints or links can miss these objects? To answer these questions, we will consider the 4 *dof* robot $|\phi_1 \ \ \theta_2 \ \ \theta_3 \ \ \theta_4|$ with simple base orientation $\mathbf{p}_1 = [0 \ \ 1 \ \ 0 \ \ 0]$ and $\boldsymbol{\nu} = [-1 \ \ 0 \ \ 0]^\#$, which has as forward kinematic solutions for its joints

$$\mathbf{u}_3 = \begin{bmatrix} -l_2\cos\phi_1 \cdot \sin\theta_2 \\ -l_2\sin\phi_1 \cdot \sin\theta_2 \\ l_1 + l_2\cos\theta_2 \\ 1 \end{bmatrix} \tag{8.14}$$

$$\mathbf{u}_4 = \begin{bmatrix} -\cos\phi_1\{l_2\sin\theta_2 + l_3\sin(\theta_2+\theta_3)\} \\ -\sin\phi_1\{l_2\sin\theta_2 + l_3\sin(\theta_2+\theta_3)\} \\ l_1 + l_2\cos\theta_2 + l_3\cos(\theta_2+\theta_3) \\ 1 \end{bmatrix} \tag{8.15}$$

$$\mathbf{u}_5 = \begin{bmatrix} -\cos\phi_1\{l_2\sin\theta_2 + l_3\sin(\theta_2+\theta_3) + l_4\sin(\theta_2+\theta_3+\theta_4)\} \\ -\sin\phi_1\{l_2\sin\theta_2 + l_3\sin(\theta_2+\theta_3) + l_4\sin(\theta_2+\theta_3+\theta_4)\} \\ l_1 + l_2\cos\theta_2 + l_3\cos(\theta_2+\theta_3) + l_4\cos(\theta_2+\theta_3+\theta_4) \\ 1 \end{bmatrix} \tag{8.16}$$

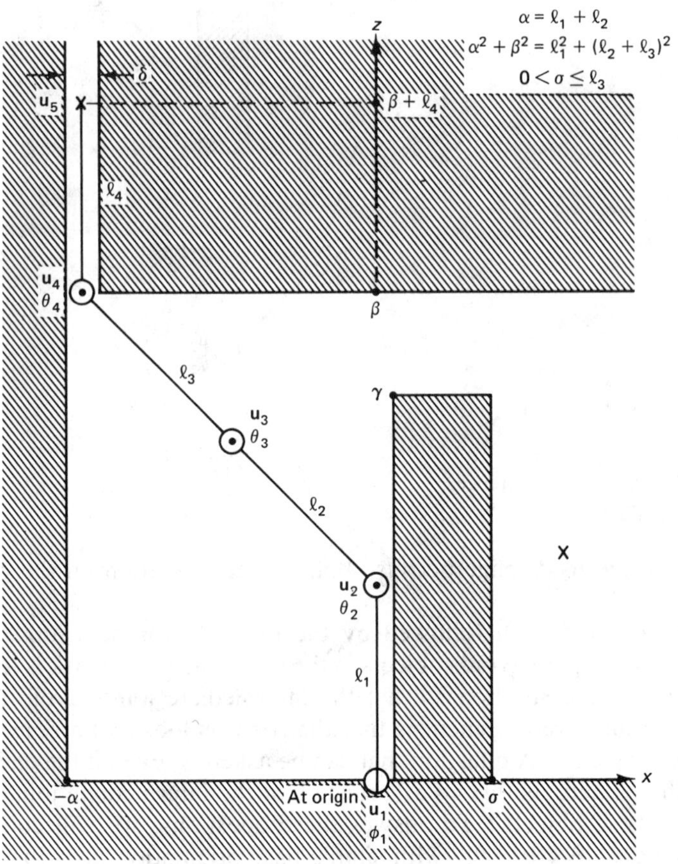

Figure 8.9

Suppose the restrictions of the workplace are shown as the section in Figure 8.9; we assume that this is a section on the xz-plane, so ϕ_1 can be set to zero to simplify the calculations. The base of the robot is at the origin. The restrictions can be written as follows:

1. $x_4 \geq -\alpha$ and $x_5 \geq -\alpha$ in order to avoid the left-hand vertical sidewall.
2. $z_4 < \beta$ if $x_5 \geq \delta - \alpha$ and $z_5 < \beta$ if $x_5 \leq \delta - \alpha$ in order to avoid the top wall.
3. If $z_5 > \beta$, then $-\alpha \leq x_4 \leq \delta - \alpha$ and $-\alpha \leq x_5 - \alpha$ in order to slide in the narrow vertical slot.

The vertical pillar with $\gamma < l_1 + l_2$ to the right of the robot imposes the following restrictions:

4. $\theta_2 \geq 0$ in order for l_2 to avoid the pillar.

5. If $x_4 > 0$, than $\theta_3 \approx -90°$ in order for l_3 to avoid striking the top of the pillar.
6. If $0 \le x_5 \le \sigma$ then $z_5 > \gamma$, or if $z_5 < \gamma$ then $x_5 > \sigma$ in order for l_4 to avoid striking the pillar.

Suppose the end effector is required to travel between the two positions marked with X in Figure 8.9. In order to extricate link l_4 from the slot, we must satisfy conditions 1, 2, and 3. Once l_4 is outside the slot, θ_3 and θ_4 can be decreased to move $\mathbf{u}_5$ to the right (decreasing x). θ_2 can be decreased while still ensuring that the z-values of $\mathbf{u}_4$ and $\mathbf{u}_5$ are less than β, and $\theta_2 \ge 0$. Care must be taken that l_3 does not strike the pillar (which can occur when $\theta_3 < -90°$).

This simple example illustrates the complexity of the problem. The complete specification of the geometric restrictions imposed by the workplace is often a daunting task, and this is only the start of the work necessary to generate a solution [Asada and By 1985]. It is possible to make a loose robot no longer appear loose with the use of penalty functions [Fu, Gozalves, and Lee 1987]; we have freedom in the choice of one or more joints when satisfying the position space. Suppose such joints are chosen by some weighting functions. As the joints of a robot approach an obstacle, a penalty function is added to each weighting function to move the robot away from the obstacle. This subject deserves an extensive study, not this oversimplified sketch.

The workplace and the tasks to be performed within the workplace should be designed for automatic assembly. In assembly work, in particular, it is poor practice to replace a human with a robot, since the resulting tasks may be and usually are difficult for a robot. The best automated assembly procedures are designed from scratch with automated assembly in mind. Replace screws with snap connectors. Every fastener that is fitted from within a convex object rather than face mounted causes problems. All flexible automation devices, including robots, are not replacements for Ethel and Sybil; rather they are part of a progression that starts in the planning process for a product and ends in the manufactured item being loaded on a truck.

Product evaluation and design are beyond the scope of this work, but the engineer who is concerned with robotics must also be concerned with CAD/CAM/CAE/CIM and all the other methodologies with acronyms that mean product design and production, material acquisition and inventory control, quality control and testing (including statistical evaluations), and yes, even marketing. The *C* in all these acronyms stands for computer, and all data bases should be computer-controlled. The common data base facilitates common control of all processes including robots, encouraging remote control of robots, which is the overall objective of the methods developed in this work.

8.9 EXAMPLES

1. Approximate the piecewise continuous function $\begin{bmatrix} \tau \\ 1 \\ 0 \\ 1 \end{bmatrix}$, $0 \le \tau \le 1$, and $\begin{bmatrix} 1 \\ 2-\tau \\ 0 \\ 1 \end{bmatrix}$, $1 \le \tau \le 2$, using parabolic blending on seven points such that the execution time is 1 s.

The function is contained in the xy-plane, and we choose the data points $(0, 1)$, $(0.5 - \delta/2, 1)$, $(1 - \delta, 1)$, $(1, 1)$, $(1, 1 - \delta)$, $(1, 0.5 - \delta/2)$, and $(1, 0)$, where δ is small. The distances between the points are $d_1 = d_2 = (1 - \delta)/2$, $d_3 = d_4 = \delta$, and $d_5 = d_6 = (1 - \delta)/2$.

A divided difference table for the value x value of these points is

0	0			
		1		
$\frac{1-\delta}{2}$	$\frac{1-\delta}{2}$		0	$P_1(\tau) = 0 + (\tau - 0)\left\{1 + \left(\tau - \frac{1-\delta}{2}\right)0\right\} = \tau$
		1		
$1-\delta$	$1-\delta$		0	$P_2(\tau) = \frac{1-\delta}{2} + \left(\tau - \frac{1-\delta}{2}\right)\{1 + (\tau - 1 + \delta)0\} = \tau$
		1		
1	1		$\frac{-1}{2\delta}$	$P_3(\tau) = 1 - \delta + (\tau - 1 + \delta)\left\{1 - (\tau - 1)\frac{1}{2\delta}\right\} = \frac{\tau+1}{2} - \frac{(\tau-1)^2}{2\delta}$
		0		
$1+\delta$	1		0	$P_4(\tau) = 1 + (\tau - 1)\{0 + (\tau - 1 - \delta)0\} = 1$
		0		
$\frac{3+\delta}{2}$	1		0	$P_5(\tau) = 1$
		0		
2	1			

The first column is τ and the second is x. In equation (8.11) we set $0 \le s_i \le d_i$, where $s_i = \tau - d_i - d_{i-1} - \cdots - d_1$. For $(1 - \delta)/2 \le \tau \le 1 - \delta$, $s_1 = \tau - (1 - \delta)/2$, so

$$Q_1(\tau) = P_1(\tau)\left(1 - \frac{2\tau - 1 + \delta}{1 - \delta}\right) + P_2(\tau)\,\frac{2\tau - 1 + \delta}{1 - \delta} = \tau$$

For $1 - \delta \le \tau \le 1$,

$$Q_2(\tau) = \tau\left(1 - \frac{\tau - 1 + \delta}{\delta}\right) + \left(\frac{\tau + 1}{2} - \frac{(\tau - 1)^2}{2\delta}\right)\frac{\tau - 1 + \delta}{\delta}$$

$$= \frac{\tau}{\delta}(1 - \tau) + \left(\frac{\tau + 1}{2} - \frac{(\tau - 1)^2}{2\delta}\right)\left(1 + \frac{\tau - 1}{\delta}\right)$$

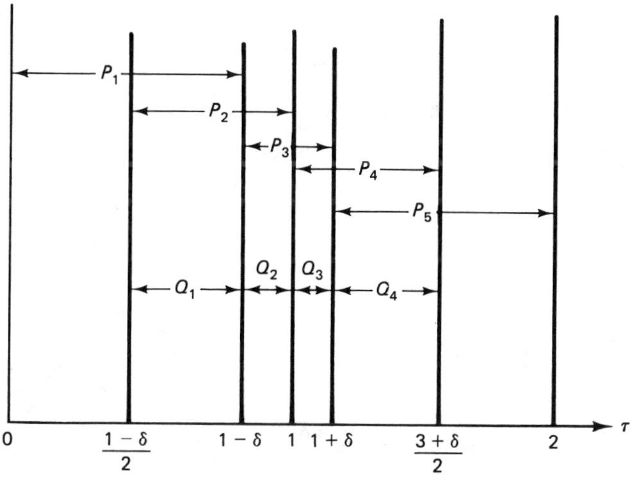

Figure 8.10

For $1 \le \tau \le 1 + \delta$,

$$Q_3(\tau) = \left\{\frac{\tau + 1}{2} - \frac{(\tau - 1)^2}{2\delta}\right\}\left(1 - \frac{\tau - 1}{\delta}\right) + \frac{\tau - 1}{\delta}$$

$$Q_4(\tau) = 1 \qquad \text{for all } \tau \ge \delta$$

The range of values of τ for which the $P_i(\tau)$ and $Q_i(\tau)$ are valid is shown in Figure 8.10. Although x and y are skew symmetric, the $Q_i(\tau)$ are different. Repeating the calculations for y gives the following

0	1			
		0		
$\frac{1-\delta}{2}$	1		0	$P_1(\tau) = 1$
		0		
$1-\delta$	1		0	$P_2(\tau) = 1$
		0		
1	1		$\frac{-1}{2\delta}$	$P_3(\tau) = 1 + (\tau - 1 + \delta)\left\{0 - (\tau - 1)\frac{1}{2\delta}\right\} = \frac{3-\tau}{2} - \frac{(\tau-1)^2}{2\delta}$
		-1		
$1+\delta$	$1-\delta$		0	$P_4(\tau) = 1 + (\tau - 1)\{-1 + (\tau - 1 - \delta)0\} = 2 - \tau$
		-1		
$\frac{3+\delta}{2}$	$\frac{1-\delta}{2}$		0	$P_5(\tau) = 1 - \delta + (\tau - 1 - \delta)\left\{-1 + \left(\tau - \frac{3+\delta}{2}\right)0\right\} = 2 - \tau$
		-1		
2	0			

For $(1 - \delta)/2 \le \tau \le 1 - \delta$, $Q_1(\tau) = 1$. For $1 - \delta \le \tau \le 1$,

$$Q_2(\tau) = \left\{1 - \left(\frac{\tau - 1 + \delta}{\delta}\right)\right\} + \left\{\frac{3 - \tau}{2} - \frac{(\tau - 1)^2}{2\delta}\right\}\left(\frac{\tau - 1 + \delta}{\delta}\right)$$

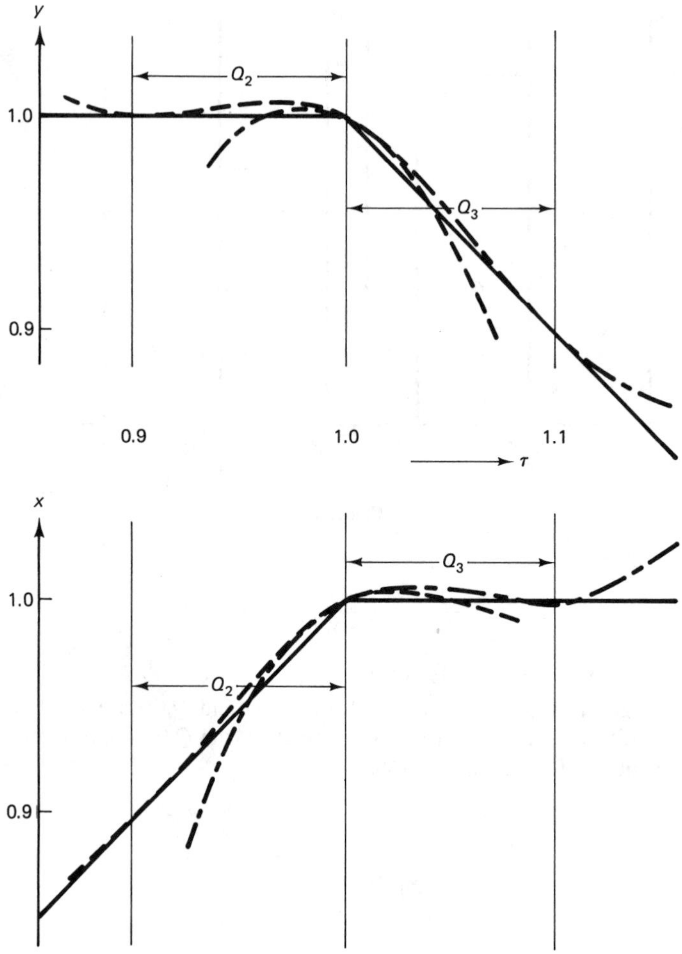

Figure 8.11 An example of parabolic blending

For $1 \leq \tau \leq 1 + \delta$,

$$Q_3(\tau) = \left\{\frac{3 - \tau}{2} - \frac{(\tau - 1)^2}{2\delta}\right)\left\{1 - \left(\frac{\tau - 1}{\delta}\right)\right\} + (2 - \tau)\left(\frac{\tau - 1}{\delta}\right)$$

For $1 + \delta \leq \tau \leq 2$, $Q_4(\tau) = 2 - \tau$.

The $Q_2(\tau)$ and $Q_3(\tau)$ approximations to $x(\tau)$ and $y(\tau)$ are shown for $\delta = 0.1$ in Figure 8.11, where the dashed line indicates Q_2 and the dotted line, Q_3. The quality of the approximation can be appreciated from Figure 8.12, which shows $y(\tau)$ against $x(\tau)$.. Notice the match of both location and position with the original function at points (x, y) given by (0.9, 1) and (1, 0.9). Since the desired execution time is 1 s, at which $\tau = 2$, we set $t = \tau/2$.

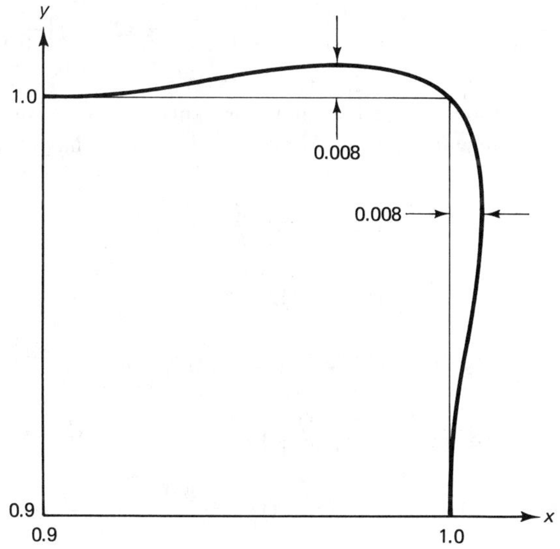

Figure 8.12 Expanded display of $x \vee y$

2. The 3 *dof* SCARA robot $| \theta\theta \times \sigma |$ with unity link lengths has its $\mathbf{u}_3$ position defined by the $x(t)$ and $y(t)$ of the first example. The base orientation and position of the robot is given by

$$\mathbf{p}_1 = [0 \quad 1 \quad 0 \quad 0], \qquad \boldsymbol{\nu} = \begin{bmatrix} 0 \\ 0 \\ 1 \end{bmatrix}, \quad \text{and} \quad \mathbf{u}_1 = \begin{bmatrix} 0 \\ 0 \\ 1 \\ 1 \end{bmatrix}$$

Defined $\theta_1(\tau)$ and $\theta_2(\tau)$ for the complete trajectory with $\delta = 0.1$. Suppose this trajectory is to be approximated by four-point Bezier curves for the time ranges 0 to 0.45, 0.45 to 0.5, 0.5 to 0.55, and 0.55 to 1, such that the initial and final velocities of the end effector are zero and there is no discontinuity in velocity anywhere on the trajectory. Determine the joint space Bezier points to execute this trajectory.

In the range $0 \leq \tau \leq 0.9$, $x(t) = \tau$, $y(\tau) = 1$. In the range $0.9 < \tau \leq 1$,

$$x(\tau) = \frac{\tau}{0.1}(1 - \tau) + \left\{\frac{\tau + 1}{2} - \frac{(\tau - 1)^2}{0.2}\right\}\left(1 + \frac{\tau - 1}{0.2}\right)$$

and

$$y(\tau) = \left\{1 - \left\{\frac{\tau - 0.9}{0.1}\right\}\right\} + \left\{\frac{3 - \tau}{2} - \frac{(\tau - 1)^2}{0.2}\right\}\left\{\frac{\tau - 0.9}{0.1}\right\}.$$

In the range $1 < \tau \leq 1.1$,

$$x(\tau) = \left\{\frac{\tau + 1}{2} - \frac{(\tau - 1)^2}{0.2}\right\}\left(1 - \frac{\tau - 1}{0.1}\right\} + \frac{\tau - 1}{0.1}$$

and $$y(\tau) = \left\{\frac{3-\tau}{2} - \frac{(\tau-1)^2}{0.2}\right\}\left\{1 - \left\{\frac{\tau-1}{0.1}\right\}\right\} + (2-\tau)\left\{\frac{\tau-1}{0.1}\right\}.$$

For $\tau > 1.1$, $x(\tau) = 1$ and $y(\tau) = 2 - \tau$.

Since the execution time of 1 second represents $\tau = 2$, we use τ as the running parameter over the range $0 \le \tau \le 1$. At $\tau = 0$ and $\tau = 2$ we have zero velocity. At $\tau = 0.9$.

$$\frac{d}{d\tau}x(\tau) = 1 \text{ and } \frac{d}{d\tau}y(\tau) = 0$$

At $\tau = 1.1$, $\frac{d}{d\tau}x(\tau) = 0$ and $\frac{d}{d\tau}y(\tau) = -1$.

At $\tau = 1$, $$\frac{d}{d\tau}x(\tau) = 10 - 20\tau + \left\{\frac{1}{2} - 10(\tau-1)\right\}\{1 + 10(\tau - 1)\}$$

$$+ \left\{\frac{\tau+1}{2} - \frac{(\tau-1)^2}{0.2}\right\} 10 = -10 + \frac{1}{2} + 10 = \frac{1}{2}.$$

and $$\frac{d}{d\tau}y(\tau) = -10 + \left\{\frac{1}{2} - 10(\tau-1)\right\}\frac{\tau-0.9}{0.1} + \left(\frac{3-\tau}{2} - \frac{\tau-1}{0.2}\right) 10$$

$$= -10 + \frac{1}{2} + 10 = \frac{1}{2}.$$

Suppose each Bezier function uses points (x_1, y_1), (x_2, y_2), (x_3, y_3) and (x_4, y_4). At $\tau = 0.9$ the speed requirement imposes a restriction on the last two Bezier points, i.e. $3(x_4 - x_3) = 1$ or $x_3 = x_4 - \frac{1}{3} = 0.5667$. The Bezier approximation over the range $0 \le \tau \le 0.9$ uses points (0, 1), (0, 1), (0.5667, 1) and (0.9, 1). Similarly, for $1.1 \le \tau \le 2$ the points are (1, 0.9), (1, 0.5667), (1, 0) and (1, 0). The speed requirement at $\tau = 1$ requires $3(x_4 - x_3) = \frac{1}{2}$, so $x_3 = \frac{5}{6}$, and $3(y_4 - y_3) = -\frac{1}{2}$, so $y_3 = \frac{7}{6}$ and the Bezier points for $0.9 \le \tau \le 1$ are (0.9, 1), (1.233, 1), $(\frac{5}{6}, \frac{7}{6})$ and (1, 1). Similarly, for $1 \le \tau \le 1.1$ the points are (1, 1), $(\frac{7}{6}, \frac{5}{6})$, (1, 1.233), (1, 0.9).

Since for this robot $x = \beta$, $y = \alpha$ and $z = -\sigma_3$, the IKS as determined from Section 6.3.5 is

$$\theta_1 = \sin^{-1}\left\{\frac{1}{2}\sqrt{x^2+y^2}\right\} - \tan^{-1}\frac{x}{y} \text{ if } y \ne 0 \text{ or}$$

$$\theta_1 = \cos^{-1}\left\{\frac{1}{2}\sqrt{x^2+y^2}\right\} + \tan^{-1}\frac{y}{x} \text{ if } x \ne 0,$$

with $$\theta_2 = \cos^{-1}\{x - \cos\theta_1\} - \theta_1,\ \sigma_3 = 1.$$

Notice that $\cos^{-1}(\cos\theta_i)$ is not necessarily equal to θ_1; it can also be $\pi - \theta_1$ or $-\theta_1$, and the choice must be logically correct. The position space and joint space Bezier points are presented in the following table:

x	0	0.5667	0.9	1.223	$\frac{5}{6}$	1	$\frac{7}{6}$	1	1	1	1
y	1	1	1	1	$\frac{7}{6}$	1	$\frac{5}{6}$	1.233	0.9	0.5667	0
θ_1	30	5.54	0.29	1.45	10.25	0	−8.67	13.50	−5.74	−25.38	−60
θ_2	120	109.84	95.45	75.65	88.41	90	88.41	74.92	95.45	109.84	120
$0 \to 0.9$	1 & 2	3	4								
$0.9 \to 1$			1	2	3	4					
$1 \to 1.1$						1	2	3	4		
$1.1 \to 2$									1	2	3 & 4

The bottom four rows in the table identify the set of points used for each Bezier curve. The actual position space trajectory using these points will be different to the trajectories found in the first worked example, except at the end points $\tau = 0, 0.9, 1, 1.1$ and 2, since execution is in joint space and the mapping to position space is nonlinear.

---

3. The 3 *dof* revolute robot $|\lambda\theta\theta|$ with link lengths of 1 m is required to travel between

$$\begin{bmatrix} 0 \\ 1 \\ 1 \\ 1 \end{bmatrix} \text{ and } \begin{bmatrix} 1 \\ 0 \\ 0 \\ 1 \end{bmatrix}$$

The base position and orientation is $\mathbf{u}_l = [0 \quad 0 \quad 0 \quad 1]^{\#}$, $\mathbf{p}_l = [0 \quad 1 \quad 0 \quad 0]$ and $\boldsymbol{\nu} = [0 \quad 0 \quad 1]^{\#}$. Determine the deviation from a straight line trajectory if the path chosen is the polynomial of order m defined by equation (8.2), where $m = 1, 2, 3,$ and 4.

We see that we have the simple base orientation discussed in Section 6.3.1 so

$$\lambda_1 = \tan^{-1}\frac{y}{x},$$

$$a = \frac{x}{\cos\lambda_1} \text{ if } \cos\lambda_1 \neq 0 \text{ or } \quad a = \frac{y}{\sin\lambda_1} \text{ if } \sin\lambda_1 \neq 0,\ b = z - 1.$$

$$\theta_2 = \sin^{-1}\left\{\frac{1}{2}\sqrt{a^2 + b^2}\right\} - \tan^{-1}\frac{b}{a} \text{ if } a \neq 0 \text{ or}$$

$$\theta_2 = \cos^{-1}\left\{\frac{1}{2}\sqrt{x^2 + y^2}\right\} + \tan^{-1}\frac{a}{b} \text{ if } b \neq 0,$$

with $\quad \theta_3 = \sin^{-1}\{a - \sin\theta_2\} - \theta_2.$

The points used in the application of equation (8.2) are

$$\begin{bmatrix} 0 \\ 1 \\ 1 \\ 1 \end{bmatrix} \text{ and } \begin{bmatrix} 1 \\ 0 \\ 0 \\ 1 \end{bmatrix} \text{ for } m = 1,$$

$$\begin{bmatrix} 0 \\ 1 \\ 1 \\ 1 \end{bmatrix}, \begin{bmatrix} 0 \\ 1 \\ 1 \\ 1 \end{bmatrix} + \beta \begin{bmatrix} 1 \\ 0 \\ 0 \\ 1 \end{bmatrix} \text{ and } \begin{bmatrix} 1 \\ 0 \\ 0 \\ 1 \end{bmatrix},$$

where $\beta = \frac{1}{2}$ for $m = 2$,
$\beta = \frac{1}{3}$ and $\frac{2}{3}$ for $m = 3$, and
$\beta = \frac{1}{4}, \frac{1}{2}$, and $\frac{3}{4}$ for $m = 4$.

The joint space values for these position space points are

	β values						
	0	$\frac{1}{4}$	$\frac{1}{3}$	$\frac{1}{2}$	$\frac{2}{3}$	$\frac{3}{4}$	1
λ_1	90.00	71.57	63.43	45.00	26.59	18.44	0
θ_2	30.00	42.04	48.19	60.92	71.79	76.51	90.00
θ_3	120.00	131.01	131.84	128.68	119.98	113.97	90.00

The straight line is $\begin{bmatrix} \beta \\ 1-\beta \\ 1-\beta \\ 1 \end{bmatrix}$, and the distance D from a point on the position space trajectory to this straight line is given by

$$D^2 = (x - \beta)^2 + (y - 1 + \beta)^2 + (z - 1 + \beta)^2.$$

$$\frac{\partial D^2}{\partial \alpha} = -2(x - \beta) + 2(y - 1 + \beta) + 2(z - 1 + \beta) = 0$$

for maximum or minimum, so

$$\beta = (2 + x - y - z)/3.$$

A C program that performs these calculations is:

```
        /* STRAIGHT LINE APPROXIMATIONS FOR rpp */
#include <math.h>
#include <stdio.h>
#define PI 3.141582654
main()
{                                          /* x, y, z are position space */
double x,y,z,j1[6],j2[6],j3[6]; /*j1,j2 and j3 are joint values at knot points*/
double a,b,h,hh,hhh,hhhh;
double lamda1,theta2,theta3,alpha,beta,D,rad;
double x0=0.,x1=1.,y0=1.,y1=0.,z0=1.,z1=0.;
int i,m;

rad = 180./PI;
for (m=1; m <= 4; m++)                  /* number of points = m+1 */
  {
  h = 1./(double)m; hh=h*h; hhh=h*hh; hhhh=hh*hh;
  for (i=0; i <= m; i++)
    {
    alpha = i; alpha = alpha*h;
    x=x0+alpha*(x1-x0); y=y0+alpha*(y1-y0); z=z0+alpha*(z1-z0);
    if (x*x >= 0.00001) j1[i] = atan(y/x);
      else j1[i] = PI/2.;
    if(cos(j1[i])*cos(j1[i]) >= 0.00001) a = x/cos(j1[i]);
      else a = y/sin(j1[i]);
    b = z-1.;
    if(a*a >= 0.00001)j2[i] = asin(0.5*sqrt(a*a+b*b)) - atan(b/a);
      else j2[i] = acos(0.5*sqrt(a*a+b*b)) + atan(a/b);
    j3[i] = PI-asin(a-sin(j2[i]))-j2[i];
    printf("\nalpha=%5.3f j1=%5.3f j2=%5.3f j3=%5.3f a=%5.3f b=%5.3f m=%d",
            alpha,j1[i],j2[i],j3[i],a,b,m);
    }

  printf("\n\nalpha lamda1 theta2 theta3   x      y      z      D");
  for (i=0; i<=20; i++)
    {
    alpha = (double)i/20.;
    lamda1=0.; theta2=0.; theta3=0.;
    D = 0.;
    if(m==4)D=(alpha-3.*h)*(j1[4]-4.*j1[3]+6.*j1[2]-4.*j1[1]+j1[0])/(24.*hhhh);
    if(m>=3)D=(alpha-2.*h)*((j1[3]-3.*j1[2]+3.*j1[1]-j1[0])/(6.*hhh)+D);
    if(m >= 2) D=(alpha-h)*((j1[2]-2.*j1[1]+j1[0])/(2.*hh)+D);
    lamda1=j1[0]+alpha*((j1[1]-j1[0])/h+D);
    D = 0.;
    if(m==4)D=(alpha-3.*h)*(j2[4]-4.*j2[3]+6.*j2[2]-4.*j2[1]+j2[0])/(24.*hhhh);
    if(m>=3)D=(alpha-2.*h)*((j2[3]-3.*j2[2]+3.*j2[1]-j2[0])/(6.*hhh)+D);
    if(m >= 2) D=(alpha-h)*((j2[2]-2.*j2[1]+j2[0])/(2.*hh)+D);
    theta2=j2[0]+alpha*((j2[1]-j2[0])/h+D);
    D = 0.;
    if(m==4)D=(alpha-3.*h)*(j3[4]-4.*j3[3]+6.*j3[2]-4.*j3[1]+j3[0])/(24.*hhhh);
    if(m>=3)D=(alpha-2.*h)*((j3[3]-3.*j3[2]+3.*j3[1]-j3[0])/(6.*hhh)+D);
    if(m >= 2) D=(alpha-h)*((j3[2]-2.*j3[1]+j3[0])/(2.*hh)+D);
    theta3=j3[0]+alpha*((j3[1]-j3[0])/h+D);
    x = cos(lamda1)*(sin(theta2)+sin(theta2+theta3));
```

```
    y = sin(lamda1)*(sin(theta2)+sin(theta2+theta3));
    z = 1.+cos(theta2)+cos(theta2+theta3);
    beta = (2.+x-y-z)/3.;
    D = sqrt((x-beta)*(x-beta)+(y-1.+beta)*(y-1.+beta)+(z-1.+beta)*(z-1.+beta));
    printf("\n %5.3f %6.2f %6.2f %6.2f",alpha,rad*lamda1,rad*theta2,rad*theta3);
    printf(" %6.4f %6.4f %6.4f %8.6f",x,y,z,D);
    }
    printf("\n");
  }
exit();
}
```

A plot of D for $m = 1, 2, 3$ and 4 as a function of α is shown in Figure 8.13. We have plotted D with implied sign so that D does not appear discontinuous in deriva-

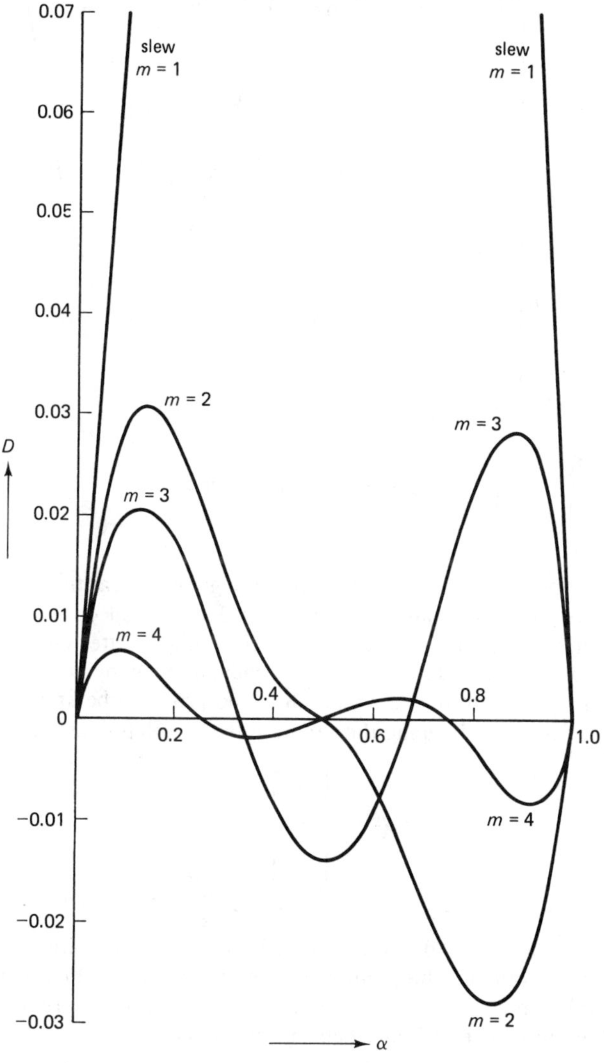

Figure 8.13 Obstacle avoidance

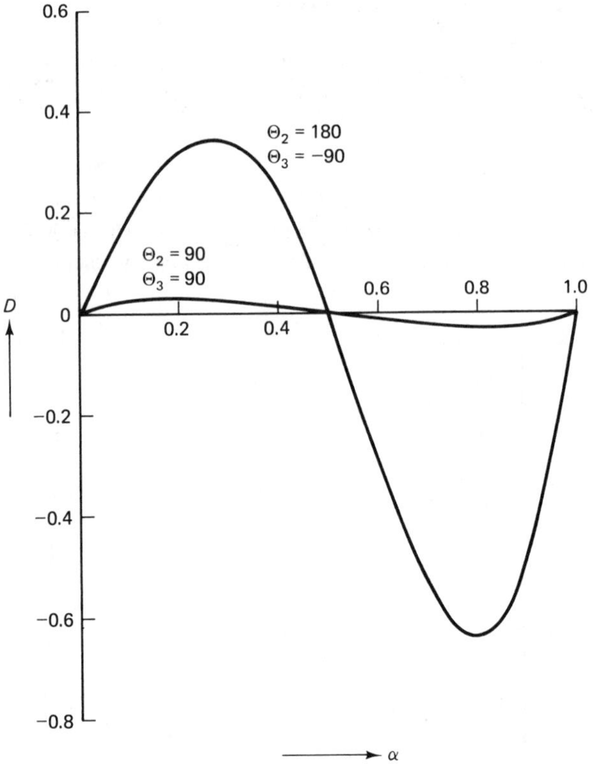

Figure 8.14 A camera image

tive at the knot points. Notice that the errors are greatest near the end points of the trajectories (at $\alpha = 0.14$, 0.12 and 0.09 for $m = 2$, 3 and 4 respectively) and smaller in the central section of the trajectory. This implies that a better choice for the knot points for $m = 3$ and 4 would be closer to the ends; moving the knot points from $\alpha =$ 0.25 and 0.75 for $m = 3$ to $\alpha = 0.2$ and 0.8 would probably be an improvement.

It was noticed in debugging the program that a solution at the end point

$$\begin{bmatrix} 1 \\ 0 \\ 0 \\ 1 \end{bmatrix}$$

could also be $\lambda_1 = 0$, $\theta_2 = 180°$, and $\theta_3 = -90°$. If these values are chosen D will be much larger. For example, the plots of D (solution shown in Figure 8.13 and with this new end value) for $m = 2$ are shown in Figure 8.14. It is apparent that care must be exercised in the choice of joint angles in the IKS.

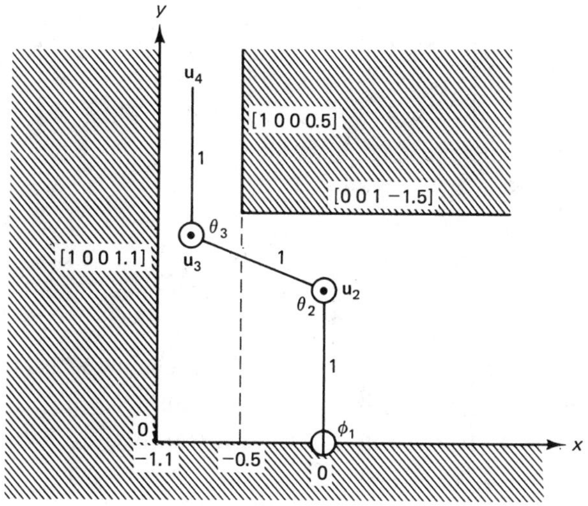

Figure 8.15 Obstacle avoidance

4. The 3 *dof* robot $|\phi\theta|$ is in the workplace shown in Figure 8.15. l_2 is vertical. $\mathbf{u}_4$ is required to move back and forth between the two points

$$\begin{bmatrix} -1 \\ 0 \\ 2 \\ 1 \end{bmatrix} \text{ and } \begin{bmatrix} 1 \\ 0 \\ 0 \\ 1 \end{bmatrix}$$

so the cycle time is 2 seconds with a pause at each end point of 0.25 seconds. Determine a suitable path to perform the required task.

The starting and end points of the path are

$$\begin{bmatrix} 180 \\ 90 \\ -90 \end{bmatrix} \text{ and } \begin{bmatrix} 0 \\ 90 \\ 90 \end{bmatrix}$$

respectively. Suppose we use the three point Bezier curve with intermediate point

$$\begin{bmatrix} 180 \\ 90 + \theta \\ -90 - \theta \end{bmatrix},$$

then

$$P(t) = (1 - t)^2 \begin{bmatrix} 180 \\ 90 \\ -90 \end{bmatrix} + 2t(1 - t) \begin{bmatrix} 180 \\ 90 + \theta \\ -90 - \theta \end{bmatrix} + t^2 \begin{bmatrix} 0 \\ 90 \\ 90 \end{bmatrix}$$

$$= \begin{bmatrix} (1 - t^2)180 \\ 90 + 2t(1 - t)\,\theta \\ -45 - 2t(1 - t)\,\theta \end{bmatrix}$$

$$P(0.5) = \begin{bmatrix} 135 \\ 90 + \theta/2 \\ -45 - \theta/2 \end{bmatrix}.$$

If $\theta = 90$, $\theta_2 = 135$ and $\theta_3 = 90$, so at $t = 0.5$ and from equation 5.7 (suitably modified)

$$\mathbf{u}_4 = \begin{bmatrix} \cos\phi_1\{\sin\theta_2 + \sin(\theta_2 + \theta_3)\} \\ \sin\phi_1\{\sin\theta_2 + \sin(\theta_2 + \theta_3)\} \\ 1 + \cos\theta_2 + \cos(\theta_2 + \theta_3) \\ 1 \end{bmatrix} = \begin{bmatrix} -0.7071\{0.7071 + 0.7071\} \\ 0.7071\{0.7071 + 0.7071\} \\ 1 - 0.7071 + 0.7071 \\ 1 \end{bmatrix} = \begin{bmatrix} -1 \\ 1 \\ 1 \\ 1 \end{bmatrix}$$

which avoids the obstacle at this point. It is also evident that the obstacle is avoided at all points along the trajectory. The trajectory planned must be executed in 0.75 seconds, so we multiply t in the Bezier curve by 0.75 to get real time. The trajectory for the return path uses these same Bezier points in reverse order.

---

5. Repeat the example 4, but this time design the trajectories so the starting and ending velocities are zero, and the end effector accelerates and decelerates uniformly with no abrupt changes in velocity.

Suppose we use a five point Bezier curve where the first pair of points are the same, the last pair of points are the same, and the middle point is the one selected in the last example, then

$$P(t) = (1 - t)^3(1 + 3t) \begin{bmatrix} 180 \\ 90 \\ -90 \end{bmatrix} + 6t^2(1 - t)^2 \begin{bmatrix} 180 \\ 180 \\ -180 \end{bmatrix} + t^3(4 - 3t) \begin{bmatrix} 0 \\ 90 \\ 90 \end{bmatrix}, \quad 0 \le t \le 1.$$

$$P(0.5) = 0.3125 \begin{bmatrix} 180 \\ 90 \\ -90 \end{bmatrix} + 0.375 \begin{bmatrix} 180 \\ 180 \\ -180 \end{bmatrix} + 0.3125 \begin{bmatrix} 0 \\ 90 \\ 90 \end{bmatrix} = \begin{bmatrix} 123.75 \\ 123.75 \\ -67.50 \end{bmatrix}.$$

$$\mathbf{u}_4 = \begin{bmatrix} \cos\lambda_1\{\sin\theta_2 + \sin(\theta_2 + \theta_3)\} \\ \sin\lambda_1\{\sin\theta_2 + \sin(\theta_2 + \theta_3)\} \\ 1 + \cos\theta_2 + \cos(\theta_2 + \theta_3) \\ 1 \end{bmatrix} = \begin{bmatrix} -0.2904\{0.9569 + 0.8315\} \\ 0.9569\{0.9569 + 0.8315\} \\ 1 - 0.2904 + 0.5555 \\ 1 \end{bmatrix}$$

$$= \begin{bmatrix} -0.5194 \\ 1.7113 \\ 1.2652 \\ 1 \end{bmatrix}$$

---

6. An *XY* inspection table uses a camera tied to an image-processing system. The table can operate over the range $0 \leq X \leq 0.4$ m, $0 \leq Y \leq 0.4$ m. The computer controls an electrical circuit that can send voltages v_x and v_y to drive the DC servomotors that drive the axes of the table to move the camera. The voltages v_x and v_y permissible are in the range $-1 \leq v_x \leq 1$ and $-1 \leq v_y \leq 1$. The acceleration of the camera in each axis is 4 m/s²-V; that is, the acceleration A_x in the X direction is $4v_x$ m/s².

The camera starts at rest in position (0, 0.2) and must come to rest at point (0.4, 0.4) in the shortest possible time such that the X and Y motors start and finish at the same time. Find the applied voltages to each servomotor as a function of time. What is the maximum velocity of the robot along the path?

If the inspection table is to be used to track paths in PC boards, explain some difficulties it could encounter in path recognition, tracking, and so on.

$A_x = 4v_x$, $A_y = 4v_y$. The table is required to move 0.4 in the x-direction and 0.2 in the y-direction. The speed of a motor is given by Newton's law of motion, $s_f = s_i + At$, where s_f is the final velocity, s_i is the initial velocity, and A is the applied (constant) acceleration. The minimum time of motion for the motor driving the X-axis to move the camera to the correct position will occur when the acceleration is always at a maximum, positive or negative. That is, the applied voltages are a maximum (positive or negative). The maximum acceleration is ± 4 m/s². The distance traveled when starting at rest and accelerating for t seconds is $d = At^2/2$. Accelerating for half the distance at this rate gives $0.2 = 4t^2/2$, or $t = \sqrt{0.1} = 0.316$ s. Thus the camera will travel to the correct position in $2 \times 0.316 = 0.632$ s.

The Y-axis motor is required to complete its travel in 0.632 s. We have many choices regarding how to achieve this. Suppose we choose the acceleration to be constant, positive or negative as A_y. A_y is applied over distance 0.1 for 0.316 s, so $0.1 = A_y 0.316^2/2$ and $A_y = 0.2$. The applied voltage on the Y motor is 0.5 for 0.316 s, followed by -0.5 for 0.316 s.

The maximum velocity in the X direction is given by $At = 4 \times 0.316 = 1.264$ m/s. The maximum velocity in the Y direction is given by $0.2 \times 0.316 = 0.632$ m/s. The maximum velocity of the camera is the vector addition of these velocities, that is, $\{1.264^2 + 0.632^2\}^{1/2} = \sqrt{2}$ m/s.

To use the *XY* table to track paths in PC boards presents the following difficulties:

1. The lighting is critical. It creates shadows, reflections, or glare from polished surfaces. The camera is usually insensitive to color, and different colors can produce the same gray level in the image. Color filtering can be used to separate gray levels. Variations in ambient light should be minimized.
2. If the camera is tracking a path, it cannot stop instantly and at a bend can overshoot and lose the path.
3. Tracking by areas requires an accurate measure of position. This is difficult to achieve without some highly accurate vernier or encoder scheme. Alternatively, the camera could be taught to recognize landmarks on the PC board.
4. The distance of the board (populated or unpopulated) must remain essentially the same in order for the camera to remain in focus. If there is a significant difference in distance from the camera, an automatic focusing scheme will be

required. In this case the depth of field could be used in the inspection process—for example, focus on the top of a DIP to check its orientation and then focus at the board level to check at the board level for defects.

---

7. A camera is mounted with pinhole at $\mathbf{w} = [2 \quad 2 \quad 2 \quad 1]^{\#}$ and with film plane at $\mathbf{p} = [1 \quad 1 \quad 1 \quad -6.25]/\sqrt{3}$. Assume that vertical in the camera is Cartesian vertical. The center line of a conveyor belt is the x-axis, and the belt moves at a velocity of 0.4 m/s. At $t = 0$ the picture shown in Figure 8.16 is of an object on the conveyor belt. Make a best guess of the shape and location of the object at $t = 0$. A 4 *dof* SCARA robot $|\,\theta\theta\underline{\psi}\sigma\,|$ with a parallel jaw gripper is positioned with $\mathbf{u}_1 = \begin{bmatrix} 1 \\ 0.25 \\ 1 \\ 1 \end{bmatrix}$,

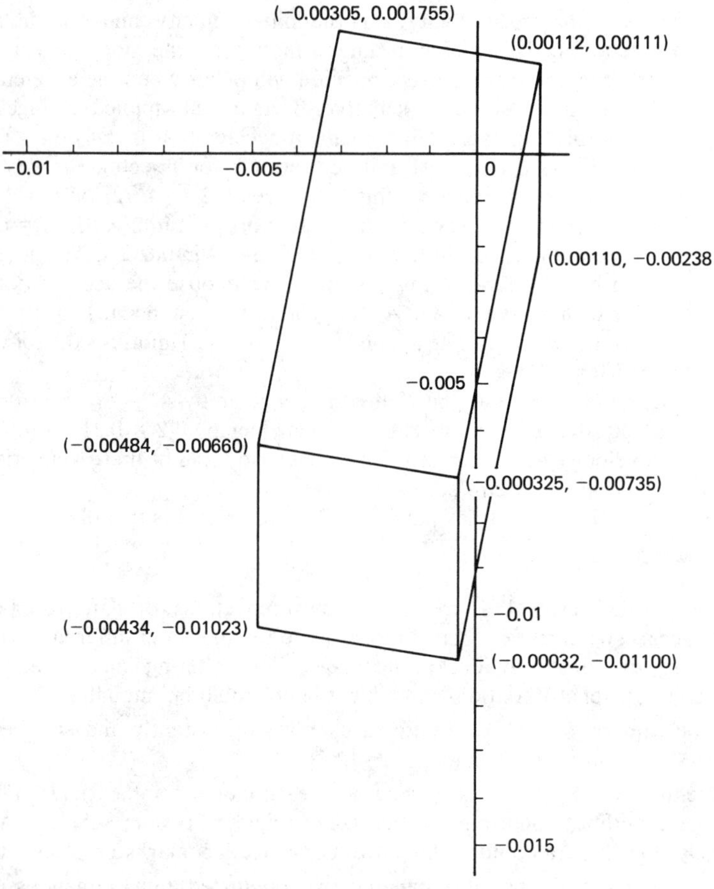

Figure 8.16 Camera image

$\boldsymbol{\nu} = \begin{bmatrix} 1 \\ 0 \\ 0 \\ 0 \end{bmatrix}$, $\mathbf{p}_l = [0 \quad 0 \quad -1 \quad 1]$, and $l_1 = l_2 = 1$, and σ_4 is a lead screw that can extend from 0.5 to 1 m. Assuming that the robot is at datum at $t = 0$, task-plan the robot so that it can pick up the object.

The shape shown in Figure 8.16 appears to be a box, and it is reasonable to assume that the box is regular in section. Further, the object lies on plane $[0 \quad 0 \quad 1 \quad 0]$, so the three image points $(-0.00434, -0.0102)$, $(-0.00032, -0.0110)$, and $(0.00112, 0.00111)$ must correspond to object points on that plane. The line through the true spatial position of each of these three image point and pinhole intersect $[0 \quad 0 \quad 1 \quad 0]$ at a corner point of the object. The true spatial position of these image points is determined by reversing the procedure discussed in Section 2.5. The axial point on the film plane is given by finding the point of intersection of the vector $[1 \quad 1 \quad 1]^{\#}$, the direction of the vector partially defining the film plane, from $\mathbf{w}$ to $\mathbf{p}$, so

$$\mathbf{v}_{ax} = \begin{bmatrix} 2 \\ 2 \\ 2 \\ 1 \end{bmatrix} - \alpha \begin{bmatrix} 1 \\ 1 \\ 1 \\ 0 \end{bmatrix} \quad \text{and} \quad \mathbf{p}\mathbf{v}_{ax} = [1 \quad 1 \quad 1 \quad -6.25] \left\{ \begin{bmatrix} 2 \\ 2 \\ 2 \\ 1 \end{bmatrix} - \alpha \begin{bmatrix} 1 \\ 1 \\ 1 \\ 0 \end{bmatrix} \right\} = 0$$

Also, $-0.25 - 3\alpha = 0$, and $\alpha = -0.08333$, so

$$\mathbf{v}_{ax} = \begin{bmatrix} 2.08333 \\ 2.08333 \\ 2.08333 \\ 1 \end{bmatrix}$$

An axial point on the object side of $\mathbf{w}$ is $\begin{bmatrix} 0 \\ 0 \\ 0 \\ 1 \end{bmatrix}$, and a point vertically above this point when imaged will give the direction of the ordinate in the film plane. Using equation (2.31) we get

$$\begin{bmatrix} 0 \\ 0 \\ 1 \\ 1 \end{bmatrix} - \frac{[1 \quad 1 \quad 1 \quad -6.25] \begin{bmatrix} 0 \\ 0 \\ 1 \\ 1 \end{bmatrix}}{[1 \quad 1 \quad 1] \begin{bmatrix} 2 \\ 2 \\ 1 \end{bmatrix}} \begin{bmatrix} 2 \\ 2 \\ 1 \\ 0 \end{bmatrix} = \begin{bmatrix} 2.100 \\ 2.100 \\ 2.050 \\ 1 \end{bmatrix}$$

$$\mathbf{v}_{\text{ord}} = \begin{bmatrix} 2.100 - 2.08333 \\ 2.100 - 2.08333 \\ 2.050 - 2.08333 \end{bmatrix} = \begin{bmatrix} 0.01667 \\ 0.01667 \\ -0.03333 \end{bmatrix} \equiv \begin{bmatrix} 0.4083 \\ 0.4083 \\ -0.8164 \end{bmatrix}$$

$$\mathbf{v}_{\text{abs}} = \frac{1}{\sqrt{3}} \begin{bmatrix} 0.4083 \\ 0.4083 \\ -0.8164 \end{bmatrix} \times \begin{bmatrix} 1 \\ 1 \\ 1 \end{bmatrix} = \begin{bmatrix} 0.7071 \\ -0.7071 \\ 0 \end{bmatrix}$$

For the three image points (−0.00434, −0.01023), (−0.00032, −0.01100), and (0.00110, −0.00238), we find their actual position in Cartesian space as $\mathbf{v} = \mathbf{v}_{ax} +$ (abscissa)$\mathbf{v}_{\text{abs}}$ + (ordinate)$\mathbf{v}_{\text{ord}}$, so the three points are

$$\begin{bmatrix} 2.08333 \\ 2.08333 \\ 2.08333 \\ 1 \end{bmatrix} - 0.00434 \begin{bmatrix} 0.7071 \\ -0.7071 \\ 0 \\ 0 \end{bmatrix} - 0.01023 \begin{bmatrix} 0.4083 \\ 0.4083 \\ -0.8164 \\ 0 \end{bmatrix} = \begin{bmatrix} 2.0761 \\ 2.0822 \\ 2.0916 \\ 1 \end{bmatrix}$$

$$\begin{bmatrix} 2.08333 \\ 2.08333 \\ 2.08333 \\ 1 \end{bmatrix} - 0.00032 \begin{bmatrix} 0.7071 \\ -0.7071 \\ 0 \\ 0 \end{bmatrix} - 0.01100 \begin{bmatrix} 0.4083 \\ 0.4083 \\ -0.8164 \\ 0 \end{bmatrix} = \begin{bmatrix} 2.0786 \\ 2.0791 \\ 2.0923 \\ 1 \end{bmatrix}$$

$$\begin{bmatrix} 2.08333 \\ 2.08333 \\ 2.08333 \\ 1 \end{bmatrix} + 0.00110 \begin{bmatrix} 0.7071 \\ -0.7071 \\ 0 \\ 0 \end{bmatrix} - 0.00238 \begin{bmatrix} 0.4083 \\ 0.4083 \\ -0.8164 \\ 0 \end{bmatrix} = \begin{bmatrix} 2.0831 \\ 2.0816 \\ 2.0852 \\ 1 \end{bmatrix}$$

The line joining each image point in its true spatial position with $\mathbf{w}$ intersects [0 0 1 0] at

$$[0 \quad 0 \quad 1 \quad 0] \left\{ \begin{bmatrix} 2.0761 \\ 2.0822 \\ 2.0916 \\ 1 \end{bmatrix} - \alpha \begin{bmatrix} 0.0761 \\ 0.0822 \\ 0.0916 \\ 0 \end{bmatrix} \right\} = 0$$

so $\alpha = 22.834$. The point is $\begin{bmatrix} 0.3384 \\ 0.2052 \\ 0 \\ 0 \end{bmatrix}$.

For $\begin{bmatrix} 2.0786 \\ 2.0791 \\ 2.0923 \\ 1 \end{bmatrix}$, $\alpha = 22.668$, so the point is $\begin{bmatrix} 0.2969 \\ 0.2860 \\ 0 \\ 1 \end{bmatrix}$.

For $\begin{bmatrix} 2.0831 \\ 2.0816 \\ 2.0852 \\ 1 \end{bmatrix}$, $\alpha = 24.474$, so the point is $\begin{bmatrix} 0.0493 \\ 0.0845 \\ 0 \\ 1 \end{bmatrix}$.

The angle formed by the edges at $\begin{bmatrix} 0.3384 \\ 0.2052 \\ 0 \\ 1 \end{bmatrix}$ is

$$\cos^{-1} \begin{bmatrix} 0.3384 - 0.2969 \\ 0.2052 - 0.2860 \\ 0 \end{bmatrix} \cdot \begin{bmatrix} 0.0493 - 0.2969 \\ 0.0845 - 0.2860 \\ 0 \end{bmatrix}$$

$$= \cos^{-1} \begin{bmatrix} 0.042 \\ -0.081 \\ 0 \end{bmatrix} \cdot \begin{bmatrix} -0.248 \\ -0.201 \\ 0 \end{bmatrix} = \cos^{-1} 0.00587 = 89.6^\circ$$

close enough to 90° to assume that the corner is a right angle. If we assume that the base is rectangular, the fourth point is at

$$\begin{bmatrix} 0.0493 + 0.0415 \\ 0.0845 - 0.8080 \\ 0 \\ 1 \end{bmatrix} = \begin{bmatrix} 0.0908 \\ 0.0037 \\ 0 \\ 1 \end{bmatrix}$$

and the width and length of the object are $\sqrt{0.042^2 + 0.081^2} = 0.0912$ and $\sqrt{0.248^2 + 0.201^2} = 0.319$. The orientation of the object with respect to the x-axis is $\tan^{-1} \dfrac{0.201}{0.248} = 39.0^\circ$.

Since we assumed a regular object, we can assume that the points (−0.00484, −0.00660), (−0.000325, −0.00735), and (0.00112, 0.00111) are vertically above (−0.00434, −0.01023), (−0.00032, −0.01100), and (0.00110, −0.00238). The plane containing the three points (−0.00484, −0.00660), (−0.000325, −0.00735), and (0.00112, 0.00111) is given in part by the vector

$$\begin{bmatrix} -0.00484 + 0.000323 \\ -0.00660 + 0.00735 \\ 0 \end{bmatrix} \times \begin{bmatrix} -0.00484 - 0.00112 \\ -0.00660 - 0.00111 \\ 0 \end{bmatrix}$$

$$= \begin{bmatrix} -0.00452 \\ 0.00075 \\ 0 \end{bmatrix} \times \begin{bmatrix} -0.00596 \\ -0.00771 \\ 0 \end{bmatrix} = \begin{bmatrix} 0 \\ 0 \\ 0.0000393 \end{bmatrix} \equiv \begin{bmatrix} 0 \\ 0 \\ 1 \end{bmatrix}$$

The centroid of the base of the object is at the intersection of a diagonal and at half the height:

$$\begin{bmatrix} \dfrac{0.0493 + 0.3384}{2} \\ \dfrac{0.0845 + 0.2052}{2} \\ 0 \\ 1 \end{bmatrix} = \begin{bmatrix} 0.1939 \\ 0.1449 \\ 0 \\ 1 \end{bmatrix}$$

For corner point (0.00112, 0.00111), we have

$$\mathbf{v} = \begin{bmatrix} 2.08333 \\ 2.08333 \\ 2.08333 \\ 1 \end{bmatrix} + 0.00112 \begin{bmatrix} 0.7071 \\ -0.7071 \\ 0 \\ 0 \end{bmatrix} + 0.00111 \begin{bmatrix} 0.4083 \\ 0.4083 \\ -0.8164 \\ 0 \end{bmatrix} = \begin{bmatrix} 2.0846 \\ 2.0830 \\ 2.0832 \\ 1 \end{bmatrix}$$

The object point lies along line

$$\begin{bmatrix} 2 \\ 2 \\ 2 \\ 1 \end{bmatrix} + \alpha \begin{bmatrix} 0.0846 \\ 0.0830 \\ 0.0832 \\ 0 \end{bmatrix}$$

If we assume that this line is meant to intersect the vertical line from the point below, a line defined by $\begin{bmatrix} 0.0493 \\ 0.0845 \\ z \\ 1 \end{bmatrix}$, then the distance D between the two lines is given by

$$D^2 = \{2 + 0.0846\alpha - 0.0493\}^2 + \{2 + 0.830\alpha - 0.0845\}^2 + \{2 + 0.0832\alpha - z\}^2$$

$$= \{1.9507 - 0.0846\alpha\}^2 + \{1.9155 + 0.0830\alpha\}^2 + \{2 + 0.0832\alpha - z\}^2$$

$$\frac{\partial D^2}{\partial \alpha} = 0.1692\{1.9507 - 0.0846\alpha\} + 0.1660\{1.9155 + 0.0830\alpha\}$$

$$+ 0.1664\{2 + 0.0832\alpha - z\} = 0 \quad \text{at maximum or minimum}$$

$$\frac{\partial D^2}{\partial z} = -2\{2 + 0.0832\alpha - z\} = 0 \quad \text{at maximum or minimum}$$

$$0.3301 + 0.3180 + 0.3328 + (0.01431 + 0.01378 + 0.01384)\alpha - 0.1664z = 0$$

$$2 + 0.0832\alpha - z = 0$$

$$0.9809 + 0.04193\alpha - 0.1664z = 0$$

or

$$\begin{aligned} 5.8948 + 0.2520\alpha - z &= 0 \\ 2 \quad\quad + 0.832\alpha - z &= 0 \\ \hline 3.8948 + 0.1688\alpha \quad\quad &= 0 \end{aligned}$$

so $\alpha = -23.073$ and $z = 0.080$. Thus, we have a rectangular object of height about 0.08 whose centroid on the conveyor belt is at

$$\begin{bmatrix} 0.194 + 0.1t \\ 0.145 \\ 0.040 \\ 1 \end{bmatrix}$$

and whose principal axis is defined by the vector

$$\begin{bmatrix} 0.2969 - 0.0493 \\ 0.2860 - 0.0845 \\ 0 \end{bmatrix} \equiv \begin{bmatrix} 0.776 \\ 0.631 \\ 0 \end{bmatrix}$$

We are now ready to plan the motions of the robot. We assume that the centroid of the opening of the end effector will coincide with the centroid of the object when the gripper is closed and the object lifted. The first two axes must move $\mathbf{u}_3$ above the centroid of the object before the lead screw lowers the end effector around the object. The object analysis must be complete before the object passes out of reach of the robot. Assume that the processing takes less than 2.5 s, so that the robot is at

$$\mathbf{u}_4 = \begin{bmatrix} 0.194 + 0.4t \\ 0.145 \\ 0.040 + l \\ 1 \end{bmatrix}$$

for times of about 2.5 s. The IKS equations for the SCARA, based on equations (6.25) and (6.26), are

$$\sin\left(\theta_1 + \tan^{-1}\frac{\bar{x}}{\bar{y}}\right) = \frac{1}{2}\sqrt{\bar{x}^2 + \bar{y}^2}; \qquad \theta_2 = \cos^{-1}(x - \cos\theta_1) - \theta_1$$

where $\bar{x} = x - 1$ and $\bar{y} = y + 0.25$. Thus driving the end effector along the straight line defined by $\mathbf{u}_4$ requires

$$\sin\left\{\theta_1(t) + \tan^{-1}\frac{0.4t - 0.806}{0.395}\right\} = \frac{1}{2}\sqrt{(0.4t - 0.806)^2 + 0.395^2}$$

$$\theta_2(t) = \cos^{-1}\{0.4t - 0.806 - \cos\theta_1(t)\} - \theta_1$$

At the time of 2.5 s,

$$\sin\left\{\theta_1(2.5) + \tan^{-1}\frac{1 - 0.806}{0.395}\right\} = \frac{1}{2}\sqrt{(1 - 0.806)^2 + 0.395^2}$$

so $\theta_1(2.5) = -13.448$ and $\theta_2(2.5) = \cos^{-1}\{1 - 0.806 - \cos 13.448\} + 13.448 = -127.68$. We drive the actuators for θ_1 and θ_2 with the four-point Bezier function

$$(1 - \alpha)^3\theta_1(t) + 3\alpha(1 - \alpha)^2\theta_1(t) + 3\alpha^2(1 - \alpha)\theta_1\left(t + \frac{2T}{3}\right) + \alpha^3\theta_1(t + T)$$

over the time interval 2.5 to $2.5 + T$. At time 2.5, θ_1 and θ_2 are stationary, but after $2.5 + 2T/3$ the velocity is approximately constant. The upper limit T depends on the achievable joint torques (see the determination of torque space in Chapter 9), and the object on the conveyor belt must be lifted before time $2.5 + T$, where $\theta_1(2.5 + T)$ and $\theta_2(2.5 + T)$ must be inside the work envelope of the robot. We choose $T = 2$. At $2.5 + 4/3 = 3.833$ s,

$$\sin\left\{\theta_1(3.833) + \tan^{-1}\frac{0.727}{0.395}\right\} = \frac{1}{2}\sqrt{(0.727)^2 + 0.395^2}$$

so $\theta_1(3.833) = 90.08°$ and $\theta_2(3.833) = \cos^{-1}\{0.727 - \cos 90.08\} - 90.088 = -46.84°$. At 4.5 s,

$$\sin\left\{\theta_1(4.5) + \tan^{-1}\frac{0.994}{0.395}\right\} = \frac{1}{2}\sqrt{0.994^2 + 0.395^2}, \text{so } \theta_1(4.5) = 79.34°$$

$$\theta_2(4.5) = \cos^{-1}\{0.994 - \cos 79.34\} - 79.34 = -43.34°$$

At time 2.5 the motions of the crank joint $\underline{\psi}_3$ and the lead screw for σ_4 commence. The parallel jaw gripper opening at 2.5 s must be greater than the width of the object 0.0912; we set this opening at 0.20 so the chance of an unwanted collision is reduced. At about time $t + T/2$ we require $\sigma_4 = 1 - 0.040 = 0.960$ and the gripper closed. For $t \geq 2.5$ s we require $\underline{\psi}_3 = 39 - \theta_1 - \theta_2$. After the gripper is closed, σ_4 is shortened and the object is lifted.

EXERCISES FOR CHAPTER 8

1. The 3 *dof* robot $|\ \lambda_1\ 0\ |\ \theta_2\ 2\ |\ \theta_3\ 2\ |$ has base orientation and position $\mathbf{u}_1 = \begin{bmatrix} 0 \\ 0 \\ 0 \\ 1 \end{bmatrix}$, $\mathbf{p}_1 = [0\ \ 0\ \ 1\ \ 0]$, and $\boldsymbol{\nu} = \begin{bmatrix} 1 \\ 0 \\ 0 \end{bmatrix}$. Determine the slew path between points $\begin{bmatrix} 4 \\ 0 \\ 0 \\ 1 \end{bmatrix}$ and $\begin{bmatrix} 0 \\ 1 \\ 3 \\ 1 \end{bmatrix}$. Find $\mathbf{u}_4$ at the midpoint of the slew path and determine the deviation of $\mathbf{u}_4$ from the midpoint of the straight line between the position space end points.

---

2. Repeat problem 1 when the endpoints of the slew path are $\begin{bmatrix} 0 \\ 0 \\ 0 \\ 1 \end{bmatrix}$ and $\begin{bmatrix} 3 \\ 0 \\ 0 \\ 1 \end{bmatrix}$.

---

3. Solve problem 1 for a quadratic approximation to a straight-line path between the endpoints; i.e., map the endpoints and the midpoint between them from position

to joint space and find the quadratic functions for each joint. Determine the error at the quarter points.

---

4. Solve problem 2 for a quadratic approximation to a straight-line path between the given endpoints.

---

5. Solve problem 2 for a cubic approximation to a straight-line path between the given endpoints—i.e., using the endpoints $\mathbf{u}_a$ and $\mathbf{u}_b$ and the points $2\mathbf{u}_a/3 + \mathbf{u}_b/3$ and $\mathbf{u}_a/3 + 2\mathbf{u}_b/3$.

---

6. The 3 *dof* revolute robot $|\ \lambda_1\ 0.2\ |\ \theta_2\ 0.2\ |\ \theta_3\ 0.2\ |$ has base orientation and position

$$\mathbf{u}_1 = \begin{bmatrix} 0 \\ 0 \\ 0 \\ 1 \end{bmatrix}, \qquad \mathbf{p}_1 = [0 \quad 1 \quad 0 \quad 0], \qquad \boldsymbol{\nu} = \begin{bmatrix} 0 \\ 0 \\ 1 \end{bmatrix}$$

$\mathbf{u}_4$ starts at point $\mathbf{v}_1 = \begin{bmatrix} 0.1 \\ 0 \\ 0 \\ 1 \end{bmatrix}$ and moves to point $\mathbf{v}_2 = \begin{bmatrix} 0 \\ 0.2 \\ 0.2 \\ 1 \end{bmatrix}$.

(a) Define the joint status for the end effector at $\mathbf{v}_1$ and $\mathbf{v}_2$.
(b) The end effector follows a slew path between the endpoints. Determine the midpoint of the path followed by $\mathbf{u}_4$.
(c) A camera sits with pinhole at $\begin{bmatrix} 1 \\ 1 \\ 1 \\ 1 \end{bmatrix}$ and film plane distance 0.2 behind the pinhole. The axis of the camera passes through the origin. Find the image of the robot on the film plane at the endpoints of the path.

---

7. The end effector of the 3 *dof* robot $|\ \phi\theta\sigma\ |$ is required to move in a straight line between points $[1 \quad \frac{1}{2} \quad 0 \quad 1]^\#$ and $[0 \quad 1 \quad 1 \quad 1]^\#$. The base of the robot is at the origin and $\mathbf{p}_1 = [0 \quad 1 \quad 0 \quad 0]$, $\boldsymbol{\nu} = [-1 \quad 0 \quad 0]$, and the link lengths are unity. Determine the joint status of the robot at the endpoints of the straight line and the middle point.

---

8. The 3 *dof* placeable robot $|\ \lambda\theta\theta\ |$ has link lengths of 10 in. and sits with its base at the origin. The end effector is required to move in a straight-line path from point $[10 \quad 10 \quad 10 \quad 1]^\#$ to point $[0 \quad 5 \quad 5 \quad 1]^\#$. Define the movement of the joints as the end effector follows this path.

---

9. The 3 *dof* revolute robot $|\ \lambda\theta\theta\ |$ with link lengths of unity, base at the origin, and $\mathbf{p}_1 = [0 \quad 1 \quad 0 \quad 0]$, $\boldsymbol{\nu} = [0 \quad 0 \quad 1]$, executes a path that is within 0.05 of a straight-line path between points $[1 \ \ \tfrac{1}{2} \ \ 1 \ \ 1]^\#$ and $[0 \ \ 0 \ \ 0 \ \ 1]^\#$. Determine the polynomial function in joint space to satisfy this criterion.

---

10. The robot $\mid \lambda\theta\theta\theta \mid$ with unity link lengths whose base is at the origin and first link is vertical is required to move its end effector along the straight-line trajectory

$$\begin{bmatrix} \alpha \\ 2-\alpha \\ 0 \\ 1 \end{bmatrix}, \qquad 0 \le \alpha \le 2$$

The last link of the robot must be kept vertical along the path. Approximate the required trajectory by a joint space quadratic that passes through the two end points and the middle point of the path. What are the position space errors at the points $\alpha = 0.5$ and $\alpha = 1.5$?

---

11. Repeat problem 10 with the endpoints $\begin{bmatrix} 2 \\ 0 \\ 0 \\ 1 \end{bmatrix}$ and $\begin{bmatrix} 0 \\ 1 \\ 1 \\ 1 \end{bmatrix}$ for the straight-line path.

---

12. The 3 *dof* revolute robot $\mid \lambda\theta\theta \mid$ with link lengths of unity, base at the origin, and $\mathbf{p}_1 = [0 \ \ 1 \ \ 0 \ \ 0]$, $\boldsymbol{\nu} = [0 \ \ 0 \ \ 1]^\#$, executes a path that is within 0.1 of a straight-line path between points $[1 \ \ 1 \ \ 1 \ \ 1]^\#$ and $[0 \ \ 0 \ \ 0 \ \ 1]^\#$ over the time period of 1 s. The starting and ending velocities are zero. Determine the Bezier function in joint space to satisfy this criterion.

---

13. Repeat problem 12 with the endpoints $\begin{bmatrix} 3 \\ 0 \\ 1 \\ 1 \end{bmatrix}$ and $\begin{bmatrix} 0 \\ 1 \\ 0 \\ 1 \end{bmatrix}$ for the straight-line path.

---

14. The 5 *dof* directable robot $\mid \lambda\theta\theta\lambda\theta \mid$ is required to be able to insert pegs into holes drilled perpendicular to the surface of a hemisphere of diameter 1 m that rests on a horizontal surface. The base of the robot is on the horizontal surface and the first link is vertical. The center of the hemisphere is at the origin. Determine the minimum link lengths of the robot and the location of the base of the robot. Assume that there are no limitations on the angles of the joints and that the size of the joints and width of the links is zero.

---

15. The 3 *dof* robot $\mid \lambda_1 \ 1 \mid \theta_2 \ 1 \mid \theta_3 \ 1 \mid$ with base orientation and position given by

$$\mathbf{u}_1 = \begin{bmatrix} 0 \\ 0 \\ 0 \\ 1 \end{bmatrix}, \qquad \mathbf{p}_1 = [0 \ \ 1 \ \ 0 \ \ 0], \qquad \boldsymbol{\nu} = \begin{bmatrix} 0 \\ 0 \\ 1 \end{bmatrix}$$

is required to travel between points $\begin{bmatrix} 0 \\ 0 \\ 2.8 \\ 1 \end{bmatrix}$ and $\begin{bmatrix} 1.8 \\ 0 \\ 1 \\ 1 \end{bmatrix}$. An obstacle that is a flat surface given by plane $[1.4 \quad 0 \quad 1.4 \quad -4]$. Does the slew path between the two points interfere with this obstacle? If it does, define some path to be executed with constant speed between the two points that will not interfere.

---

16. Repeat the second part of problem 15 assuming the path to be executed starts and ends with zero velocity and the acceleration is limited to 1 m/s^2.

---

17. A 3 *dof* SCARA robot with usual base orientation has its base at $\begin{bmatrix} 0 \\ 0 \\ 1 \\ 1 \end{bmatrix}$. Its major link lengths are $l_1 = l_2 = 1$. Its end effector is required to move between points $\begin{bmatrix} 1.8 \\ 0 \\ 0.5 \\ 1 \end{bmatrix}$ and $\begin{bmatrix} 0 \\ 1.8 \\ 0 \\ 1 \end{bmatrix}$, but a slew path will interfere with a thin vertical barrier defined by points

$$\begin{bmatrix} 1 + \alpha \\ 1 + \alpha \\ z \\ 1 \end{bmatrix}, \qquad 0 \le \alpha < \infty, \qquad \text{all } z$$

Determine a time function for θ_1 and θ_2 so that the robot can transit between the two points without collision in 1 second.

---

18. Repeat the prior exercise assuming that the velocity at the endpoints of the path is zero.

---

19. The polar robot $|\lambda\theta\sigma|$ has base orientation and position $\mathbf{p}_1 = [1 \quad 0 \quad 0 \quad 0]$, $\boldsymbol{\nu} = \begin{bmatrix} 0 \\ 0 \\ 1 \end{bmatrix}$, and $\mathbf{u}_1 = \begin{bmatrix} 0 \\ 0 \\ 0 \\ 1 \end{bmatrix}$. Plan a trajectory between points $\begin{bmatrix} 0.5 \\ 0.5 \\ 1.1 \\ 1 \end{bmatrix}$ and $\begin{bmatrix} -0.7 \\ 0 \\ 0.9 \\ 1 \end{bmatrix}$ to avoid a ceiling given by $[0 \quad 0 \quad 1 \quad -1.2]$. The velocity at the endpoints must be zero and the execution time must be 1 s.

---

20. A series of machine parts in the shape of rectangular bars $0.1 \times 0.15 \times 0.25$ m in size is moving along a horizontal conveyor belt whose center line is

$[1 \quad y \quad 0 \quad 1]^{\#}$. An image-processing scheme senses when a hole is missing in a bar and determines the centroid, major axis, and orientation of the defective bar. The 5 *dof* robot $| \lambda\theta\theta\theta\lambda |$ with base at the origin and parallel jaw gripper is required to pick up defective bars and place them in a tray for reworking. Task plan the robot, assuming that the conveyor belt is moving at 1 m/s and that information on the defective bar is received before the bar passes the point $y = -0.5$. Assume that l_1 to l_4 are unity, and $l_5 = 0$. Further, it can be assumed that the centroids of the objects can vary by ± 0.5 from the center line of the conveyor belt.

CHAPTER NINE

Dynamics

The subject of dynamic analysis of robots has appeared very complicated. The equations are messy, and the methodology is advanced and (for many) obscure. However, as seen in this chapter, dynamic analysis need not be difficult. The difference is the point-plane method of analysis, which was shown to provide simpler kinematic analysis than standard formulations involving the Denavit-Hartenberg notation.

We start from first principles, considering spring-mass-dashpot systems, and move into loaded kinematic chains that characterize typical robots. An *n dof* manipulator with distributed loads on the links and point loads at the joints is initially modeled with point loads, and the vector force on each point load is computed. The component of each force that acts on each downlink joint is computed to determine the torque at each joint. The method is the simple application of two cookbook formulas. The method is similar to what others [Craig 1986; Fu, Gozalves, and Lee 1987] call the iterative Newton-Euler method, but we avoid the term iterative, since it implies not in closed form, whereas we present a closed-form method. The modifications necessary to model a system with distributed loads is discussed, as is the degenerate nature of what is called torque space.

9.1 FRICTION, MASS, AND FORCE IN MECHANICAL SYSTEMS

A simple spring-mass-dashpot system is shown in Figure 9.1. A dashpot is also called a damper and provides a viscous frictional force; its action is similar to the shock absorbers in an automobile. The units of measure we use are the international MKS (meter, m; kilogram, kg; second, s) system. Forces are in newtons, where 1 N is the force required to accelerate a mass of 1 kg by 1 m/s^2, so the units of a Newton are $m\text{-}kg/s^2$. The acceleration g due to gravity is 9.80616 m/s^2, or 32.172 ft/s^2. In British units, mass is measured in slugs and force in pounds, so a pound is defined as the force required to accelerate a mass of 1 slug by 1 ft/s^2. Sometimes confusion is created under the British system by defining mass in pounds, where 1 lb mass = 1/32.172 = 0.03108 slugs. There is no such confusion in the MKS system. The spring constant K is in N/m. The dashpot exerts a frictional force proportional to the rate of change of position $y(t)$, so F is in N-s/m.

Newton's laws of motion can be stated as follows:

1. Every body continues at rest or uniform motion in a straight line except when changed by the algebraic sum of applied forces.
2. $f = Ma$.
3. For every action, there is an equal and opposite reaction elsewhere.

Using Newton's third law on the system of Figure 9.1 with externally applied force f we have

$$f = M\frac{d^2y}{dt^2} + F\frac{dy}{dt} + Ky \tag{9.1}$$

Under the condition that F/M is small and M is large, the system shown in Figure 9.1 can be used as a seismograph (an instrument to measure

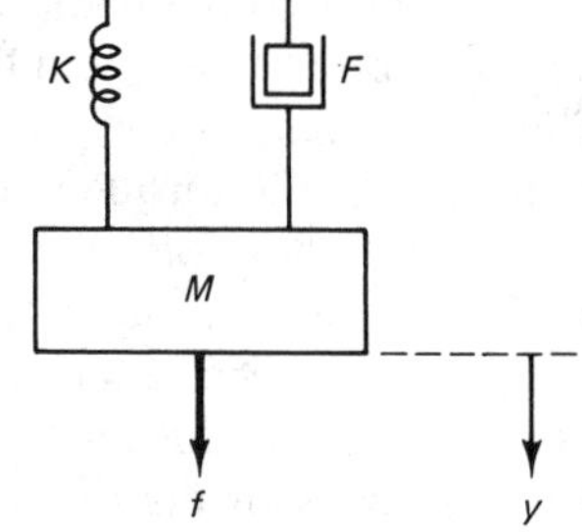

Figure 9.1 A translational mechanical system

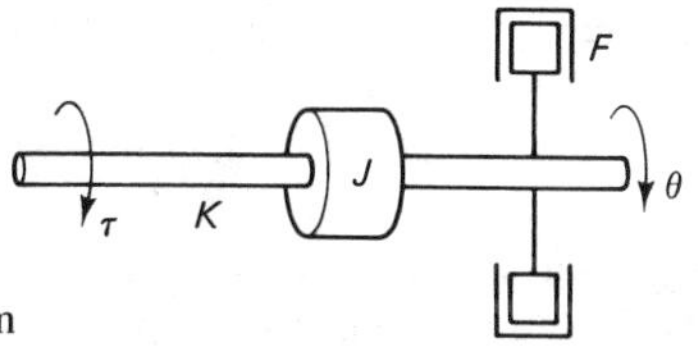

Figure 9.2 A rotational mechanical system

Earth tremors). The case of the instrument (attached to the crosshatched top piece in Figure 9.1) is solidly mounted on the ground. A scale attached to the case is read with a pointer mounted on the mass. A tremor causes the case to move, but the mass is slow to follow such movements, and the pointer records the relative magnitude of the tremor. Instead, if the spring constant K is large and the input frequency is small compared to $\sqrt{(K/M)}$, then the device can be used to measure acceleration, with application on aircraft.

The rotational equivalent of the spring-mass-dashpot system is shown in Figure 9.2. J is the moment of inertia of the mass, known simply as the inertia, measured in kg-m^2. The rotational equivalent of the spring is the torsion spring governed by the equation $\tau(t) = K\theta(t)$, where K has the dimensions N-m/rad. Recall that radians are dimensionless. The complete rotational system is described in the time domain by

$$\tau = J\frac{d^2\theta}{dt^2} + F\frac{d\theta}{dt} + K\theta \tag{9.2}$$

Work is the application of force over distance, so it has the dimensions of N-m = kg-m^2/s^2. Power is the rate of doing work in N-m/s = kg-m^2/s^3. The energy of a body is its capacity to do work. Potential energy is the ability of the body to do work based on its position, so the potential energy of a mass of M Kg at rest l meters above datum is Mgl kg-m^2/s^2. The kinetic energy of the mass moving at velocity v m/s is $Mv^2/2$ kg-m^2/s^2. If the mass is a point mass rotating in a circle of radius r at angular velocity ω rad/s, the kinetic energy is $\omega^2r^2M/2$ kg-m^2/s^2. The law of conservation of energy states that energy is neither created nor destroyed but can be converted from one form to another—for instance, kinetic to potential energy and heat, light, or sound. Energy is constant, and as we shall see, mass is constant for our systems: Einstein's famous equation $E = mc^2$ does not apply in our Newtonian world.

Example:

A point mass M is attached by a string of length r to a pivot point. At $t < 0$ the system is at rest, and at $t = 0$ the system is given kinetic energy E, so the mass rotates in vertical plane $\mathbf{p} = [0 \quad 1 \quad 0 \quad 0]$. When M is traveling vertically upward, the string is cut. Determine the velocities and energies at all salient points on the path followed by M. Assume there is no friction or losses in the system.

The kinetic energy E causes M to rotate at radius r and angular velocity ω_0,

$$E = \frac{1}{2} M\omega_0^2 r^2, \quad \text{so} \quad \omega_0^2 = \frac{2E}{Mr^2}$$

Since the energy in the system is unaltered by position, the velocity ω_1 when the body reaches the same height as the pivot is given by

$$\frac{1}{2} M\omega_1^2 r^2 = E - Mgr, \quad \text{so} \quad \omega_1^2 = \frac{2E}{Mr^2} - \frac{2g}{r}$$

When the string is cut, the vertical velocity is $\omega_1 r$. M is then in free flight and reaches its apogee in time t given by

$$0 = \omega_1 r - gt, \quad \text{so} \quad t = \frac{\omega_1 r}{g}$$

Also, M has traveled the distance

$$s = \omega_1 rt - \frac{1}{2} gt^2 = \frac{\omega_1^2 r^2}{2g}$$

since the string was cut. The height of M above its starting position is $s + r$, so the potential energy above the datum position is

$$M(s + r)g = M\left(\frac{\omega_1^2 r^2}{2g} + r\right) g = E, \quad \text{or} \quad s + r = \frac{E}{Mg}$$

The mass falls and reaches the same height as the datum point at time t, given by

$$s + r = \frac{1}{2} gt^2, \quad \text{so} \quad t^2 = \frac{2}{g}(s + r) = \frac{2E}{Mg^2}$$

The velocity at this point is $v = gt = \sqrt{2E/M}$, where the body has kinetic energy $\frac{1}{2}Mv^2 = E$. We have shown that the energies at all salient points are identical.

When M is held by the string, it is subject to an acceleration toward the pivot point given by $\omega^2 r$, so the centrifugal force is $M\omega^2 r$. Since this force does no work, it is ignored.

The center of gravity of any rigid body can be found by suspending the body from a point and determining a vertical through the body, repeating the procedure at another point. The center of gravity is the unique point of intersection of any two such verticals provided they are not coincident. For a point mass the center of gravity is the location of the point mass. For the uniform rod of uniform density the center of gravity is the center point of the rod. For a body comprised of n point masses M_i located at u_i in Cartesian space, the center of gravity is given by

$$\frac{\sum_{i=1}^{n} M_i u_i}{\sum_{i=1}^{n} M_i}$$

To a first-order approximation, any rigid body of total mass $\overline{M}$ can be reduced to a single mass $\overline{M}$ at its center of gravity. That this is an approximation can be seen by considering the kinetic energy of a thin bar of length l and total distributed mass m connected at one end to a pitch joint. The kinetic energy of the bar as it rotates at angular velocity ω can be found by integrating the kinetic energies in the incremental mass particles, i.e.

$$E = \int_0^l \frac{m}{2l}\, \omega^2 x^2\, dx = \frac{\omega^2}{2} I, \text{ where } I = \frac{ml^2}{3}$$

where I is known as the moment of inertia. Alternatively, if we consider mass m at the center of gravity, the kinetic energy is

$$\frac{\omega^2}{2}\left(\frac{ml^2}{2}\right)$$

Further, a solid body with three dimensional mass distribution requires integration (performed in one variable x above) in three mutually orthogonal directions. However, we can model a distributed rigid mass by two or more point masses and achieve any assigned accuracy. We defer discussion of other than point masses to a later section and take some comfort in this assumption by noting that gravity forces are significant in a typical robot, and the center of gravity point model is exact for gravitational forces.

An n *dof* robot with distributed mass in each link, point loads due to its actuator masses, and end effector loading can be reduced to an unloaded kinematic chain with m masses at m centers of gravity as shown in Figure 9.3, where $m \le n$ and each mass is fixed with respect to its corresponding link. For example, the complete dynamic system for the 6 *dof* revolute, orientable robot $|\,\lambda\theta\theta\lambda\theta\lambda\,|$ reduces to four masses $\overline{M}_{i+1}$ at $\bar{\mathbf{u}}_{i+1}$, $i = 1, \ldots, 4$, where $|\,\bar{\mathbf{u}}_{i+1} - \bar{\mathbf{u}}_i\,| = \bar{l}_i$, the radius of the center of gravity of $\overline{M}_{i+1}$. If each link l_i has point load M_{i+1} at $\mathbf{u}_{i+1}$ and total uniform distributed load of m_i, we reduce the load on the link to point loads $\overline{M}_{i+1} = m_i + M_{i+1}$ at center of gravity $\bar{l}_i$ where

$$\bar{l}_i = l_i \frac{\dfrac{m_i}{2} + M_{i+1}}{M_{i+1} + m_i}$$

For a roll joint at $\mathbf{u}_i$ we have the point load of $\overline{M}_{i+1} = M_i + m_i + M_{i+1} + m_{i+1}$ at

$$\bar{l}_i = \frac{l_i\left(\dfrac{m_i}{2} + M_i\right) + \left\{l_i(m_{i+1} + M_{i+1}) + l_{i+1}\left(\dfrac{m_{i+1}}{2} + M_{i+1}\right)\right\}}{\overline{M}_{i+1}}$$

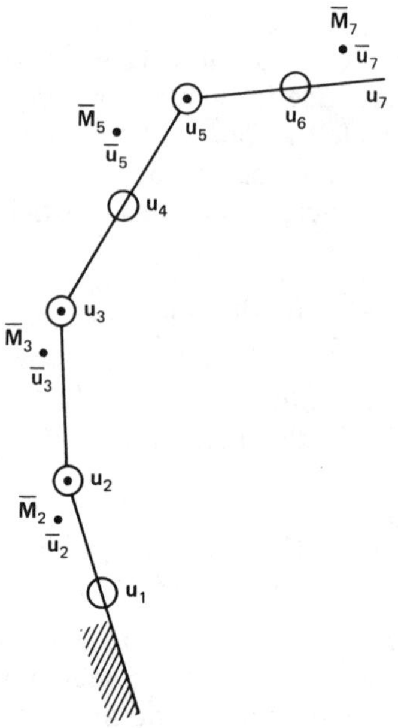

Figure 9.3 Point mass model of an orientable robot

For the crank joint ψ at $\mathbf{u}_i$ with equivalent point loads of $\overline{M}_i$ and $\overline{M}_{i+1}$ at $\overline{\mathbf{u}}_i$ and $\overline{\mathbf{u}}_{i+1}$. The total loading effect is as a point load $\overline{M}_i + \overline{M}_{i+1}$ at

$$\frac{\overline{\mathbf{u}}_i\overline{M}_i + \overline{\mathbf{u}}_{i+1}\overline{M}_{i+1}}{\overline{M}_i + \overline{M}_{i+1}}$$

Notice that this location is on neither planes of motion of the uplink and downlink of the crank joint. For the system with the cylindrical joint ϕ at $\mathbf{u}_i$, the equivalent single point load is as given for the crank joint, but in this case the point load is on both planes of motion.

The joints are considered to have hard, non-compliant bearing surfaces that can handle any bearing load. The bearing loads will produce greater point pressures when eccentric loads (such as are produced by loads following a crank joint) exist. We assume that the mechanical integrity of the bearing is upheld and neglect bearing failure. However, it should be recognized that the greater the pressure on the bearing and gear system the more the wear, and ultimately the accuracy of the robot is prejudiced.

The viscous frictional force F at a joint is a function of the bearings at the joint, the gearing, and to a lesser extent the eccentricity of the load at the joint; we assume no Coulomb friction. There is no eccentricity when all forces lie within the plane of motion $\mathbf{p} = [a \quad b \quad c \quad d]$: this will occur only when gravity forces are contained in $\mathbf{p}$ so $\mathbf{p}$ is vertical and $c = 0$.

9.2 LAGRANGIAN ANALYSIS OF SIMPLE KINEMATIC CHAINS

Lagrangian dynamics is based on a determination of the energy contained in a system. The Lagrangian E_L is defined as $E_L = E_K - E_P$, where E_K is the total kinetic energy and E_P is the total potential energy of the system [Greenwood 1965; Asada 1986]. These energies can be expressed in any convenient coordinate system [Paul 1981b]. Here we choose the Cartesian coordinate system. Throughout this section we use base orientation and position of $\mathbf{u}_1$ given by

$$\boldsymbol{\nu} = \begin{bmatrix} 0 \\ 0 \\ 1 \end{bmatrix}, \mathbf{p}_1 = [0 \quad 1 \quad 0 \quad 0] \text{ and } \mathbf{u}_1 = \begin{bmatrix} 0 \\ 0 \\ 0 \\ 1 \end{bmatrix}$$

Consider a mass M at $\mathbf{u}_2$ connected by a link of length l_1 connected to a pitch joint at the origin as shown in Figure 9.4. The mass is at point

$$\mathbf{u}_2 = \begin{bmatrix} -l_1 \sin\theta \\ 0 \\ l_1 \cos\theta \\ 1 \end{bmatrix}, \text{ so } V^2 = \dot{x}_2^2 + \dot{z}_2^2 = l_1^2\dot{\theta}_1^2(\cos^2\theta_1 + \sin^2\theta_1) = l_1^2\dot{\theta}_1^2,$$

where V is the velocity of $\mathbf{u}_2$. The kinetic energy of M is $E_K = \frac{1}{2}Ml_1^2\dot{\theta}_1^2$. The potential energy of M is $Mgl_1\cos\theta_1$. The Lagrangian is

$$E_L = E_K - E_P = \tfrac{1}{2}Ml_1^2\dot{\theta}_1^2 - Mgl_1\cos\theta_1.$$

The general form of the torque equations is given by [Greenwood 1965; Fu 1987]

$$\tau_i = \frac{d}{dt}\frac{\partial E_L}{\partial \dot{\theta}_i} - \frac{\partial E_L}{\partial \theta_i} \tag{9.3}$$

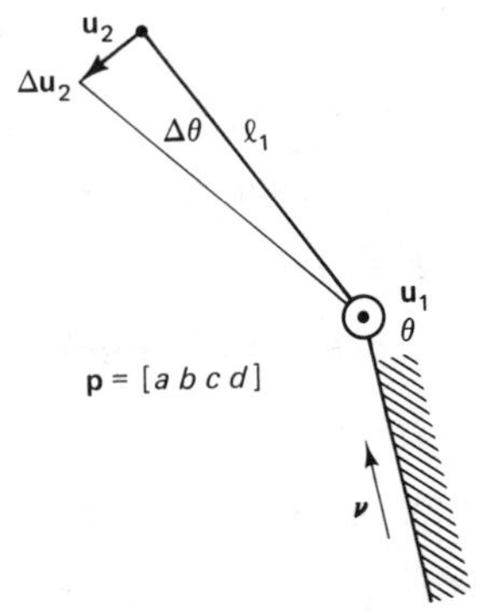

Figure 9.4 A pitch joint

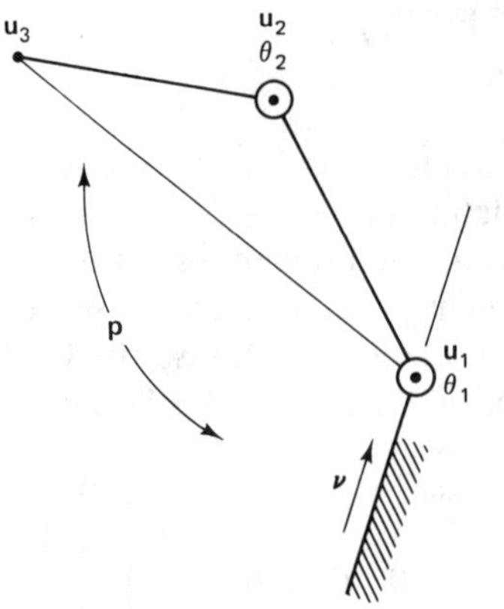

Figure 9.5 Two pitch joints

so
$$\frac{\partial E_L}{\partial \dot{\theta}_1} = Ml_1^2\dot{\theta}_1;\ \frac{d}{dt}Ml_1^2\dot{\theta}_1 = Ml_1^2\ddot{\theta};\ \frac{\partial E_L}{\partial \theta_1} = Mgl_1 \sin\theta_1.$$

$$\tau = Ml_1^2\ddot{\theta}_1 - Mgl_1\sin\theta_1.$$

Given the 2 *dof* system | $\theta\theta$ | shown in Figure 9.5, assume there is a point mass M_2 at $\mathbf{u}_2$ and a point mass M_3 at $\mathbf{u}_3$, $\mathbf{p} = [0\ \ 1\ \ 0\ \ 0]$ and $\boldsymbol{\nu} = [0\ \ 0\ \ 1]^{\#}$. The kinetic and potential energies for M_2 are

$$E_{K2} = \tfrac{1}{2}M_2l_1^2\dot{\theta}_1^2;\ E_{P2} = M_2gl_1\cos\theta_1.$$

M_3 is at
$$x_3 = l_1\sin\theta_1 + l_2\sin(\theta_1 + \theta_2),$$
$$z_3 = l_1\cos\theta_1 + l_2\cos(\theta_1 + \theta_2),$$

so
$$\dot{x}_3 = l_1\dot{\theta}_1\cos\theta_1 + l_2(\dot{\theta}_1 + \dot{\theta}_2)\cos(\theta_1 + \theta_2)$$

and
$$\dot{z}_3 = -l_1\dot{\theta}_1\sin\theta_1 - l_2(\dot{\theta}_1 + \dot{\theta}_2)\sin(\theta_1 + \theta_2)$$

The velocity V_3 of M_3 is given by

$$\begin{aligned} V_3^2 &= \dot{x}_3^2 + \dot{z}_3^2 \\ &= \{l_1\dot{\theta}_1\cos\theta_1 + l_2(\dot{\theta}_1 + \dot{\theta}_2)\cos(\theta_1 + \theta_2)\}^2 \\ &\quad + \{l_1\dot{\theta}_1\sin\theta_1 + l_2(\dot{\theta}_1 + \dot{\theta}_2)\sin(\theta_1 + \theta_2)\}^2 \\ &= l_1^2\dot{\theta}_1^2 + l_2^2(\dot{\theta}_1 + \dot{\theta}_2)^2 + 2l_1l_2\dot{\theta}_1(\dot{\theta}_1 + \dot{\theta}_2) \\ &\quad \{\cos\theta_1\cos(\theta_1 + \theta_2) + \sin\theta_1\sin(\theta_1 + \theta_2)\} \\ &= l_1^2\dot{\theta}_1^2 + l_2^2(\dot{\theta}_1 + \dot{\theta}_2)^2 + 2l_1l_2\dot{\theta}_1(\dot{\theta}_1 + \dot{\theta}_2)\cos\theta_2 \end{aligned}$$

so the kinetic energy of M_3 is

$$E_{K3} = \tfrac{1}{2}M_3l_1^2\dot{\theta}_1^2 + \tfrac{1}{2}M_3l_2^2(\dot{\theta}_1 + \dot{\theta}_2)^2 + M_3l_1l_2\dot{\theta}_1(\dot{\theta}_1 + \dot{\theta}_2)\cos\theta_2$$

The potential energy of M_3 is

$$E_{P3} = M_3 g\{l_1 \cos\theta_1 + l_2 \cos(\theta_1 + \theta_2)\}$$

The Lagrangian is given by

$$\begin{aligned} E_L &= E_{K2} + E_{K3} - E_{P2} - E_{P3} \\ &= \tfrac{1}{2}(M_2 + M_3)l_1^2\dot{\theta}_1^2 + \tfrac{1}{2}M_3 l_2^2(\dot{\theta}_1 + \dot{\theta}_2)^2 + M_3 l_2 l_3 \dot{\theta}_1(\dot{\theta}_1 + \dot{\theta}_2)\cos\theta_2 \\ &\quad - M_2 g l_1 \cos\theta_1 - M_3 g\{l_1\cos\theta_1 + l_2\cos(\theta_1 + \theta_2)\} \end{aligned} \tag{9.5}$$

The torque equations are

$$\tau_1 = \frac{d}{dt}\frac{\partial E_L}{\partial \dot{\theta}_1} - \frac{\partial E_L}{\partial \theta_1}$$

so

$$\frac{\partial E_L}{\partial \dot{\theta}_1} = (M_2 + M_3)l_1^2\dot{\theta}_1 + M_3 l_2^2(\dot{\theta}_1 + \dot{\theta}_2) + M_3 l_1 l_2(2\dot{\theta}_1 + \dot{\theta}_2)\cos\theta_2$$

$$\begin{aligned} \frac{d}{dt}\frac{\partial E_L}{\partial \dot{\theta}_1} &= (M_2 + M_3)l_1^2\ddot{\theta}_1 + M_3 l_2^2(\ddot{\theta}_1 + \ddot{\theta}_2) + M_3 l_1 l_2(2\ddot{\theta}_1 + \ddot{\theta}_2)cos\,\theta_2 \\ &\quad - M_3 l_1 l_2(2\dot{\theta}_1 + \dot{\theta}_2)\dot{\theta}_2 \sin\theta_2 \\ &= M_2 l_1^2\ddot{\theta}_1 + M_3(l_1^2 + l_2^2 + 2l_2 l_2\cos\theta_2)\ddot{\theta}_1 + M_3 l_2(l_2 + l_1\cos\theta_2)\ddot{\theta}_2 \\ &\quad - M_3 l_1 l_2(2\dot{\theta}_1 + \dot{\theta}_2)\dot{\theta}_2\sin\theta_2 \end{aligned}$$

$$\frac{\partial E_L}{\partial \theta_1} = -M_2 g l_1 \sin\theta_1 - M_3 g\{l_1\cos\theta_1 + l_2\cos(\theta_1 + \theta_2)\}$$

$$\begin{aligned} \tau_1 &= M_2 l_1^2\ddot{\theta}_1 + M_3(l_1^2 + l_2^2 + 2l_2 l_2\cos\theta_2)\ddot{\theta}_1 + M_3 l_2(l_2 + l_1\cos\theta_2)\ddot{\theta}_2 \\ &\quad - M_3 l_1 l_2(2\dot{\theta}_1 + \dot{\theta}_2)\dot{\theta}_2 \sin\theta_2 + M_2 g l_1 \sin\theta_1 + M_3 g\{l_1\cos\theta_1 \\ &\quad + l_2\cos(\theta_1 + \theta_2)\} \end{aligned} \tag{9.6}$$

$$\tau_2 = \frac{d}{dt}\frac{\partial E_L}{\partial \dot{\theta}_2} - \frac{\partial E_L}{\partial \theta_2}$$

$$\frac{\partial E_L}{\partial \dot{\theta}_2} = M_3 l_2^2(\dot{\theta}_1 + \dot{\theta}_2) + M_3 l_1 l_2 \dot{\theta}_1\cos\theta_2$$

$$\frac{d}{dt}\frac{\partial E_L}{\partial \dot{\theta}_2} = M_3 l_2^2(\ddot{\theta}_1 + \ddot{\theta}_2) + M_3 l_1 l_2(\ddot{\theta}_1\cos\theta_1 - \dot{\theta}_1\dot{\theta}_2\sin\theta_2)$$

$$\frac{\partial E_L}{\partial \theta_2} = -M_3 l_1 l_2 \dot{\theta}_1(\dot{\theta}_1 + \dot{\theta}_2)\sin\theta_2 + M_3 g l_2 \sin(\theta_1 + \theta_2)$$

$$\tau_2 = M_3 l_2^2(\ddot{\theta}_1 + \ddot{\theta}_2) + M_3 l_1 l_2\{\ddot{\theta}_1 \cos\theta_2 - \dot{\theta}_1\dot{\theta}_2 \sin\theta_2 + \dot{\theta}_1(\dot{\theta}_1 + \dot{\theta}_2)\sin\theta_2\}$$
$$- M_3 g l_2 \sin(\theta_1 + \theta_2)$$
$$= M_3 l_2^2(\ddot{\theta}_1 + \ddot{\theta}_2) + M_3 l_1 l_2(\ddot{\theta}_1 \cos\theta_2 + \dot{\theta}_1^2 \sin\theta_2) - M_3 g l_2 \sin(\theta_1 + \theta_2) \tag{9.7}$$

Given the 2 *dof* kinematic chain $| \lambda\theta |$ as shown in Figure 9.6. Assume that there is a point mass M at $\mathbf{u}_3$.

$$\begin{bmatrix} x_3 \\ y_3 \\ z_3 \end{bmatrix} = \begin{bmatrix} -l_2 \sin\theta_2 \cos\lambda_1 \\ -l_2 \sin\theta_2 \sin\lambda_1 \\ l_1 + l_2 \cos\theta_2 \end{bmatrix}$$

so

$$\dot{x}_3 = -l_2\dot{\theta}_2 \cos\theta_2 \cos\lambda_1 - l_2\dot{\lambda}_1 \sin\theta_2 \sin\lambda_1$$
$$\dot{y}_3 = -l_2\dot{\theta}_2 \cos\theta_2 \sin\lambda_1 - l_2\dot{\lambda}_1 \sin\theta_2 \cos\lambda_1$$
$$\dot{z}_3 = -l_2\dot{\theta}_2 \sin\theta_2$$

The velocity V_3 of M is given by

$$V_3^2 = l_2^2\{\{\dot{\theta}_2 \cos\theta_2 \cos\lambda_1 - \dot{\lambda}_1 \sin\theta_2 \sin\lambda_1\}^2$$
$$+ \{\dot{\theta}_2 \cos\theta_2 \sin\lambda_1 + \dot{\lambda}_1 \sin\theta_2 \cos\lambda_1\}^2 + \{\dot{\theta}_2 \sin\theta_2\}^2\}$$
$$= l_2^2\{\dot{\theta}_2^2 \cos^2\theta_2 \cos^2\lambda_1 + \lambda_1^2 \sin^2\theta_2 \sin^2\lambda_1 + \theta_2^2 \cos\theta_2^2 \sin^2\lambda_1$$
$$+ \lambda_1^2 \sin^2\theta_2 \cos^2\lambda_1 + \theta_2^2 \sin^2\theta_2\}$$
$$= l_2^2\{\theta_2^2 + \dot{\lambda}_1^2 \sin^2\theta_2\}$$

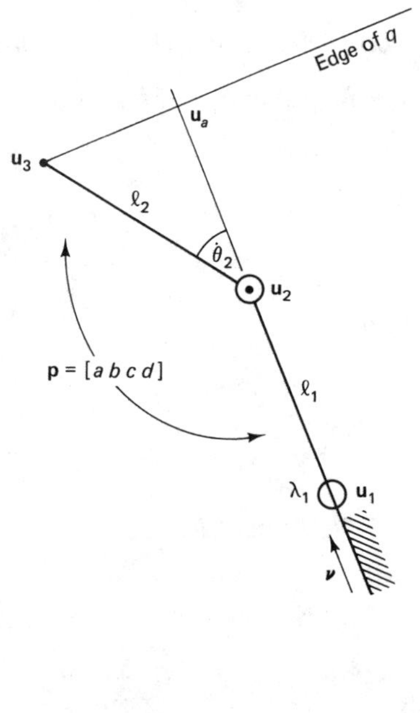

Figure 9.6 Accelerations on $\mathbf{u}_3$ with two pitch joints

The kinetic energy of M is $E_{K3} = \frac{1}{2}Ml_2^2\{\dot{\theta}_2^2 + \dot{\lambda}_1^2\sin^2\theta_2\}$. The potential energy of M is $E_{P3} = Mg(l_1 + l_2\cos\theta_2)$. The Lagrangian is

$$E_L = E_{K3} - E_{P3} = \frac{1}{2}Ml_2^2\{\dot{\theta}_2^2 + \dot{\lambda}_1^2\sin^2\theta_2\} - Mg(l_1 + l_2\cos\theta_2) \qquad (9.8)$$

$$\tau_1 = \frac{d}{dt}\frac{\partial E_L}{\partial \dot{\lambda}_1} - \frac{\partial E_L}{\partial \lambda_1}$$

so

$$\frac{\partial E_L}{\partial \dot{\lambda}_1} = Ml_2^2\dot{\lambda}_1\sin^2\theta_2; \qquad \frac{d}{dt}\frac{\partial E_L}{\partial \dot{\lambda}_1} = Ml_2^2\{\ddot{\lambda}_1\sin^2\theta_2 + 2\dot{\lambda}_1\dot{\theta}_2\sin\theta_2\cos\theta_2\}$$

$$\frac{\partial E_L}{\partial \lambda_1} = 0$$

$$\tau_1 = Ml_2^2\{\ddot{\lambda}_1\sin^2\theta_2 + 2\dot{\lambda}_1\dot{\theta}_2\sin\theta_2\cos\theta_2\} = Ml_2^2\{\ddot{\lambda}_1\sin^2\lambda_1 + \dot{\lambda}_2\dot{\theta}_2\sin 2\theta_2\} \qquad (9.9)$$

$$\tau_2 = \frac{d}{dt}\frac{\partial E_L}{\partial \dot{\theta}_2} - \frac{\partial E_L}{\partial \theta_2}$$

so

$$\frac{\partial E_L}{\partial \dot{\theta}_2} = Ml_2^2\dot{\theta}_1, \qquad \frac{d}{dt}\frac{\partial E_L}{\partial \dot{\theta}_2} = Ml_2^2\ddot{\theta}_2$$

$$\frac{\partial E_L}{\partial \theta_2} = Ml_2^2\dot{\lambda}_1^2\sin\theta_2\cos\theta_2 - Mgl_2\sin\theta_2$$

$$\tau_2 = Ml_2(l_2\ddot{\theta}_2 + l_2\dot{\lambda}_1^2\sin\theta_2\cos\theta_2 - g\sin\theta_2) \qquad (9.10)$$

9.3 ACCELERATION OF A PARTICLE AT THE END OF A KINEMATIC CHAIN

The advantage of the Lagrangian formulation is its generality. Another seeming advantage is the scalar nature of the calculations—that is, the energies are scalar quantities. However, the derivatives required become complex, and the formulation is too cumbersome to use in practical cases. Even for the simple systems $|\,\theta\theta\,|$ and $|\,\lambda\theta\,|$ considered in the prior section, we used simple base orientations to ease the calculations. In this section we determine the acceleration of a particle at the end of kinematic chains in terms of the point-plane method and assume general orientations.

Suppose $[a \quad b \quad c]^\#$ is a vector pointing into the paper; this convention is used in this limited context to avoid an excessive number of negative signs in subsequent derivations. $\mathbf{p}$ will be used as a plane $[a \quad b \quad c \quad d]$ and as a vector $[a \quad b \quad c]^\#$, and the context resolves the ambiguity. If a vector $\boldsymbol{\mu}$ is rotated by 90° in $\mathbf{p}$, then the resulting vector is given by $\boldsymbol{\mu} \times \mathbf{p}$. Rotating by 90° once more gives $(\boldsymbol{\mu} \times \mathbf{p}) \times \mathbf{p} = -\boldsymbol{\mu}$.

Consider a particle at the end of a link of length l attached to a pitch joint, as shown in Figure 9.4. We use the notation that $\mathbf{u}_2 = \mathbf{u}_1 + \mathbf{u}_{21}$, where $\mathbf{u}_{21}$ is a vector, and assume that $\mathbf{u}_1$ and $\boldsymbol{\nu}$ are fixed. A change of $\Delta\theta$ in θ results in a change $\Delta\mathbf{u}_2 = \Delta\theta\mathbf{u}_{21} \times \mathbf{p}$ in $\mathbf{u}_2$; notice the change in $\mathbf{u}_2$ is in the direction given by rotating $\mathbf{u}_{21}$ by 90° in $\mathbf{p}$ [Luh, Walker, and Paul 1980b; Balafoutis, Patel, and Misra 1988]. Thus, the velocity $\dot{\mathbf{u}}_2$ of $\mathbf{u}_2$ is

$$\dot{\mathbf{u}}_2 = \frac{d\theta}{dt}\frac{\partial}{\partial\theta}\mathbf{u}_2 = \dot{\theta}\frac{\partial}{\partial\theta}\mathbf{u}_{21} = \dot{\theta}\mathbf{u}_{21} \times \mathbf{p}$$

and

$$\bar{\mathbf{u}}_2 = \frac{d}{dt}(\dot{\theta}\mathbf{u}_{21} \times \mathbf{p}) = \ddot{\theta}\mathbf{u}_{21} \times \mathbf{p} - \dot{\theta}^2\mathbf{u}_{21} \tag{9.11}$$

We are using the total derivative rule that

$$\frac{d}{dt}f(\theta_1, \theta_2, \ldots, \theta_n) = \sum_{i=1}^{n} \dot{\theta}_1 \frac{\partial}{\partial\theta_i} f(\theta_1, \theta_2, \ldots, \theta_n)$$

where f is a function only of $\theta_1, \theta_2, \ldots$ and θ_n. Notice that the double cross product $(\mathbf{u}_{21} \times \mathbf{p}) \times \mathbf{p} = -\mathbf{u}_{21}$, but this is only true, since $\mathbf{u}_{21}$ and $\mathbf{p}$ are orthogonal.

Consider the motion of a particle at $\mathbf{u}_3$, which is $\mathbf{u}_{31} = \mathbf{u}_{21} + \mathbf{u}_{32}$ in the system of two pitch joints shown in Figure 9.5.

$$\begin{aligned}
\dot{\mathbf{u}}_3 &= \dot{\theta}_1\frac{\partial}{\partial\theta_1}\mathbf{u}_{31} + \dot{\theta}_2\frac{\partial}{\partial\theta_2}\mathbf{u}_{31} = \dot{\theta}_1\frac{\partial}{\partial\theta_1}\mathbf{u}_{31} + \dot{\theta}_2\frac{\partial}{\partial\theta_2}\mathbf{u}_{32} \\
&= \dot{\theta}_1\mathbf{u}_{31} \times \mathbf{p} + \dot{\theta}_2\mathbf{u}_{32} \times \mathbf{p} \\
\ddot{\mathbf{u}}_3 &= \frac{d}{dt}\dot{\mathbf{u}}_{31} = \frac{d}{dt}\{\dot{\theta}_1\mathbf{u}_{31} \times \mathbf{p} + \dot{\theta}_2\mathbf{u}_{32} \times \mathbf{p}\} \\
&= \ddot{\theta}_1\mathbf{u}_{31} \times \mathbf{p} + \ddot{\theta}_2\mathbf{u}_{32} \times \mathbf{p} + \dot{\theta}_1\frac{\partial}{\partial\theta_1}\{\dot{\theta}_1\mathbf{u}_{31} \times \mathbf{p} + \dot{\theta}_2\mathbf{u}_{32} \times \mathbf{p}\} \\
&\quad + \dot{\theta}_2\frac{\partial}{\partial\theta_2}\{\dot{\theta}_1\mathbf{u}_{31} \times \mathbf{p} + \dot{\theta}_2\mathbf{u}_{32} \times \mathbf{p}\} \\
&= \ddot{\theta}_1\mathbf{u}_{31} \times \mathbf{p} + \ddot{\theta}_2\mathbf{u}_{32} \times \mathbf{p} - \dot{\theta}_1^2\mathbf{u}_{31} - \dot{\theta}_2^2\mathbf{u}_{32} \\
&\quad + \dot{\theta}_1\dot{\theta}_2\left(\frac{\partial}{\partial\theta_1}\mathbf{u}_{32} \times \mathbf{p} + \frac{\partial}{\partial\theta_2}\mathbf{u}_{31} \times \mathbf{p}\right) \\
&= \ddot{\theta}_1\mathbf{u}_{31} \times \mathbf{p} + \ddot{\theta}_2\mathbf{u}_{32} \times \mathbf{p} - \dot{\theta}_1^2\mathbf{u}_{31} - \dot{\theta}_2^2\mathbf{u}_{32} \\
&\quad + \dot{\theta}_1\dot{\theta}_2\left(\frac{\partial}{\partial\theta_1}(\mathbf{u}_{31} - \mathbf{u}_{21}) \times \mathbf{p} + \frac{\partial}{\partial\theta_2}(\mathbf{u}_{21} + \mathbf{u}_{32}) \times \mathbf{p}\right) \\
&\qquad\qquad\qquad\qquad \dot{\theta}_1\dot{\theta}_2(-\mathbf{u}_{31} \\
&= \ddot{\theta}_1\mathbf{u}_{31} \times \mathbf{p} + \ddot{\theta}_2\mathbf{u}_{32} \times \mathbf{p} - \dot{\theta}_1^2\mathbf{u}_{31} - \dot{\theta}_2^2\mathbf{u}_{32} + \dot{\theta}_1\dot{\theta}_2(-\mathbf{u}_{31}\ \mathbf{u}_{21} - \mathbf{u}_{32}) \cdot \\
&= \ddot{\theta}_1\mathbf{u}_{31} \times \mathbf{p} + \ddot{\theta}_2\mathbf{u}_{32} \times \mathbf{p} - \dot{\theta}_1^2\mathbf{u}_{31} - \dot{\theta}_2^2\mathbf{u}_{32} - 2\dot{\theta}_1\dot{\theta}_2\mathbf{u}_{32}
\end{aligned} \tag{9.12}$$

The directions of these accelerations are shown in Figure 9.7.

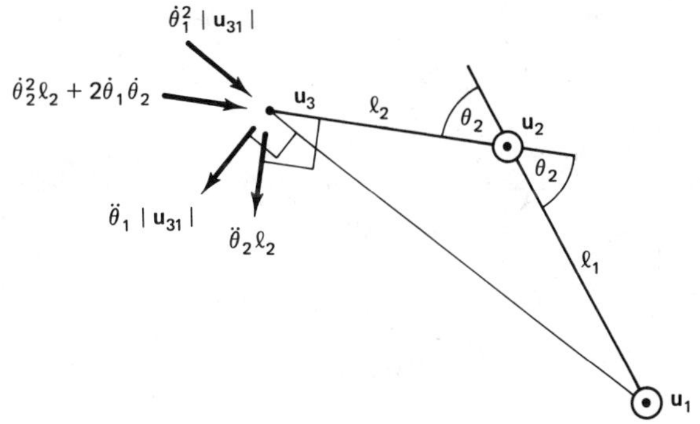

Figure 9.7 The system $| \lambda\theta |$

Consider next the system $| \lambda\theta |$ shown in Figure 9.6. The axis of rotation for λ_1 is given by $\boldsymbol{\nu}$. The particle is at $\mathbf{u}_3 = \mathbf{u}_2 + \mathbf{u}_{32}$, so

$$\dot{\mathbf{u}}_3 = \frac{d}{dt}\mathbf{u}_{31} = \frac{d}{dt}(\mathbf{u}_{21} + \mathbf{u}_{32}) = 0 + \dot{\theta}_2 \frac{\partial}{\partial \theta_2}\mathbf{u}_{32} + \dot{\lambda}_1 \frac{\partial}{\partial \lambda_1}\mathbf{u}_{32}$$

$$= \dot{\theta}_2 \mathbf{u}_{32} \times \mathbf{p} + \dot{\lambda}_1 \mathbf{u}_{32} \times \boldsymbol{\nu}$$

$$\ddot{\mathbf{u}}_3 = \frac{d}{dt}\dot{\mathbf{u}}_{31} = \frac{d}{dt}\{\dot{\theta}_2 \mathbf{u}_{32} \times \mathbf{p} + \dot{\lambda}_1 \mathbf{u}_{32} \times \boldsymbol{\nu}\}$$

$$= \ddot{\theta}_2 \mathbf{u}_{32} \times \mathbf{p} + \ddot{\lambda}_1 \mathbf{u}_{32} \times \boldsymbol{\nu} + \dot{\theta}_2 \frac{\partial}{\partial \theta_2}\{\dot{\theta}_2 \mathbf{u}_{32} \times \mathbf{p} + \dot{\lambda}_1 \mathbf{u}_{32} \times \boldsymbol{\nu}\}$$

$$+ \dot{\lambda}_1 \frac{\partial}{\partial \lambda_1}\{\dot{\theta}_2 \mathbf{u}_{32} \times \mathbf{p} + \dot{\lambda}_1 \mathbf{u}_{32} \times \boldsymbol{\nu}\}$$

$$= \ddot{\theta}_2 \mathbf{u}_{32} \times \mathbf{p} + \ddot{\lambda}_1 \mathbf{u}_{32} \times \boldsymbol{\nu} - \dot{\theta}_2^2 \mathbf{u}_{32} + \dot{\lambda}_1^2 (\mathbf{u}_{32} \times \boldsymbol{\nu}) \times \boldsymbol{\nu}$$

$$+ \dot{\theta}_2 \dot{\lambda}_1 \left\{\frac{\partial}{\partial \lambda_1}\mathbf{u}_{32} \times \mathbf{p} + \frac{\partial}{\partial \theta_2}\mathbf{u}_{32} \times \boldsymbol{\nu}\right\}$$

$$= \ddot{\theta}_2 \mathbf{u}_{32} \times \mathbf{p} + \ddot{\lambda}_1 \mathbf{u}_{32} \times \boldsymbol{\nu} - \dot{\theta}_2^2 \mathbf{u}_{32} + \dot{\lambda}_1^2 (\mathbf{u}_{32} \times \boldsymbol{\nu}) \times \boldsymbol{\nu} + \dot{\theta}_2 \dot{\lambda}_1 (\mathbf{u}_{32} \times \mathbf{p}) \times \boldsymbol{\nu} \tag{9.13}$$

since $\frac{\partial}{\partial \theta_2}\mathbf{u}_{32} \times \boldsymbol{\nu} = 0$. Notice from Figure 9.6 that $(\mathbf{u}_{32} \times \boldsymbol{\nu}) \times \boldsymbol{\nu} = -\mathbf{u}_{3a}$. The directions of the components of the acceleration are shown in Figure 9.8.

A system in which a particle at the end of a chain of links does not lie on the first plane of motion is $| \theta\lambda\theta |$ and is shown in Figure 9.9; $\mathbf{u}_4$ is not on $\mathbf{p}_1$. For this system

$$\dot{\mathbf{u}}_4 = \frac{d}{dt}\mathbf{u}_{41} = \dot{\theta}_1 \frac{\partial}{\partial \theta_1}\mathbf{u}_{41} + \dot{\lambda}_2 \frac{\partial}{\partial \lambda_2}\mathbf{u}_{41} + \dot{\theta}_3 \frac{\partial}{\partial \theta_3}\mathbf{u}_{41}$$

$$= \dot{\theta}_1 \mathbf{u}_{41} \times \mathbf{p}_1 + \dot{\lambda}_2 \mathbf{u}_{42} \times \boldsymbol{\nu} + \dot{\theta}_3 \mathbf{u}_{43} \times \mathbf{p}_2$$

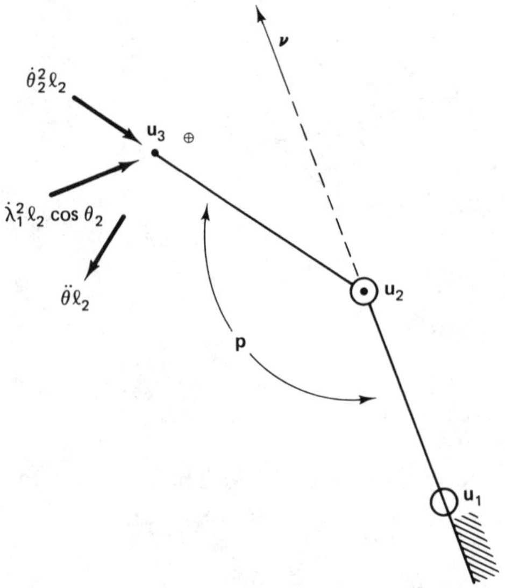

Figure 9.8 Acceleration on $\mathbf{u}_3$ for $|\,\lambda\theta\,|$

$$
\begin{aligned}
\ddot{\mathbf{u}}_4 &= \frac{d}{dt}\{\dot{\theta}_1\mathbf{u}_{41} \times \mathbf{p}_1 + \dot{\lambda}_2\mathbf{u}_{42} \times \boldsymbol{\nu} + \dot{\theta}_3\mathbf{u}_{43} \times \mathbf{p}_2\} \\
&= \ddot{\theta}_1\mathbf{u}_{41} \times \mathbf{p}_1 + \ddot{\lambda}_2\mathbf{u}_{42} \times \boldsymbol{\nu} + \ddot{\theta}_3\mathbf{u}_{43} \times \mathbf{p}_2 \\
&\quad + \dot{\theta}_1 \frac{\partial}{\partial\theta_1}\{\dot{\theta}_1\mathbf{u}_{41} \times \mathbf{p}_1 + \dot{\lambda}_2\mathbf{u}_{42} \times \boldsymbol{\nu} + \dot{\theta}_3\mathbf{u}_{43} \times \mathbf{p}_2\} \\
&\quad + \dot{\lambda}_2 \frac{\partial}{\partial\lambda_2}\{\dot{\theta}_1\mathbf{u}_{41} \times \mathbf{p}_1 + \dot{\lambda}_2\mathbf{u}_{42} \times \boldsymbol{\nu} + \dot{\theta}_3\mathbf{u}_{43} \times \mathbf{p}_2\} \\
&\quad + \dot{\theta}_3 \frac{\partial}{\partial\theta_3}\{\dot{\theta}_1\mathbf{u}_{41} \times \mathbf{p}_1 + \dot{\lambda}_2\mathbf{u}_{42} \times \boldsymbol{\nu} + \dot{\theta}_3\mathbf{u}_{43} \times \mathbf{p}_2\} \\
&= \ddot{\theta}_1\mathbf{u}_{41} \times \mathbf{p}_1 + \ddot{\lambda}_2\mathbf{u}_{42} \times \boldsymbol{\nu} + \ddot{\theta}_3\mathbf{u}_{43} \times \mathbf{p}_2 + \dot{\theta}_1^2(\mathbf{u}_{41} \times \mathbf{p}_1) \times \mathbf{p}_1 \\
&\quad + \dot{\lambda}_2^2(\mathbf{u}_{42} \times \boldsymbol{\nu}) \times \boldsymbol{\nu} - \dot{\theta}_3^2\mathbf{u}_{43} \\
&\quad + \dot{\theta}_1 \frac{\partial}{\partial\theta_1}\{\dot{\lambda}_2\mathbf{u}_{42} \times \boldsymbol{\nu} + \dot{\theta}_3\mathbf{u}_{43} \times \mathbf{p}_2\} + \dot{\lambda}_2 \frac{\partial}{\partial\lambda_2}\{\dot{\theta}_1\mathbf{u}_{41} \times \mathbf{p}_1 + \dot{\theta}_3\mathbf{u}_{43} \times \mathbf{p}_2\} \\
&\quad + \dot{\theta}_3 \frac{\partial}{\partial\theta_3}\{\dot{\theta}_1\mathbf{u}_{41} \times \mathbf{p}_1 + \dot{\lambda}_2\mathbf{u}_{42} \times \boldsymbol{\nu}\} \\
&= \ddot{\theta}_1\mathbf{u}_{41} \times \mathbf{p}_1 + \ddot{\lambda}_2\mathbf{u}_{42} \times \boldsymbol{\nu} + \ddot{\theta}_3\mathbf{u}_{43} \times \mathbf{p}_2 + \dot{\theta}_1^2(\mathbf{u}_{41} \times \mathbf{p}_1) \times \mathbf{p}_1 \\
&\quad + \dot{\lambda}_2^2(\mathbf{u}_{42} \times \boldsymbol{\nu}) \times \boldsymbol{\nu} - \dot{\theta}_3^2\mathbf{u}_{43} \\
&\quad + \dot{\theta}_1\dot{\theta}_3\{(\mathbf{u}_{43} \times \mathbf{p}_2) \times \mathbf{p}_1 + (\mathbf{u}_{41} \times \mathbf{p}_1) \times \mathbf{p}_2\} \\
&\quad + \dot{\lambda}_2\{\dot{\theta}_1\mathbf{u}_{41} \times \mathbf{p}_1 \times \boldsymbol{\nu} + \dot{\theta}_3(\mathbf{u}_{43} \times \mathbf{p}_2) \times \boldsymbol{\nu}\} \qquad (9.14)
\end{aligned}
$$

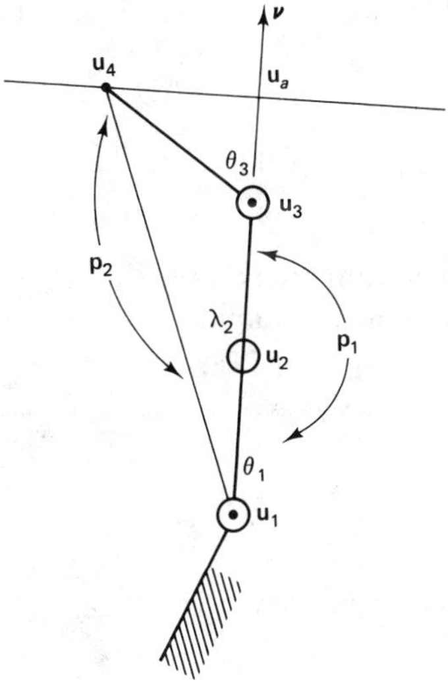

Figure 9.9 The system $|\ \theta\lambda\theta\ |$

For the 2 *dof* system $|\ \theta\sigma\ |$ shown in Figure 9.10,

$$\dot{\mathbf{u}}_2 = \frac{d}{dt}\mathbf{u}_{21} = \dot{\theta}_1 \frac{\partial}{\partial\theta_1}\mathbf{u}_{21} + \dot{\mathbf{u}}_{21} = \dot{\theta}_1\mathbf{u}_{21} \times \mathbf{p} + \dot{\mathbf{u}}_{21}$$

where $\dot{\mathbf{u}}_{21}$ is the rate of change of the length of the sliding link with respect to time.

$$\ddot{\mathbf{u}}_2 = \frac{d}{dt}\{\dot{\theta}_1\mathbf{u}_{21} \times \mathbf{p} + \dot{\mathbf{u}}_{21}\} = \frac{d}{dt}\{\dot{\theta}_1\mathbf{u}_{21} \times \mathbf{p}\} + \ddot{\mathbf{u}}_{21}$$

$$= \ddot{\theta}_1\mathbf{u}_{21} \times \mathbf{p} + \dot{\theta}_1^2 \frac{\partial}{\partial\theta_1}\mathbf{u}_{21} \times \mathbf{p} + \dot{\theta}_1\mathbf{u}_{21}\frac{\partial}{\partial\sigma_2}\mathbf{u}_{21} \times \mathbf{p} + \ddot{\mathbf{u}}_{21}$$

$$= \ddot{\theta}_1\mathbf{u}_{21} \times \mathbf{p} - \dot{\theta}_1^2\mathbf{u}_{21} + \dot{\theta}_1\dot{\mathbf{u}}_{21} \times \mathbf{p} + \ddot{\mathbf{u}}_{21} \tag{9.15}$$

The determination of the acceleration of a particle at the end of a kinematic chain can be codified into a simple methodology. Consider an n

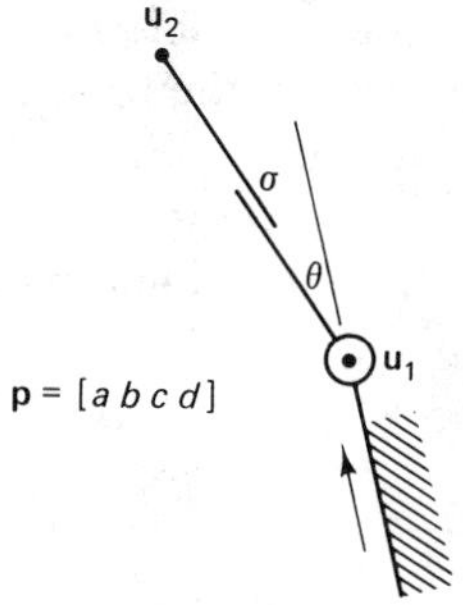

Figure 9.10 The system $|\ \theta\sigma\ |$

dof kinematic chain of revolute joints in which the axes of rotation are $\mathbf{\Omega}_i$, $i = 1, \ldots, n$. For example, with the 6 *dof* orientable robot

$$\begin{array}{c} \bar{\mathbf{p}}_2^2 \qquad\qquad \bar{\mathbf{p}}_3^1 \qquad \bar{\mathbf{p}}_4^0 \\ |\,\lambda_1 l_1\,|\,\theta_2 l_2\,|\,\theta_3 l_3\,|\,\lambda_4 l_4\,|\,\theta_5 l_5\,|\,\lambda_6 l_6\,| \\ \mathbf{u}_1 \quad \mathbf{u}_2 \quad \mathbf{u}_3 \quad \mathbf{u}_4 \quad \mathbf{u}_5 \quad \mathbf{u}_6 \quad \mathbf{u}_7 \end{array} \tag{9.16}$$

the axes of rotation are $\mathbf{\Omega}_1 = \mathbf{u}_{21}/l_1$, $\mathbf{\Omega}_2 = \mathbf{\Omega}_3 = \bar{\mathbf{p}}_2^2$, $\mathbf{\Omega}_4 = \mathbf{u}_{54}/l_4$, $\mathbf{\Omega}_5 = \bar{\mathbf{p}}_3^1$, and $\mathbf{\Omega}_6 = \mathbf{u}_{76}/l_6$. Further, suppose the set of joints is Θ_i, $i = 1, \ldots, n$; then

$$\dot{\mathbf{u}}_k = \sum_{i=1}^{k-1} \dot{\Theta}_i \frac{\partial}{\partial \theta_i} \mathbf{u}_{k1} = \sum_{i=1}^{k-1} \dot{\Theta}_i \frac{\partial}{\partial \theta_i} \mathbf{u}_{ki} = \sum_{i=1}^{k-1} \dot{\Theta}_i \mathbf{u}_{ki} \times \mathbf{\Omega}_i$$

and

$$\begin{aligned} \ddot{\mathbf{u}}_k &= \sum_{i=1}^{k-1} \ddot{\Theta}_1 \mathbf{u}_{ki} \times \mathbf{\Omega}_i + \sum_{j=1}^{k-1} \frac{\partial}{\partial \Theta_j} \left\{ \sum_{i=1}^{k-1} \dot{\Theta}_i \mathbf{u}_{ki} \times \mathbf{\Omega}_i \right\} \\ &= \sum_{i=1}^{k-1} \ddot{\Theta}_i \mathbf{u}_{ki} \times \mathbf{\Omega}_i + \sum_{j=1}^{k-1} \frac{\partial}{\partial \Theta_j} \left\{ \sum_{i=1}^{k-1} \dot{\Theta}_i (\mathbf{u}_{kj} - \mathbf{u}_{ij}) \times \mathbf{\Omega}_i \right\} \end{aligned}$$

In this expression we notice that $\dfrac{\partial}{\partial \Theta_j} \mathbf{u}_{ij} = 0$ if $i \le j$, so

$$\begin{aligned} \ddot{\mathbf{u}}_k &= \sum_{i=1}^{k-1} \ddot{\Theta}_i \mathbf{u}_{ki} \times \mathbf{\Omega}_i + \sum_{j=1}^{k-1} \frac{\partial}{\partial \Theta_j} \left\{ \sum_{i=1}^{k-1} \dot{\Theta}_i \mathbf{u}_{kj} \times \mathbf{\Omega}_i \right\} - \sum_{j=1}^{k-2} \frac{\partial}{\partial \Theta_j} \left\{ \sum_{i=j+1}^{k-1} \dot{\Theta}_i \mathbf{u}_{ij} \times \mathbf{\Omega}_i \right\} \\ &= \sum_{i=1}^{k-1} \ddot{\Theta}_i \mathbf{u}_{ki} \times \mathbf{\Omega}_i + \sum_{j=1}^{k-1} \dot{\Theta}_j \left\{ \sum_{i=1}^{k-1} \dot{\Theta}_i \mathbf{u}_{kj} \times \mathbf{\Omega}_i \right\} \times \mathbf{\Omega}_j \\ &\quad - \sum_{j=1}^{k-2} \dot{\Theta}_j \left\{ \sum_{i=j+1}^{k-1} \dot{\Theta}_i \mathbf{u}_{ij} \times \mathbf{\Omega}_i \right\} \times \mathbf{\Omega}_j \end{aligned} \tag{9.17}$$

$k = 2, \ldots, n + 1$. The units of $\ddot{\mathbf{u}}_k$ are m/s^2 or ft/s^2. Equation (9.17) holds for any robot that can be modeled by kinematically simple revolute joints, and the revolute joints of all commercial robots are kinematically simple. Thus, the acceleration of a particle of mass $\overline{M}_k$ at point $\bar{\mathbf{u}}_{k+1}$ can be determined for every point mass of the robot. We are now ready to calculate the forces on these point masses.

If one or more sliding links occur in a chain of revolute joints, equation (9.17) still holds, with the following modifications. For sliding link $\mathbf{u}_{l+1,l}$ add terms

$$\ddot{\mathbf{u}}_{l+1,l} + \sum_{i=1}^{l} \dot{\Theta}_1 \dot{\mathbf{u}}_{l+1,l} \times \mathbf{\Omega}_1$$

to the acceleration $\ddot{\mathbf{u}}_k$.

9.4 FORCES AND TORQUES

Newton's second law states that the force f required to accelerate a body of mass M by $\ddot{\mathbf{u}}$ m/s^2 is given by $f = M\ddot{\mathbf{u}}$. In the prior section we determined the acceleration $\bar{\mathbf{u}}$ of a particle in a kinematic chain at point $\mathbf{u}$ as a vector in Cartesian space. If the particle has mass M, the force $\mathbf{f}$ that must be applied to this mass is given by

$$\mathbf{f} = M\ddot{\mathbf{u}} + Mg\begin{bmatrix} 0 \\ 0 \\ 1 \end{bmatrix} \quad \text{newtons} \tag{9.18}$$

An n *dof* robot with point loads and distributed loads can be modeled by an unloaded kinematic chain with m point masses at known points $\bar{\mathbf{u}}_1$; it is seen in Section 9.6 that provided the links of the robot can be classified as slender (the length of the link much greater than its maximum cross-sectional dimension—in practice a ratio greater than 5 is usually good enough to classify a member as slender), then any n *dof* robot can be modeled with n point loads M_k at the joints and n point loads m_k elsewhere on the each link. The acceleration of each mass can be found from equation (9.17), so the force on mass m_k is given by

$$\mathbf{f}_{mk} = m_k\left\{\sum_{i=1}^{k-1}\ddot{\Theta}_i\bar{\mathbf{u}}_{ki}\times\boldsymbol{\Omega}_i + \sum_{j=1}^{k-1}\dot{\Theta}_j\left\{\sum_{i=1}^{k-1}\dot{\Theta}_i\bar{\mathbf{u}}_{kj}\times\boldsymbol{\Omega}_i\right\}\times\boldsymbol{\Omega}_j\right.$$
$$\left. - \sum_{j=1}^{k-2}\dot{\Theta}_j\left\{\sum_{i=j+1}^{k-1}\dot{\Theta}_i\mathbf{u}_{ij}\times\boldsymbol{\Omega}_i\right\}\times\boldsymbol{\Omega}_j + \begin{bmatrix} 0 \\ 0 \\ g \end{bmatrix}\right\} \tag{9.19}$$

where $\mathbf{u}_{k-1} \le \bar{\mathbf{u}}_{ki} \le \mathbf{u}_k$, $k = 2, \ldots, n+1$. Some of these $2n$ vector forces $\mathbf{f}_{Mk}$ or $\mathbf{f}_{mk}$ will be zero. For example, in the 6 *dof* orientable robot given by expression (9.16), the point masses are m_1, M_2, . . . , m_6, and M_7, but the forces on m_1 and M_2 do no work, so we will not calculate the vector force acting on them. Further, some point masses can be combined depending on the configuration of the robot, so the number of masses to be considered in determining the torques at the joints can be small.

The component of the vector force $\mathbf{f}_{mk}$ at $\bar{\mathbf{u}}_k$ that produces torque at $\mathbf{u}_i$ whose axis is $\boldsymbol{\Omega}_i$ is given by $\mathbf{f}_{mk} \cdot \bar{\mathbf{u}}_{ki} \times \boldsymbol{\Omega}_i / |\bar{\mathbf{u}}_{ki}|$, and this force produces torque $\mathbf{f}_{mk} \cdot \bar{\mathbf{u}}_{ki} \times \boldsymbol{\Omega}_i$. Recall, vector cross-multiplication is associative, so $\mathbf{f}_{mk} \cdot \bar{\mathbf{u}}_{ki} \times \boldsymbol{\Omega}_i \equiv \mathbf{f}_{mk} \cdot (\bar{\mathbf{u}}_{ki} \times \boldsymbol{\Omega}_i)$. The total torque produced at $\mathbf{u}_i$ is

$$\tau_i = \sum_{k=i+1}^{n+1}\mathbf{f}_{mk}\cdot\bar{\mathbf{u}}_{ki}\times\boldsymbol{\Omega}_i \quad \text{newton-meters} \tag{9.20}$$

We have applied d'Alembert's principle [Greenwood 1965], which is a slightly different form of Newton's second law. d'Alembert's principle states that the external forces balance the internal forces in a system. τ_i is the only external force that balances moments occurring at joint i due to forces on distal masses. We treat the forces in this context as static forces. Notice that the torques at distal joints produce no moment at joint i and so are ignored; joints $i + 1, i + 2, \ldots, n$ are frozen as far as the torque at joint i is concerned.

The methodology for determining torques in this point mass model is as follows:

1. Determine the acceleration of each point mass m_k and M_k.
2. Determine the total vector force $\mathbf{f}_{mk}$ and $\mathbf{f}_{Mk}$ on each m_k and M_k, respectively.
3. Sum the components of $\mathbf{f}_{mk}$ and $\mathbf{f}_{Mk}$ that contribute to torque at $\mathbf{u}_i$.

9.5 THE EFFECT OF DISTRIBUTED MASSES

In the prior section we presented two equations that enable the dynamic analysis of any unloaded kinematic chains with point loads at specified locations to be performed. However, actual robots have distributed loads as well as loads that can, with accuracy, be considered as point loads. Here we will consider the effect of distributed loads in a kinematic chain.

Suppose we have an incremental mass Δm (an infinitesimal segment $\Delta\chi$ long on a slender link of length l and total mass m) in a loaded kinematic chain. A slender link has a small cross section, so we can neglect inertial effects on any cross-sectional axis. Substituting this incremental mass in equation (9.19) and inserting this force in equation (9.20) results in an incremental torque $\Delta\tau_i$. To find the total torque on τ_i requires the integration of the incremental torques as

$$\int_0^l \Delta\tau_i \, d\chi$$

Since $\Delta\tau_i$ has terms in χ^0, χ, and χ^2, the integration produces terms in χ, $\chi^2/2$, and $\chi^3/3$. These terms are intermingled and difficult to simplify and codify. We illustrate with an example. Consider the 3 *dof* revolute robot, unloaded apart from a uniform distributed mass m on link l_3. Further, we omit the gravity terms for simplicity. The torque at $\mathbf{u}_1$ due to m is

$$\tau_1 = \frac{m}{l_3}\int_0^{l_3} \left(\ddot{\lambda}_1\left(\mathbf{u}_{31} + \frac{\mathbf{u}_{43}}{l_3}\chi\right) \times \boldsymbol{\nu} + \ddot{\theta}_2\left(\mathbf{u}_{32} + \frac{\mathbf{u}_{43}}{l_3}\chi\right) \times \mathbf{p} + \ddot{\theta}_3 \frac{\mathbf{u}_{43}}{l_3}\chi \times \mathbf{p}\right.$$

$$+ \dot{\lambda}_1^2\left\{\left(\mathbf{u}_{31} + \frac{\mathbf{u}_{43}}{l_3}\chi\right) \times \boldsymbol{\nu}\right\} \times \boldsymbol{\nu} - \dot{\theta}_2^2\left(\mathbf{u}_{32} + \frac{\mathbf{u}_{43}}{l_3}\chi\right)$$

$$- \dot{\theta}_3^2 \frac{\mathbf{u}_{43}}{l_3} \chi - 2\dot{\theta}_3\dot{\theta}_2 \frac{\mathbf{u}_{43}}{l_3} \chi - \dot{\lambda}_1(\dot{\theta}_2 + \dot{\theta}_3)\left\{\left(\mathbf{u}_{31} + \frac{\mathbf{u}_{43}}{l_3}\chi\right) \times \mathbf{p}\right\} \times \boldsymbol{\nu}\Big)$$

$$\cdot \left(\mathbf{u}_{31} + \frac{\mathbf{u}_{43}}{l_3}\chi\right) \times \boldsymbol{\nu}\, d\chi$$

$$= m \Big[\{\ddot{\lambda}_1\mathbf{u}_{31} \times \boldsymbol{\nu} + \ddot{\theta}_2\mathbf{u}_{32} \times \mathbf{p} + \dot{\lambda}_1^2(\mathbf{u}_{31} \times \boldsymbol{\nu}) \times \boldsymbol{\nu} - \dot{\theta}_2^2\mathbf{u}_{32}$$

$$- \dot{\lambda}_1(\dot{\theta}_2 + \dot{\theta}_3)\mathbf{u}_{31} \times \mathbf{p} \times \boldsymbol{\nu}\} \cdot \mathbf{u}_{31} \times \boldsymbol{\nu} + \frac{1}{2}\{\ddot{\lambda}_1\mathbf{u}_{31} \times \boldsymbol{\nu} + \ddot{\theta}_2\mathbf{u}_{32} \times \mathbf{p}$$

$$+ \dot{\lambda}_1^2(\mathbf{u}_{31} \times \boldsymbol{\nu}) \times \boldsymbol{\nu} - \dot{\theta}_2^2\mathbf{u}_{32} - \dot{\lambda}_1(\dot{\theta}_2 + \dot{\theta}_3)\mathbf{u}_{31} \times \mathbf{p} \times \boldsymbol{\nu}\} \cdot \mathbf{u}_{43} \times \boldsymbol{\nu}$$

$$+ \frac{1}{2}\{\ddot{\lambda}_1\mathbf{u}_{43} \times \boldsymbol{\nu} + \ddot{\theta}_2\mathbf{u}_{43} \times \mathbf{p} + \ddot{\theta}_3\mathbf{u}_{43} \times \mathbf{p} + \dot{\lambda}_1^2(\mathbf{u}_{43} \times \boldsymbol{\nu}) \times \boldsymbol{\nu} - \dot{\theta}_2^2\mathbf{u}_{43}$$

$$- \dot{\theta}_3^2\mathbf{u}_{43} - 2\dot{\theta}_2\dot{\theta}_3\mathbf{u}_{43} - \dot{\lambda}_1(\dot{\theta}_2 + \dot{\theta}_3)\mathbf{u}_{43} \times \mathbf{p} \times \boldsymbol{\nu}\} \cdot \mathbf{u}_{31} \times \boldsymbol{\nu}$$

$$+ \frac{1}{3}\{\ddot{\lambda}_1\mathbf{u}_{43} \times \boldsymbol{\nu} + \ddot{\theta}_2\mathbf{u}_{43} \times \mathbf{p} + \ddot{\theta}_3\mathbf{u}_{43} \times \mathbf{p} + \dot{\lambda}_1^2(\mathbf{u}_{43} \times \boldsymbol{\nu}) \times \boldsymbol{\nu} - \dot{\theta}_2^2\mathbf{u}_{43}$$

$$- \dot{\theta}_3^2\mathbf{u}_{43} - 2\dot{\theta}_2\dot{\theta}_3\mathbf{u}_{43} - \dot{\lambda}_1(\dot{\theta}_2 + \dot{\theta}_3)\mathbf{u}_{43} \times \mathbf{p} \times \boldsymbol{\nu}\} \cdot \mathbf{u}_{43} \times \boldsymbol{\nu}\Big]$$

$$= m \Big[\ddot{\lambda}_1 \Big(\mathbf{u}_{31} \times \boldsymbol{\nu} \cdot \mathbf{u}_{31} \times \boldsymbol{\nu} + \mathbf{u}_{31} \times \boldsymbol{\nu} \cdot \mathbf{u}_{43} \times \boldsymbol{\nu} + \frac{1}{3}\mathbf{u}_{43} \times \boldsymbol{\nu} \cdot \mathbf{u}_{43} \times \boldsymbol{\nu}\Big)$$

$$- \dot{\lambda}_1(\dot{\theta}_2 + \dot{\theta}_3)\Big\{\mathbf{u}_{31} \times \mathbf{p} \times \boldsymbol{\nu} \cdot \mathbf{u}_{31} \times \boldsymbol{\nu} + \frac{1}{2}\mathbf{u}_{31} \times \mathbf{p} \times \boldsymbol{\nu} \cdot \mathbf{u}_{43} \times \boldsymbol{\nu}$$

$$+ \frac{1}{2}\mathbf{u}_{43} \times \mathbf{p} \times \boldsymbol{\nu} \cdot \mathbf{u}_{31} \times \boldsymbol{\nu} + \frac{1}{3}\mathbf{u}_{43} \times \mathbf{p} \times \boldsymbol{\nu} \cdot \mathbf{u}_{43} \times \boldsymbol{\nu}\Big\}\Big]$$

$$= m \Big[\ddot{\lambda}_1 \Big(\mathbf{u}_{41} \times \boldsymbol{\nu} \cdot \mathbf{u}_{31} \times \boldsymbol{\nu} + \frac{1}{3}\mathbf{u}_{43} \times \boldsymbol{\nu} \cdot \mathbf{u}_{43} \times \boldsymbol{\nu}\Big)$$

$$- \dot{\lambda}_1(\dot{\theta}_2 + \dot{\theta}_3)\Big\{\Big(\mathbf{u}_{31} + \frac{1}{2}\mathbf{u}_{43}\Big) \times \mathbf{p} \times \boldsymbol{\nu} \cdot \mathbf{u}_{31} \times \boldsymbol{\nu}$$

$$+ \Big(\frac{1}{2}\mathbf{u}_{31} + \frac{1}{3}\mathbf{u}_{43}\Big) \times \mathbf{p} \times \boldsymbol{\nu} \cdot \mathbf{u}_{43} \times \boldsymbol{\nu}\Big\}\Big] \tag{9.21}$$

The form of the terms in $\ddot{\lambda}_1$ are fairly simple to codify, but a general methodology for determining the form of the cross (Coriolis) terms is not clear. Thus, we see how to calculate the torques when distributed loads are present but not how to simplify the procedure. As a result, systems containing distributed loads require calculations that couple equations (9.19) and (9.20) with a considerable expansion in computation. It is better to model the system with point loads, find the total vector force on each mass, and then apply equation (9.20) to find the torque at each joint rather than to attempt the preceding calculations. This is considered in the next section.

9.6 POINT LOAD APPROXIMATIONS TO DISTRIBUTED LOAD SYSTEMS

Consider a pitch joint at $\mathbf{u}_1$, to which uniformly distributed, loaded link l is attached, where the load per unit length is m/l. The end of the link is at $\mathbf{u}_2$. The torque at the joint is derived from equations (9.11) and (9.20) as

$$\tau = \int_0^l \frac{m}{l}\left(\ddot{\theta}\mathbf{u}_{21} \times \mathbf{p}\,\frac{\chi}{l} - \dot{\theta}^2\mathbf{u}_{21}\,\frac{\chi}{l} + \begin{bmatrix} 0 \\ 0 \\ g \end{bmatrix}\right) \cdot \left(\mathbf{u}_{21} \times \mathbf{p}\,\frac{\chi}{l}\right) d\chi$$

$$= \frac{m}{l}\int_0^l \left\{\ddot{\theta}(\mathbf{u}_{21} \times \mathbf{p}) \cdot (\mathbf{u}_{21} \times \mathbf{p})\,\frac{\chi^2}{l^2} + \begin{bmatrix} 0 \\ 0 \\ g \end{bmatrix} \cdot (\mathbf{u}_{21} \times \mathbf{p})\,\frac{\chi}{l}\right\} d\chi$$

$$= \frac{m\ddot{\theta}}{3}(\mathbf{u}_{21} \times \mathbf{p}) \cdot (\mathbf{u}_{21} \times \mathbf{p}) + \frac{m}{2}\begin{bmatrix} 0 \\ 0 \\ g \end{bmatrix}(\mathbf{u}_{21} \times \mathbf{p})$$

$$= \frac{m\ddot{\theta}l^2}{3} + \frac{m}{2}\begin{bmatrix} 0 \\ 0 \\ g \end{bmatrix}(\mathbf{u}_{21} \times \mathbf{p})$$

where $I = ml^2/3$ is the inertia of the link with respect to the joint. Thus, as far as this single-pitch joint system is concerned, we can replace the uniformly distributed load m by a point load m at $l/\sqrt{3}$ (point $l/\sqrt{3}$ is known as the *center of gyration*) as far as the term in $\ddot{\theta}$ is concerned and by a point load m at the halfway point along the link as far as the gravitational forces are concerned; this point is the center of gravity. In fact, any nonslender, nonuniform distributed load can be replaced by a single point mass at the center of gyration and a mass at the center of gravity as far as the torque at a single-pitch joint is concerned.

Suppose a three-dimensional link is connected to a pitch joint and the incremental particle of mass is given by $\Delta m(\alpha, \beta, \gamma)$, where α is measured along the length l of the link, β is measured along its width, and γ is measured along its depth. The length is considered in the direction given by the two joints connected by the link l, where an end effector is also considered a joint—that is, the length is along the link axis. The width is considered in the plane of motion of the link axis and orthogonal to the link axis. The depth is in the direction given by $\mathbf{p}$, the plane of motion of the link. The joint is at $\alpha = \beta = \gamma = 0$. The total mass of the body is given by

$$\int_{\gamma_1}^{\gamma_2}\int_{\beta_1}^{\beta_2}\int_{\alpha_1}^{\alpha_2} \Delta m(\alpha, \beta, \gamma)\, d\alpha\, d\beta\, d\gamma = m$$

where the limits of integration are the extremities of the body and the order of integration is immaterial. As an example, consider a rectanguloid of total mass m and with uniform density and dimensions $0 \le \alpha \le \alpha_2$, $0 \le \beta \le \beta_2$ and $0 \le \gamma \le \gamma_2$; then

$$\Delta m(\alpha, \beta, \gamma) = \frac{m}{\alpha_2\beta_2\gamma_2}, \quad \text{so} \quad \int_0^{\gamma_2}\int_0^{\beta_2}\int_0^{\alpha_2} \frac{m}{\alpha_2\beta_2\gamma_2}\, d\alpha\, d\beta\, d\gamma = m$$

For a joint at $\mathbf{u}_i$, the center of gravity of a three-dimensional body forming link l_i is at

$$\frac{1}{m}\int_{\alpha_1}^{\alpha_2}\left\{\int_{\beta_1}^{\beta_2}\int_{\gamma_1}^{\gamma_2} \Delta m(\alpha, \beta, \gamma)\beta\gamma\, d\beta\, d\gamma\right\}\alpha\, d\alpha = l_{g\alpha} \text{ along the link axis,}$$

$$\frac{1}{m}\int_{\beta_1}^{\beta_2}\left\{\int_{\alpha_1}^{\alpha_2}\int_{\gamma_1}^{\gamma_2} \Delta m(\alpha, \beta, \gamma)\alpha\gamma\, d\beta\right\}\beta\, d\beta = l_{g\beta} \text{ radial to the link axis and in}$$
the plane,

$$\frac{1}{m}\int_{\gamma_1}^{\gamma_2}\left\{\int_{\alpha_1}^{\alpha_2}\int_{\beta_1}^{\beta_2} \Delta m(\alpha, \beta, \gamma)\alpha\beta\, d\beta\right\}\gamma\, d\gamma = l_{g\gamma} \text{ in the direction of the vector } \mathbf{p}.$$

The distance of the center of gravity from the joint is $\sqrt{l_{g\alpha}^2 + l_{g\beta}^2 + l_{g\gamma}^2}$. If the link is slender, $l_{g\beta} = l_{g\gamma} = 0$, and the center of gravity is at $l_{g\alpha}$. Inertia I is defined as the sum of the product of all mass particles and the distance from the axis of rotation squared, so

$$I = \int_{\gamma_1}^{\gamma_2}\int_{\beta_1}^{\beta_2}\int_{\alpha_1}^{\alpha_2} \Delta m(\alpha, \beta, \gamma)(\alpha^2 + \beta^2 + \gamma^2)\, d\alpha\, d\beta\, d\gamma = ml_1^2$$

where l_1 is the distance of the center of gyration from the joint.

The effect of the point masses at two different locations (center of gravity and the center of gyration) is unfortunate but can be avoided with the two-point mass model shown in Figure 9.11. Since we require equivalence, we have $m_a + m_b = m$. Since the centers of gravity of the two systems are

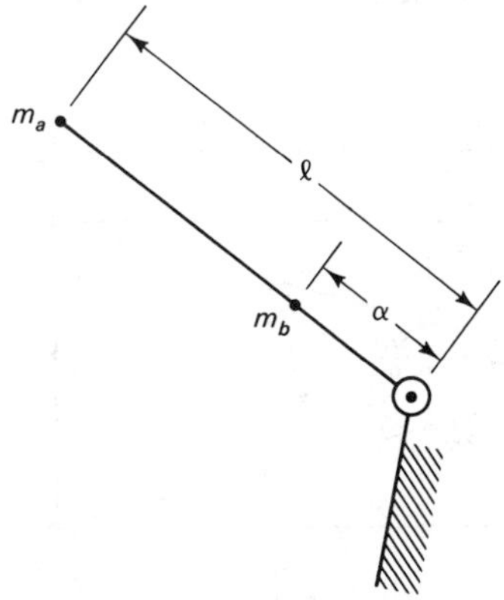

Figure 9.11 A point mass approximation to a distributed load

the same, then

$$m_a l + m_b \alpha = \frac{m}{2}$$

Since the inertia is required to be the same, we have

$$m_a l^2 + m_b \alpha^2 = \frac{m}{3} l^2$$

The solution of these three simultaneous equations gives

$$m_a = \tfrac{1}{4}, \qquad m_b = \tfrac{3}{4}, \quad \text{and} \quad \alpha = \tfrac{1}{3}$$

Any slender member can be modeled in this manner. For example, suppose the link of length l has incremental mass $(2m/l)(l - \chi/l)$, so its total mass is

$$\int_0^l \frac{2m}{l}\left(l - \frac{\chi}{l}\right) d\chi = m$$

To match centers of gravity we have

$$m_a l + m_b \alpha = \int_0^l \frac{2m}{l}\left(l - \frac{\chi}{l}\right) \chi \, d\chi = \frac{m}{3} l$$

To match inertias we have

$$m_a l^2 + m_b \alpha^2 = \int_0^l \frac{2m}{l}\left(l - \frac{\chi}{l}\right) \chi^2 \, d\chi = \frac{m}{6} l^2$$

The solution of these equations is $m_a = m/9$, $m_b = 8m/9$, and $\alpha = l/4$. This two-mass model works extremely well for slender links in which the length to maximum width is large. When this is not the case, we must model with more than two masses. For example, suppose the link of mass m has a flat rectangular cross section, as shown in Figure 9.12; then a good four-point mass model becomes as shown. If it is a rectanguloid, an eight-point mass model is used. If the triangular member previously analyzed has the dimensions shown in Figure 9.13, the inertia is found by integrating the incremen-

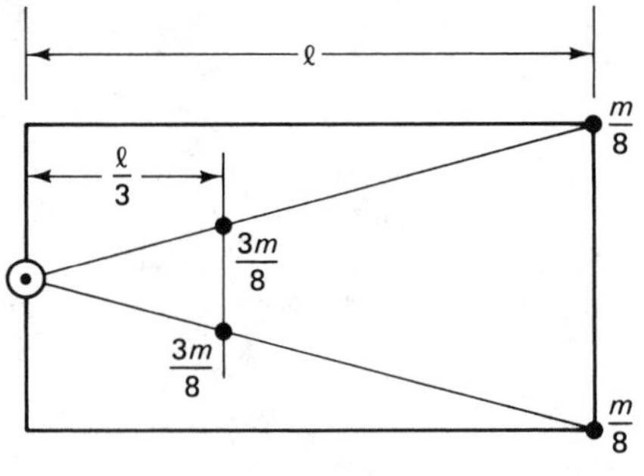

Figure 9.12 A point mass model for a rectangular link

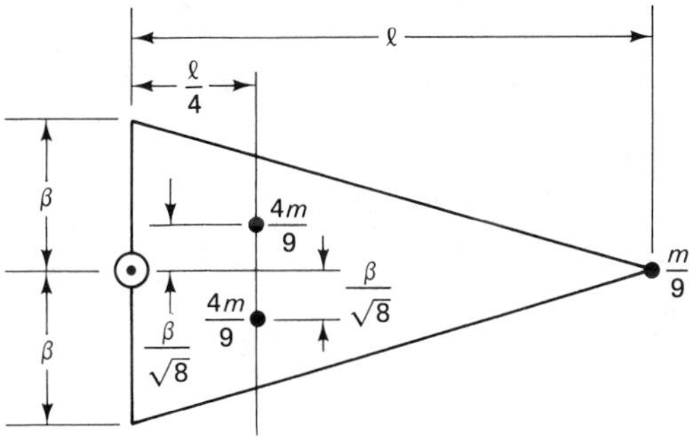

Figure 9.13 A point mass model of a triangular member

tal inertias as

$$I = 2\int_0^{\beta} \frac{2m}{\beta^2}\,\chi\left(\frac{\chi}{2}\right)^2 d\chi = \frac{1}{4}\,m\beta^2$$

The equivalent point mass model would place masses at $\pm\beta/\sqrt{8}$ on either side of the axis, as shown in Figure 9.13.

It should be emphasized that the point mass models presented are exact. That is, no approximation is involved.

9.7 DYNAMIC ANALYSIS OF POINT-MASS ROBOTS

We saw in the prior sections that point-mass models are exact for distributed load systems. Here we use these point-mass models in the dynamic analysis of typical robots. For the 3 *dof* revolute robot $|\,\lambda\theta\theta\,|$ shown in Figure 9.14, we assume uniformly distributed loads of m_i on l_i and point loads M_j at $\underline{\mathbf{u}}_j$, $i = 1, 2, 3$ and $j = 2, 3, 4$. The equivalent point masses in the model are $\overline{M}_{i+1} = m_i/4 + M_{i+1}$ at $\mathbf{u}_i$ and $3m_i/4$ at $\mathbf{u}_i + (\mathbf{u}_{i+1} - \mathbf{u}_i)/3$, $i = 1, 2, 3$. Only four of these point masses do work, so $3m_1/4$ and $\overline{M}_2 = M_2 + M_1/4$, which do no work, are ignored. Applying equation (9.19),

$$\mathbf{f}_{M4} = \overline{M}_4 \left\{ \sum_{i=1}^{3} \ddot{\Theta}_i \mathbf{u}_{4i} \times \boldsymbol{\Omega}_i + \sum_{j=1}^{3} \dot{\Theta}_j \left\{ \sum_{i=1}^{3} \dot{\Theta}_i \mathbf{u}_{4j} \times \boldsymbol{\Omega}_i \right\} \times \boldsymbol{\Omega}_j \right.$$

$$\left. - \sum_{j=1} \dot{\Theta}_j \left\{ \sum_{i=j+1}^{3} \dot{\Theta}_i \mathbf{u}_{ij} \times \boldsymbol{\Omega}_i \right\} \times \boldsymbol{\Omega}_j + \begin{bmatrix} 0 \\ 0 \\ g \end{bmatrix} \right\}$$

In this equation we have $\boldsymbol{\Omega}_1 = \boldsymbol{\nu}$, $\boldsymbol{\Omega}_2 = \boldsymbol{\Omega}_3 = \mathbf{p}$, $\Theta_1 = \lambda_1$, $\Theta_2 = \theta_2$, and $\Theta_3 = \theta_3$.

$$\mathbf{f}_{M4} = \overline{M}_4 \Bigg\{ \ddot{\lambda}_1 \mathbf{u}_{41} \times \boldsymbol{\nu} + \ddot{\theta}_2 \mathbf{u}_{42} \times \mathbf{p} + \ddot{\theta}_3 \mathbf{u}_{43} \times \mathbf{p}$$

$$+ \dot{\lambda}_1\{\dot{\lambda}_1 \mathbf{u}_{41} \times \boldsymbol{\nu} + (\dot{\theta}_2 + \dot{\theta}_3)\mathbf{u}_{41} \times \mathbf{p}\} \times \boldsymbol{\nu}$$

$$+ \dot{\theta}_2\{\dot{\lambda}_1 \mathbf{u}_{42} \times \boldsymbol{\nu} + (\dot{\theta}_2 + \dot{\theta}_3)\mathbf{u}_{42} \times \mathbf{p}\} \times \mathbf{p}$$

$$+ \dot{\theta}_3\{\dot{\lambda}_1 \mathbf{u}_{43} \times \boldsymbol{\nu} + (\dot{\theta}_2 + \dot{\theta}_3)\mathbf{u}_{43} \times \mathbf{p}\} \times \mathbf{p}$$

$$- \dot{\lambda}_1\{(\dot{\theta}_2 \mathbf{u}_{21} + \dot{\theta}_3 \mathbf{u}_{31}) \times \mathbf{p}\} \times \boldsymbol{\nu} - \dot{\theta}_2\{\dot{\theta}_3 \mathbf{u}_{32} \times \mathbf{p}\} \times \mathbf{p} + \begin{bmatrix} 0 \\ 0 \\ g \end{bmatrix}\Bigg\}$$

$$= \overline{M}_4 \Bigg\{ \ddot{\lambda}_1 \mathbf{u}_{41} \times \boldsymbol{\nu} + \ddot{\theta}_2 \mathbf{u}_{42} \times \mathbf{p} + \ddot{\theta}_3 \mathbf{u}_{43} \times \mathbf{p} + \dot{\lambda}_1^2(\mathbf{u}_{41} \times \boldsymbol{\nu}) \times \boldsymbol{\nu}$$

$$- \dot{\theta}_2^2 \mathbf{u}_{42} - \dot{\theta}_3^2 \mathbf{u}_{43} - \dot{\theta}_3 \dot{\theta}_2(\mathbf{u}_{42} + \mathbf{u}_{43} - \mathbf{u}_{32})$$

$$+ \dot{\lambda}_1(\dot{\theta}_2 + \dot{\theta}_3)(\mathbf{u}_{41} \times \mathbf{p}) \times \boldsymbol{\nu} - \dot{\lambda}_1\{(\dot{\theta}_2 \mathbf{u}_{21} + \dot{\theta}_3 \mathbf{u}_{31}) \times \mathbf{p}\} \times \boldsymbol{\nu} + \begin{bmatrix} 0 \\ 0 \\ g \end{bmatrix}\Bigg\}$$

$$= \overline{M}_4 \Bigg\{ \ddot{\lambda}_1 \mathbf{u}_{41} \times \boldsymbol{\nu} + \ddot{\theta}_2 \mathbf{u}_{42} \times \mathbf{p} + \ddot{\theta}_3 \mathbf{u}_{43} \times \mathbf{p} + \dot{\lambda}_1^2(\mathbf{u}_{41} \times \boldsymbol{\nu}) \times \boldsymbol{\nu}$$

$$- \dot{\theta}_2^2 \mathbf{u}_{42} - \dot{\theta}_3^2 \mathbf{u}_{43} - 2\dot{\theta}_3 \dot{\theta}_2 \mathbf{u}_{43}$$

$$- \dot{\lambda}_1(\dot{\theta}_2 \mathbf{u}_{42} \times \mathbf{p} + \dot{\theta}_3 \mathbf{u}_{41} \times \mathbf{p}) \times \boldsymbol{\nu} + \begin{bmatrix} 0 \\ 0 \\ g \end{bmatrix}\Bigg\} \tag{9.22}$$

which is consistent with equation (9.12) when $\lambda_1 = 0$ and is consistent with equation (9.13) when $\theta_3 = 0$. To find the force due to point load $3m_3/4$ at $\overline{\mathbf{u}}_4 = \mathbf{u}_3 + (\mathbf{u}_4 - \mathbf{u}_3)/3$, we modify equation (9.22).

$$\mathbf{f}_{m3} = \frac{3m_3}{4} \Bigg\{ \ddot{\lambda}_1 \overline{\mathbf{u}}_{41} \times \boldsymbol{\nu} + \ddot{\theta}_2 \overline{\mathbf{u}}_{42} \times \mathbf{p} + \ddot{\theta}_3 \overline{\mathbf{u}}_{43} \times \mathbf{p} + \dot{\lambda}_1^2(\overline{\mathbf{u}}_{41} \times \boldsymbol{\nu}) \times \boldsymbol{\nu}$$

$$- \dot{\theta}_2^2 \overline{\mathbf{u}}_{42} - \dot{\theta}_3^2 \overline{\mathbf{u}}_{43} - 2\dot{\theta}_3 \dot{\theta}_2 \overline{\mathbf{u}}_{43}$$

$$- \dot{\lambda}_1(\dot{\theta}_2 \overline{\mathbf{u}}_{42} \times \mathbf{p} + \dot{\theta}_3 \overline{\mathbf{u}}_{41} \times \mathbf{p}) \times \boldsymbol{\nu} + \begin{bmatrix} 0 \\ 0 \\ g \end{bmatrix}\Bigg\} \tag{9.23}$$

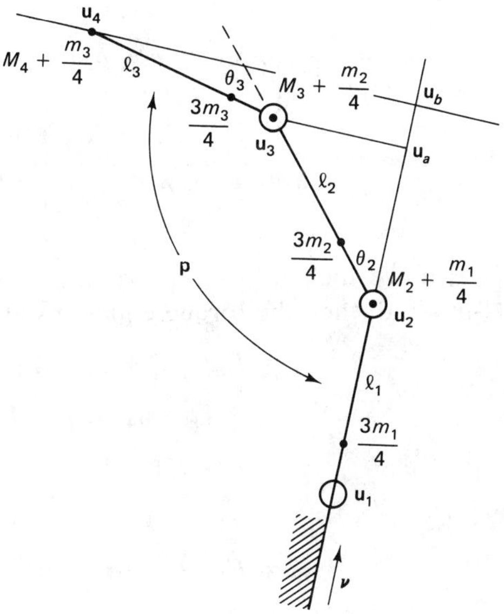

Figure 9.14 The 3 *dof* revolute robot

$$\mathbf{f}_{M3} = \overline{M}_3 \left\{ \ddot{\lambda}_1 \mathbf{u}_{31} \times \boldsymbol{\nu} + \ddot{\theta}_2 \mathbf{u}_{32} \times \mathbf{p} + \dot{\lambda}_1 \{ \dot{\lambda}_1 \mathbf{u}_{31} \times \boldsymbol{\nu} + \dot{\theta}_2 \mathbf{u}_{31} \times \mathbf{p} \} \times \boldsymbol{\nu} \right.$$

$$\left. + \dot{\theta}_2 \{ \dot{\lambda}_1 \mathbf{u}_{32} \times \boldsymbol{\nu} + \dot{\theta}_2 \mathbf{u}_{32} \times \mathbf{p} \} \times \mathbf{p} - \dot{\lambda}_1 \{ \dot{\theta}_2 \mathbf{u}_{21} \times \mathbf{p} \} \times \boldsymbol{\nu} + \begin{bmatrix} 0 \\ 0 \\ g \end{bmatrix} \right\}$$

$$= \overline{M}_3 \left\{ \ddot{\lambda}_1 \mathbf{u}_{31} \times \boldsymbol{\nu} + \ddot{\theta}_2 \mathbf{u}_{32} \times \mathbf{p} + \dot{\lambda}_1^2 (\mathbf{u}_{31} \times \boldsymbol{\nu}) \times \boldsymbol{\nu} - \dot{\theta}_2^2 \mathbf{u}_{32} \right.$$

$$\left. + \dot{\lambda}_1 \dot{\theta}_2 \{ (\mathbf{u}_{31} \times \mathbf{p}) \times \boldsymbol{\nu} + (\mathbf{u}_{32} \times \boldsymbol{\nu}) \mathbf{p} - \{ \mathbf{u}_{21} \times \mathbf{p}) \times \boldsymbol{\nu} \} + \begin{bmatrix} 0 \\ 0 \\ g \end{bmatrix} \right\}$$

$$= \overline{M}_3 \left\{ \ddot{\lambda}_1 \mathbf{u}_{31} \times \boldsymbol{\nu} + \ddot{\theta}_2 \mathbf{u}_{32} \times \mathbf{p} + \dot{\lambda}_1^2 (\mathbf{u}_{31} \times \boldsymbol{\nu}) \times \boldsymbol{\nu} - \dot{\theta}_2^2 \mathbf{u}_{32} \right.$$

$$\left. + \dot{\lambda}_1 \dot{\theta}_2 (\mathbf{u}_{32} \times \mathbf{p}) \times \boldsymbol{\nu} + \begin{bmatrix} 0 \\ 0 \\ g \end{bmatrix} \right\} \tag{9.24}$$

which is consistent with equation (9.13).

$$\mathbf{f}_{m2} = \frac{3m_2}{4}\left\{\ddot{\lambda}_1\bar{\mathbf{u}}_{31} \times \boldsymbol{\nu} + \ddot{\theta}_2\bar{\mathbf{u}}_{32} \times \mathbf{p} + \dot{\lambda}_1^2(\bar{\mathbf{u}}_{31} \times \boldsymbol{\nu}) \times \boldsymbol{\nu} - \dot{\theta}_2^2\bar{\mathbf{u}}_{32} \right.$$

$$\left. + \dot{\lambda}_1\dot{\theta}_2(\bar{\mathbf{u}}_{32} \times \mathbf{p}) \times \boldsymbol{\nu} + \begin{bmatrix} 0 \\ 0 \\ g \end{bmatrix}\right\} \tag{9.25}$$

If we include viscous friction F_i at joint $\mathbf{u}_i$, where F_i is defined in N-m-s/rad, then the torque equations are given by

$$\tau_1 = F_1 \mid \dot{\lambda}_1 \mid + \mathbf{f}_{m2} \cdot \bar{\mathbf{u}}_{31} \times \boldsymbol{\nu} + \mathbf{f}_{M3} \cdot \mathbf{u}_{31} \times \boldsymbol{\nu}$$

$$+ \mathbf{f}_{m3} \cdot \bar{\mathbf{u}}_{41} \times \boldsymbol{\nu} + \mathbf{f}_{M4} \cdot \mathbf{u}_{41} \times \boldsymbol{\nu} \tag{9.26}$$

$$\tau_2 = F_2 \mid \dot{\theta}_2 \mid + \mathbf{f}_{m2} \cdot \bar{\mathbf{u}}_{32} \times \mathbf{p} + \mathbf{f}_{M3} \cdot \mathbf{u}_{32} \times \mathbf{p}$$

$$+ \mathbf{f}_{m3} \cdot \bar{\mathbf{u}}_{42} \times \mathbf{p} + \mathbf{f}_{M4} \cdot \mathbf{u}_{42} \times \mathbf{p} \tag{9.27}$$

$$\tau_3 = F_3 \mid \dot{\theta}_3 \mid + \mathbf{f}_{m3} \cdot \bar{\mathbf{u}}_{43} \times \mathbf{p} + \mathbf{f}_{M4} \cdot \mathbf{u}_{43} \times \mathbf{p} \tag{9.28}$$

The set of joint torques $[\tau_1 \quad \tau_2 \quad \tau_3]$ are collectively known as *torque space*. For any robot[1] the order of position space, joint space, and torque space are the same. In Chapter 5 we saw how to convert from joint space to position space. In Chapter 6 we saw how to convert from position space into joint space. In Chapter 8 we planned trajectories in position space as time functions and obtained time functions for the joint space. Here we see how to obtain torque space from the position space and joint space and so can convert from position space to torque space. It is not clear how to perform the inverse (from torque to position space) and not clear that such an operation would be useful even if it could be performed. The sequence of operations from task planning to torque space is shown in Figure 9.15.

We can use equations (9.22) through (9.28) in the analysis of the 6 *dof* orientable robot $\mid \lambda\theta\theta\lambda\theta\lambda \mid$. The only other masses to consider are $\overline{M}_7 = (m_5 + m_6)/4 + M_7$ at $\mathbf{u}_7$ and $3(m_5 + m_6)/4$ at $\bar{\mathbf{u}}_7 = \mathbf{u}_5 + \mathbf{u}_{75}/3$; we assume here that the distributed load per unit length is the same on the two links l_5 and l_6. Further, we modify the two point masses at $\mathbf{u}_5$ and $\mathbf{u}_3 + \mathbf{u}_{53}/3$ to $(m_3 + m_4)/4 + M_5$ and $3(m_3 + m_4)/4$, respectively, under the same assumption. The two additional forces needed are

$$\mathbf{f}_{M7} = \overline{M}_7\left\{\sum_{i=1}^{6} \ddot{\Theta}_i\mathbf{u}_{7i} \times \boldsymbol{\Omega}_i + \sum_{j=1}^{6} \dot{\Theta}_j\left\{\sum_{i=1}^{6} \dot{\Theta}_i\mathbf{u}_{7j} \times \boldsymbol{\Omega}_i\right\} \times \boldsymbol{\Omega}_j \right.$$

$$\left. - \sum_{j=1}^{5} \dot{\Theta}_j\left\{\sum_{i=j+1}^{6} \dot{\Theta}_i\mathbf{u}_{ij} \times \boldsymbol{\Omega}_i\right\} \times \boldsymbol{\Omega}_j + \begin{bmatrix} 0 \\ 0 \\ g \end{bmatrix}\right\} \tag{9.29}$$

[1] We add the restriction that we consider only nonloose robots with a consistent position space—that is, a position space that can be satisfied by the robot configuration.

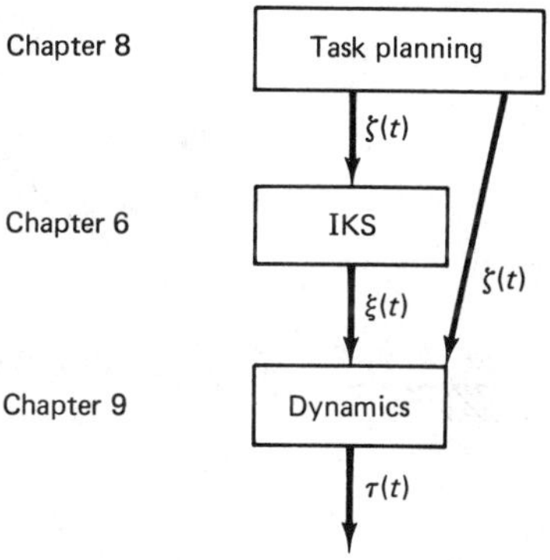

Figure 9.15 From task planning to torque space

and

$$\mathbf{f}_{m5} = \frac{3(m_3 + m_4)}{4}\left\{\sum_{i=1}^{6} \ddot{\Theta}_i \bar{\mathbf{u}}_{7i} \times \boldsymbol{\Omega}_i + \sum_{j=1}^{6} \dot{\Theta}_j \left\{\sum_{i=1}^{6} \dot{\Theta}_i \bar{\mathbf{u}}_{7j} \times \boldsymbol{\Omega}_i\right\} \times \boldsymbol{\Omega}_j \right.$$

$$\left. - \sum_{j=1}^{5} \dot{\Theta}_j \left\{\sum_{i=j+1}^{6} \dot{\Theta}_i \bar{\mathbf{u}}_{ij} \times \boldsymbol{\Omega}_i\right\} \times \boldsymbol{\Omega}_j + \begin{bmatrix} 0 \\ 0 \\ g \end{bmatrix}\right\} \tag{9.30}$$

The efficiency of the methodology presented here will now be assessed. We assume as our starting point the position and joint space, and so the set of joint axes. The dominant computational expense is contained in the vector cross-multiplications. For the 3 *dof* robot | $\lambda\theta\theta$ |, there are 38 cross-multiplications. Each cross-multiplication requires 6 multiplications and 3 additions. For the 6 *dof* robot | $\lambda\theta\theta\lambda\theta\lambda$ |, if no simplifications are performed[2] on equation (9.26), an additional 156 cross-multiplications are required. In general, equation (9.19) applied to the mass on the last link of an n *dof* robot requires $n + 2n^2$ cross-multiplications. This computational expense compares favorably with the recursive Newton-Euler method [Fu, Gozalves, and Lee 1987] but less efficient than the derivations of Balafoutis et al. [Balafoutis, Patel, and Misra 1988]. However, when logical simplifications are made using this method, as was done to obtain equations (9.25), (9.26), and (9.27), this method is the most efficient. The Newton-Euler method is considerably more efficient than Lagrangian analysis [Fu, Gozalves, and Lee 1987]. However, the main advantage of this method is its simplicity: Equations (9.19) and (9.20) can be used cookbook-style for any configuration robot without need for analysis or thought.

For the 4 *dof* SCARA robot | $\theta\theta\psi\sigma$ |, we can place mass $\overline{M} = m_3 + M_4$ on link l_3 to lie at $\mathbf{u}_3$, since its effect on the two pitch joints will be unaltered

[2] One of the exercises at the end of this chapter concerns the simplification for this particular case.

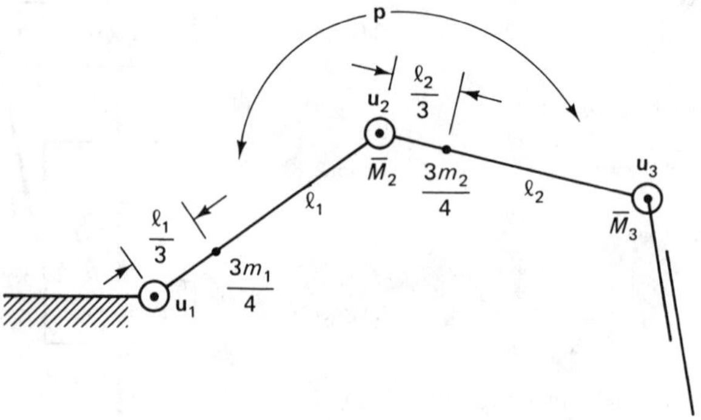

Figure 9.16 Point mass in the SCARA robot

by this artifice. The effective masses on l_2 are $\overline{M}_3 = m_2/4 + M_3 + m_3 + M_4$ at $\mathbf{u}_3$ and $3m_2/4$ at $\mathbf{u}_2 + \mathbf{u}_{32}/3$. The point masses are shown for this robot in Figure 9.16. The two vector forces are

$$\begin{aligned}\mathbf{f}_{M3} = \overline{M}_3 \Bigg\{ & \ddot{\theta}_1\mathbf{u}_{31} \times \mathbf{p} + \ddot{\theta}_2\mathbf{u}_{31} \times \mathbf{p} \\ & + \{\dot{\theta}_1^2\mathbf{u}_{31} \times \mathbf{p} + \dot{\theta}_1\dot{\theta}_2\mathbf{u}_{31} \times \mathbf{p} + \dot{\theta}_2\dot{\theta}_1 \times \mathbf{p} + \dot{\theta}_2^2\mathbf{u}_{32} \times \mathbf{p}\} \times \mathbf{p} \\ & - \dot{\theta}_1\dot{\theta}_2(\mathbf{u}_{21} \times \mathbf{p}) \times \mathbf{p} + \begin{bmatrix} 0 \\ 0 \\ g \end{bmatrix}\Bigg\} \\ = \overline{M}_3 \Bigg\{ & \ddot{\theta}_1\mathbf{u}_{31} \times \mathbf{p} + \ddot{\theta}_2\mathbf{u}_{31} \times \mathbf{p} - \dot{\theta}_1^2\mathbf{u}_{31} - \dot{\theta}_2\mathbf{u}_{32} - 2\dot{\theta}_1\dot{\theta}_2\mathbf{u}_{32} + \begin{bmatrix} 0 \\ 0 \\ g \end{bmatrix}\Bigg\}\end{aligned} \tag{9.31}$$

$$\mathbf{f}_{m2} = \frac{3m_2}{4}\left\{\ddot{\theta}_1\bar{\mathbf{u}}_{31} \times \mathbf{p} + \ddot{\theta}_2\bar{\mathbf{u}}_{31} \times \mathbf{p} - \dot{\theta}_1^2\bar{\mathbf{u}}_{31} - \dot{\theta}_2\bar{\mathbf{u}}_{32} - 2\dot{\theta}_1\dot{\theta}_2\bar{\mathbf{u}}_{32} + \begin{bmatrix} 0 \\ 0 \\ g \end{bmatrix}\right\}$$

$$\text{where } \bar{\mathbf{u}}_{31} = \mathbf{u}_{21} + \frac{\mathbf{u}_{32}}{3} \tag{9.32}$$

$$\begin{aligned}\mathbf{f}_{M2} &= \overline{M}_2\left\{\ddot{\theta}_1\mathbf{u}_{21} \times \mathbf{p} + \dot{\theta}_1^2(\mathbf{u}_{21} \times \mathbf{p}) \times \mathbf{p} + \begin{bmatrix} 0 \\ 0 \\ g \end{bmatrix}\right\} \\ &= \overline{M}_2\left\{\ddot{\theta}_1\mathbf{u}_{21} \times \mathbf{p} - \dot{\theta}_1^2\mathbf{u}_{21} + \begin{bmatrix} 0 \\ 0 \\ g \end{bmatrix}\right\}\end{aligned} \tag{9.33}$$

$$\mathbf{f}_{m1} = \frac{m_1}{4}\{\ddot{\theta}_1\mathbf{u}_{21} \times \mathbf{p} - \dot{\theta}_1^2\mathbf{u}_{21}\} + \frac{3m_1}{4}\left[\begin{matrix}0\\0\\g\end{matrix}\right]\Bigg\} \tag{9.34}$$

so

$$\tau_1 = F_1 \mid \dot{\theta}_2 \mid + \tfrac{1}{3}\,\mathbf{f}_{m1} \cdot \mathbf{u}_{21} \times \mathbf{p} + \mathbf{f}_{M2} \cdot \mathbf{u}_{21} \times \mathbf{p}$$

$$+ \mathbf{f}_{m2} \cdot \bar{\mathbf{u}}_{31} \times \mathbf{p} + \mathbf{f}_{M3} \cdot \mathbf{u}_{31} \times \mathbf{p} \tag{9.35}$$

$$\tau_2 = F_2 \mid \dot{\theta}_2 \mid + \tfrac{1}{3}\,\mathbf{f}_{m2} \cdot \mathbf{u}_{32} \times \mathbf{p} + \mathbf{f}_{M3} \cdot \mathbf{u}_{32} \times \mathbf{p} \tag{9.36}$$

9.8 THE DEFECTIVE NATURE OF TORQUE SPACE

We end this chapter with a disclaimer. We find it difficult to design in torque space, since we have no efficient system for determining either position space or joint space from torque space. To plan a trajectory using torque space is an open, iterative procedure, since equation (9.19) defines a set of coupled, second-order, nonlinear ordinary differential equations that have no analytic solution. For example, if we wish to minimize the execution time for a robot in moving between two position space points, the ideal method would be to determine the joint most likely to be torque-limited and drive that joint with maximum actuator acceleration, adjusting the torques at the other joints to follow the prescribed path. This is not possible in closed form, using the methodologies developed in this chapter. Instead, we can monitor the torques at specific points along the path and change the path or execution time so that the resulting torques are feasible; this itself is an iterative procedure.

9.9 EXAMPLES

1. Suppose the system shown in Figure 9.17 is at rest. At $t = 0$ the string is cut by the scissors. What happens to the masses?

This is an initial condition problem. For $t < 0$, the system is stationary, the frictional forces are zero, and the top spring exerts force Mg, where g is the acceleration due to gravity. When the string is cut, the bottom mass (with associated spring and dashpot) is in free flight and accelerates at $g = 9.80616$ m/s^2. The top spring-mass dashpot system is then treated as a stand-alone single-mass system with an initial condition given by $K_1 y(0) = M_2 g$, where $y(0)$ is the initial displacement. Thus, $y(0) = M_2 g/K_1$. The system is then described by

$$M_1 \frac{d^2y}{dt^2} + F_1 \frac{dy}{dt} + K_1 y = 0$$

in the time domain, or

$$M_1 s^2 Y - M_1 s y(0) - M_1 y'(0) + F_1 s Y - F_1 y(0) + K_1 Y = 0$$

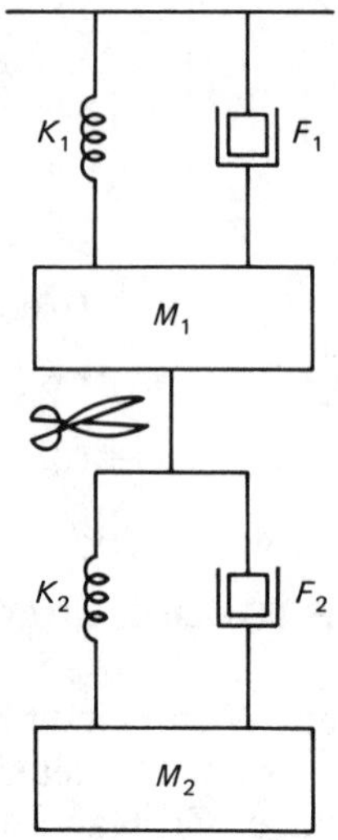

Figure 9.17 A two-mass system

in the frequency domain. Entering the known initial conditions, we get

$$Y = \frac{(M_1 s + F_1)g/K_1}{M_1 s^2 + F_1 s + K_1}$$

The system response to the step input is a damped oscillation if $4M_1K_1 > F_1^2$ or an exponential response otherwise.

---

2. Suppose a point mass is held at distance l from a frictionless pivot point at the origin so it rotates in $\mathbf{p} = [0 \quad 0 \quad 1 \quad 0]$. Suppose at $\theta = \pi$ rad the angular velocity is ω. Indicate how to determine the motion of M.

The torque equation for the system as given by equation (9.4) is $0 = Ml\ddot{\theta} - Mgl \sin \theta$, where M is the mass. Taking the Laplace transform gives

$$\Theta = \frac{Ml^2\{s\theta(0) + \theta'(0)\}}{Ml^2s^2 - \dfrac{Mgl}{s^2 + 1}} = \frac{(s^2 + 1)(s\pi + \omega)}{s^2(s^2 + 1) - \dfrac{g}{l}}$$

A partial fraction expansion enables the inverse Laplace transform to be performed, giving the time domain solution.

An alternative method is to integrate numerically, starting with one or more known initial condition points. Each integration step produces a new initial condition point for the next integration step. A time-varying input applied to a time-dependent system can be treated in the time domain using the convolution integral. If this integral is evaluated up to some time point t_0, we can always define the system as $\dot{\mathbf{y}} = \mathbf{f}(\mathbf{y}, \mathbf{x})$ and apply a numeric integration formula starting at $t = t_0$ with known $\mathbf{y}_0$. $\dot{\mathbf{y}} = d\mathbf{y}/dt$. The nonlinear equation $\dot{\mathbf{y}} = \mathbf{f}(\mathbf{y}, \mathbf{x})$ can be obtained for any physical system containing one or more differential equations of any order: the linear equivalent of this equation is $\dot{\mathbf{y}} = \mathbf{A}\mathbf{y} + \mathbf{B}\mathbf{x}(t)$, where $\mathbf{A}$ and $\mathbf{B}$ are matrices.

The simplest numeric integration formula is Euler's formula, $\mathbf{y}_{n+1} = \mathbf{y}_n + h\mathbf{f}(\mathbf{y}_n, \mathbf{x}_n)$. In fact, all numeric integration processes that start with a single set of initial conditions are forced to start with Euler's formula. For example, one of the most popular numeric integration systems is the fourth-order Runge-Kutta process. This system applies Euler's formula to generate another point so the next formula can use this point. After a total of four integration formulas are applied, the next time point is chosen as a weighted sum of the four answers obtained. One such fourth-order Runge-Kutta process is

$$\left.\begin{aligned} &\boldsymbol{\phi}_1 = h\mathbf{f}(\mathbf{y}_n, \mathbf{x}_n), && \boldsymbol{\phi}_2 = h\mathbf{f}\left(\mathbf{y}_n + \frac{1}{2}\boldsymbol{\phi}_1, \mathbf{x}_n + \frac{h}{2}I\right) \\ &\boldsymbol{\phi}_3 = h\mathbf{f}\left(\mathbf{y}_n + \frac{1}{2}\boldsymbol{\phi}_2, \mathbf{x}_n + \frac{h}{2}I\right), && \boldsymbol{\phi} = h\mathbf{f}(\mathbf{y}_n + \boldsymbol{\phi}_3, \mathbf{x}_n + hI) \\ &\mathbf{y}_{n+1} = \mathbf{y}_n + \frac{1}{6}(\boldsymbol{\phi}_1 + 2\boldsymbol{\phi}_2 + 2\boldsymbol{\phi}_3 + \boldsymbol{\phi}_4) && \end{aligned}\right\} \tag{9.37}$$

Another popular numeric integration system is the Mth-order predictor-corrector method that requires N initial conditions; typically $M = 4$ and $N = 4$. One such method based on Adams formula [Ralston and Rabinowitz 1978] is

$$\begin{aligned} \mathbf{y}_p &= \mathbf{y}_{n-1} + \frac{h}{3}\{8\mathbf{f}(\mathbf{y}_n, \mathbf{x}_n) - 5\mathbf{f}(\mathbf{y}_{n-1}, \mathbf{x}_{n-1}) \\ &\quad + 4\mathbf{f}(\mathbf{y}_{n-2}, \mathbf{x}_{n-2}) - \mathbf{f}(\mathbf{y}_{n-3}, \mathbf{x}_{n-3})\} \\ \mathbf{y}_{n+1} &= \mathbf{y}_{n-1} + \frac{h}{3}\{\mathbf{f}(\mathbf{y}_p, \mathbf{x}_{n+1}) + 4\mathbf{f}(\mathbf{y}_n, \mathbf{x}_n) + \mathbf{f}(\mathbf{y}_{n-1}, \mathbf{x}_{n-1})\} \end{aligned} \tag{9.38}$$

We need the four points $(\mathbf{y}_{n-3}, \mathbf{x}_{n-3})$, $(\mathbf{y}_{n-2}, \mathbf{x}_{n-2})$, $(\mathbf{y}_{n-1}, \mathbf{x}_{n-1})$, and $(\mathbf{y}_n, \mathbf{x}_n)$ to apply the method. If one initial condition point is given, three more can be generated by another method (the fourth-order Runge-Kutta, for example). The advantages of the predictor-corrector method over a Runge-Kutta method are the following:

1. The difference between the predicted value $\mathbf{y}_p$ and $\mathbf{y}_{n+1}$ is a known measure of the error of $\mathbf{y}_{n+1}$.
2. There is one calculation of the function $\mathbf{f}(\mathbf{y}, \mathbf{x})$ for one complete step, unlike four such calculations for the Runge-Kutta method.

The methodology of iterating the time point by h is the same regardless of the numeric integration method, so we discuss the process with Euler's formula. The integration proceeds from initial conditions with chosen h. h must be small enough to preserve numeric stability and to maintain accuracy. Consider the simple system $dy/dt = -\lambda y$, where $\lambda > 0$, so

$$y_{n+1} = y_n + h\frac{dy_n}{dt} = (1 - h\lambda)y_n$$

The usual solution of a difference equation (an equation with discrete points such as y_n and y_{n+1}) is $y_n = \rho^n$, so $\rho^{n+1} = (1 - h\lambda)\rho^n$, or $\rho = 1 - h\lambda$. The stability of a difference equation requires all its roots to lie on or within the unit circle—that is, $|\rho| \leq 1$ for all roots. Typically the difference equation for a fourth-order correct integration formula applied to the simple differential equation just used is a polynomial of order N in ρ, where N is usually about 4. We can determine stability of higher-order difference equations using z-transform techniques [Kuo 1987] or using the Schur test [Marden 1966]. For the simple application of Euler's formula just given, we require $|1-h\lambda| \leq 1$, or $0 \leq h\lambda \leq 2$. λ is the negative of the eigenvalue (natural frequency) of the differential equation. System stability requires that $\lambda \geq 0$. Numeric stability requires $h \leq 2/\lambda$. For a system with two or more eigenvalues, we choose the magnitude of the largest eigenvalue as λ.

The fourth-order Runge-Kutta system given by equations (9.37) has the stability requirement $0 \leq h\lambda \leq 2.785295$. Substituting $\theta_1 = \theta$ and $\dot{\theta}_1 = \theta_2$, we get

$$\dot{\boldsymbol{\theta}} = \mathbf{f}(\boldsymbol{\theta}) = \begin{bmatrix} \theta_2 \\ \frac{g}{l}\sin\theta_1 \end{bmatrix}$$

and so

$$\boldsymbol{\phi}_1 = h\begin{bmatrix} \theta_2 \\ \frac{g}{l}\sin\theta_1 \end{bmatrix}$$

$$\boldsymbol{\phi}_2 = h\begin{bmatrix} \theta_2 + \frac{hg}{2l}\sin\theta_1 \\ \frac{g}{l}\sin\left(\theta_1 + \frac{h}{2}\theta_2\right) \end{bmatrix}$$

$$\boldsymbol{\phi}_3 = h\begin{bmatrix} \theta_2 + \frac{hg}{2l}\sin\left(\theta_1 + \frac{h}{2}\theta_2\right) \\ \frac{g}{l}\sin\left\{\theta_1 + \frac{h}{2}\left(\theta_2 + \frac{hg}{2l}\sin\theta_1\right)\right\} \end{bmatrix}$$

$$\boldsymbol{\phi}_4 = h\begin{bmatrix} \theta_2 + \frac{hg}{2l}\sin\left\{\theta_1 + \frac{h}{2}\theta_2\left(\theta_2 + \frac{hg}{2l}\sin\theta_1\right)\right\} \\ \frac{g}{l}\sin\left(\theta_1 + \frac{h}{2}\left\{\theta_2 + \frac{hg}{2l}\sin\left(\theta_1 + \frac{h}{2}\theta_2\right)\right\}\right) \end{bmatrix}$$

and $\boldsymbol{\theta}_{n+1} = \boldsymbol{\theta}_n + \frac{1}{6}(\boldsymbol{\phi}_1 + 2\boldsymbol{\phi}_2 + 2\boldsymbol{\phi}_3 + \boldsymbol{\phi}_4)$.

---

3. Use equation (9.17) to verify the acceleration at $\mathbf{u}_3$ given by equation (9.12) for the two-pitch joint system.

We use $k = 3$ and $\boldsymbol{\Omega}_1 = \boldsymbol{\Omega}_2 = \mathbf{p}$ in equation (9.17):

$$\begin{aligned}
\ddot{\mathbf{u}}_3 &= \sum_{i=1}^{2} \ddot{\Theta}_i \mathbf{u}_{3i} \times \mathbf{p} + \sum_{j=1}^{2} \dot{\theta}_j \left\{ \sum_{i=1}^{2} \dot{\Theta}_i \mathbf{u}_{3j} \times \mathbf{p} \right\} \times \mathbf{p} - \sum_{j=1}^{1} \dot{\theta}_j \left\{ \sum_{i=j+1}^{2} \dot{\theta}_i \mathbf{u}_{ij} \times \mathbf{p} \right\} \times \mathbf{p} \\
&= \ddot{\theta}_1 \mathbf{u}_{31} \times \mathbf{p} + \ddot{\theta}_2 \mathbf{u}_{32} \times \mathbf{p} + \dot{\theta}_1\{\dot{\theta}_1 \mathbf{u}_{31} \times \mathbf{p} + \dot{\theta}_2 \mathbf{u}_{31} \times \mathbf{p}\} \times \mathbf{p} \\
&\quad + \dot{\theta}_2\{\dot{\theta}_1 \mathbf{u}_{32} \times \mathbf{p} + \dot{\theta}_2 \mathbf{u}_{32} \times \mathbf{p}\} \times \mathbf{p} - \dot{\theta}_1\{\dot{\theta}_2 \mathbf{u}_{21} \times \mathbf{p}\} \times \mathbf{p} \\
&= \ddot{\theta}_1 \mathbf{u}_{31} \times \mathbf{p} + \ddot{\theta}_2 \mathbf{u}_{32} \times \mathbf{p} - \dot{\theta}_1^2 \mathbf{u}_{31} - \dot{\theta}_2^2 \mathbf{u}_{32} + \dot{\theta}_1\dot{\theta}_2(\mathbf{u}_{31} + \mathbf{u}_{32} - \mathbf{u}_{21}) \times \mathbf{p} \times \mathbf{p} \\
&= \ddot{\theta}_1 \mathbf{u}_{31} \times \mathbf{p} + \ddot{\theta}_2 \mathbf{u}_{32} \times \mathbf{p} - \dot{\theta}_1^2 \mathbf{u}_{31} - \dot{\theta}_2^2 \mathbf{u}_{32} - 2\dot{\theta}_1\dot{\theta}_2 \mathbf{u}_{32}
\end{aligned}$$

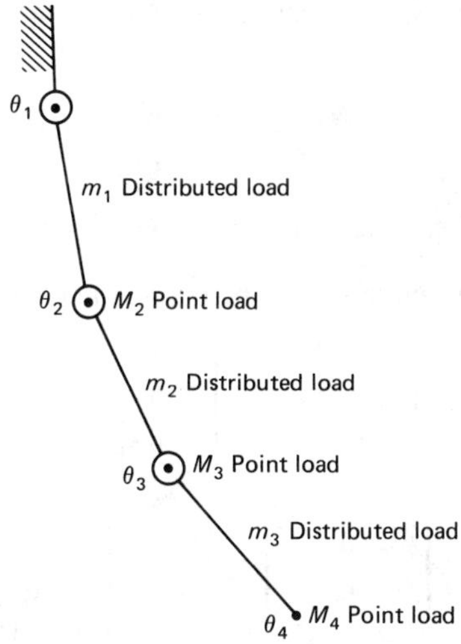

Figure 9.18 A compound pendulum

4. Why is it not possible to use the methods developed in this chapter to analyze the compound pendulum shown in Figure 9.18?

The answer is to be found in Figure 9.15 and the disclaimer at the end of the chapter. We use position space and joint space to determine the joint torques, so the motion of the joints is predetermined. We cannot go backwards and determine the joint space or position space from the torque space. In the compound pendulum, the torques at the joint are zero, and the motion of the pendulum depends on the initial conditions. The last example in this section indicates the difficulty of designing with torques even for a single-pitch joint.

---

5. Obtain the torque equations for the 3 *dof* polar robot $|\lambda\theta\sigma|$ for general base orientation $\mathbf{p}_1$ and $\boldsymbol{\nu}$. Assume point masses only at the joint locations.

Without loss of generality we can assume $\mathbf{u}_1$ is the origin. The mass M_3 at $\mathbf{u}_3$ is the only mass that does work in this system. Adding $\ddot{\mathbf{u}}_{32} + \dot{\lambda}_1\dot{\mathbf{u}}_{32} \times \boldsymbol{\nu} + \dot{\theta}_2\dot{\mathbf{u}}_{32} \times \mathbf{p}$ to equation (9.24) to take into account the sliding link gives

$$\mathbf{f}_{M3} = M_3\left\{\ddot{\lambda}\overline{\mathbf{u}}_{31} \times \boldsymbol{\nu} + \ddot{\theta}_2\overline{\mathbf{u}}_{32} \times \mathbf{p} + \dot{\lambda}_1^2(\overline{\mathbf{u}}_{31} \times \boldsymbol{\nu}) \times \boldsymbol{\nu} - \dot{\theta}_2^2\overline{\mathbf{u}}_{32} + \dot{\lambda}_1\dot{\theta}_2(\overline{\mathbf{u}}_{32} \times \mathbf{p}) \times \boldsymbol{\nu}\right.$$

$$\left. + \ddot{\mathbf{u}}_{32} + \dot{\lambda}_1\dot{\mathbf{u}}_{32} \times \boldsymbol{\nu} + \dot{\theta}_2\dot{\mathbf{u}}_{32} \times \mathbf{p} - \begin{bmatrix} 0 \\ 0 \\ g \end{bmatrix}\right\}$$

The torque equations are $\tau_1 = \mathbf{f}_{M3} \cdot \overline{\mathbf{u}}_{31} \times \boldsymbol{\nu}$ and $\tau_2 = \mathbf{f}_{M3} \cdot \overline{\mathbf{u}}_{32} \times \mathbf{p}$.

---

6. For the two-pitch joint system $|\theta\theta|$ with unity link lengths whose base orientation is given by $\boldsymbol{\nu} = [0 \quad 0 \quad 1]^{\#}$ and $\mathbf{p} = [0 \quad 1 \quad 0 \quad 0]$, assume that the only mass is 1 kg at $\mathbf{u}_3$. Suppose $\mathbf{u}_3$ follows the four-point Bezier curve

$$(1 - t)^2(1 + 2t)\begin{bmatrix} -1 \\ 0 \\ 0 \end{bmatrix} + t^2(3 - 2t)\begin{bmatrix} -1 \\ 0 \\ 1 \end{bmatrix}, \qquad 0 \le t \le 1$$

If the actual execution time is 0.5 s, obtain a time function for the torques.

The acceleration of the mass is known directly from the Bezier function

$$\mathbf{u}_3 = (1 - 3t^2 + 2t^3)\begin{bmatrix} -1 \\ 0 \\ 0 \end{bmatrix} + (3t^2 - 2t^3)\begin{bmatrix} -1 \\ 0 \\ 1 \end{bmatrix} = \begin{bmatrix} -1 \\ 0 \\ 3t^2 - 2t^3 \end{bmatrix}$$

so

$$\dot{\mathbf{u}}_3 = \begin{bmatrix} 0 \\ 0 \\ 6t - 6t^2 \end{bmatrix}, \qquad \ddot{\mathbf{u}}_3 = \begin{bmatrix} 0 \\ 0 \\ 6 - 12t \end{bmatrix}$$

The force on the mass is

$$\begin{bmatrix} 0 \\ 0 \\ -6 + 12t - 9.80616 \end{bmatrix} = \begin{bmatrix} 0 \\ 0 \\ 12t - 15.80616 \end{bmatrix} \text{ N}$$

From equation (9.20) the torques are

$$\tau_1 = \mathbf{f}_M \cdot \mathbf{u}_{31} \times \mathbf{p} = \begin{bmatrix} 0 \\ 0 \\ 15.80616 - 12t \end{bmatrix} \cdot \begin{bmatrix} -1 \\ 0 \\ 3t^2 - 2t^3 \end{bmatrix} \times \begin{bmatrix} 0 \\ 1 \\ 0 \end{bmatrix}$$

$$= 16t - 15.80616$$

and

$$\tau_2 = \mathbf{f}_M \cdot \mathbf{u}_{32} \times \mathbf{p} = \begin{bmatrix} 0 \\ 0 \\ 15.80616 - 12t \end{bmatrix} \cdot \begin{bmatrix} -1 + \sin\theta_1 \\ 0 \\ \cos(\theta_1 + \theta_2) \end{bmatrix} \times \begin{bmatrix} 0 \\ 1 \\ 0 \end{bmatrix}$$

$$= (12t - 15.80616)(1 - \sin\theta_1)$$

since

$$\mathbf{u}_{31} = \begin{bmatrix} -\sin\theta_1 - \sin(\theta_1 + \theta_2) \\ 0 \\ \cos\theta_1 + \cos(\theta_1 + \theta_2) \end{bmatrix} = \begin{bmatrix} -1 \\ 0 \\ \cos\theta_1 + \cos(\theta_1 + \theta_2) \end{bmatrix}$$

and

$$\mathbf{u}_{32} = \mathbf{u}_{31} - \mathbf{u}_{21} = \begin{bmatrix} -1 + \sin\theta_1 \\ 0 \\ \cos(\theta_1 + \theta_2) \end{bmatrix}$$

7. Consider mass M, stationary at time $t = 0$, to be at the end of a link of length l rotating in $[1 \quad 0 \quad 0 \quad 0]$. The joint is at $[0 \quad 0 \quad 0 \quad 1]^{\#}$ and $\mathbf{u}_2 = [0 \quad 1 \quad 0 \quad 1]^{\#}$. The maximum torque permissible is 20 N-m. If the mass M is 1 kg, what is the minimum time required to move the mass to point $[0 \quad 0 \quad 1 \quad 1]^{\#}$ with zero velocity at this endpoint?

Recall that the chapter ended with a disclaimer that torque is not a true space. We now reformulate the equations for this simple case so that torque can be used as a space.

$$\ddot{\mathbf{u}}_2 = \ddot{\theta}\mathbf{u}_{21} \times \mathbf{p} - \dot{\theta}^2\mathbf{u}_{21} = \ddot{\theta}\begin{bmatrix} 0 \\ -\sin\theta \\ \cos\theta \end{bmatrix} \times \begin{bmatrix} 1 \\ 0 \\ 0 \end{bmatrix} - \dot{\theta}^2\begin{bmatrix} 0 \\ -\sin\theta \\ \cos\theta \end{bmatrix}$$

$$= \ddot{\theta}\begin{bmatrix} 0 \\ \cos\theta \\ \sin\theta \end{bmatrix} - \dot{\theta}^2\begin{bmatrix} 0 \\ -\sin\theta \\ \cos\theta \end{bmatrix}$$

$$\mathbf{f}_M = \left\{\ddot{\theta}\begin{bmatrix} 0 \\ \cos\theta \\ \sin\theta \end{bmatrix} - \dot{\theta}^2\begin{bmatrix} 0 \\ -\sin\theta \\ \cos\theta \end{bmatrix} - \begin{bmatrix} 0 \\ 0 \\ 9.80616 \end{bmatrix}\right\}$$

$$\tau = \mathbf{f}_M \cdot \mathbf{u}_{21} \times \mathbf{p} = \left\{\ddot{\theta}\begin{bmatrix} 0 \\ \cos\theta \\ \sin\theta \end{bmatrix} - \dot{\theta}^2\begin{bmatrix} 0 \\ -\sin\theta \\ \cos\theta \end{bmatrix} - \begin{bmatrix} 0 \\ 0 \\ 9.80616 \end{bmatrix}\right\}$$

$$\cdot\left\{\begin{bmatrix} 0 \\ -\sin\theta \\ \cos\theta \end{bmatrix} \times \begin{bmatrix} 1 \\ 0 \\ 0 \end{bmatrix}\right\}$$

$$= \left\{\ddot{\theta}\begin{bmatrix} 0 \\ \cos\theta \\ \sin\theta \end{bmatrix} - \dot{\theta}^2\begin{bmatrix} 0 \\ -\sin\theta \\ \cos\theta \end{bmatrix} - \begin{bmatrix} 0 \\ 0 \\ 9.80616 \end{bmatrix}\right\} \cdot \begin{bmatrix} 0 \\ \cos\theta \\ \sin\theta \end{bmatrix}$$

$$= (\ddot{\theta} - 9.80616 \sin\theta)$$

$$\ddot{\theta} = \tau + 9.80616 \sin\theta$$

Suppose we approximate $\dot{\theta}$ and θ by the first-order recursive formulas

$$\dot{\theta}_{i+1} = \dot{\theta}_i + h\ddot{\theta}_i = \dot{\theta}_i + h(\tau + 9.80616 \sin\theta_i); \qquad \theta_{i+1} = \theta_i + h\dot{\theta}_i$$

where h is the time step between points. τ is a function of time. Recall that the ideal acceleration characteristic as shown in Figure 8.4 indicates that we drive the joint with maximum positive torque for part of the way and then drive with the maximum negative torque for the rest of the way. For $h = 0.05$ the iterative formulas become

$$\dot{\theta}_{i+1} = \dot{\theta}_i \pm 1 + 0.4903 \sin\theta_i; \qquad \theta_{i+1} = \theta_i + 0.05\dot{\theta}_i$$

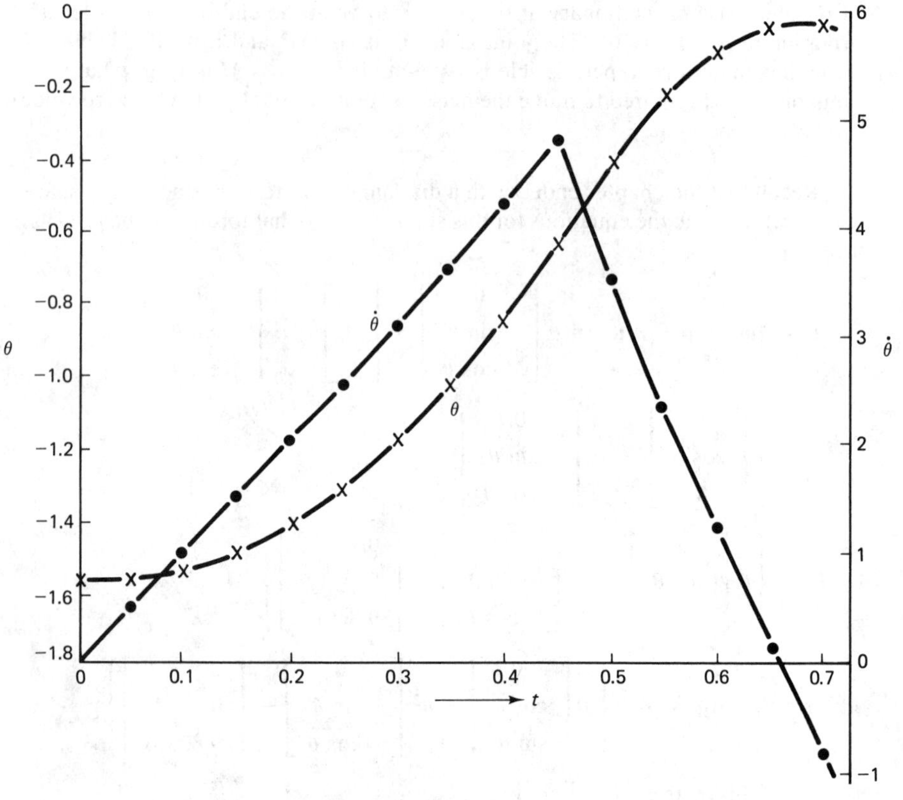

Figure 9.19 Motion of a mass subject to maximum torque

and we construct the following table with positive torque until indicated:

t	θ	$\dot{\theta}$
0	−1.571	0
0.05	−1.571	0.510
0.1	−1.546	1.019
0.15	−1.491	1.527
0.2	−1.415	2.038
0.25	−1.313	2.554
0.3	−1.185	3.080
0.35	−1.031	3.626
0.4	−0.850	4.205
0.45	−0.640	4.837
0.5*	−0.398	3.544
0.55	−0.221	2.354
0.6	−0.103	1.247
0.65	−0.041	0.197
0.7	−0.031	−0.823

* Negative torque applied here.

A plot of the angle and the velocity is shown in Figure 9.19 for the two trajectories. When negative torque is applied at 0.5 s the final goal is not quite achieved, indicating that the positive torque must be continued slightly longer.

Comparing the asymmetry of the shape of the velocity $\dot{\theta}$ in Figure 9.19 with the idealized velocity characteristic of Figure 8.4, we see that gravity is significant. The average acceleration with positive torque is 4.837/0.45 = 10.75 rad/s^2, whereas the average acceleration with negative torque is 4.837/0.21 = 23.0. In a typical robot, gravity is a more significant force than the forces of acceleration. However, in the future, as robots are made faster, the influence of gravity will reduce.

EXERCISES FOR CHAPTER 9

1. A point mass is held by a weightless string of length l and swung in a horizontal plane at angular velocity ω rad/s. Gravity causes the mass to droop with respect to this horizontal plane. Calculate the amount of droop as a function of l and ω. Assume zero friction and no wind drag.

---

2. Two masses M_1 and M_2 are attached by a spring with spring constant K newtons per meter and rest on a horizontal surface with coefficient of friction 0.1. For $t < 0$, the system is at rest. At $t = 0$ a force of x newtons is applied to M_1 in a direction toward M_2. Determine the equations of motion of the masses, including the initial conditions. Indicate a computationally feasible method of solution.

---

3. A pitch joint at the origin in Cartesian space connects link l, at the end of which mass M is attached. l moves in normalized plane $\mathbf{p} = [a \quad b \quad c \quad 0]$. For $t < 0$, the mass is held stationary at

$$\begin{bmatrix} \dfrac{bl}{\sqrt{a^2 + b^2}} \\ \dfrac{-al}{\sqrt{a^2 + b^2}} \\ 0 \\ 1 \end{bmatrix}$$

Viscous frictional force F N-m-s/rad occurs at the joint. Determine the differential equation that describes the motion of the mass for $t \geq 0$ if the mass is released at $t = 0$.

---

4. Find the acceleration of a particle at the end of kinematic chain

$$| \lambda_1 \, l_1 \, | \, (+) \, l_2 \, |$$

in terms of the motion of joint λ_1. If the particle has mass M, find an expression for the torque at the joint.

---

5. Find the acceleration of a particle at the end of kinematic chain

$$| \lambda_1 \; l_1 \; | \; (\times) \; l_2 \; |$$

in terms of the motion of joint λ_1. If the particle has mass M, find an expression for the torque at the joint.

---

6. Find the acceleration of a particle at the end of kinematic chain

$$| \; \theta_1 \; l_1 \; | \; (+) \; \theta_2 l_2 \; |$$

in terms of the motions of joints θ_1 and θ_2. If the particle has mass M, find expressions for the torque at the joints.

---

7. Find the acceleration of a particle at the end of kinematic chain

$$| \; \theta_1 \; l_1 \; | \; (\times) \; \theta_2 l_2 \; |$$

in terms of the motions of joints θ_1 and θ_2. If the particle has mass M, find an expression for the torque at the joint.

---

8. Find the acceleration of a particle at $\mathbf{u}_3$ for

$$\begin{array}{cccc} | \; \sigma_1 & | \; \theta_2 & l_2 & | \\ \mathbf{u}_1 & \mathbf{u}_2 & & \mathbf{u}_3 \end{array}$$

Assume that σ_1 lies along vector $\boldsymbol{\nu}$ and moves in $\mathbf{p}$.

---

9. Find the acceleration of a particle at the end of $| \; \sigma + \sigma \; |$, where σ_1 lies on $\boldsymbol{\nu}$ and moves in $\mathbf{p}$.

---

10. Find the acceleration of a particle at the end of $| \; \sigma + \sigma + \sigma \; |$, where σ_1 lies on $\boldsymbol{\nu}$ and moves in $\mathbf{p}$.

---

11. Find the acceleration of a particle at the end of $| \; \sigma + \sigma \times \sigma \; |$, where σ_1 lies on $\boldsymbol{\nu}$ and moves in $\mathbf{p}$.

---

12. Find the acceleration of a particle at the end of the kinematic chain $| \; \phi\phi\phi \; |$ with unity link lengths. Assume the standard base orientation $\boldsymbol{\nu}$ and $\mathbf{p}_1$.

---

13. Verify equation (9.17) when $n = 1$ and the joint is pitch.

---

14. Verify equation (9.17) when $n = 2$ and the joints are pitch.

---

15. With a uniformly distributed total load m on l_2 in $| \; \lambda\theta \; |$ but the system unloaded otherwise, find the torque on λ_1. Compare the result with that obtained from the two-point mass model.

---

16. Calculate the dynamic equations for the 2 *dof* manipulator $| \; \theta\gamma \; |$ when $\boldsymbol{\nu} = [1 \;\; 0 \;\; 0]^{\#}$ and $\mathbf{p}_1 = [0 \;\; 1 \;\; 0 \;\; 0]$. Assume unity link lengths, uniformly distrib-

uted link masses of m kilogram per unit length, point load of M_2 at $\mathbf{u}_2$, and point load M_3 at $\mathbf{u}_3$. Ignore gravity.

---

17. Repeat problem 16 for $| \, \theta\phi \, |$.

---

18. The system $| \, \theta\psi \, |$ has distributed loads of m kilogram per meter along its links of length l_1 and l_2. Given the position and joint space, calculate the torque required at the base joint.

---

19. The system $| \, \theta\tilde{\psi} \, |$ has distributed loads of m kilogram per meter along its links of length l_1 and l_2. Given the position and joint space, calculate the torque required at the base joint. Ignore gravity.

---

20. Find the torque on θ_1 in the system $| \, \theta\lambda\theta \, |$ when the only load is a point load M at $\mathbf{u}_4$. Assume that the link lengths are unity.

---

21. A compound pendulum is composed of 2 pitch joints at $\mathbf{u}_1$ and $\mathbf{u}_2$ with links of length l_1 and l_2. There are point masses M_1 and M_2 at $\mathbf{u}_2$ and $\mathbf{u}_3$. The links move in plane [0 1 0 0]. If $\boldsymbol{\nu} = [0 \;\; 0 \;\; -1]^{\#}$, write the torques at the joints (which will be zero) in terms of joint space. Is it possible to invert these equations—i.e., write the joint space explicitly?

---

22. Consider the mechanism

$$\begin{array}{l}\mathbf{p}_1\bar{\mathbf{p}}_2^2 \\ \;\; | \, \lambda 0 \, | \, \theta l_2 \, | \, \theta l_3 \, | \\ \;\; \mathbf{u}_1 \quad \mathbf{u}_2 \quad \mathbf{u}_3 \quad \mathbf{u}_4 \end{array} \quad \text{with} \quad \mathbf{u}_1 = \begin{bmatrix} 0 \\ 0 \\ 0 \\ 1 \end{bmatrix}, \qquad \boldsymbol{\nu} = \begin{bmatrix} 0 \\ 0 \\ -1 \end{bmatrix}$$

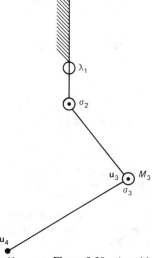

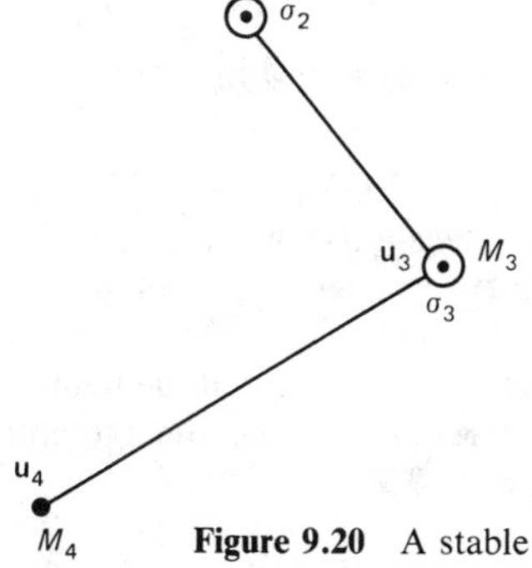

Figure 9.20 A stable state for a driven mechanism

and $\mathbf{p}_l = [0 \quad 1 \quad 0 \quad 0]$. λ_1 is driven at constant velocity, and no torque is applied at the other two joints. There is evidence that it is possible to find nontrivial values for λ_1, θ_2, and θ_3 to produce a stable equilibrium state. Write the equations for the system and investigate the possibility of such an equilibrium state. *Hint:* At equilibrium, $\ddot{\lambda}_1 = \ddot{\theta}_2 = \dot{\theta}_2 = \ddot{\theta}_3 = \dot{\theta}_3 = 0$.

---

23. Determine the point load equivalent for the flat triangular member shown in the accompanying figure. Assume that the total mass is m and the member has uniform thickness.

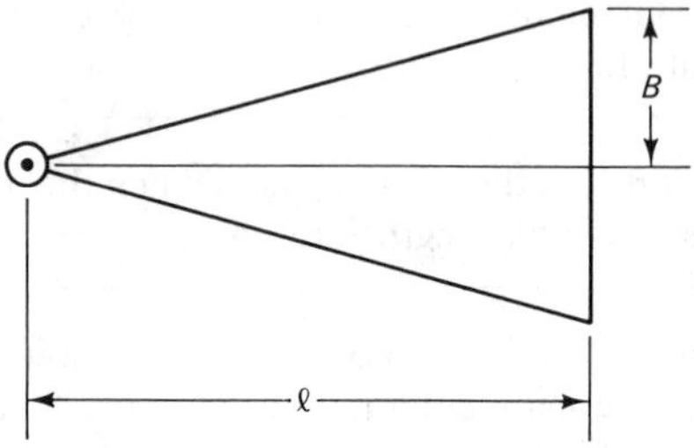

Figure 9.21 A triangular member

---

24. Suppose a uniformly distributed load m on a straight link of length l is approximated by a chain of n equally spaced point loads with equal mass spaced $l/(n + 1)$ apart. Further suppose that the center of mass is at the center. Determine the inertia with respect to the end of the link and compare it to the inertia $I = ml^2/3$ for the distributed load. Solve for $n = 2$, 3, and 4 and generalize to produce an expression for the inertia with a general n.

---

25. Simplify equation (9.29) specifying the force on a mass at the end of kinematic chain $|\ \lambda\theta\theta\lambda\theta\lambda\ |$. What is the number of cross-multiplications in this result?

---

26. Obtain the equations for the torques at the joints for the 3 *dof* cylindrical robot

$$\begin{array}{c} \mathbf{p} \\ |\ \lambda_1\ \sigma_2\ |\ (+)\ \sigma_3\ | \\ \mathbf{u}_1 \qquad \mathbf{u}_2 \qquad\qquad \mathbf{u}_3 \end{array}$$

assuming point masses M_2 at $\mathbf{u}_2$ and M_3 at $\mathbf{u}_3$ in terms of $\mathbf{p}$ and $\boldsymbol{\nu}$, where $\boldsymbol{\nu}$ is the direction of the first link.

---

27. Two pitch joints lie along the z-axis when $\theta_1 = 0$ and $\theta_2 = 0$. The plane of motion is $[0 \quad 1 \quad 0 \quad 0]$ and $\mathbf{u}_1$ is at the origin. A point mass of M kilograms is at $\mathbf{u}_3$, and $l_1 = l_2 = 1$ m.

(a) Write the expression for the torques at the joints as a function of joint space.

(b) Determine the torque required at the joints to hold the mass motionless for $0 \leq \theta_1 \leq 90°$ and for $\theta_2 = 0°$.

(c) If the system is required to move over the slew path $\theta_1(t) = 90t/T$ from $t = 0$ to $t = T$ seconds, where θ_2 is kept at (or close to) zero, for what range of T will the applied torques be essentially the same as in (b).

(d) At the lower limit of T, when the torques start deviating from their static values, what term in the torque expression is most significant in the deviation from the static torques?

CHAPTER TEN

Actuators and Their Control

In Chapter 9 a methodology was developed for converting joint space to torque space. In particular, we showed how to calculate the bottom block in Figure 9.15. Here we address the problem of determining the internal workings of this particular block—that is, how to take the joint space information and drive the joints to produce the required time functions for the joint torque.

The devices that drive the joints of the robot are known as *actuators*. Most discussions of actuators use desired angular position of the actuator shaft as input and actual angular position as the output [Kuo 1987; Ogata 1970]; angular position is determined using optical encoders. We introduce actuators using this format but briefly consider coupled systems of joints and the output torque of each actuator. Most actuator theory uses linear models in the frequency domain, but since the dynamic equations governing practical robots are highly nonlinear, we limit our analysis to the time domain.

The principal classes of actuators are hydraulic, pneumatic, and electric, of which the most diverse and important are the electric actuators. The characteristics of common actuators and their control are discussed. Another critical consideration in the control systems we consider in this chapter is stability. In open-loop systems, stability is not a problem, but since the actuators are employed in closed-loop systems, some of the fundamental techniques used in feedback control systems are sketched [Snyder 1985]. A controller is the closed-loop system around the actuator, and we outline its characteristics here.

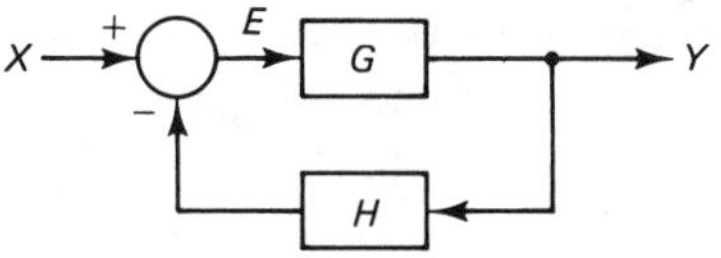

Figure 10.1 The classic single-loop feedback system

We can barely scratch the surface of the control of actuators. However, with the information presented here, the reader should be in a better position to assess the limitations and requirements imposed by the actuators on the path-planning process and the more extensive subject of work-cell design.

Actuators are part of an automatic control system [Kuo 1987; Ogata 1970], and basic knowledge in the classic theories of control systems is a prerequisite to a study of the design of actuators. A block diagram of a simple feedback control system is shown in Figure 10.1. The signal flow through the blocks is indicated by the arrows [Kuo 1987], and the input impedance of each block is assumed infinite, while the output impedance is assumed to be zero. Here the input X is compared at a summing point to the feedback signal passing through measuring element H to produce error signal E. G is often called the plant or the process. The frequency domain equations for this control system are $E = X - HY$, and $Y = EG = (X - HY)G$, so

$$\frac{Y}{X} = \frac{G}{1 + GH} \tag{10.1}$$

10.1 SHAFT ENCODERS

The feedback elements H in actuators are most commonly the encoders. A transducer that converts the position of a rotating shaft into another form is called an encoder. Although a rack-and-pinion gear performs this function (it converts from the rotational pinion to a linear rack), we do not normally classify this as an encoder. More commonly, the encoder produces an electrical output. A simplified version of a potentiometer encoder is shown in Figure 10.2. The position of the wiper is the position of the shaft, and the wiper makes electrical contact with a resistive wire wound on the perimeter of a circle. As the shaft and wiper change position, the resistance changes.

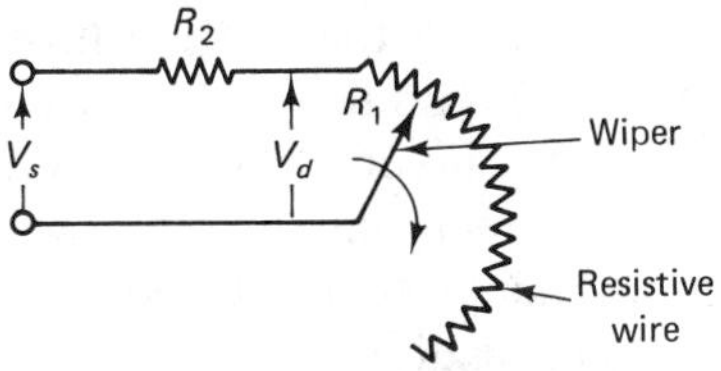

Figure 10.2 A resistive shaft encoder

The resistance of the potentiometer can be converted to voltage using series resistance R_2 and applied voltage V_s, so the voltage across the potentiometer is $R_1V_s/(R_1 + R_2)$. Commercial potentiometers may be more sophisticated, but the transduction of shaft position to analog electrical signal is the same. The device suffers from wear and noise in the contact between the wiper and the resistive wire and errors in the analog output.

The most common encoders in use today are optical encoders. These use an opaque disc attached to the shaft of the motor with holes or slits cut through the opaque layer to allow light from a light source on one side of the disc to strike a sensor on the other side. The sensor is typically a photovoltaic cell (which converts light energy to electricity), a phototransistor, or a photodiode. More sophisticated encoders use collimated light (produced by passing the light through a lens to produce parallel rays) and/or a stationary section of disc with holes or slits similar to that in the rotating disc. The purpose of these devices is to make finer resolution possible.

Optical encoders can be absolute or relative. Absolute encoders can read the position of the shaft at any time and are useful in cases where power outages or disruptions are expected. However, absolute encoders are expensive and more bulky than relative encoders. Relative encoders give shaft position relative to a prior known position and are susceptible to loss of positional information during a power outage. However, the chances of a power outage are small, and most commercial robots are designed to enter a home-position mode before commencing work: The joints are driven to a designated datum position and the optical encoders are reset at this position. Most commercial robots use relative encoders.

The relative shaft encoder can be considered a physical realization of delta modulation. In a delta modulator, the waveshape $W(t)$ at $t = t_i + h \equiv t_{i+1}$ is achieved by adding δ_i (with value $-\delta$, 0 or δ, where δ is fixed) to the value at t_i, i.e. $W(t_{i+1}) = W(t_i) + \delta_i$. The delta modulator and the relative shaft encoder can be thought as acting as numeric integrators.

Due to edge effects, the sensed light is received with the waveshapes shown in Figure 10.3(a). Notice that it is not possible to tell if the rotation is clockwise or counterclockwise. Usually, two light sources and detectors are set up in quadrature (90° out of phase), so that the waveshapes received are as shown in Figure 10.3(b). Notice that as detector B changes state from 0 to 1, detector A is in the 1 position if the shaft rotation is clockwise and in the 0 position if the shaft rotation is counterclockwise. It is a simple logic task to sum the rising pulses of B multiplied by -1 if A is in the 0 state. The total number of pulses together with the starting datum position gives the actual status of the output shaft of the motor, plus or minus the quantization error. This error is inversely proportional to the resolution of the encoder, so the more slits per unit length, the more accurate the encoder. Most electrical motors operate in the 2500-rpm (revolutions per minute) range, so a substantial gearing down is required. The response speed of the optoelectrical part of the system is far faster than that of the mechanical components,

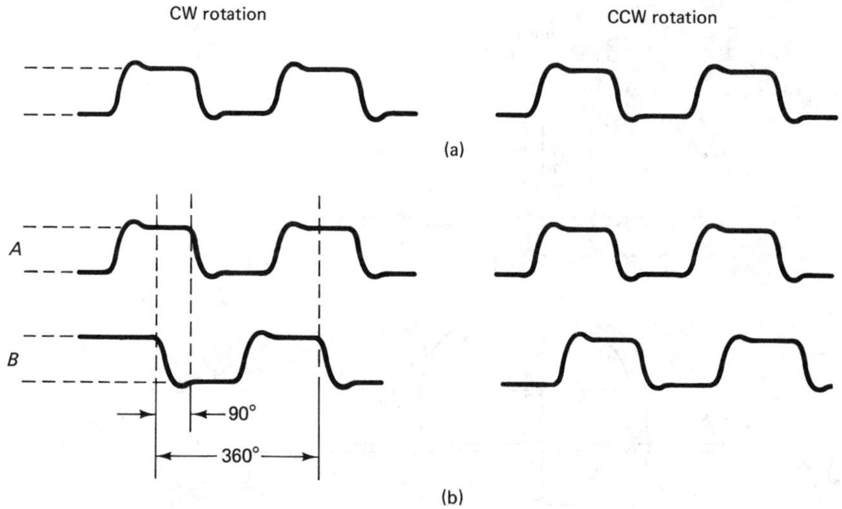

Figure 10.3 Quadrature encoder outputs

so the optical encoder can be attached to the nonreduced shaft of the motor, thus improving the effective resolution of the motor by a factor equal to the gear reduction ratio.

10.2 HYDRAULIC DRIVES

Hydraulic actuators were the dominant method of driving manipulators until recently. The reasons for this are as follows:

1. Most early robots were designed to carry heavy payloads.
2. Hydraulic valves permit the motive force to be remote from the joints.

A diagram of a hydraulic actuator is shown in Figure 10.4 [Ogata 1970]. If the pilot valve moves to the right by x meters, the high-pressure oil moves at rate Q kilograms per second down port I and moves the power cylinder to the right by y meters. The instantaneous pressures in the left and the right sides of the power cylinder are P_1 and P_2 kilogram per square meter^2, respectively. Let $P_1 - P_2 = \Delta P$ and the normal operating point for ΔP be $\Delta \overline{P}$ at $x = \bar{x}$. $Q = f(x, \Delta P)$ is a nonlinear function of x and ΔP, which can be linearized by a Taylor series about an operating point. A Taylor series expansion of $g(x)$ about operating point x_0 is

$$g(x_0 + h) = g(x_0) + h \left.\frac{dg(x)}{dx}\right|_{x=x_0} + \left.\frac{h^2}{2!}\frac{d^2g(x)}{dx^2}\right|_{x=x_0} + \left.\frac{h^3}{3!}\frac{d^3g(x)}{dx^3}\right|_{x=x_0} + \cdots + L(x) \tag{10.2}$$

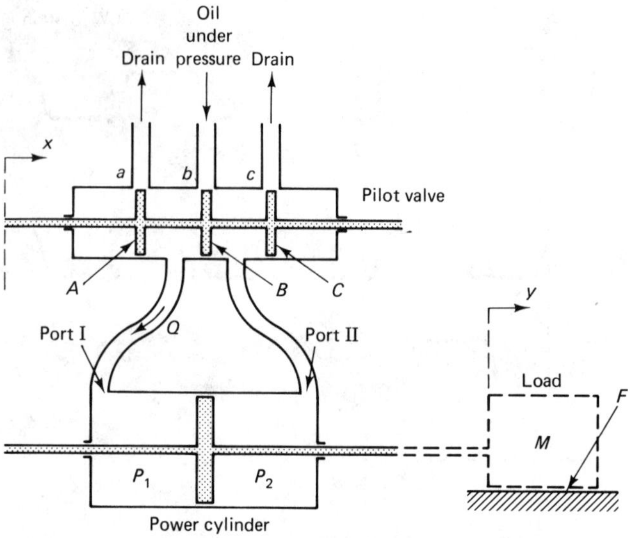

Figure 10.4 A hydraulic drive system

and for $Q(x, \Delta P)$ about operating point $(\bar{x}, \Delta\bar{P})$ the first-order approximation (truncating the Taylor series after the first two terms) becomes

$$Q - \bar{Q} = k_1(x - \bar{x}) - k_2(\Delta P - \Delta\bar{P}) \tag{10.3}$$

where

$$k_1 = \left.\frac{\partial Q}{\partial x}\right|_{x=\bar{x},\ \Delta P=\Delta\bar{P}} \quad \text{and} \quad k_2 = \left.\frac{\partial Q}{\partial \Delta P}\right|_{x=\bar{x},\ \Delta P=\Delta\bar{P}}$$

where the operating point is often $\bar{x} = 0$ and $\Delta\bar{P} = 0$, so

$$Q = k_1 x - k_2 \Delta P \tag{10.4}$$

If the area of the power cylinder piston is A and oil density is ρ kilograms per cubic meter, then

$$Q = A\rho \frac{dy}{dt}$$

Combining this with equation (10.3) gives

$$f = A\Delta P = \frac{A}{k_2}\left\{k_1 x - A\rho \frac{dy}{dt}\right\} \tag{10.5}$$

where f is the force generated by the power piston. If f moves mass M subject to a frictional force F, as shown in Figure 10.1, then

$$f = M\frac{d^2y}{dt^2} + F\frac{dy}{dt}$$

or

$$M \frac{d^2y}{dt^2} + \left(F + \frac{A^2\rho}{k_2}\right) \frac{dy}{dt} = \frac{Ak_1}{k_2} x \tag{10.6}$$

which has the same form as the spring-mass-dashpot system of Figure 10.1 and equation (9.1) with the spring constant zero. If M is small enough to have little effect on the response of the system, then the rate of change of y is proportional to the pilot valve displacement x.

The work involved in moving the pilot valve by distance x is small compared to the work produced, so the gain of the system (output/input) is large. Hydraulic systems can produce large forces or torques to drive robotic joints, so robots with hydraulic actuators are often used in heavy industry. The hydraulic pump and oil reservoir can be placed close to but not on the manipulator linkages themselves, and the reinforced hoses run up the linkages to drive the joints. An optical encoder controls the motion of the pilot valve. Usually the pilot valve is driven by a lead screw, which is discussed in Section 10.10 and shown in Figure 10.30.

10.3 PNEUMATIC ACTUATORS

The structure of the pneumatic actuator shown in Figure 10.5 is similar to that of the hydraulic actuator. These are the principal differences:

1. The purge valves release the compressed air to the environment, whereas the drains in the hydraulic system return the oil at low pres-

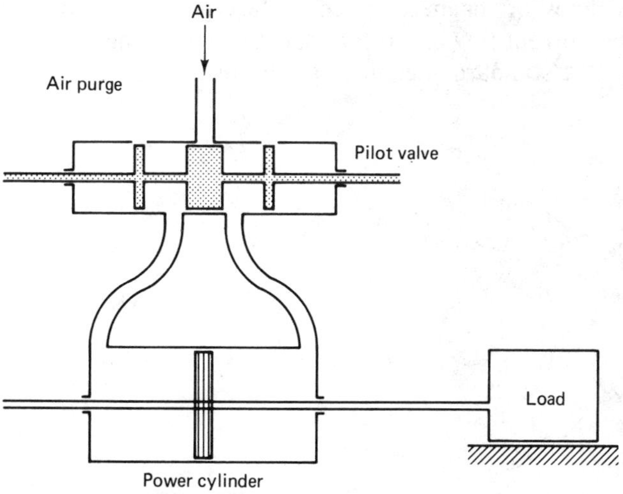

Figure 10.5 A pneumatic drive system

sure to a sump, from which a hydraulic pump draws oil to pressurize and feed the input to the pilot valve.

2. The mass to be moved has cushioned constraints that limit the motion of the power cylinder. These constraints are manually adjustable, and the pneumatic action causes the mass to bang against an end stop. Hence we say that a pneumatic robot is a bang-bang robot.

The mathematical specification of a pneumatic system is difficult to obtain, primarily because air is a compressible fluid. The response of the power piston is a function of the differential air pressure on either side of the piston, which is itself a function of the lengths, areas, number of bends, and obstructions in the air flow passages from the pressure tank to the power cylinder, where the opening of the orifice to the pressure tank produced by the motion of x is the most important determinant. Usually, the pressure in the air tank is adjusted as required to produce acceptable motion characteristics.

10.4 DC MOTORS

The DC motor is used more than any other electric motor as robotic actuator. All electric motors operate by passing current through conductors that lie in a magnetic field. The detailed subject of electric motor design, including windings, pole shape, and placement, and so on, is complex and beyond the scope of this work. However, certain principles can be introduced that can help characterize the performance of electrically operated robot actuators.

A current passing through an electrical conductor produces magnetic flux (called lines of magnetic force or flux lines) around the conductor. Suppose the current flowing in a conductor is into the paper, as shown in Figure 10.6 (the standard specification for current flow is + for current into

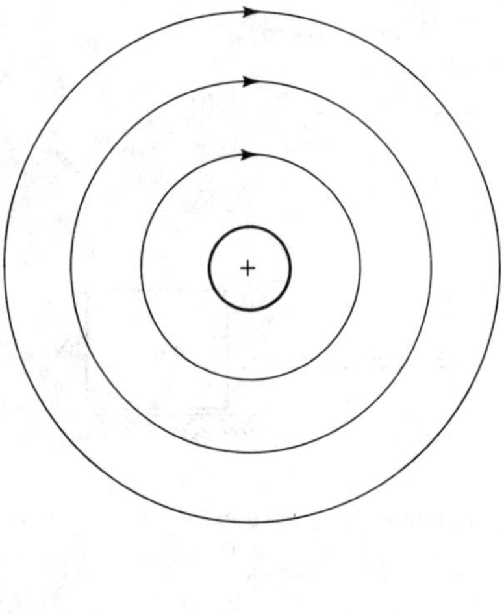

Figure 10.6 The lines of flux around a current-carrying conductor

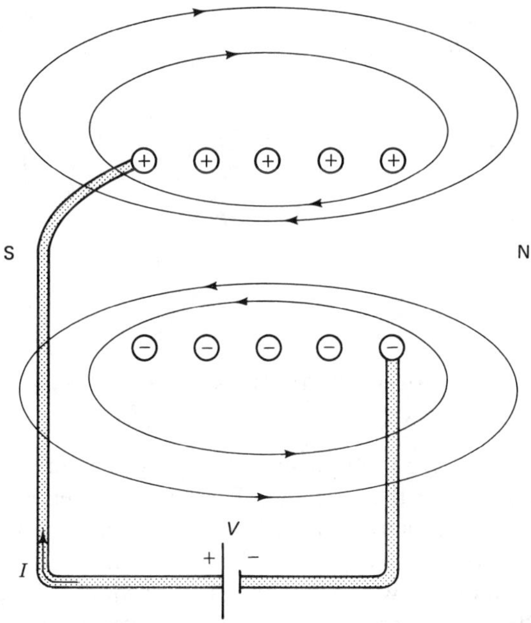

Figure 10.7 The magnetic field around a current-carrying spiral

the paper and − for current flow out of the paper), then the flux lines produced are as shown. The direction of flux is given by the right hand rule: partially curl the fingers, point the thumb in the direction of the current and the curling fingers give the direction of the lines of flux.

Suppose the conductor-carrying current is bent into a spiral, as shown in cross section in Figure 10.7; if we assume that the number of turns is large, the magnetic effect of the lead-in and lead-out conductors can be considered negligible. A property of flux lines is that they cannot cross each other, and the interference of the flux produced by one part of the conductor with another deforms the flux lines. The resulting flux lines are shown in Figure 10.7. These flux lines have the same shape as produced by a bar magnet, where the north and south poles are marked N and S in the figure. That is, the current-carrying spiral conductor behaves as a bar magnet. However, it has a great advantage over the bar magnet, since the current can be altered to change the intensity and direction of the magnetic flux. Further, the current can be controlled automatically as part of a feedback system, a situation that occurs for the DC motors in all robotic actuators.

A property of magnets is that like poles repel and unlike poles attract, so a current-carrying spiral in a magnetic field will result in a force between spiral and magnet. This force can be employed to change electrical energy into motion (mechanical energy) and is the principle of operation of all electric motors.

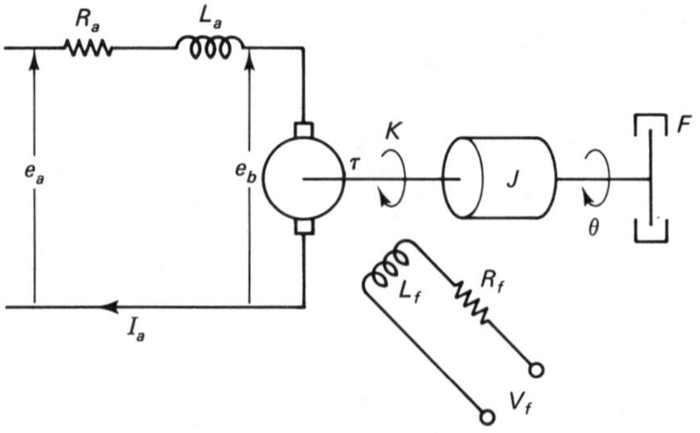

Figure 10.8 The DC motor and load

In DC motors there are two sets of windings. One set is in the armature, the rotating part of the machine. Electrical connection is made to the armature with soft conductors called *brushes* that contact axial copper conductors on the armature shaft, called the *commutator*. The other set comprises the field windings, which are fixed and produce a magnetic field that interacts with the field produced by the armature windings. A simplified electrical circuit that describes the DC motor is shown in Figure 10.8. J is the moment of inertia of the armature and its connected load. F is the frictional force of the armature and load. Subscripts a and f designate armature and field, respectively.

The magnetic field produced by the field coil will be quite small unless iron is added. Just as the current in an electrical circuit is inversely proportional to the resistance in that circuit, the flux in a magnetic circuit (the flux lines are always closed) is inversely proportional to what is known as the magnetic reluctance of the circuit. Air has a high reluctance, but iron has a low reluctance. The field and armature windings are embedded in iron, and the air gap between the iron (this iron produces the poles, so-called because of their shape and function) in the field windings and the armature is small, so the overall magnetic reluctance is low and the flux in the air gap is high. This makes the conversion from electrical to mechanical energy more efficient in all types of electric motor. The iron is close to but electrically insulated from the current-carrying conductors.

Typically the armature resistance is small, and increasing I_a sufficiently to satisfy the equation while leaving V_a untouched will burn the motor out. A series resistor in the armature circuit prevents burnout, as does reducing V_a. We assume that R_a is the sum of armature and series resistance. Increasing R_a while leaving V_a unaltered gives a nonzero torque at $\omega = 0$. If V_a is reduced to zero, both the speed and the torque collapse. Typical speed-torque curves as functions of R_a and V_a are shown in Figure 10.9, where we assume the simple condition $K = 1$ and, nominally, $R_a = 1$ and $V_a = 1$.

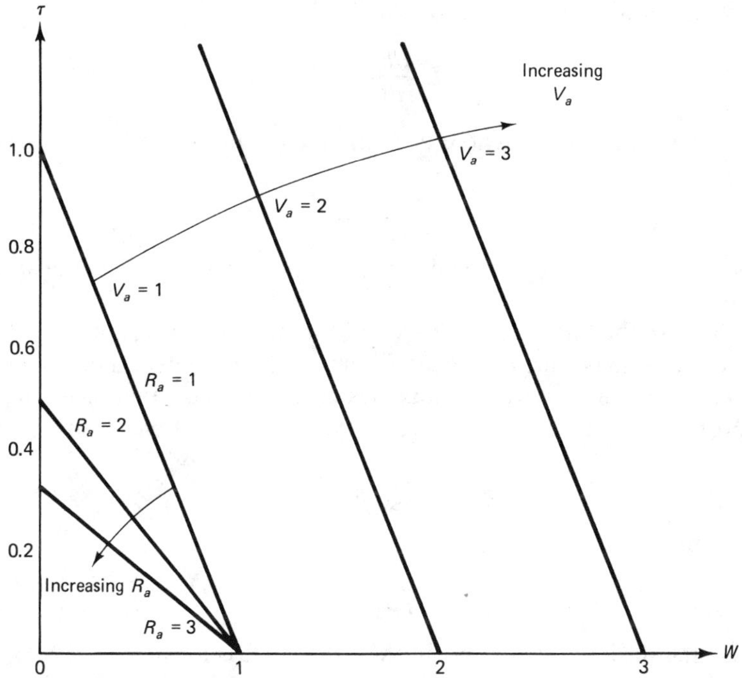

Figure 10.9 Speed-torque characteristics of the DC motor

One way of varying DC voltage V_a is to replace it with a pulse of width δ over cycle Δ seconds, as shown in Figure 10.10. If the frequency $1/(2\pi\Delta)$ cycles/second is sufficiently high, the inductive load imposed by the DC motor integrates and removes all variations in V_a, so the motor sees an effective input voltage of $V_a = \overline{V}\delta/\Delta$. The controller (Section 9.3) of the DC motor controls δ and so linearly controls V_a. We say that the controller employs pulse-width modulation.

The time domain equations governing the electrical part of the system

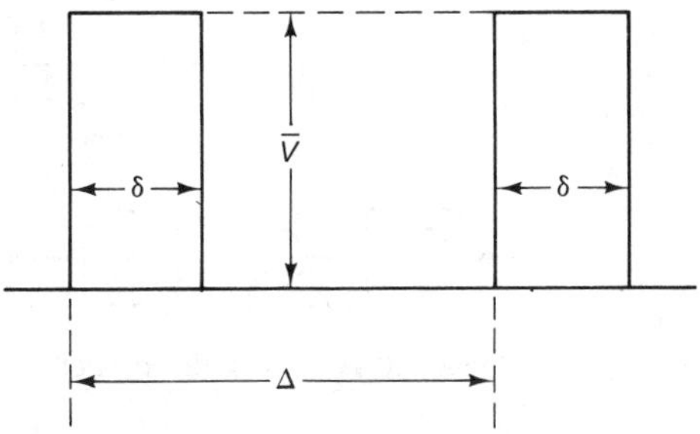

Figure 10.10 Pulse-width modulation

are [Fitzgerald, Kingsley, and Kusko 1971]

$$e_a = L_a \frac{di_a}{dt} + i_a R_a + e_b \tag{10.7}$$

for the armature circuit, where e_b is the back emf induced in the armature windings.

$$\tau = k\phi i_a \tag{10.8}$$

since the torque τ is proportional to the flux in the air gap ϕ and the armature current.

Suppose the flux in the air gap is constant [Ogata 1970], as would occur if the field windings were replaced by permanent magnets. Then the motor is said to be armature controlled, flux ϕ is essentially constant ($k\phi = K_1$), and e_b is proportional to the angular velocity of the output shaft, so

$$e_b = K_2 \frac{d\theta}{dt} \tag{10.9}$$

and

$$\tau = K_1 i_a \tag{10.10}$$

Suppose the desired angular position of the shaft is θ_d, and $e_a = K_3(\theta_d - \theta)$. Then we can construct the block diagram shown in Figure 10.11. The dimensions of the constants are kg/m-A for K_1 and volts for K_2 and K_3. The meaning of the blocks DE1 and DE2 is that the output is the solution to a differential solution. The input to DE1 is

$$L_a \frac{di_a}{dt} + i_a R_a$$

and its output is i_a. The input of DE2 is τ and its output is θ.

It is common to analyze the system assuming a simple, inertial load on the system, so

$$J \frac{d^2\theta}{dt^2} + F \frac{d\theta}{dt} + K\theta = \tau \tag{10.11}$$

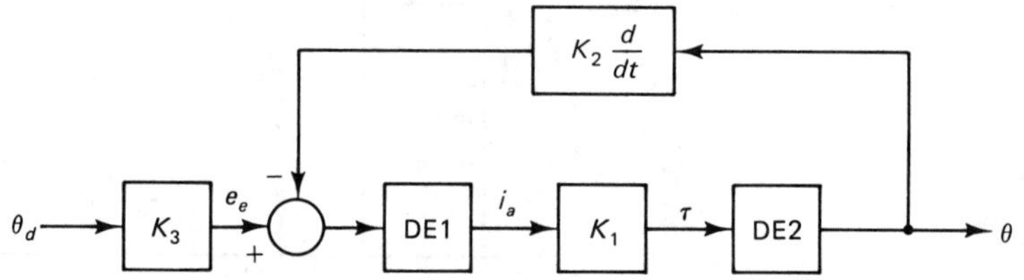

Figure 10.11 Block diagram of an armature-controlled DC motor

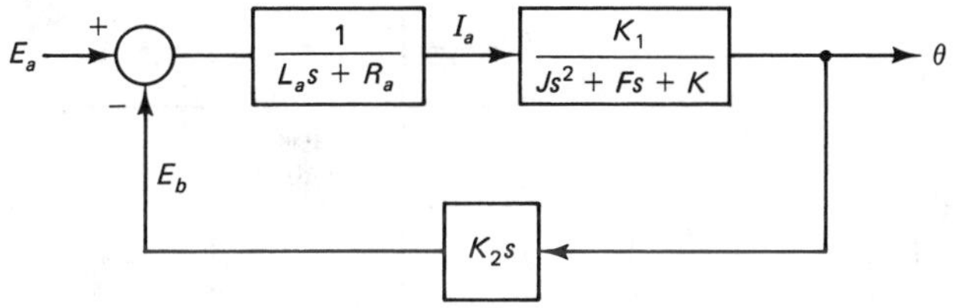

Figure 10.12 Block diagram of an armature-controlled DC motor

and the equations become linear, permitting the Laplace transform to be taken to produce equations that can be put in block diagram form. (See Figure 10.12.) We can replace this linear system with a single block G, whose input is armature voltage and whose output is the actual shaft angle. Since the load on the motor is nonlinear, we cannot analyze the system using Laplace transforms, as discussed. Instead we classify the nonlinear block in the time domain by symbol g.

If the input of a joint actuator is desired position and the actual position is determined by an optical encoder, the block diagram of the position-controlled actuator becomes as shown in Figure 10.13: $e_a = K_3(\theta_d - \theta)$. The shaft position controls the armature voltage e_a, which produces current i_a, which develops torque τ, but we cannot determine θ from τ. We can, however, determine torque space from joint space, and reformulating the block diagram to be in closed-form-analyzable form produces Figure 10.14. It is not clear that this form can be used in the design or control process. Henceforth, all motors will be modeled by nonlinear block g with the controlling parameter (armature voltage for the armature-controlled DC motor, field voltage for the field-controlled DC motor discussed next, and so on) as input and shaft position θ as output.

The DC motor can be controlled via its field windings with the armature current constant [Fitzgerald, Kingsley, and Kusko 1971]. For this connection and with position control the operating equations are

$$\tau = K_1 i_f; \qquad L_f \frac{di_f}{dt} + R_f i_f = e_f; \qquad e_f = K_3(\theta_d - \theta) \tag{10.12}$$

The advantage of field control over armature control is the lower power level required in the control circuit. However, the disadvantages of the field-

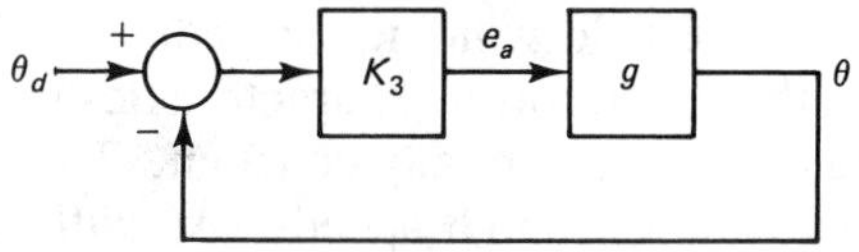

Figure 10.13 Position control of an armature-controlled DC motor

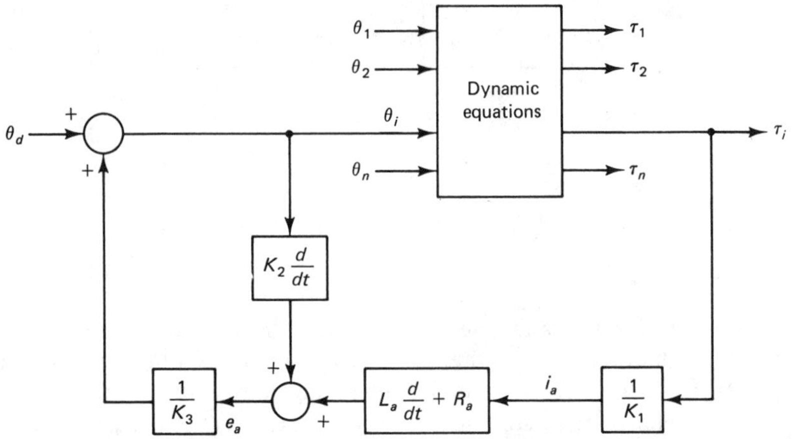

Figure 10.14 Block diagram of DC motor for use with kinematic linkages

controlled motor include the following:

1. The requirement that the armature current remain fixed requires a current power source, not a simple requirement to satisfy.
2. There is a lack of damping, which must be provided by the motor and load, whereas in the armature-controlled motor the back emf provides damping.
3. The heat losses caused by the constant armature current could cause armature burnout.
4. The time constants are large, so the response may be slow.

The field windings are often connected to the armature windings. The usual connection is to place them in parallel (shunt), and the motor is then called a shunt motor; it is common to have a variable resistor in series with the field windings, and we will take this into account as R_f. Under the assumption that L_a is small, the speed-torque characteristics of the motor are shown in Figure 10.15(a) for increasing R_f and in 10.15(b) for increasing R_a; $\omega = d\theta/dt$. The speed is not a strong function of the shaft torque. The equations governing the system (even when we ignore centrifugal and gravitational forces and cross-coupling effects between the joints) are nonlinear, since $\tau = K_1 i_f i_a$, and no attempt will be made to reduce this type of motor to a block diagram description.

If the field windings are connected in series with the armature windings, the speed-torque characteristics become as shown in Figure 10.16: The product $\omega\tau$ is essentially constant. This produces a large starting torque but a low torque at high speeds. As with the shunt motor, the equations are nonlinear, and no attempt will be made to produce a block diagram.

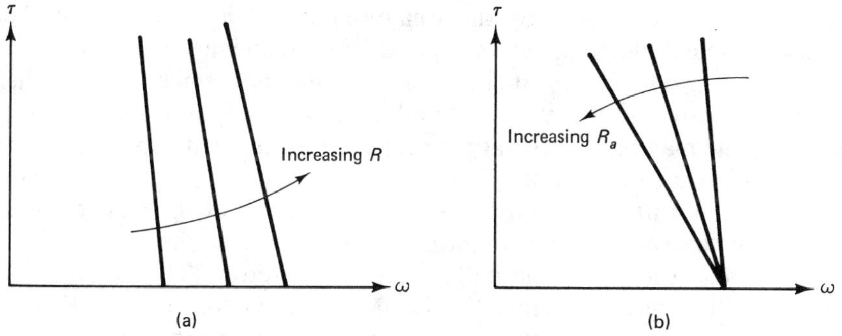

Figure 10.15 Speed-torque characteristics of shunt DC motor

We have assumed throughout that the initial conditions are zero when taking Laplace transforms. This does not negate the technique if the initial conditions are not zero. Frequency domain specifications are often used under steady-state conditions for which the initial conditions are ignored. However, the frequency domain specification can readily be converted to a state variable description [Kuo 1987], and state variable descriptions use the initial conditions. We avoided the problems of taking series of convolution integrals by converting to polynomial descriptions in s, performing as many calculations as possible in s before reconverting (if necessary) to the time domain.

The preceding discussion of DC motors barely scratches the surface. The design of such motors is an ongoing task absorbing the energies of many engineers. The versatility of such motors for robotic actuators has accelerated that attention. It is possible to do such things as shape the poles, construct the windings, and design the control circuitry to produce specialized responses. However, for our purposes, knowledge of the speed-torque

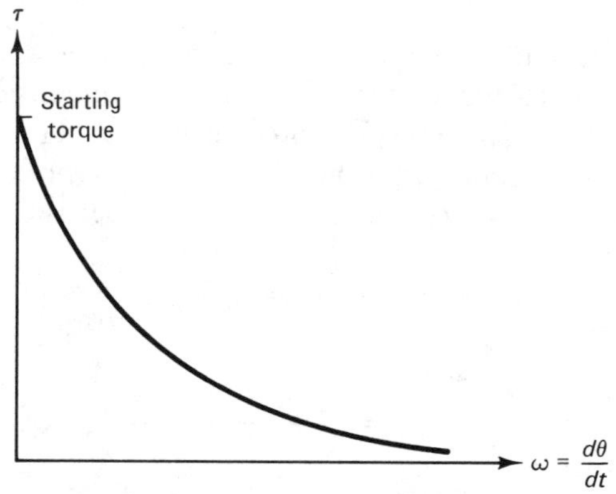

Figure 10.16 Speed-torque characteristics of series DC motor

characteristics and (at least for the armature-controlled and field-controlled motors) the block diagrams are adequate for a preliminary analysis.

It may appear that DC motors are unsatisfactory, since when $\omega = d\theta/dt = 0$, τ is also essentially zero. The load on the output shaft of a motor caused by the manipulator linkages produces a torque that varies according to the position and orientation of the linkages[1]. This load torque could cause the linkages to drift, moving the position θ of the output shaft. The optical encoders on this shaft indicate to the control circuitry that the motor is out of position, and an input on the motor corrects the error. The process continues of drifting and correcting; this is often called motor hunting. Hunting can be corrected electrically with a compensator (see Section 10.8.3) or mechanically by frictional forces in the gear trains (see Section 10.10).

10.5 AC MOTORS

In the DC motor it is possible for the stationary field windings (called the stator) to be replaced by permanent magnets. In AC machines the field windings are driven by alternating current, so the effect is the same as rotating magnets surrounding the rotor (the name used in AC motors to describe the armature). Often the rotor conductors are connected at the ends to produce what is called a squirrel cage because of its shape [Fitzgerald, Kingsley, and Kusko 1971]; the cage is filled with iron to increase the flux in the air gap by several orders of magnitude. There are no electrical connections to the squirrel cage. When this rotating field cuts the rotor windings, current is induced in the rotor, the fields interact, and a torque is produced. The torque is zero when the rotor is moving at the same angular velocity as the rotating field produced by the field windings. A typical squirrel-cage induction motor is almost a constant-speed device with speed-torque characteristics: From no load to full load, the speed may drop by 5%.

If ω_s is the speed of the stator in radians per second and ω_r is the speed of the rotor in radians per second, S is the slip (dimensionless), f is the frequency of the electrical power source in cycles per second, P is the number of poles on the stator, n_s is the speed of the stator field in radians per second, and n_r is the speed of the rotor in radians per second, then

$$S = \frac{\omega_s - \omega_r}{\omega_s} = \frac{n_s - n_r}{n_s} \tag{10.13}$$

where

$$n_s = \frac{120f}{P} \text{ rpm} \tag{10.14}$$

[1] Typical payloads are less than 3% of the weight of the linkages and attached drives and so are rarely a significant factor.

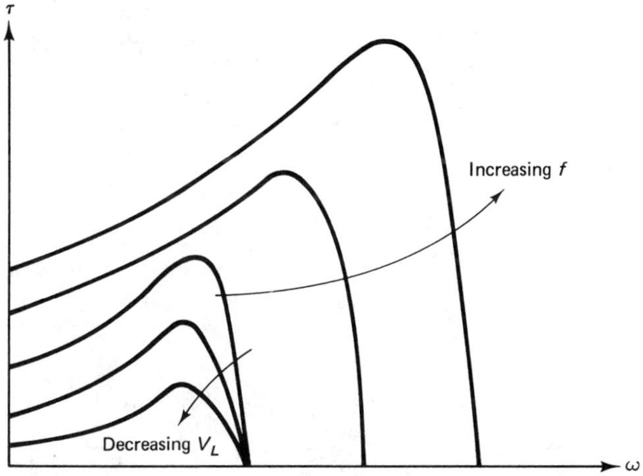

Figure 10.17 Speed-torque characteristics of induction motor

and

$$\omega_s = \frac{2\pi n_s}{60} \tag{10.15}$$

The speed-torque curves of the induction motor are shown in Figure 10.17, illustrating the effect of increasing f and decreasing the line voltage V_L applied to the stator. It is apparent that the most effective speed control for this motor is varying the incoming frequency. Frequency control is provided using a rectifier (which changes the AC fixed frequency line power into DC power) and an invertor (which converts DC power into AC power at a controllable frequency). The input to the invertor is provided by the motor controller.

A brushless DC motor is a special type of AC motor gaining popularity as a replacement for small DC motors. It is a three-phase synchronous AC machine with a permanent magnet rotor whose rotor speed is a direct function of the input frequency [Krause 1986]. The input frequency can be a direct function of a DC input voltage, and with this DC voltage as input, the control characteristics of the motor are similar to a DC motor without brushes—hence the name.

An AC fractional horsepower motor used most often for instrument control is the two-phase AC servomotor [Ogata 1970] shown in Figure 10.18(a). The fixed-phase field is continually exited with a reference voltage $v_f = v_f \sin \omega t$ in the frequency range 60 to 1000 Hz. The control phase is $v_c = V_c \sin(\omega t + 90)$. The stator windings are 90° apart. Change the sign of V_c and the motor changes direction. The motor is usually designed with low inertia, and it has high torque at low speed. We assume that changes in V_c

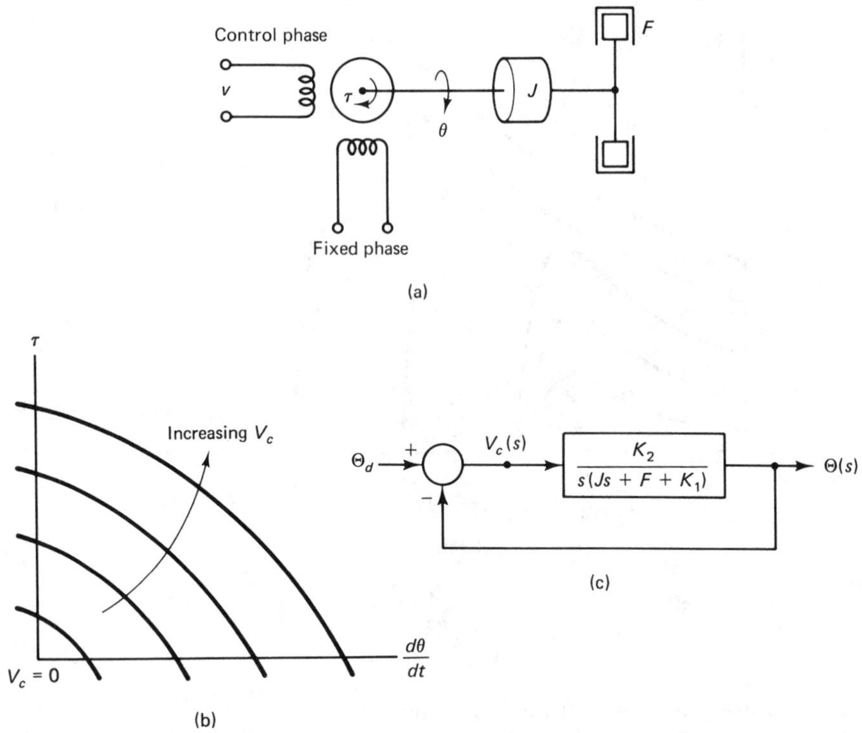

Figure 10.18 The two-phase servomotor

are slow compared to ω; then the speed torque characteristic is as shown in Figure 10.18(b). As a first approximation, we assume the lines in Figure 10.18(b) are parallel and the distance between them is proportional to the change in V_c, so the motor has the characteristics defined by

$$\tau = -K\frac{d\theta}{dt} + K_2V_c; \qquad V_c = k_3(\theta_d - \theta).$$

A block diagram realization of these equations is shown in Figure 10.18(c).

A motor with a flat torque characteristic over all operating speeds is the hysteresis motor. Hysteresis is the phenomenon in which the magnetic character of iron is a function of how it was magnetized before. Place an unmagnetized iron bar in a magnetic field and remove the iron, and the bar is magnetized. However, if the bar is then replaced in reverse in the magnetic field and then removed, its magnetic characteristics will be somewhat different. This is called *hysteresis,* and it causes energy losses in the iron of a motor. Soft iron is poor at retaining magnetization. Hard steel can retain magnetization much better than soft iron, but it is more difficult to magnetize in the first place. Hysteresis is far more pronounced in hard steel than in soft iron.

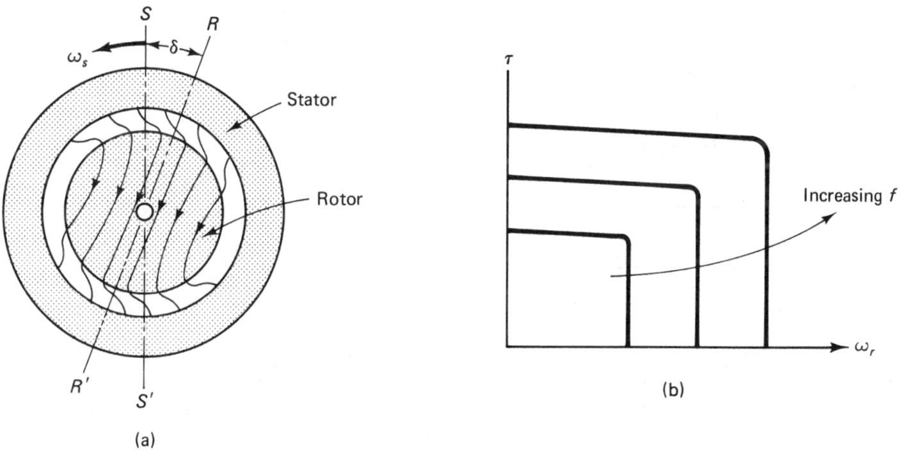

Figure 10.19 The hysteresis motor

In a hysteresis motor the rotor is made out of hard steel without windings. The windings and iron in the stator are designed to produce a sinusoidal rotating flux in the air gap that magnetizes the steel. The effects of hysteresis are such as to cause the magnetization of the steel to lag the rotating field. A simplified diagram of the hysteresis motor [Fitzgerald, Kingsley, and Kusko 1971] and its speed-torque characteristics as a function of w_s are shown in Figure 10.19. There are some specialized robotic applications that can require such a machine.

10.6 STEPPER MOTORS

The problem with the electric motors described in Sections 10.4 and 10.5 is that the position of the output shaft of each motor is not accurately known as a function of the electrical inputs. Positional information is usually obtained from optical encoders, and such encoders are relatively expensive and delicate. A motor in which position is always known as a function of the electrical input is a stepper motor [Fitzgerald, Kingsley, and Kusko 1971].

A stepper motor rotates a specific number of degrees for each electrical pulse received. Stepper motors vary in size from minute to multiple horsepower and can accept input pulses at rates up to about 1200 pulses per second. The construction of a simple four-phase stepper motor is shown in Figure 10.20. The rotor assumes angles of $45i^0$, where i is an integer, including zero. The windings are excited in the sequence N_a, $N_a + N_b$, N_b, $N_b + N_c$, N_c, $N_c + N_d$, N_d, and $N_d + N_a$. The permanent magnet rotor is locked in place by the magnetic field windings. If the current through these windings is removed, the locking torque is also removed and the rotor is free to move.

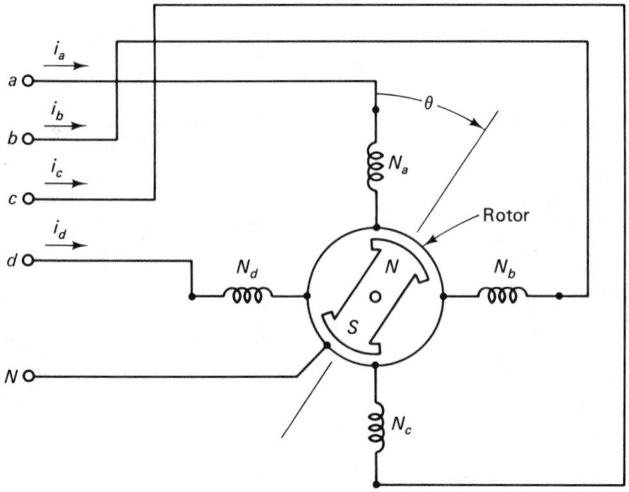

Figure 10.20 Elementary diagram of four-phase stepper motor

The position of the output shaft of the stepper motor connected to the rotor is algebraic sum of the pulses times the angle of rotation per pulse from the assigned datum. If the pulse rate to the control system that drives the field windings is above the capability of the stepper motor to respond, the motor groans and refuses to move. The energy absorbed by the stepper motor when energized but quiescent (not moving) is fairly high, so heat buildup must be considered.

10.7 STABILITY

Given some system, linear or nonlinear, we inquire as to its stability. For the linear system $\mathbf{y}' = A\mathbf{y}$, stability is a function of the eigenvalues of A: The real parts of the eigenvalues must all be negative for the system to be stable. Determination of eigenvalues is computationally expensive. If A is negative definite, the system will be stable. We can test for negative definiteness by checking the signs of the leading principle minors [Gantmacher 1959] (this is Sylvester's criterion). If the characteristic polynomial of A is known, the efficient Routh test [Routh 1877; 1959] can be used. If the system is nonlinear, such as $\mathbf{y}' = \mathbf{f}(\mathrm{y}, t)$, we can do the following:

1. Linearize the system at a point (Taylor series or describing function methods [Ogata 1970]) and use the linear model to describe the system.
2. Use information at a series of points to estimate the value of the function at the next point—this is the technique employed in numeric integration.

3. Find the conditions for forcing $\mathbf{f}(y)$ to zero, which is then an equilibrium point of $\mathbf{y}' = \mathbf{f}(y)$.
4. Apply a Liapunov function to the system about a known equilibrium point to test for asymptotic stability.

10.8 TRACKING ERROR AND COMPENSATION

The robotic joint actuator and its control system should possess the following properties:

1. Have a low tracking error
2. Have no overshoot or oscillation
3. Have low sensitivity to variations such as temperature or humidity
4. Be insensitive to noise in the system

Techniques for improving these factors are considered in this section.

The most common system for basic study of transient response is the second-order system

$$\frac{Y(s)}{X(s)} = \frac{\omega_n}{s^2 + 2\xi\omega_n s + \omega_n^2} \tag{10.17}$$

where ξ is the damping factor and ω_n is the undamped natural frequency. With input $x(t)$ a unit step, performing a partial fraction expansion and taking the inverse Laplace transform gives the response

$$y(t) = 1 + \frac{e^{-\xi\omega_n t}}{\sqrt{1-\xi^2}} \sin\left(\omega_n \sqrt{1-\xi^2}\, t - \tan^{-1}\frac{\sqrt{1-\xi^2}}{-\xi}\right), \qquad t \geq 0 \tag{10.18}$$

The two poles of equation (10.17) are at

$$-\xi\omega_n \pm j\omega_n\sqrt{1-\xi^2} \tag{10.19}$$

and these poles are real and identical when $\xi = 1$, a condition known as critical damping. When $\xi = 0$, the unit step results in $y(t) = 1 - \cos \omega_n t$, an oscillatory response. $y(t)$ as a function of $\omega_n t$ is shown for various damping factors in Figure 10.21, showing damped oscillations for $0 < \xi < 1$ and slow, nonoscillatory response for $1 < \xi < \infty$. Taking the derivative of $y(t)$ in equation (10.18) gives

$$\begin{aligned} \frac{dv(t)}{dt} &= \frac{\omega_n e^{-\xi\omega_n t}}{\sqrt{1-\xi^2}} \left\{ -\xi \sin\left(\omega_n\sqrt{1-\xi^2}\, t - \tan^{-1}\frac{\sqrt{1-\xi^2}}{-\xi}\right) \right. \\ &\quad \left. + \sqrt{1-\xi^2} \cos\left(\omega_n\sqrt{1-\xi^2}\, t - \tan^{-1}\frac{\sqrt{1-\xi^2}}{-\xi}\right)\right\} \\ &= \frac{\omega_n e^{-\xi\omega_n t}}{\sqrt{1-\xi^2}} \sin \omega_n\sqrt{1-\xi^2}\, t, \qquad t \geq 0. \end{aligned} \tag{10.20}$$

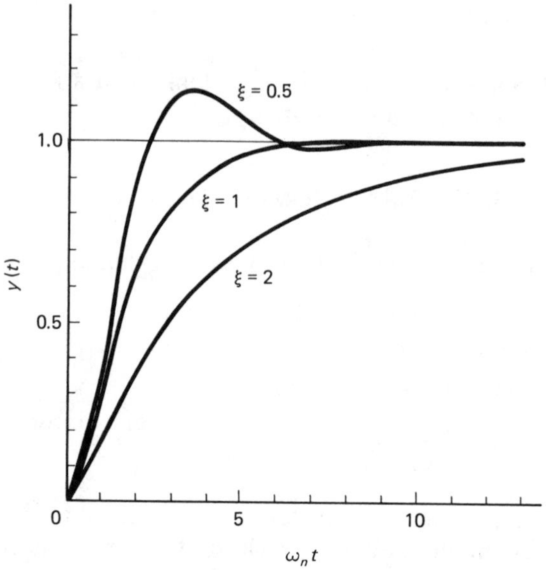

Figure 10.21 Response of a second-order system as a function of damping

At a maximum or minimum, $dy(t)/dt = 0$; then $t = \omega$ and $\omega_n\sqrt{1 - \xi^2}\, t = n\pi$, $n = 0, 1, 2, \ldots$, so

$$t = \frac{n\pi}{\omega_n\sqrt{1 - \xi^2}} \tag{10.21}$$

The first maximum value occurs at $n = 1$, so the maximum overshoot is

$$t_{\text{max}} = \frac{\pi}{\omega_n\sqrt{1 - \xi^2}} \tag{10.22}$$

Linear or quasilinear first- and second-order systems are amenable to analysis such as this. Higher-order and nonlinear systems are more difficult to analyze, and we are usually forced to resort to numerical approximations.

10.8.1 Tracking Errors

The tracking error is $e(t)$ in the time domain and $E = E(s)$ in the frequency domain, and from Figure 10.1 we have

$$E = \frac{X}{1 + HG} \tag{10.23}$$

The time domain error for the second-order system defined by equation (10.17) and subject to a step input is shown in Figure 10.22 for $\xi = 2$, 1, and 0.5. This error $e(t)$ starts at 1 and, provided $\xi > 0$, becomes zero at $t \rightarrow \infty$.

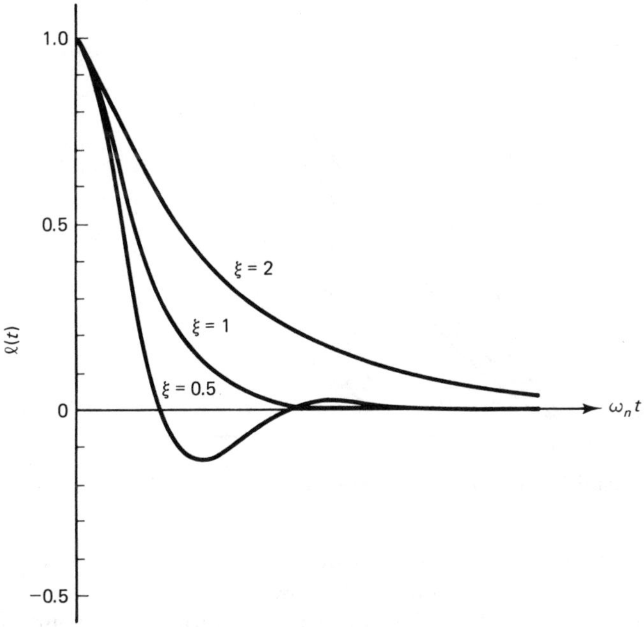

Figure 10.22 The tracking error due to a step input

Of interest is the steady-state error

$$e_{ss} = \lim_{t\to\infty} e(t) = \lim_{s\to 0} sE(s) = \lim_{s\to 0} \frac{sX}{1 + HG} \tag{10.24}$$

Suppose

$$HG = \frac{K(s + a_1)(s + a_2)\cdots(s + a_m)}{s^j(s + b_1)(s + b_2)\cdots(s + b_n)} \tag{10.25}$$

where j is said to determine the type of the system. We say that this HG is type j. Suppose the input $r(t)$ is a step function $\alpha u(t)$, so $X = \alpha/s$, and

$$e_{ss} = \lim_{s\to 0} \frac{\alpha}{1 + HG} = \frac{\alpha}{1 + \lim_{s\to 0} HG} = \frac{\alpha}{1 + K_p} \tag{10.26}$$

where K_p is called the *step error constant;* we want K_p to be large in order to make e_{ss} small, and this requires $j \geq 1$, so the system must be type 1 or higher. That is, in order for the system to follow a step input, one or more poles of E must exist at the origin.

Next, suppose the input $x(t)$ is a ramp function αt; then $X = \alpha/s^2$, and

$$e_{ss} = \lim_{s\to 0} \frac{\alpha}{s + sHG} = \lim_{s\to 0} \frac{\alpha}{sHG} = \frac{\alpha}{K_v} \tag{10.27}$$

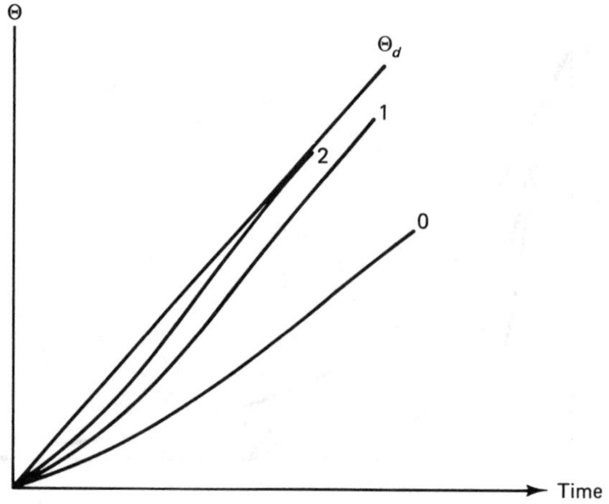

Figure 10.23 Typical responses of type 0, 1, and 2 to a ramp input

where K_v is called the *ramp error constant*. For $e_{ss} = 0$, $K_v = \infty$, and so $j \geq 2$. For $j = 0$, the tracking error becomes unbounded, and for $j = 1$, $e_{ss} = \alpha/K_v$. Sketches of the tracking errors $e(t)$ to a ramp input for type 0, 1, and 2 systems are shown in Figure 10.23.

Suppose the initial conditions are zero; then [Ogata 1970]

$$e_{\text{ramp}}(t) = \int_0^t e_{\text{step}}(t)\, dt$$

so the instantaneous error e_{ramp} due to a ramp input is the integral of the step error. Stating this another way, the ramp error at t is the area under the curve $e_{\text{step}} = 1 - y(t)$ from $t = 0$ to time t.

In summary, the form of HG determines the tracking error of the control system and should be a critical consideration in the overall design.

10.8.2 Compensators

Suppose the actuator and its control system with transfer function G is shown to exhibit undesirable characteristics. We may have a pair of complex conjugate poles that can be excited by the given input, leading to ringing or overshoot in the response. If there is no direct way of modifying G, a compensator can be employed to appear to move these poles. A *compensator* is a device with a transfer function containing one or more poles and one or more zeros that is placed in series with G or placed as a feedback element across G. It is also possible to have both forms of compensation simultaneously. Suppose G has pole-zero pattern, as shown in Figure 10.24(a). The compensator with pole-zero pattern shown in Figure 10.24(b) will result in an effective pole-zero pattern, as shown in Figure 10.24(c), provided the

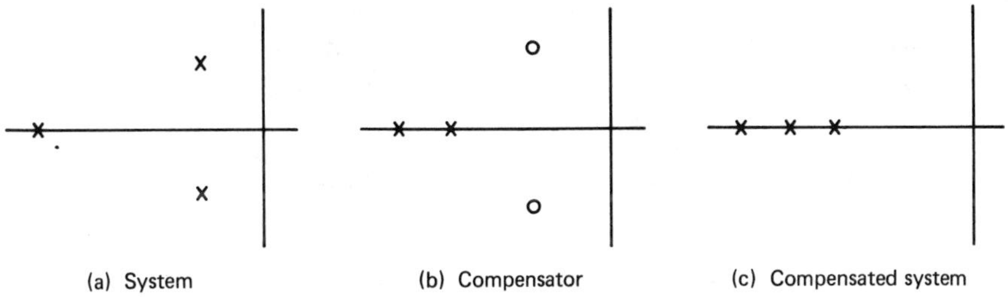

Figure 10.24 Root movement under compensation

zeros of the compensator are sufficiently close to the corresponding poles of G.

A compensator cannot remove instability. If a pole(s) exists in the right-hand s-plane, the use of a compensator with zero(s) lying on top of the poles will not stop the output from becoming unbounded. The reason is that it is impossible to position any singularity with the infinitesimal accuracy required, and even the smallest variation permits the unstable poles to draw and build up energy from the power source. It is rare to find a conditionally stable system that can be made unconditionally stable with a compensator.

If the compensator for the motor is properly designed with integral control (that causes the motor to increase output torque until all movement stops in the absence of a control input) in addition to proportional control (always needed so the output tracks the input) and some derivative control (to enable the system to respond rapidly to an input), the resulting proportional-integral-derivative (PID) compensator can eliminate hunting while still making the system responsive and free from overshoot or oscillation [Kuo 1987]. A system using the PID compensator is shown in block diagram form in Figure 10.25. If we set

$$\frac{K_p}{K_d} = 2\xi\omega_n, \qquad \frac{K_i}{K_d} = \omega_n^2$$

then the resulting forward gain is $K_dK/(1 + K_dK)$. This means that the output is proportional to the input, and ideal situation for the control system.

Suppose in addition to position control we need to limit the joint velocity. Velocity can be sensed with a DC generator whose output voltage is

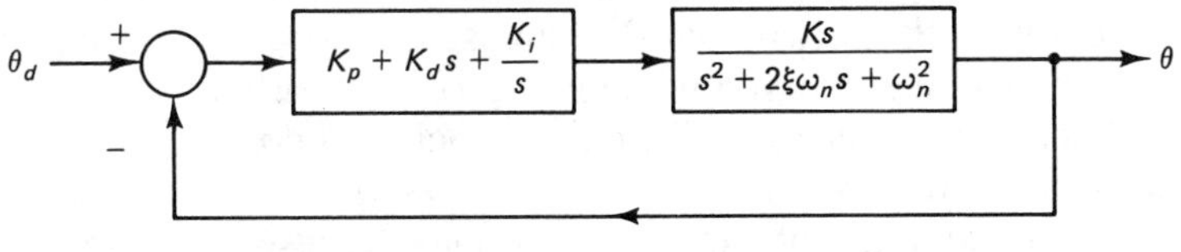

Figure 10.25 A PID compensator in a position-controlled motor

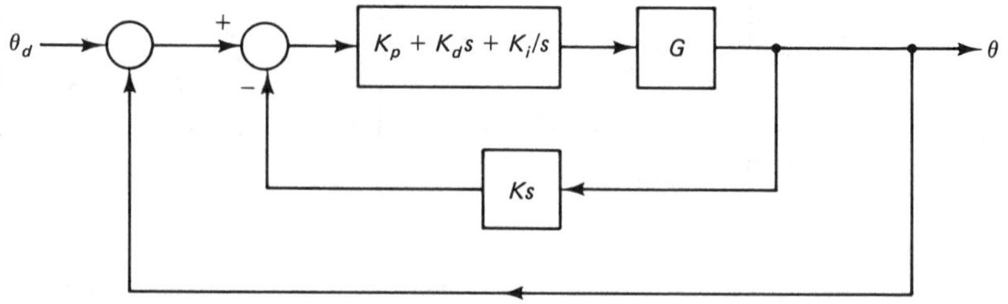

Figure 10.26 A position- and velocity-controlled system

proportional to the angular velocity. A block diagram with both position and velocity control is shown in Figure 10.26.

In the brief analysis given, it is assumed that the actuator is considered in isolation with constant inertial loads. We know from the dynamic analysis of Chapter 9 that this is not the case. In practice the loads on actuator i are inertial, centrifugal, and gravitational and are a function of the following:

1. The joint status
2. The payload (usually an insignificant factor)
3. The forces exerted by the motion of joints $1, 2, \ldots, i-1, i+1, \ldots, n$

We generally consider fixing the values of K_p, K_d, and K_i for some standard set of conditions for which the resulting motion is critically damped (fast response with no overshoot with respect to some type of input—step, ramp, or some other). A time-dependent factor can be added to the PID compensator to take into account changing inertial loads caused by the motion of other joints [Craig 1986; Fu, Gozalves, and Lee 1987]; this time-dependent factor provides cross-coupling between the actuators. The determination of these cross-coupling factors is largely empirical.

10.8.3 Controllers

The control system that surrounds the actuator and enables desired joint space-time functions to be realized is called the *controller*. The controllers in most older robots are of the point-to-point variety. Thus the IKS of the robot is solved at a pair of points, and an approximate slew solution is executed between these two points. Suppose the two joint space points are $\xi 1$ and $\xi 2$; then joint j is driven with input $\xi(t) = \xi 1 + t\{\xi 2 - \xi 1\}$, where $0 \leq t \leq 1$ is a linear ramp function. The output tracks but does not exactly match the input—hence the use of the term *approximate*.

The two endpoints of the slew path may represent a segment of the overall path. Execution of these individual slew paths can cause difficulties, since although the overall path is piecewise continuous in position, its velocity is (in theory) discontinuous. In practice, the finite torques possible in the actuators may smooth out some discontinuities. The jerkiness can also be minimized by nesting the solutions—that is, starting the next path before the previous path is completed—and the result can look like that of a continuous path solution. In fact, what some manufacturers call continuous path controllers are often point-to-point controllers with nesting.

The method of choice is to use a continuous path or contouring controller. Most use the joint space-time function as input without modification, and time delays in the response cause tracking errors. These tracking errors can be minimized by increasing the order of the system; see Section 10.10.1. The tracking errors can also be reduced using preview or feedforward control [Yoshimoto and Wakatsuki 1984; Yoshimoto and Suliuchi 1985]. The method is used when the required path is predictive.

Tracking errors can also be caused by computational errors, sensory errors, and looseness or slop in the mechanical linkages. Computational errors are rarely a problem, since the word length of most computers is more than adequate to provide a solution with negligible computational error. Chopping errors occur in the optical encoders, since angular position is translated to binary information, but the resolution is usually more than adequate, and these errors are negligible. Looseness or slop are part of the accuracy and repeatability [Veitschegger and Wu 1986] specified by the robot manufacturer but are not applicable when the robot is worn. Users of robots must be aware that accuracy may deteriorate with age. It may be prudent to relegate an older robot to a task with less critical tolerances. Tracking errors are the principal source of error during the trajectory execution, and slop can be a problem when the manipulator is stationary.

10.9 PATH CONTROL

Once a task has been planned in Cartesian space, the objective is to translate this information into robot motion. Suppose the robot has n joints; then each of these joints must be driven individually as time functions θ_d, where the θ_d are the inputs to the systems, such as in Figure 10.26. Here we will discuss the path planning and execution using the models developed in this chapter.

Suppose a series of straight-line trajectories is required in position space, where the velocity at the end points of each trajectory are zero and the velocities follow closely the ideal velocity characteristics shown in Figure 8.4. We will use the four-point Bezier curve with the first set of points the same and the last set of points the same. In position space, and with ζ_k replacing x_0 and x_1 and ζ_{k+1} replacing x_2 and x_3 in equation (7.22), we have

$$\zeta(t) = (1 - t)^2(1 + 2t)\zeta_k + t^2(3 - 2t)\zeta_{k+1} \qquad (10.28)$$

$$k \leq t \leq k + 1, \qquad k = 1, 2, \ldots, m$$

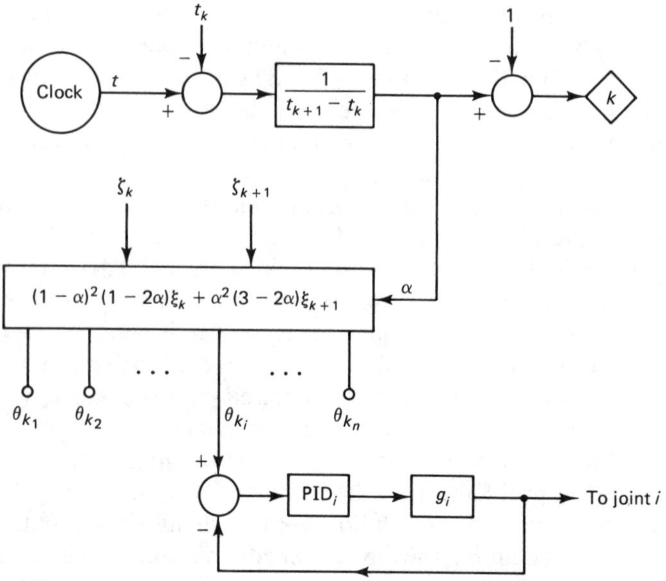

Figure 10.27 Straight-line trajectory control

The given initial conditions are time points $t_1, t_2, \ldots, t_m$, where $t_1 < t_2 < \cdots < t_m$, and $\zeta_1, \zeta_2, \ldots, \zeta_m$, where $\zeta = \zeta_k$ at $t = t_k$. The sequencing and control layout for driving the joints are shown in Figure 10.27, and the clock starts only after the IKS for $\xi_1, \xi_2, \ldots, \xi_i$ have been calculated, where i is at least 2. Provided the time required to calculate a ξ_i is less than the actual time required to execute a line segment, it is possible to have $i = 2$; otherwise, a longer set of ξ_i must be calculated and stored for instant use. To be avoided is the possibility that a joint space point is not available at time of execution. The diamond box iterates the enscribed variable, in this case k. If the input signal τ is less than zero, nothing occurs. When $\tau \geq 0$, k is increased by 1. When this occurs, input τ becomes negative and k is unaltered until τ again becomes positive. With this protocol, a complete segment of path is executed with ξ_k and ξ_{k-1} unaltered. The calculations for the joint conditions, shown as θk_i, $i = 1, 2, \ldots, n$ for the n *dof* robot, occur without noticeable time delay, and θk_i is fed to the input of the feedback control system describing the actuator, its compensator, and the optical encoder feedback. In this way, the joints move in concert and in delayed (by the time it takes to store a sequence of ξ_i) real time.

Consider next the execution of a curved trajectory in which position and velocity are continuous functions of time at all points. We assume that position space is given at ζ_i, $i = 1, 2, \ldots, m$, where velocity is zero at ζ_1 and at ζ_m. Using the Newton interpolating polynomial discussed in Section 7.3, the polynomial passing though the three points (t_1, x_1), (t_2, x_2), (t_3, x_3) is given by

$$P(t) = x_1 + (t - t_1)(\overline{t_1, t_2} + (t - t_2)\overline{t_1, t_2, t_3})$$
$$= x_1 + t^2\overline{t_1, t_2, t_3} + t\{\overline{t_1, t_2} - t_1\overline{t_1, t_2, t_3} - t_2\overline{t_1, t_2, t_3}\}$$
$$- t_1\overline{t_1, t_2} + t_1t_2\overline{t_1, t_2, t_3} \tag{10.29}$$

and

$$\frac{dP(t)}{dt} = 2t\,\overline{t_1, t_2, t_3} + \overline{t_1, t_2} - t_1\,\overline{t_1, t_2, t_3} - t_2\,\overline{t_1, t_2, t_3}$$

so at $t = t_2$ the slope is

$$(t_2 - t_1)\overline{t_1, t_2, t_3} + \overline{t_1, t_2} \tag{10.30}$$

We calculate the slopes χ_1 at points $i = 2, 3, \ldots, m - 1$, where χ_1 has the dimension of n for the n *dof* robot. The ordered set of position space points used in each Bezier curve and joint space trajectory function in Table 10.1, where the IKS of ζ_i is ξ_i, of $\zeta_i + 3\chi_i$ is ξ_{i1} and $0 \le \alpha \le 1$.

Executing successively the Bezier functions with fixed clock rate produces a position space trajectory that starts at rest, finishes at rest, and passes through all intermediate points with no discontinuity in velocity at those intermediate points. The velocity along the trajectory is controlled by the spacing of the points, and the slopes at the points are independent of this spacing. The protocol for controlling joint actuators is shown in Figure 10.28. Two or more IKS sequences must be calculated in advance before the clock is started and the trajectory commences.

Efficient operation may require minimum time delays [Huang and McClamroch 1988]. The robot under discussion must be synchronized with other robots and other flexible automation devices in a work cell. IKS calculations are the usual cause of the majority of time delays that are not part of the actual execution function. Minimizing such time delays is desirable, and the efficient programming of the canonical IKS equations developed in Chapter 6 is an important part of timed operations.

TABLE 10.1

Position Space Points	Joint Space Function	Time Interval
$\zeta_1, \zeta_1, \zeta_2 - 3\chi_2, \zeta_2$	$(1 - \alpha)^3\xi_1 + 3\alpha(1 - \alpha)^2\xi_{11} + 3\alpha^2(1 - \alpha)\xi_{12} + \alpha^3\xi_2$	$t_1 \le t \le t_2$
$\zeta_2, \zeta_2 + 3\chi_2, \zeta_3 - 3\chi_3, \zeta_3$	$(1 - \alpha)^3\xi_2 + 3\alpha(1 - \alpha)^2\xi_{21} + 3\alpha^2(1 - \alpha)\xi_{22} + \alpha^3\xi_3$	$t_2 \le t \le t_3$
$\vdots$	$\vdots$	$\vdots$
$\zeta_{m-1}, \zeta_{m-1} + 3\chi_{m-1}, \zeta_m, \zeta_m$	$(1 - \alpha)^3\xi_{m-1} + 3\alpha(1 - \alpha)^2\xi_{m-1,1} + 3\alpha^2(1 - \alpha)\xi_{m-1,2} + \alpha^3\xi_m$	$t_{m-1} \le t \le t_m$

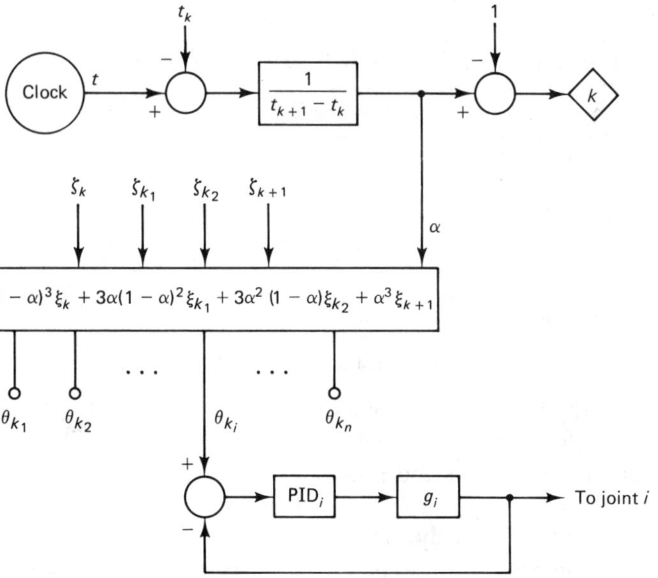

Figure 10.28 Splined-path trajectory control

10.10 TRANSMISSIONS AND GEAR TRAINS

Transmissions in a robotic sense transmit power from an actuator to the specific joint. In most cases a gear train is part of this transmission. Gear trains are assemblies of gears that transform motion. The transformation can be in speed or direction or from rotational to translational or vice versa. The output of most electrical motors requires a speed-reduction gear train to reduce this speed to a motion suitable for application to the joints of a robot. The joint displacement limits of a typical revolute joint are $45° \leq \theta \leq 360°$. Some joints can have unlimited motion, such as a roll joint when no actuators are mounted on the distal side (uplink side of the joint). Of all revolute joints the roll joint tends to be designed with the greatest freedom of motion.

On the other hand, the drive shafts of stepper motors can be connected directly to the robot joint—that is, the body of the motor is mounted on the downlink side of the joint and the drive shaft is directly connected to the uplink side of the joint. The outputs of hydraulic or pneumatic drives can often be directly connected to the joints of the robot.

Friction is present in the most efficient gear train, and typically the friction in a reduction gearing precludes driving the train in reverse. A benefit of this is that the problem of hunting (when input torque is reduced and the weight of the manipulator causes the output shaft of the motor to drift) in electrical motors is eliminated. Hunting is a particular problem in direct-drive systems—those in which the output shaft of the motor is directly connected to a robotic joint without reduction gearing. Most motor drives use reduction gearing. Even the best gear drives contain friction and most reduction (output speed lower than the input gear speed) gear trains cannot be driven in reverse.

Consider the power screw in which threads are the bearing surface between the driven screw and the nut. Suppose the screw pitch is p meters per revolution, the applied torque is T newton-meters μ is the coefficient of friction between the screw threads, where $0 < \mu < 1$, β is the thread angle for Acme or unified threads (the common threads with sawtooth cross section), and d is the diameter of the screw. Then screw friction F is given by [Groover et al. 1986]

$$F = \frac{2T\left(\pi - \dfrac{\mu p}{d \cos \beta}\right)}{p + \dfrac{\mu \pi d}{\cos \beta}}$$

If the threads are square, then $\beta = 0$, and the expression reduces to

$$F = \frac{2T\left(\pi - \dfrac{\mu p}{d}\right)}{p + \mu \pi d} \cong \frac{2\pi}{p + \mu \pi d} T \quad \text{or} \quad \frac{Fd}{T} \cong \frac{2\pi}{\dfrac{p}{d} + \mu \pi} > 0.5$$

so it is impossible to drive the system in reverse.

Wear is an ever-present problem in gear trains. The accuracy of the manipulator worsens with use due to wear in the gear trains, and the problem may not be correctable with the positional information from the optical encoders. Recognizing this some manufacturers (ADEPT, for example) advertise the benefits of their direct-drive systems.

A gear system that can translate rotational motion into translational motion or vice versa is the rack and pinion shown in Figure 10.29. On the

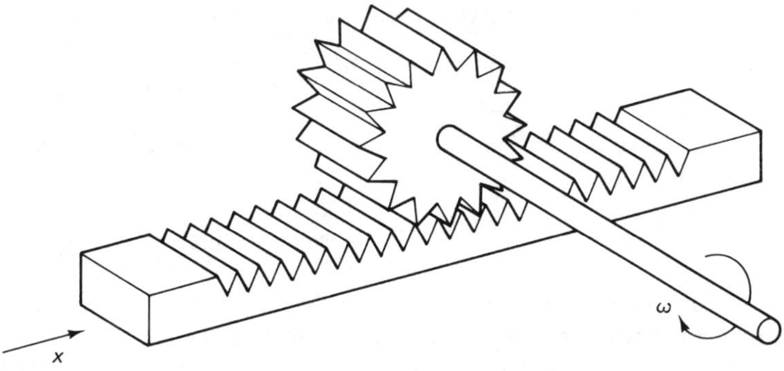

Figure 10.29 A rack-and-pinion gear for translating motion

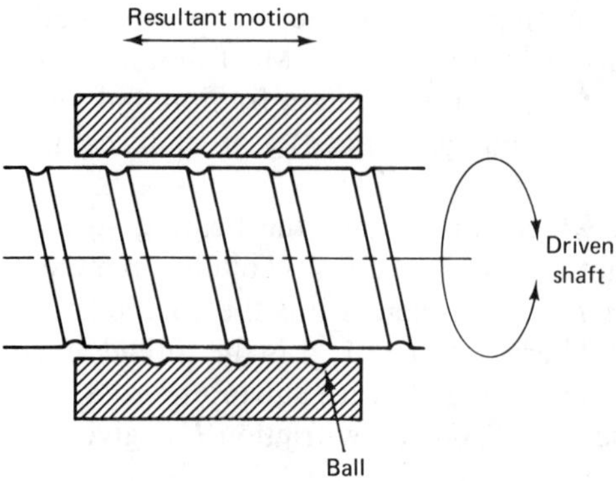

Figure 10.30 The lead screw

other hand, the fast and accurate lead screw [Kafrissen and Stephans 1984] shown in Figure 10.30 can convert rotational to linear motion only. This is similar to the way that a nut can drive a bolt but not vice versa.

Gear trains with reductions of 50 or more are common, and such gear trains are subject to wear and slop (backlash or deadband, where reversing torque on the output shaft produces a nonlinear response. A gear-reduction system that virtually eliminates slop is the harmonic gear. A *harmonic gear* has three parts, an elliptic rotor, which is the driven input part, a rigid circular splined gear with n inside teeth, which is the output, and a floating, flexible splined cylinder with $n - 2$ outside teeth. The flexible part is known as the flexspline. The rotor contacts the flexspine at two locations 180° apart and deforms it so its teeth mesh with the teeth in the outer splined cylinder at these two locations. One complete rotation of the elliptical rotor causes the flexspine to deform over 360°, and the effect is to move the cylindrical output cylinder by two teeth out of n in the opposite direction. The gear reduction is $n/2$.

10.11 EXAMPLES

1. An armature-controlled DC servomotor is described by the equations

$$e_a = L_a \frac{di_a}{dt} + r_a i_a + K_2 \frac{d\theta}{dt} \quad \text{and} \quad \tau = K_1 i_a$$

The system is position-controlled, so $e_a = K_3(\theta_d - \theta)$. There is a simple inertial load such that

$$\tau = J \frac{d^2\theta}{dt^2} + F \frac{d\theta}{dt} \quad \text{newton-meters}$$

where J is the inertia and F is the viscous friction at the joint. Assuming the system starts at rest, determine the response of the system as measured by the motor shaft angle θ to a step change of 1 rad on θ_d. Integrate the differential equation using Euler's formula [Ralston, 1978]. Assume $L_a = 0.1$ H, $r_a = 1\ \Omega$, $J =$ 1 kg-m^2, and $F = 0.1$ N-m/s. Determine the value of $K_1 = K_2 = k$ that produces a critically damped response with no overshoot for various values of K_3. k is the strength of the motor.

$$K_3(\theta_d - \theta) = 0.1\frac{di_a}{dt} + i_a + k\frac{d\theta}{dt} \quad \text{and} \quad \ddot{\theta} + 0.1\dot{\theta} = ki_a$$

Eliminating i_a, we produce the single linear differential equation

$$K_3(\theta_d - \theta) = \frac{0.1}{k}\left(\frac{d^3\theta}{dt^3} + 0.1\ddot{\theta}\right) + \frac{1}{k}(\ddot{\theta} + 0.1\dot{\theta}) + k\dot{\theta}$$

so

$$10kK_3(\theta_d - \theta) = \left(\frac{d^3\theta}{dt^3} + 0.1\ddot{\theta}\right) + 10\ddot{\theta} + \dot{\theta} + 10k^2\dot{\theta}$$

or

$$\frac{d^3\theta}{dt^3} = -10.1\ddot{\theta} - (1 + 10k^2)\dot{\theta} - 10kK_3(\theta - \theta_d)$$

Using Euler's formula we get the three recursive equations

$$\theta_{i+1} = \theta_i + h\dot{\theta}_i$$

$$\dot{\theta}_{i+1} = \dot{\theta}_i + h\ddot{\theta}_i$$

$$\ddot{\theta}_{i+1} = \ddot{\theta}_i + h\frac{d^3\theta_i}{dt^3} = \ddot{\theta}_i - h[10.1\ddot{\theta}_i + (1 + 10k^2)\dot{\theta}_i + 10kK_3(\theta_i - \theta_d)]$$

These equations are easily programmed. Some critically damped cases are shown in Figure 10.31. A table of some critically damped values is as follows.

	K_3	k	90% Rise Time
1	1	1.35	2.6
2	2	1.73	1.6
	3	1.95	1.13
3	4	2.15	0.88 Nonmonotonic response
	6	2.51	0.65
4	10	3.16	0.44
	20	4.5	0.30 Ringing overshoot
	40	5.55	0.18
	100	Unstable for $k < 10.2$	

In the simulation the integration step length h was small enough to maintain numeric stability (the product of h and the magnitude of the largest eigenvalue must be less than 2) and make numeric approximation errors insignificant (since Euler's formula is first order, the error is a function of h^2). Notice the increase in oscillatory response as k_3 is increased. What is happening as k and K_3 are varied can be seen from a root-locus analysis. We take the Laplace transform of the third-order polynomial in θ and

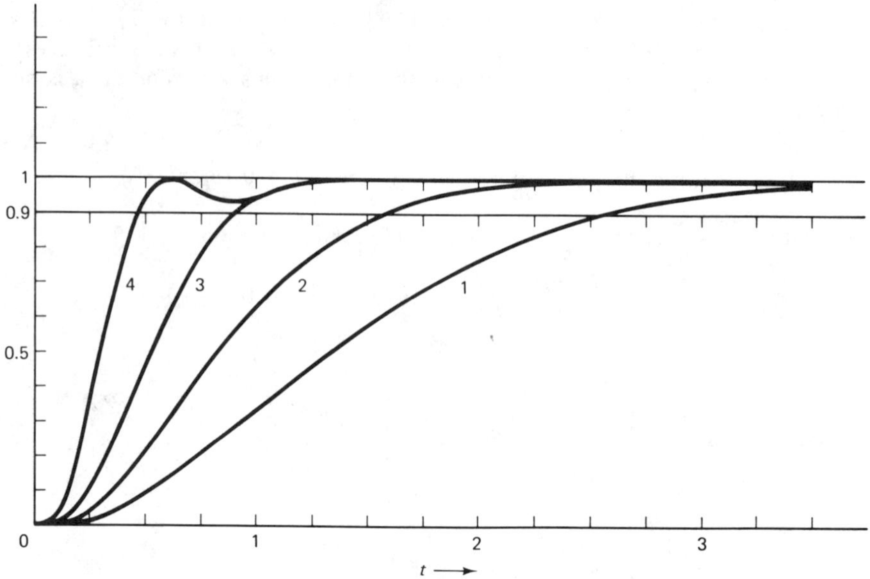

Figure 10.31 Critically Damped Responses for Position Controlled Motor.

produce the transfer function

$$\frac{\Theta}{\Theta_d} = \frac{10kK_3}{s^3 + 10.1s^2 + (1 + 10k^2)s + 10kK_3}$$

At $K_3 = 0$, the poles are at $s = 0$ and $s^2 + 10.1s + 1 + 10k^2 = 0$. The roots of this quadratic are at $s = -5.05 \pm \sqrt{24.50 - 10k^2}$, which are real for $k < 2.45$. At $k = 2.45$ the double root is at -5.05. For large k the roots are at $-5.05 \pm jk$, where $j = \sqrt{-1}$. The closed-loop poles are at $s^3 + 10.1s^2 + (1 + 10k^2)s + 10kK_3 = 0$, and the partial Routh array [Kuo, 1987] is

s^3	1	$1 + 10k^2$
s^2	10.1	$10kK_3$
s^1	$1 + 10k^2 - \frac{10}{10.1}kK_3$	

For roots on the imaginary axis in the s-plane we require the s^1 row to be zero, so

$$1 + 10k^2 - \frac{10}{10.1}kK_3 = 0, \quad \text{or} \quad 10kK_3 = 10.1(1 + 10k^2)$$

The imaginary axis roots are at $10.1s^2 + 10kK_3 = 0$, so $s = \pm j\sqrt{(1 + 10k^2)}$. To maintain stability we require the s^1 term to be positive, so

$$K_3 < \frac{1.01}{k} + 10.1k.$$

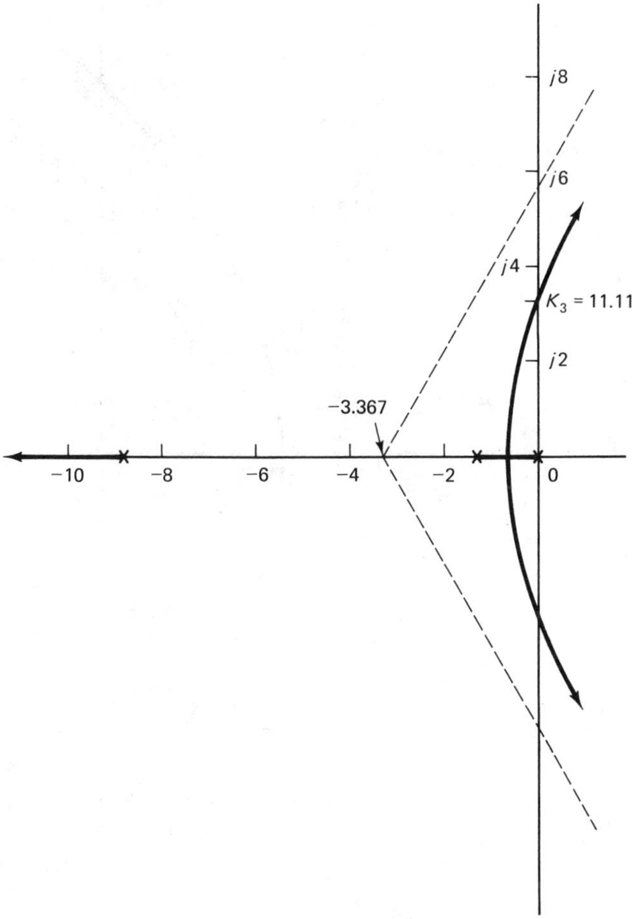

(a) Root locus for k = 1

Figure 10.32

We sketch the root locus for $k = 1$, $k = 2.45$, and $k = 10$ in Figure 10.32(a), (b) and (c). For all three cases the center of the asymptotes is at $\sigma_a = -10.1/3 = 3.367$. The location of the complex roots for $k > 2.45$ explains the oscillatory responses. For example, the poles shown in Figure 32(c) are at $-5.05 \pm j31.233$ for $k = 10$ and $K_3 = 0$, the angle of departure from the pole at $-5.05 + j31.233$ is $-9.18°$, and this pole crosses the imaginary axis at $\omega = 31.638$. It is impossible to choose a satisfactory positive K_3 to eliminate overshoot for this case.

To put the 90% rise time with respect to root locus in perspective, consider a system with a single pole at $s = -b$. The response to a unit step $u(t)$ is $(1 - e^{-bt})u(t)$, and the time to reach 90% of the steady-state value of unity is given by $e^{-bt} = 0.1$, so $bt = \log_n 10 = 2.303$. For a double pole at $s = -c$, the response to a unit step is $\{1 - c(1 + t)e^{-ct}\}u(t)$, and the time to reach 90% of the steady-state value is $ct = 3.89$. Thus, for the double-pole system to have the same 90° rise time, we require $c =$ 1.689b.

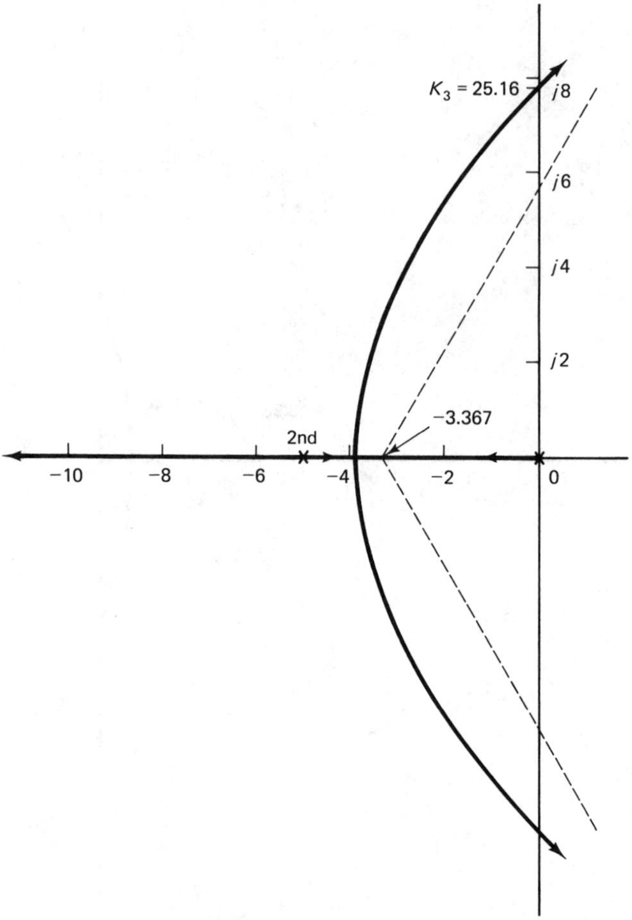

(b) Root locus for k = 2.45

Figure 10.32 *Continued*

The response time of the system is governed directly by the armature circuit, in particular the value of L_a; reducing L_a reduces the response time. We can also improve the response by adding a zero to the system (as in a PD compensator).

---

2. Use a PD compensator with transfer function given by $K(s/\alpha + 1)$ with the motor discussed in example 1 such that the response to the step input exhibits no overshoot. Assume the strength of the motor is $k = 1$. Show the effect of K and α on the response.

The block diagram for the system is shown in Figure 10.33 and the equations governing the system are given by

$$e_p = (\theta_d - \theta), \qquad e_a = K\left(\frac{d}{\alpha\, dt} + 1\right)e_p,$$

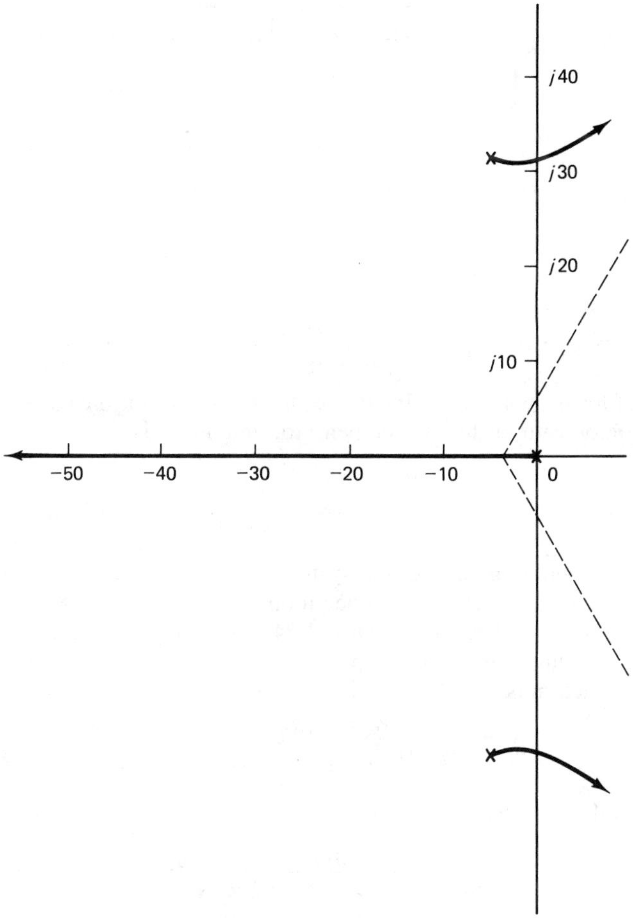

(c) Root locus for k = 10

Figure 10.32 *Continued*

$$e_a = 0.1\frac{di_a}{dt} + i_a + \frac{d\theta}{dt} \quad \text{and} \quad \theta + 0.1\theta = i_a$$

Eliminating e_p in the equation for e_a gives $e_a = \dfrac{K}{\alpha}\dfrac{d}{dt}(\theta_d - \theta) + K(\theta_d - \theta)$ and for the motor

$$e_a = 0.1\left(\frac{d^3\theta}{dt^3} + 0.1\ddot{\theta}\right) + (\ddot{\theta} + 0.1\dot{\theta}) + \dot{\theta}$$

so

$$e_a(t_i) = 0.1\frac{d^3\theta_i}{dt^3} + 1.01\ddot{\theta}_i + 1.1\dot{\theta}_i = \frac{K}{\alpha}(\dot{\theta}_d - \dot{\theta}) + K(\theta_d - \theta)$$

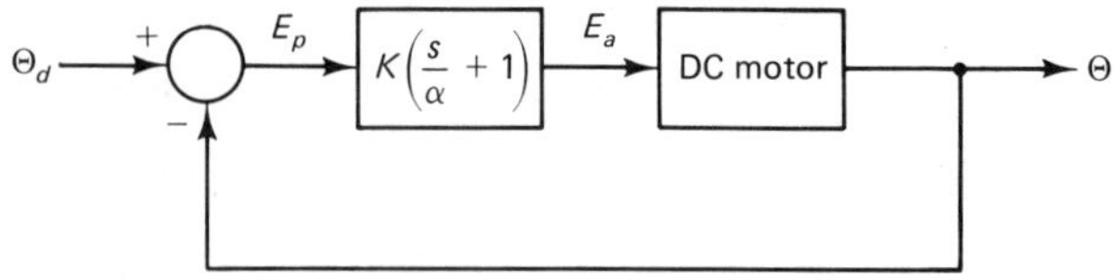

Figure 10.33 The PD-compensated DC motor

or

$$\frac{d^3\theta_i}{dt^3} = -10.1\ddot{\theta}_i - \left[1 + 10\left(1 + \frac{K}{\alpha}\right)\right]\dot{\theta}_i - 10K\theta_i + 10K\left(\frac{\dot{\theta}_d}{\alpha} + \theta_d\right)$$

We use a root locus approach to determine suitable values for the parameters K and α. The open loop gain of the PD compensator and motor is

$$G = \frac{10K\left(\frac{s}{\alpha} + 1\right)}{s(s^2 + 10.1s + 11)}$$

The sum of the open-loop poles of the system is -10.1 and potential instability (for large K) will occur if $\alpha \geq 10.1$. The open-loop poles are at 0, -8.858, and -1.242. The best response should occur when $\alpha = 1.242$ (since it then neutralizes the pole at -1.242) and K is chosen for a double pole at $s = -5.05$. With $\alpha = 1.242$, the open-loop transfer function is

$$G = \frac{10K}{s(s + 8.858)}$$

and the closed-loop transfer function is

$$\frac{10K}{s(s + 8.858) + 10K}$$

with poles at

$$s = -\frac{8.858}{2} \pm \sqrt{8.858^2 - 40K}$$

A double pole occurs when $K = 1.9616$, which makes the radical zero. Thus the best response should occur for $\alpha = 1.242$ and $K \cong 2$. To show the actual response we rely on numeric integration of the differential equations. The three recursive equations derived using Euler's equation are

$$\theta_{i+1} = \theta_i + h\dot{\theta}_i$$

$$\dot{\theta}_{i+1} = \dot{\theta}_i + h\ddot{\theta}_i$$

$$\ddot{\theta}_{i+1} = \ddot{\theta}_i + h\frac{d^3\theta_i}{dt^3} = (1 - 10.1h)\ddot{\theta}_i - h\left[\left(11 + 10\frac{K}{\alpha}\right)\dot{\theta}_i + 10K\theta_i - 10K\left(\frac{\dot{\theta}_d}{\alpha} + \theta_d\right)\right]$$

Care must be taken in modeling $\dot{\theta}_d$. For example, if $\theta_d(t) = u(t)$, a unit step commencing at $t = 0$, then $\dot{\theta}(t) = \delta(t)$, a unit impulse with area of unity. Since the

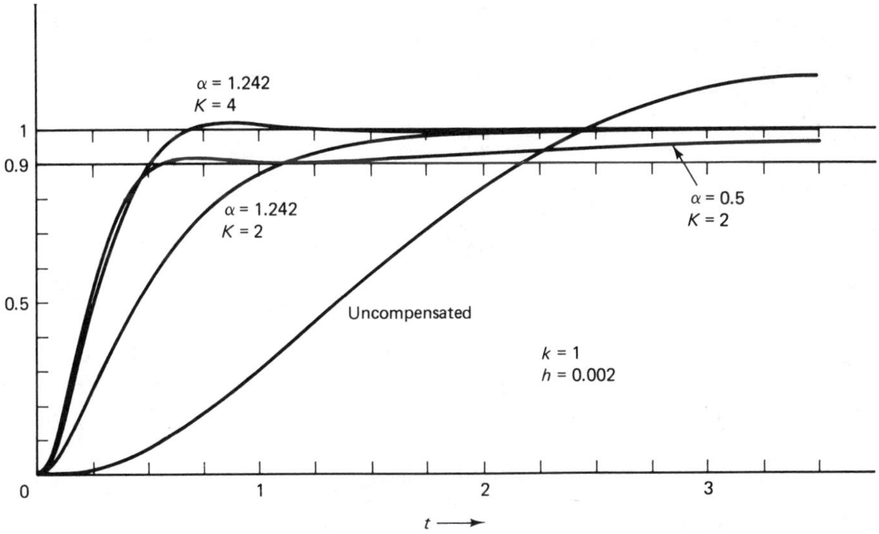

Figure 10.34 PD-compensated Responses

integration occurs over time step h, we can model $\theta(0)$ by any function such that

$$\int_0^h \dot{\theta}_d(t)\, dt = 1, \quad \text{and} \quad \dot{\theta}_d(t) = 0 \text{ outside this range}$$

For convenience we choose $\dot{\theta}_d(t) = 1/h, 0 \le t \le h$. Some responses at or close to the theoretically ideal values are shown in Figure 10.34. Notice a substantial improvement over the uncompensated motor.

---

3. Repeat example 1 but this time assume that the motor drives a pitch joint at $\begin{bmatrix} 0 \\ 0 \\ 0 \\ 1 \end{bmatrix}$.

The pitch joint drives a link of length 1 m and the distal end of the link is required to move from $\begin{bmatrix} 1 \\ 0 \\ 0 \\ 1 \end{bmatrix}$ to $\begin{bmatrix} 0 \\ 0 \\ 1 \\ 1 \end{bmatrix}$. A mass of M kilograms is at the distal end, $0 \le M \le 1$. Assume $L_a = 0.1$ H, $r_a = 1\ \Omega$, $K_1 = K_2 = k$, and the viscous friction at the joint is $F = 0.1$ n-m/s. Suppose the total inertial load at the joint is 1 kg-m^2, so $J + M = 1$, where J is inertial.

Determine the values of k and K_3 for the best response of the system for $M = 0$. What effect does shifting mass from J to M have on the response of this system, and for $M = 1$ are the previously chosen values for k and k_3 satisfactory?

The input of the system is θ_d, where $\theta_d = -1.571$ rad for $t < 0$, and $\theta_d = 0$ for $t > 0$. From worked example 7 of Chapter 9 and adding the viscous friction, we have $\tau = (J + M)\ddot{\theta} + F\dot{\theta} - 9.8062M \sin\theta$. The system equations become

$$K_3(\theta_d - \theta) = L_a \frac{di_a}{dt} + r_a i_a + K_2\dot{\theta} \quad \text{and} \quad (J + M)\ddot{\theta} + F\dot{\theta} - 9.8062M \sin\theta = K_1 i_a$$

Eliminating i_a we produce the single, nonlinear differential equation

$$K_1K_3(\theta_d - \theta) = L_a \left[(J + M)\frac{d^3\theta}{dt^3} + F\ddot{\theta} - 9.8062M\dot{\theta}\cos\theta\right]$$
$$+ r_a[(J + M)\ddot{\theta} + F\dot{\theta} - 9.8062M \sin\theta] + K_1K_2\dot{\theta}$$

or

$$L_a(J + M)\frac{d^3\theta}{dt^3} + [L_aF + r_a(J + M)]\ddot{\theta} + (r_aF + K_1K_2 - 9.8062L_aM\cos\theta)\dot{\theta}$$
$$- K_1K_3(\theta_d - \theta) - 9.8062r_aM\sin\theta = 0$$

Writing this equation in recursive form,

$$\frac{d^3\theta_{i+1}}{dt^3} = \frac{-1}{L_a(J + M)}\{[L_aF + r_a(J + M)]\ddot{\theta}_i$$
$$+ (r_aF + K_1K_2 - 9.8062L_aM\cos\theta_i)\dot{\theta}_i - K_1K_3(\theta_d - \theta_i)$$
$$- 9.8062r_aM\sin\theta_i\}$$

Inserting the known constants,

$$\frac{d^3\theta_{i+1}}{dt^3} = -10[1.1\ddot{\theta}_i + (k^2 + 0.1 - 0.98062M\cos\theta_i)\dot{\theta}_i$$
$$+ kK_3(\theta_i - \theta_d) - 9.8062M\sin\theta_i\}$$

This third-order system will be solved numerically with three recursive equations derived from Euler's formula as

$$\theta_{i+1} = \theta_i + h\dot{\theta}_i$$

$$\dot{\theta}_{i+1} = \theta_i + h\ddot{\theta}_i$$

$$\ddot{\theta}_{i+1} = \ddot{\theta}_i + h\frac{d^3\theta_i}{dt^3}$$

$$= \ddot{\theta}_i - 10h\{1.1\ddot{\theta}_i + (k^2 + 0.1 - 0.98062M\cos\theta_i)\dot{\theta}_i$$
$$+ kK_3(\theta_i - \theta_d) - 9.8062M\sin\theta_i\}$$

The initial conditions for the simulations shown in Figure 10.35 are $\theta(0) = 0$. $\dot{\theta}(0) = \ddot{\theta}(0) = 0$, and $\theta_d = -\pi/2$. We chose $h = 0.01$. Response 1 is the same as that produced in the first worked example for $K_3 = 3$ and with $M = 0$. The other responses use $M = 1$, and we see that this eccentric load produces a steady-state error that is a function of k and K_3; we require a large k and K_3 for this error to be small. However, we cannot increase the strength of the motor enough to make the steady-

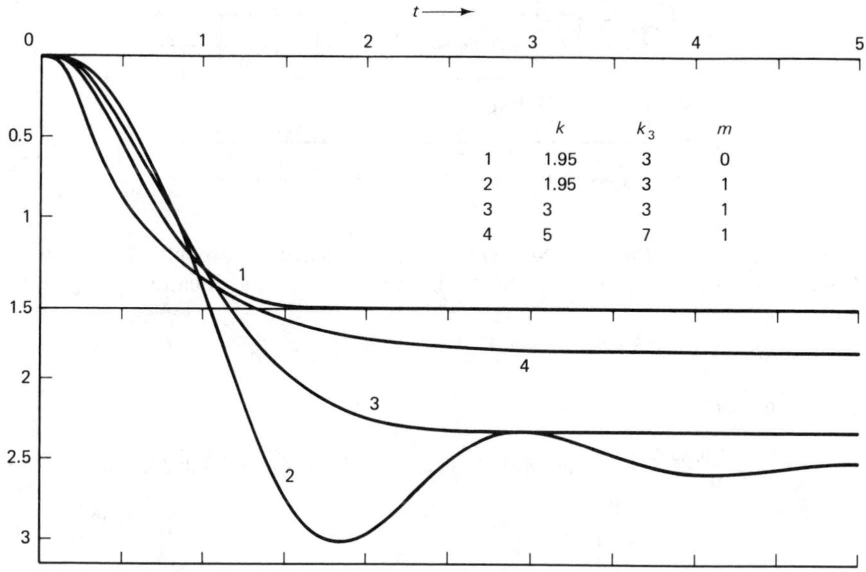

Figure 10.35 Response of Position Controlled Motor with Eccentric Load

	k	k_3	m
#1	1.95	3	0
#2	1.95	3	1
#3	3	3	1
#4	5	7	1

state error insignificant and need a series integrator to remove this steady-state error. However, the effectiveness of PD compensation indicates that we need an additional zero (not the additional pole an integrator gives) to decrease the response time. The conclusion to be drawn is that we need a PID compensator with two zeros on the negative real axis and a pole at the origin; this case is discussed in the next worked example.

---

4. Design a PID compensator to improve the response of the position-controlled motor with eccentric load M in the third worked example. Assume $r_a = 1$, $L_a = 0.1$, $J + M = 1$, $F = 0.1$, $K_1 = K_2 = 1$, and $K_3 = 1$. If the results are unsatisfactory, discuss possible ways of improving performance.

The block diagram for the system is shown in Figure 10.36.

The equations governing the system are given by

$$e_p = (\theta_d - \theta), \qquad e_a = K_P e_p + K_I \int e_p \, dt + K_D \frac{d}{dt} e_p, \qquad e_a = 0.1 \frac{di_a}{dt} + i_a + k\dot{\theta}$$

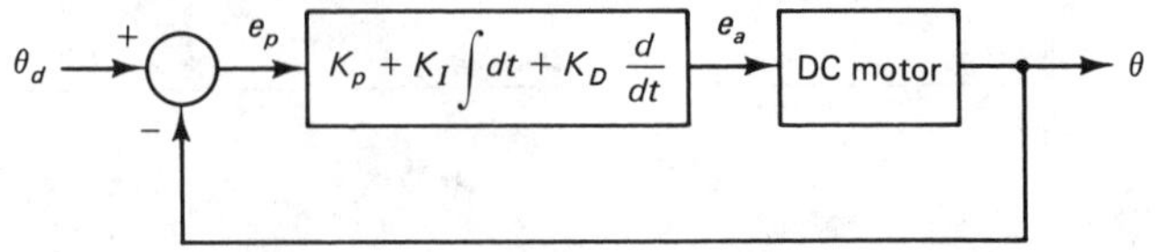

Figure 10.36 The PID-compensated DC Motor

and for the load $\ddot{\theta} + F\dot{\theta} - 9.8062M \sin\theta = ki_a$. We leave the strength of the motor as k and F as a variable in the equations. Eliminating e_p in the equation for e_a gives

$$e_a = K_P(\theta_d - \theta) + K_I \int (\theta_d - \theta)\, dt + K_D \frac{d}{dt}(\theta_d - \theta)$$

For the motor,

$$e_{a,i} = \frac{0.1}{k}\left(\frac{d^3\theta_i}{dt^3} + F\ddot{\theta}_i - 9.8062M\dot{\theta}_i \cos\theta_i\right) + \frac{1}{k}(\ddot{\theta}_i + F\dot{\theta}_i - 9.8062M \sin\theta_i) + k\dot{\theta}_i$$

$$= K_P(\theta_d - \theta) + K_I \int (\theta_d - \theta)\, dt + K_D \frac{d}{dt}(\theta_d - \theta)$$

$$\frac{d^3\theta_i}{dt^3} = -(10 + F)\ddot{\theta}_i + \dot{\theta}_i(9.8062M \cos\theta_i - 10(k^2 + F) - 10kK_D) + 98.062M \sin\theta_i$$

$$+ 10kK_P(\theta_d - \theta_i) + 10kK_I \int (\theta_d - \theta_i)\, dt + 10kK_D\dot{\theta}_d$$

Rather than determine K_P, K_I, and K_D empirically, we use a root locus approach. There seems no disadvantage to setting the two zeros at the same location; with $K_P = 2\alpha K_D$ and $K_I = \alpha^2 K_D$, then the compensator becomes $K_D(s + \alpha)^2/s$ in the frequency domain. The frequency domain block diagram is given in Figure 10.37.

In Figure 10.37,

$$T = \begin{cases} M9.8062 & \text{if } \theta \cong 0 \\ 0 & \text{if } \theta \cong \pi/2 \\ -M9.8062 & \text{if } \theta \cong \pi \end{cases}$$

Thus, when $\theta \cong 0$, the load acts like an upside-down pendulum that has the equivalent of a negative spring constant, and the uncompensated motor is unstable. When θ

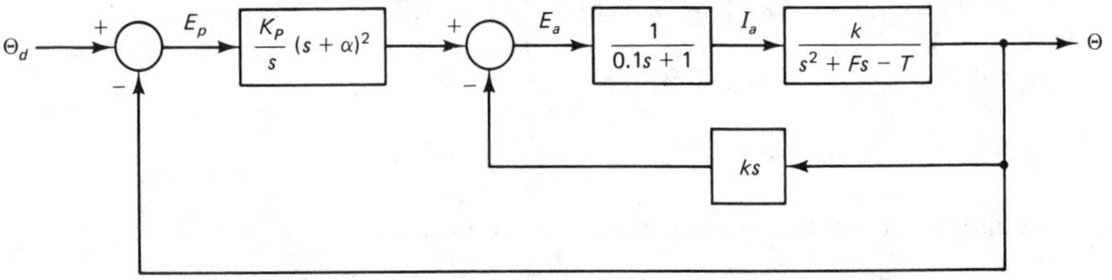

Figure 10.37 Block Diagram of PID-compensated System with Eccentric Load

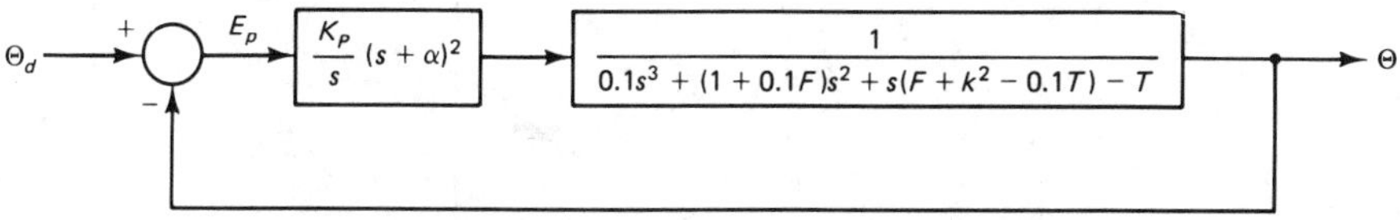

Figure 10.38 The Reduced System

$\cong \pi$, the load acts like a normal pendulum. When $\theta \cong \pi/2$, there is no spring, and the motor has a pole at the origin. We reduce the motor to a single block in Figure 10.38. The locations of the poles of the motor as a function of θ are sketched in Figure 10.39 for the case $k = 1$, $F = 0.1$, and $M = 1$. The most negative pole moves over a small range in all three cases, from $s = -8.66$ at $\theta = 0$, $s = -8.86$ at $\theta = \pi/2$, to $s = -8.96$ at $\theta = \pi$. As far as the design of the compensator is concerned, we can assume that this pole is stationary at the average value of $s = -8.83$. The load position has a strong influence on the other two poles of the motor.

We have four poles and two zeros, so the asymptotes are $\pm 90°$; they cross the real axis at σ_a, where in all cases $\sigma_a = -5.05 + \alpha$. We choose α in the range $1 \le \alpha \le 3$ and set K_p to produce closed-loop poles close to the real axis. For $\alpha = 3$ and $k = 1$, a sketch of the root locus for $0 \le K_D \le \infty$ is shown in Figure 10.40. We see that small and large K_D will be unsatisfactory.

To show the actual response, we rely on numeric integration of the differential equations. The four recursive equations derived using Euler's equation are:

$$\theta_{i+1} = \theta_i + h\dot{\theta}_i$$

$$\dot{\theta}_{i+1} = \dot{\theta}_i + h\ddot{\theta}_i$$

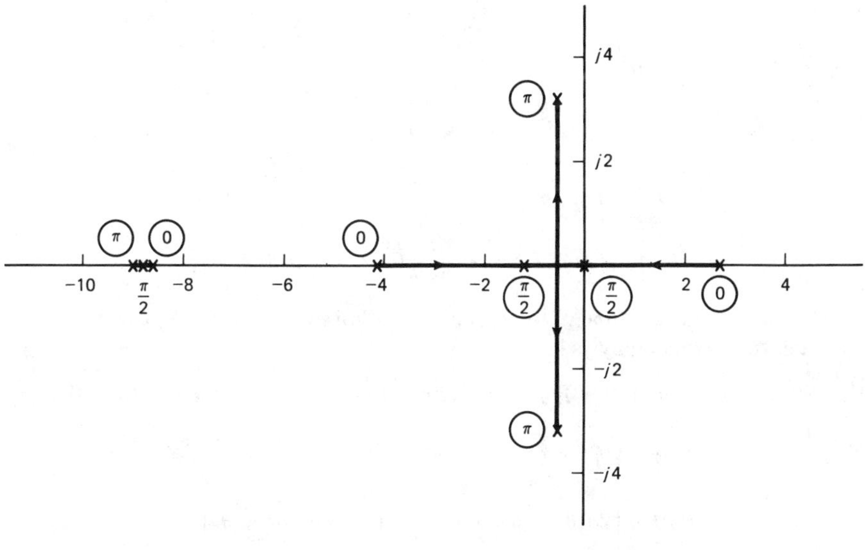

Figure 10.39 Effect of Robot Arm Position on Motor Poles

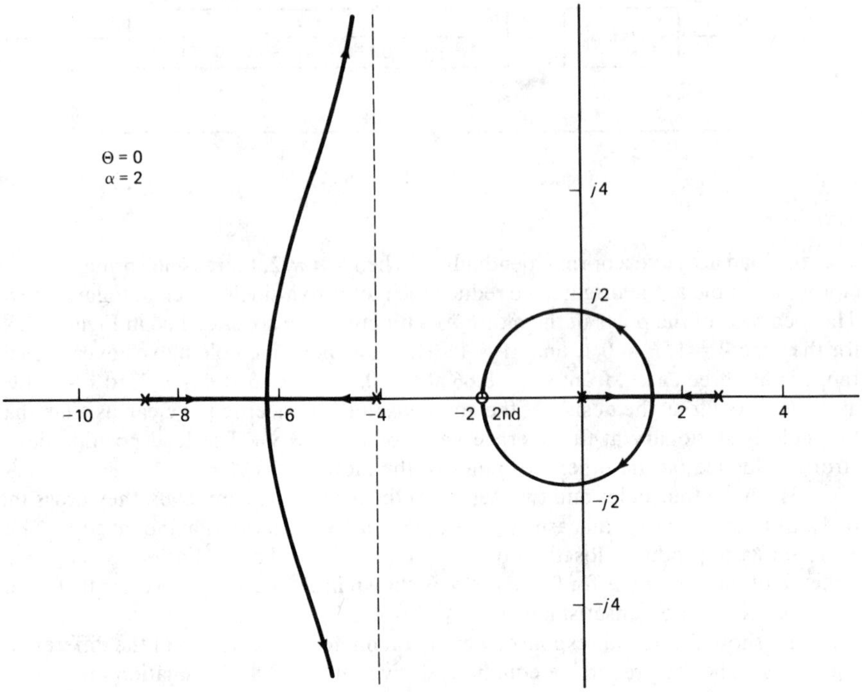

Figure 10.40 A Root Locus for Compensated Motor

$$
\begin{aligned}
\ddot{\theta}_{i+1} &= \ddot{\theta}_i + h\,\frac{d^3\theta_i}{dt^3} \\
&= [1 - h(10 + F)]\ddot{\theta}_i + h\dot{\theta}_i(9.8062M\cos\theta_i - 10(k^2 + F) - 10kK_D) \\
&\quad + 98.062Mh\sin\theta_i \\
&\quad + 10kh\left[K_P(\theta_d - \theta_i) + K_I\int_{t_0}^{t_{i+1}}(\theta_d - \theta_i)\,dt + K_D\dot{\theta}_d\right]
\end{aligned}
$$

where t_0 is the starting time of the procedure. With the values for K_I and K_D chosen as in the root locus analysis,

$$
\begin{aligned}
\ddot{\theta}_{i+1} &= [1 - h(10 + F)]\ddot{\theta}_i + h\dot{\theta}_i(9.8062M\cos\theta_i - 10(1 + F) - 10kK_D) \\
&\quad + 98.062Mh\sin\theta_i \\
&\quad + 10hK_D\left[2\alpha(\theta_d - \theta_i) + \alpha^2\int_{t_0}^{t_{i+1}}(\theta_d - \theta_i)\,dt + \dot{\theta}_d\right]
\end{aligned}
$$

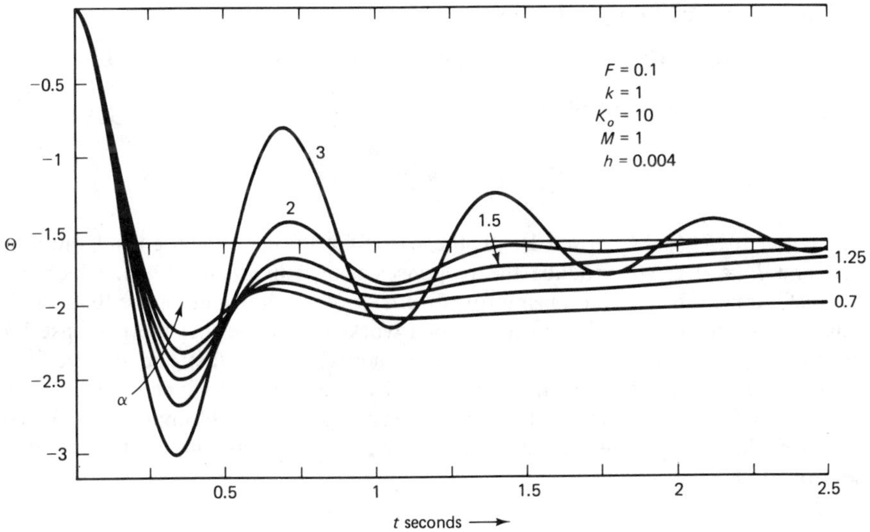

Figure 10.41 PID-compensated Motor Response with $k = 1$, $F = 0.1$

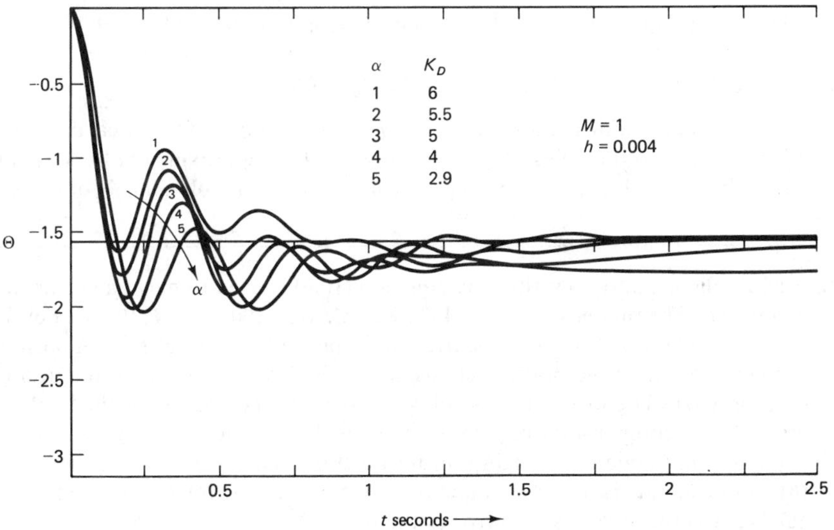

Figure 10.42 PID-compensated Motor Response with $k = 4$, $F = 0.1$

with $\theta_d(t_{0-}) = 0$ and $\theta_d(t) = \theta_d$ when $t > 0$, then $t \geq 0$,

$$\ddot{\theta}_{i+1} = [1 - (10 + F)h]\ddot{\theta}_i + h\dot{\theta}_i(9.8062M \cos \theta_i - 10(k^2 + F) - 10kK_D)$$
$$+ 98.062Mh \sin \theta_i$$
$$+ 10hkK_D\alpha \left[2(\theta_d - \theta_i) + \alpha h \sum_{j=0}^{i} (\theta_d - \theta_j)\right] + \Delta$$

where $\Delta = 10kK_D\theta_d$ if $i = 0$ and $\Delta = 0$ otherwise. For the case under consideration, $k = 1$ and $F = 0.1$, we produce the responses shown in Figure 10.41, which are unsatisfactory. The response is improved with $k = 4$, as shown in Figure 10.42 but is still unsatisfactory. Recall from the second worked example that the response for the PD-compensated system with a simple inertial load is fast and trouble-free. Perhaps the completely eccentric load $M = 1$ and $J = 0$ will never produce completely satisfactory responses. Perhaps increasing damping will improve the situation. Considerably more investigation is needed for this important armature controlled DC motor.

EXERCISES FOR CHAPTER 10

1. A robotic actuator is described by the transfer function

$$G = \frac{0.5}{s^2 + 5s + 4}$$

If this actuator is used with a position encoder and proportional compensator of gain K, determine the value of K for fastest response to a step input.

---

2. The feedback control system shown in Figure 1 has $G = \dfrac{1}{s^2 + 2\xi s + 1}$ and $H = e^{\sqrt{s}}$. At operating point $s = 0$ we set $H = 1$ and choose ξ for critical damping. What happens to the poles of the system when H is approximated by a Taylor series about $s = 0$: perform the approximation to obtain polynomials of order 1 and 2.

---

3. A hydraulic actuator described by equation (10.6) is used to move a sliding link of a robot. The oil density is $\rho = 1000$ Kg/m^3, $k_1 = 100$, $k_2 = 1$, the area of the power cylinder is $A = 0.1$ m^2, and friction on the load is $F = 2$ N-s/m. A position encoder is used so the position status $y(t)$ of the load is compared to a desired position $y_d(t)$. The load is a mass of $M = 10$ Kg at the end of a sliding link. A proportional compensator of gain K preceeds the actuator.

(a) Find the frequency domain transfer function $Y(s)/Y_d(s)$.

(b) Suppose $y_d(t)$ is a unit step and the system is at rest for $t < 0$. Find $y(t)$.

(c) What is the steady-state error of the output?

(d) If $y_d(t) = t$, find $y(t)$; if there is no upper limit on the value assumed for $y(t)$, what is the steady-state error in this case?

---

4. The equations of the position-controlled and field-controlled DC servomotor are given by equations (12), where $L_f = 1$ H, $R_f = 1\ \Omega$, and $K_1 = 1$. The motor drives an inertial load of J = 1 N-sec^2 with friction F = 0.2 N-s. Draw a block diagram for the system, where one block has E_f as input and I_f as output, the other block has I_f as input and torque as output, and the third block has torque as input and position θ as output. Determine the transfer function θ/θ_d. Comment on the range of values for K_3 to prevent overshoot of the response to a step input.

---

5. If $K_3 = 2$ in exercise 4, design a PD compensator to improve the step response of the system.

---

6. The field-controlled DC servomotor with values described in exercise 5 is required to drive massless robot arm of length 1 m with point mass $M = 1$ kg at the end. Friction in the robot joint is given by $F = 0.2$ N-s. If the joint is at the origin, the plane of motion of the robot arm is $\mathbf{p} = [1/\sqrt{2}\ \ 0\ \ 1/\sqrt{2}\ \ 0]$, and the mass M is initially at point $\mathbf{u} = [0\ \ 1\ \ 0\ \ 1]^\#$, simulate the response of the system to a $\pm\pi/2$ change in θ_d.

---

7. Improve the response of the field-controlled motor of exercise 6 by PD compensation.

---

8. The fourth worked example shows the inherently unstable nature of the eccentric loading experienced in an armature-controlled DC motor with eccentric load. In that system the frictional forces were small and the gain of the motor was chosen as 1 or 4. Investigate the system further, increasing F and k until a reasonable response to a step change in θ_d is obtained.

---

9. In the fourth worked example, the zeros of the PID compensator were chosen at the same point, $s = -\alpha$. Investigate the possibility of improvement in the response for the isolated case of a step change in θ_d from 0 to $-\pi/2$ when the system is initially at rest. Keep $k = 1$ and $F = 0.1$ and all the other parameters as chosen in that worked example.

---

10. Repeat the first worked example using a two-phase AC servomotor instead of the armature-controlled DC motor. Insert $K_1 = 10$, $K_2 = 10$, and $K_3 = 0.1$ in the dynamic equation for joint status and integrate using Euler's formula with step $h = 0.1$.

---

11. A 3 *dof* revolute robot $|\ \lambda\theta\theta\ |$ with link lengths of unity and base orientation and position given by $\mathbf{p}_1 = [0\ \ 1\ \ 0\ \ 0]$, $\boldsymbol{\nu} = [0\ \ 0\ \ 1]^\#$, $\mathbf{u}_1 = [0\ \ 0\ \ 0\ \ 1]^\#$ is driven by armature-controlled DC motors. It is required to execute a straight-line path between points $[1\ \ 1\ \ 0\ \ 1]^\#$ and $[1\ \ 0\ \ 0\ \ 1]^\#$ using the four-point Bezier curve with zero velocity at the endpoints. Indicate how to drive the robot over the required trajectory.

APPENDIX A

A Case for Cartesian Coordinates

The Cartesian coordinate space is used extensively in robotic work to the exclusion of the other three spaces—cylindrical, spherical, and polar. A particular class of robot, cylindrical axis robots, seem to lend themselves to specification in cylindrical coordinates. Another is polar. However, it will be shown that this is not the case, since a robot is useful only when it interacts with its environment, and the environment is almost always Cartesian. The problem will be seen to be caused in part by the dimensions of the coordinates in each space. In Cartesian coordinates all dimensions are length, but the cylindrical and spherical coordinate systems have mixed dimensions; the polar coordinate system can be specified such that the dimensions are consistent. However, the reference to the origin in all coordinate spaces is the problem with all but Cartesian space.

In all spaces, three independent quantities determine the space. Position in Cartesian space is given by $[x \quad y \quad z \quad 1]^{\#}$ in terms of the three orthogonal axes. The other coordinate spaces are defined in terms of x, y, and z (and vice versa) as follows.

In cylindrical coordinates, $r = \sqrt{x^2 + y^2}$ and $\phi = \tan^{-1}y/x$, so $x = r\cos\phi$, $y = r\sin\phi$, and $z = z$. In spherical coordinates,

$$\rho = \sqrt{x^2 + y^2 + z^2}, \qquad \theta = \tan^{-1}y/x, \quad \text{and} \quad \phi = \cos^{-1}z/\rho$$

so $x = \rho\sin\phi \cdot \cos\theta$, $y = \rho\sin\phi \cdot \sin\theta$, and $z = \rho\cos\phi$.
In polar coordinates, $\rho = \sqrt{x^2 + y^2 + z^2}$, $\gamma_x = \cos^{-1}x/\rho$, $\gamma_y = \cos^{-1}y/\rho$, and $\gamma_z = \cos^{-1}z/\rho$, so $x = \rho\cos\gamma_x$, $y = \rho\cos\gamma_y$, and $z = \rho\cos\gamma_z$.

The end effector of the cylindrical axis robot $\left| \phi_1 \begin{matrix} \overline{\sigma}_1 \\ \sigma_1 \end{matrix} \middle| \begin{matrix} \overline{\sigma}_2 \\ \sigma_2 \end{matrix} \right|$ in Cartesian space is

$$\begin{bmatrix} -\sigma_1 \cos \phi_1 \\ -\sigma_2 \sin \phi_1 \\ \sigma_1 \\ 1 \end{bmatrix}$$

In cylindrical space, it is

$$\begin{bmatrix} \sigma_2 \\ \phi_1 \\ \sigma_1 \\ 1 \end{bmatrix}$$

The 3 *dof* robot $| \phi_1 l_1 | \theta_2 l_2 | \theta_3 l_3 |$ has its end effector in Cartesian space at

$$\begin{bmatrix} -\cos \phi_1 \{l_2 \sin \theta_2 + l_3 \sin (\theta_2 + \theta_3)\} \\ -\sin \phi_1 \{l_2 \sin \theta_2 + l_3 \sin (\theta_2 + \theta_3)\} \\ l_1 + l_2 \cos \theta_2 + l_3 \cos (\theta_2 + \theta_3) \\ 1 \end{bmatrix}$$

and, in cylindrical space, at

$$\begin{bmatrix} l_2 \sin \theta_2 + l_3 \sin(\theta_2 + \theta_3) \\ \phi_1 \\ l_1 + l_2 \cos \theta_2 + l_3 \cos(\theta_2 + \theta_3) \\ 1 \end{bmatrix}$$

Extensive use is made of the dot and cross product in Cartesian space. These operations are inefficient to the point of ineffectiveness in other coordinate spaces. The dot and cross product must be carried out in a dimensionally consistent space: Cylindrical coordinates have the mixed dimensions of length, angle, and length. Converting the dot product from Cartesian space to cylindrical space gives

$$\begin{bmatrix} x_1 \\ y_1 \\ z_1 \end{bmatrix} \cdot \begin{bmatrix} x_2 \\ y_2 \\ z_2 \end{bmatrix} \equiv \begin{bmatrix} r_1 \cos \phi_1 \\ r_1 \sin \phi_1 \\ z_1 \end{bmatrix} \cdot \begin{bmatrix} r_2 \cos \phi_2 \\ r_2 \sin \phi_2 \\ z_2 \end{bmatrix} = r_1 r_2 \{\cos \phi_1 \cdot \cos \phi_2 + \sin \phi_1 \cdot \sin \phi_2\} + z_1 z_1$$

$$= r_1 r_2 \cos(\phi_1 + \phi_2) + z_1 z_2$$

Similarly, the cross product becomes

$$\begin{bmatrix} x_1 \\ y_1 \\ z_1 \end{bmatrix} \times \begin{bmatrix} x_2 \\ y_2 \\ z_2 \end{bmatrix} \equiv \begin{bmatrix} r_1\cos\phi_1 \\ r_1\sin\phi_1 \\ z_1 \end{bmatrix} \times \begin{bmatrix} r_2\cos\phi_2 \\ r_2\sin\phi_2 \\ z_2 \end{bmatrix}$$

$$= \begin{bmatrix} r_1z_2\sin\phi_1 - r_2z_1\sin\phi_2 \\ r_2z_1\cos\phi_2 - r_1z_2\cos\phi_1 \\ r_1r_2\{\cos\phi_1 \cdot \sin\phi_2 - \sin\phi_1 \cdot \cos\phi_2\} \end{bmatrix}$$

$$= \begin{bmatrix} r_1z_2\sin\phi_1 - r_2z_1\sin\phi_2 \\ r_2z_1\cos\phi_2 - r_1z_2\cos\phi_1 \\ r_1r_2\sin(\phi_1 - \phi_2) \end{bmatrix}$$

$$\equiv \begin{bmatrix} \{r_1^2z_2^2 + r_2^2z_1^2 - 2r_1r_2z_1z_2\cos(\phi_1 - \phi_2)\}^{1/2} \\ \tan^{-1}\left\{\dfrac{r_2z_1\cos\phi_2 - r_1z_2\cos\phi_1}{r_1z_2\sin\phi_1 - r_2z_1\sin\phi_2}\right\} \\ r_1r_2\sin(\phi_1 - \phi_2) \end{bmatrix}$$

The distance between two points is

$$\sqrt{(x_1 - x_2)^2 + (y_1 - y_2)^2 + (z_1 - z_2)^2}$$
$$= \sqrt{(r_1\cos\phi_1 - r_2\cos\phi_2)^2 + (r_1\sin\phi_1 - r_2\sin\phi_2)^2 + (z_1 - z_2)^2}$$
$$= \sqrt{r_1^2 + r_2^2 - 2r_1r_2\cos(\phi_1 - \phi_2)}$$

A line can be expressed as $\mathbf{u} + \alpha(\mathbf{v} - \mathbf{u})$, where $\mathbf{u}$ and $\mathbf{v}$ are two points on the line. In Cartesian space the line is straight. In cylindrical space the line will not be straight (except for a trivial case) and is in fact a spiral section of a line on the surface of a cone with apex at the origin and z-axis. The line is

$$\begin{bmatrix} r_1 \\ \phi_1 \\ z_1 \\ 1 \end{bmatrix} + \alpha \begin{bmatrix} r_2 - r_1 \\ \phi_2 - \phi_1 \\ z_2 - z_1 \\ 0 \end{bmatrix} = \begin{bmatrix} (1 - \alpha)\sqrt{x_1^2 + y_1^2} + \alpha\sqrt{x_2^2 + y_2^2} \\ (1 - \alpha)\tan^{-1}\dfrac{y_1}{x_1} + \alpha\tan^{-1}\dfrac{y_2}{x_2} \\ (1 - \alpha)z_1 + \alpha z_2 \\ 1 \end{bmatrix}$$

For the particular case of $x_1^2 + y_1^2 = x_2^2 + y_2^2$, r does not change, and the line becomes a section of a helix. This helix will intersect the xz- or yz-plane more than once if ϕ varies by more than 2π. Recall, the intersection of a straight line with any plane in Cartesian space is a unique point.

A surface in any coordinate space can sometimes be obtained by fixing one of the coordinates. The surface obtained by fixing $x = 0$ in Cartesian space is the yz-plane. In cylindrical space the surface obtained by

1. Fixing r is the surface of a cylinder of radius r and z-axis,
2. Fixing ϕ is the plane intersecting the z-axis,
3. Fixing z is a plane parallel to the xy-plane.

The homogeneous representation of a plane in Cartesian coordinates is of paramount importance to the methods employed to solve kinematic problems. If other coordinate systems are to be used, they must provide a similar entity with relationships similar in application to those developed for Cartesian space in Chapter 1. In particular, we need a mechanism similar to the $\mathbf{pu} = 0$ of Cartesian space to determine that point $\mathbf{u}$ is on plane $\mathbf{p}$. Suppose a surface in cylindrical coordinates is defined by $[a \quad b \quad c \quad d]$, where a and c have the dimensions of inverse length and b and d are dimensionless.

$$[a \quad b \quad c \quad d] \begin{bmatrix} r \\ \phi \\ z \\ 1 \end{bmatrix} = ar + b\phi + cz + d = 0 \quad \text{if the point is on the surface}$$

We should assume, just as we did with Cartesian space, that a linear dependency exists between a, b, c, and d, but it is not clear how to establish this dependency, particularly as there is a change in dimensions. Thus, three independent points will establish the surface.

The meaning of the origin in all coordinate spaces is crucial to the operations carried out in those spaces. A rigid body must maintain its shape when translated in position, but only Cartesian space permits this: A rigid body changes shape when its distance from the origin is changed in all but Cartesian space. A corollary of this is that a rigid body with straight-line edges will have curved edges when translated to a new location in every coordinate space but Cartesian.

To summarize, we live in a Cartesian world. What we process through our senses and understand with our brains are lines, corners, and rigid bodies whose shape remains invariant with position. Throw a baseball at someone and it is still a baseball with the same shape and size when it has changed position. Our brains decode, classify, and identify objects regardless of their position and orientation. We perform these operations easily in Cartesian space. To humans the real world is Cartesian, and we are unlikely to find uses in robotics for other spaces.

APPENDIX B

Derivations of Forward Kinematic Solutions for Revolute Robots

The plane of motion $\mathbf{p}_2 = [a_2 \quad b_2 \quad c_2 \quad 0]$ of the 3 *dof* revolute robot $|\ \lambda\theta\theta\ |$ with general base orientation is given by equation (3.6) as

$$\begin{bmatrix} a_2 \\ b_2 \\ c_2 \end{bmatrix} = \begin{bmatrix} \cos\lambda_1 & -\nu_z \sin\lambda_1 & \nu_y \sin\lambda_1 \\ \nu_z \sin\lambda_1 & \cos\lambda_1 & -\nu_x \sin\lambda_1 \\ -\nu_y \sin\lambda_1 & \nu_x \sin\lambda_1 & \cos\lambda_1 \end{bmatrix} \begin{bmatrix} a_1 \\ b_1 \\ c_1 \end{bmatrix}$$

$$= \begin{bmatrix} a_1 \cos\lambda_1 + (\nu_y c_1 - \nu_z b_1) \sin\lambda_1 \\ b_1 \cos\lambda_1 + (\nu_z a_1 - \nu_x c_1) \sin\lambda_1 \\ c_1 \cos\lambda_1 + (\nu_x b_1 - \nu_y a_1) \sin\lambda_1 \end{bmatrix} \tag{B.1}$$

$$\mathbf{u}_2 = \begin{bmatrix} l_1 \nu_x \\ l_1 \nu_y \\ l_1 \nu_z \\ 1 \end{bmatrix} \tag{B.2}$$

$\mathbf{u}_3$ is calculated from equation (3.1) as

$$\mathbf{u}_3 = \mathbf{u}_2 + l_2 \begin{bmatrix} \cos\theta_2 & -c_2 \sin\theta_2 & b_2 \sin\theta_2 & 0 \\ c_2 \sin\theta_2 & \cos\theta_2 & -a_2 \sin\theta_2 & 0 \\ -b_2 \sin\theta_2 & a_2 \sin\theta_2 & \cos\theta_2 & 0 \\ 0 & 0 & 0 & 1 \end{bmatrix} \begin{bmatrix} \nu_x \\ \nu_y \\ \nu_z \\ 0 \end{bmatrix}$$

$$= \begin{bmatrix} (l_1 + l_2\cos\theta_2)\nu_x + l_2\{\nu_y\{c_1\cos\lambda_1 + (\nu_x b_1 - \nu_y a_1)\sin\lambda_1\} \\ \qquad + \nu_z\{b_1\cos\lambda_1 + (\nu_z a_1 - \nu_x c_1)\sin\lambda_1\}\}\sin\theta_2 \\ (l_1 + l_2\cos\theta_2)\nu_y + l_2\{-\nu_x\{c_1\cos\lambda_1 + (\nu_x b_1 - \nu_y a_1)\sin\lambda_1\} \\ \qquad - \nu_z\{a_1\cos\lambda_1 + (\nu_y c_1 - \nu_z b_1)\sin\lambda_1\}\}\sin\theta_2 \\ (l_1 + l_2\cos\theta_2)\nu_z + l_2\{\nu_x\{b_1\cos\lambda_1 + (\nu_z a_1 - \nu_x c_1)\sin\lambda_1\} \\ \qquad + \nu_y\{a_1\cos\lambda_1 + (\nu_y c_1 - \nu_z b_1)\sin\lambda_1\}\}\sin\theta_2 \\ 1 \end{bmatrix}$$

$$= \begin{bmatrix} (l_1 + l_2\cos\theta_2)\nu_x - l_2\{(\nu_y c_1 - \nu_z b_1)\cos\lambda_1 - a_1\sin\lambda_1\}\sin\theta_2 \\ (l_1 + l_2\cos\theta_2)\nu_y - l_2\{(\nu_z a_1 - \nu_x c_1)\cos\lambda_1 - b_1\sin\lambda_1\}\sin\theta_2 \\ (l_1 + l_2\cos\theta_2)\nu_z - l_2\{(\nu_x b_1 - \nu_y a_1)\cos\lambda_1 - c_1\sin\lambda_1\}\sin\theta_2 \\ 1 \end{bmatrix} \tag{B.3}$$

$\mathbf{u}_4$ is calculated from equation (3.1) as

$$\mathbf{u}_4 = \mathbf{u}_3 + \frac{l_3}{l_2}\begin{bmatrix} \cos\theta_3 & -c_2\sin\theta_2 & b_2\sin\theta_3 & 0 \\ c_2\sin\theta_3 & \cos\theta_3 & -a_2\sin\theta_3 & 0 \\ -b_2\sin\theta_3 & a_2\sin\theta_3 & \cos\theta_3 & 0 \\ 0 & 0 & 0 & 1 \end{bmatrix}(\mathbf{u}_3 - \mathbf{u}_2)$$

where the matrix block on the right hand side is given by l_3 times

$$\begin{bmatrix} \cos\theta_3 & -c_2\sin\theta_3 & b_2\sin\theta_3 & 0 \\ c_2\sin\theta_3 & \cos\theta_3 & -a_2\sin\theta_3 & 0 \\ -b_2\sin\theta_3 & a_2\sin\theta_3 & \cos\theta_3 & 0 \\ 0 & 0 & 0 & 1 \end{bmatrix}$$

$$\begin{bmatrix} \nu_x\cos\theta_2 - \{(\nu_y c_1 - \nu_z b_1)\cos\lambda_1 - a_1\sin\lambda_1\}\sin\theta_2 \\ \nu_y\cos\theta_2 - \{(\nu_z a_1 - \nu_x c_1)\cos\lambda_1 - b_1\sin\lambda_1\}\sin\theta_2 \\ \nu_z\cos\theta_2 - \{(\nu_x b_1 - \nu_y a_1)\cos\lambda_1 - c_1\sin\lambda_1\}\sin\theta_2 \\ 0 \end{bmatrix}$$

$$= \begin{bmatrix} (\nu_x\cos\theta_2 - \{(\nu_y c_1 - \nu_z b_1)\cos\lambda_1 - a_1\sin\lambda_1\}\sin\theta_2)\cos\theta_3 \\ \quad + (\nu_y\cos\theta_2 - \{(\nu_z a_1 - \nu_x c_1)\cos\lambda_1 - b_1\sin\lambda_1\}\sin\theta_2)c_2\sin\theta_3 \\ \qquad - (\nu_z\cos\theta_2 - \{(\nu_x b_1 - \nu_y a_1)\cos\lambda_1 - c_1\sin\lambda_1\}\sin\theta_2)b_2\sin\theta_3 \\ -(\nu_x\cos\theta_2 - \{(\nu_y c_1 - \nu_z b_1)\cos\lambda_1 - a_1\sin\lambda_1\}\sin\theta_2)c_2\sin\theta_3 \\ \quad + (\nu_y\cos\theta_2 - \{(\nu_z a_1 - \nu_x c_1)\cos\lambda_1 - b_1\sin\lambda_1\}\sin\theta_2)\cos\theta_3 \\ \qquad + (\nu_z\cos\theta_2 - \{(\nu_x b_1 - \nu_y a_1)\cos\lambda_1 - c_1\sin\lambda_1\}\sin\theta_2)a_2\sin\theta_3 \\ (\nu_x\cos\theta_2 - \{(\nu_y c_1 - \nu_z b_1)\cos\lambda_1 - a_1\sin\lambda_1\}\sin\theta_2)b_2\sin\theta_3 \\ \quad - (\nu_y\cos\theta_2 - \{(\nu_z a_1 - \nu_x c_1)\cos\lambda_1 - b_1\sin\lambda_1\}\sin\theta_2)a_2\sin\theta_3 \\ \qquad + (\nu_z\cos\theta_2 - \{(\nu_x b_1 - \nu_y a_1)\cos\lambda_1 - c_1\sin\lambda_1\}\sin\theta_2)\cos\theta_3 \\ 0 \end{bmatrix}$$

$$
= \begin{bmatrix}
\begin{aligned}
&\{\nu_x\cos\theta_3 - (\nu_y c_2 - \nu_z b_2)\sin\theta_3\}\cos\theta_2 \\
&\qquad - \{(\nu_y c_1 - \nu_z b_1)\cos\lambda_1 - a_1\sin\lambda_1\}\sin\theta_2\cos\theta_3 \\
&\qquad + (\{(\nu_z a_1 - \nu_x c_1)\cos\lambda_1 - b_1\sin\lambda_1\}c_2 \\
&\qquad - \{(\nu_x b_1 - \nu_y a_1)\cos\lambda_1 - c_1\sin\lambda_1\}b_2)\sin\theta_2\sin\theta_3
\end{aligned} \\
\begin{aligned}
&\{\nu_y\cos\theta_3 - (\nu_z a_2 - \nu_x c_2)\sin\theta_3\}\cos\theta_2 \\
&\qquad - (\{(\nu_z a_1 - \nu_x c_1)\cos\lambda_1 - b_1\sin\lambda_1\}\sin\theta_2)\cos\theta_3 \\
&\qquad + (-\{(\nu_y c_1 - \nu_z b_1)\cos\lambda_1 - a_1\sin\lambda_1\}c_2 \\
&\qquad + \{(\nu_x b_1 - \nu_y a_1)\cos\lambda_1 - c_1\sin\lambda_1\}a_2)\sin\theta_2\sin\theta_3
\end{aligned} \\
\begin{aligned}
&\{\nu_z\cos\theta_3 + (\nu_x b_2 - \nu_y a_2)\sin\theta_3\}\cos\theta_2 \\
&\qquad - \{(\nu_x b_1 - \nu_y a_1)\cos\lambda_1 - c_1\sin\lambda_1\}\sin\theta_2\cos\theta_3 \\
&\qquad + (\{(\nu_y c_1 - \nu_z b_1)\cos\lambda_1 - a_1\sin\lambda_1\}b_2 \\
&\qquad - \{(\nu_z a_1 - \nu_x c_1)\cos\lambda_1 - b_1\sin\lambda_1\}a_2)\sin\theta_2\sin\theta_3
\end{aligned} \\
0
\end{bmatrix}
$$

$$
= \begin{bmatrix}
\begin{aligned}
&\{\nu_x\cos\theta_3 + (\nu_y c_2 - \nu_z b_2)\sin\theta_3\}\cos\theta_2 \\
&\qquad - \{(\nu_y c_1 - \nu_z b_1)\cos\lambda_1 - a_1\sin\lambda_1\}\sin\theta_2\cos\theta_3 \\
&\qquad + \{(c_1 b_2 - b_1 c_2)\sin\lambda_1 - \nu_x\cos^2\lambda_1\}\sin\theta_2\sin\theta_3
\end{aligned} \\
\begin{aligned}
&\{\nu_y\cos\theta_3 + (\nu_z a_2 - \nu_x c_2)\sin\theta_3\}\cos\theta_2 \\
&\qquad - \{(\nu_z a_1 - \nu_x c_1)\cos\lambda_1 - b_1\sin\lambda_1\}\sin\theta_2\cos\theta_3 \\
&\qquad + \{(a_1 c_2 - c_1 a_2)\sin\lambda_1 - \nu_y\cos^2\lambda_1\}\sin\theta_2\sin\theta_3
\end{aligned} \\
\begin{aligned}
&\{\nu_z\cos\theta_3 + (\nu_x b_2 - \nu_y a_2)\sin\theta_3\}\cos\theta_2 \\
&\qquad - \{(\nu_x b_1 - \nu_y a_1)\cos\lambda_1 - c_1\sin\lambda_1\}\sin\theta_2\cos\theta_3 \\
&\qquad + \{(b_1 a_2 - a_1 b_2)\sin\lambda_1 - \nu_z\cos^2\lambda_1\}\sin\theta_2\sin\theta_3
\end{aligned} \\
0
\end{bmatrix}
$$

$$
= \begin{bmatrix}
\begin{aligned}
&\{\nu_x\cos\theta_3 - (\nu_y c_1\cos\lambda_1 + \nu_y(\nu_x b_1 - \nu_y a_1)\sin\lambda_1 \\
&\quad - \nu_z b_1\cos\lambda_1 - \nu_z(\nu_z a_1 - \nu_x c_1)\sin\lambda_1)\sin\theta_3\}\cos\theta_2 \\
&\quad - \{(\nu_y c_1 - \nu_z b_1)\cos\lambda_1 - a_1\sin\lambda_1\}\sin\theta_2\cos\theta_3 \\
&\quad + \{(c_1(\nu_z a_1 - \nu_x c_1) - b_1(\nu_x b_1 - \nu_y a_1))\sin^2\lambda_1 - \nu_x\cos^2\lambda_1\}\sin\theta_2\sin\theta_3
\end{aligned} \\
\begin{aligned}
&\{\nu_y\cos\theta_3 - (\nu_z a_1\cos\lambda_1 + \nu_z(\nu_y c_1 - \nu_z b_1)\sin\lambda_1 \\
&\quad - \nu_x c_1\cos\lambda_1 - \nu_x(\nu_x b_1 - \nu_y a_1)\sin\lambda_1)\sin\theta_3\}\cos\theta_2 \\
&\quad - \{(\nu_z a_1 - \nu_x c_1)\cos\lambda_1 - b_1\sin\lambda_1\}\sin\theta_2\cos\theta_3 \\
&\quad + \{a_1(\nu_x b_1 - \nu_y a_1) - c_1(\nu_y c_1 - \nu_z b_1)\sin^2\lambda_1 - \nu_y\cos^2\lambda_1\}\sin\theta_2\sin\theta_3
\end{aligned} \\
\begin{aligned}
&\{\nu_z\cos\theta_3 - (\nu_x b_1\cos\lambda_1 + \nu_x(\nu_z a_1 - \nu_x c_1)\sin\lambda_1 \\
&\quad - \nu_y a_1\cos\lambda_1 - \nu_y(\nu_y c_1 - \nu_z b_1)\sin\lambda_1)\sin\theta_3\}\cos\theta_2 \\
&\quad - \{(\nu_x b_1 - \nu_y a_1)\cos\lambda_1 - c_1\sin\lambda_1\}\sin\theta_2\cos\theta_3 \\
&\quad + \{b_1(\nu_y c_1 - \nu_z b_1) - a_1(\nu_z a_1 - \nu_x c_1)\sin^2\lambda_1 - \nu_z\cos^2\lambda_1\}\sin\theta_2\sin\theta_3
\end{aligned} \\
0
\end{bmatrix}
$$

$$= \begin{bmatrix} \nu_x\cos\theta_3\cos\theta_2 - \{(\nu_y c_1 - \nu_z b_1)\cos\lambda_1 - a_1\sin\lambda_1\} \\ \qquad \times (\sin\theta_3\cos\theta_2 + \sin\theta_2\cos\theta_3) - \nu_x\sin\theta_2\sin\theta_3 \\ \nu_y\cos\theta_3\cos\theta_2 - \{(\nu_z a_1 - \nu_x c_1)\cos\lambda_1 - b_1\sin\lambda_1\} \\ \qquad \times (\sin\theta_3\cos\theta_2 + \sin\theta_2\cos\theta_3) - \nu_y\sin\theta_2\sin\theta_3 \\ \nu_z\cos\theta_3\cos\theta_2 - \{(\nu_x b_1 - \nu_y a_1)\cos\lambda_1 - c_1\sin\lambda_1\} \\ \qquad \times (\sin\theta_3\cos\theta_2 + \sin\theta_2\cos\theta_3) - \nu_z\sin\theta_2\sin\theta_3 \\ 0 \end{bmatrix}$$

$$= \begin{bmatrix} \nu_x\cos(\theta_2+\theta_3) - \{(\nu_y c_1 - \nu_z b_1)\cos\lambda_1 - a_1\sin\lambda_1\}\sin(\theta_2+\theta_3) \\ \nu_y\cos(\theta_2+\theta_3) - \{(\nu_z a_1 - \nu_x c_1)\cos\lambda_1 - b_1\sin\lambda_1\}\sin(\theta_2+\theta_3) \\ \nu_z\cos(\theta_2+\theta_3) - \{(\nu_x b_1 - \nu_y a_1)\cos\lambda_1 - c_1\sin\lambda_1\}\sin(\theta_2+\theta_3) \\ 0 \end{bmatrix} \tag{B.4}$$

so $\mathbf{u}_4$ is given by

$$\begin{bmatrix} \nu_x\{l_1 + l_2\cos\theta_2 + l_3\cos(\theta_2+\theta_3)\} \\ \quad - \{(\nu_y c_1 - \nu_z b_1)\cos\lambda_1 - a_1\sin\lambda_1\}\{l_2\sin\theta_2 + l_3\sin(\theta_2+\theta_3)\} \\ \nu_y\{l_1 + l_2\cos\theta_2 + l_3\cos(\theta_2+\theta_3)\} \\ \quad - \{(\nu_z a_1 - \nu_x c_1)\cos\lambda_1 - b_1\sin\lambda_1\}\{l_2\sin\theta_2 + l_3\sin(\theta_2+\theta_3)\} \\ \nu_z\{l_1 + l_2\cos\theta_2 + l_3\cos(\theta_2+\theta_3)\} \\ \quad - \{(\nu_x b_1 - \nu_y a_1)\cos\lambda_1 - c_1\sin\lambda_1\}\{l_2\sin\theta_2 + l_3\sin(\theta_2+\theta_3)\} \\ 0 \end{bmatrix} \tag{B.5}$$

It can be shown that the robot $|\ \lambda\theta\theta\theta\ |$ has

$$\mathbf{u}_5 = \begin{bmatrix} l_1 + \nu_x\{l_2\cos\theta_2 + l_3\cos(\theta_2+\theta_3) + l_4\cos(\theta_2+\theta_3+\theta_4)\} \\ \quad - \{(\nu_y c_1 - \nu_z b_1)\cos\lambda_1 - a_1\sin\lambda_1\}\{l_2\sin\theta_2 + l_3\sin(\theta_2+\theta_3) \\ \qquad + l_4\sin(\theta_2+\theta_3+\theta_4)\} \\ l_1 + \nu_y\{l_2\cos\theta_2 + l_3\cos(\theta_2+\theta_3) + l_4\cos(\theta_2+\theta_3+\theta_4)\} \\ \quad - \{(\nu_z a_1 - \nu_x c_1)\cos\lambda_1 - b_1\sin\lambda_1\}\{l_2\sin\theta_2 + l_3\sin(\theta_2+\theta_3) \\ \qquad + l_4\sin(\theta_2+\theta_3+\theta_4)\} \\ l_1 + \nu_z\{l_2\cos\theta_2 + l_3\cos(\theta_2+\theta_3) + l_4\cos(\theta_2+\theta_3+\theta_4)\} \\ \quad - \{(\nu_x b_1 - \nu_y a_1)\cos\lambda_1 - c_1\sin\lambda_1\}\{l_2\sin\theta_2 + l_3\sin(\theta_2+\theta_3) \\ \qquad + l_4\sin(\theta_2+\theta_3+\theta_4)\} \\ 0 \end{bmatrix} \tag{B.6}$$

The 4 *dof* pencilable robot $|\ \lambda\theta\theta\lambda\ |$ has the same $\mathbf{u}_5$ as $\mathbf{u}_4$ of equation (B.5) with l_3 replaced by $l_3 + l_4$. The robot has two planes of motion, $\bar{\mathbf{p}}_2^2$ and $\bar{\mathbf{p}}_3^0$, and from equation (3.7),

$$\begin{bmatrix} a_3 \\ b_3 \\ c_3 \end{bmatrix} = \begin{bmatrix} \cos\lambda_4 & \eta_z\sin\lambda_4 & -\eta_y\sin\lambda_4 \\ -\eta_z\sin\lambda_4 & \cos\lambda_4 & \eta_x\sin\lambda_4 \\ \eta_y\sin\lambda_4 & -\eta_x\sin\lambda_4 & \cos\lambda_4 \end{bmatrix} \begin{bmatrix} a_2 \\ b_2 \\ c_2 \end{bmatrix}$$

$$
= \begin{bmatrix}
\begin{array}{l}
\{a_1\cos\lambda_1 + (\nu_y c_1 - \nu_z b_1)\sin\lambda_1\}\cos\lambda_4 \\
\qquad + \{\{b_1\cos\lambda_1 + (\nu_z a_1 - \nu_x c_1)\sin\lambda_1\}\eta_z \\
\qquad - \{c_1\cos\lambda_1 + (\nu_x b_1 - \nu_y a_1)\sin\lambda_1\}\eta_y\}\sin\lambda_4
\end{array} \\
\begin{array}{l}
\{b_1\cos\lambda_1 + (\nu_z a_1 - \nu_x c_1)\sin\lambda_1\}\cos\lambda_4 \\
\qquad + \{-\{a_1\cos\lambda_1 + (\nu_y c_1 - \nu_z b_1)\sin\lambda_1\}\eta_z \\
\qquad + \{c_1\cos\lambda_1 + (\nu_x b_1 - \nu_y a_1)\sin\lambda_1\}\eta_x\}\sin\lambda_4
\end{array} \\
\begin{array}{l}
\{c_1\cos\lambda_1 + (\nu_x b_1 - \nu_y a_1)\sin\lambda_1\}\cos\lambda_4 \\
\qquad + \{\{a_1\cos\lambda_1 + (\nu_y c_1 - \nu_z b_1)\sin\lambda_1\}\eta_y \\
\qquad - \{b_1\cos\lambda_1 + (\nu_z a_1 - \nu_x c_1)\sin\lambda_1\}\eta_x\}\sin\lambda_4
\end{array}
\end{bmatrix}
$$

$$
= \begin{bmatrix}
\begin{array}{l}
\{a_1\cos\lambda_1 + (\nu_y c_1 - \nu_z b_1)\sin\lambda_1\}\cos\lambda_4 \\
\quad + \{\{b_1\cos\lambda_1 + (\nu_z a_1 - \nu_x c_1)\sin\lambda_1\}\{\nu_z\cos(\theta_2 + \theta_3) \\
\quad - \{(\nu_x b_1 - \nu_y a_1)\cos\lambda_1 - c_1\sin\lambda_1\}\sin(\theta_2 + \theta_3)\} \\
\quad - \{c_1\cos\lambda_1 + (\nu_x b_1 - \nu_y a_1)\sin\lambda_1\}\{\nu_y\cos(\theta_2 + \theta_3) \\
\quad - \{(\nu_z a_1 - \nu_x c_1)\cos\lambda_1 - b_1\sin\lambda_1\}\sin(\theta_2 + \theta_3)\}\}\sin\lambda_4
\end{array} \\
\begin{array}{l}
\{b_1\cos\lambda_1 + (\nu_z a_1 - \nu_x c_1)\sin\lambda_1\}\cos\lambda_4 \\
\quad + \{-\{a_1\cos\lambda_1 + (\nu_y c_1 - \nu_z b_1)\sin\lambda_1\}\{\nu_z\cos(\theta_2 + \theta_3) \\
\quad - \{(\nu_x b_1 - \nu_y a_1)\cos\lambda_1 - c_1\sin\lambda_1\}\sin(\theta_2 + \theta_3)\} \\
\quad + \{c_1\cos\lambda_1 + (\nu_x b_1 - \nu_y a_1)\sin\lambda_1\}\{\nu_x\cos(\theta_2 + \theta_3) \\
\quad - \{(\nu_y c_1 - \nu_z b_1)\cos\lambda_1 - a_1\sin\lambda_1\}\sin(\theta_2 + \theta_3)\}\}\sin\lambda_4
\end{array} \\
\begin{array}{l}
\{c_1\cos\lambda_1 + (\nu_x b_1 - \nu_y a_1)\sin\lambda_1\}\cos\lambda_4 \\
\quad + \{\{a_1\cos\lambda_1 + (\nu_y c_1 - \nu_z b_1)\sin\lambda_1\}\{\nu_y\cos(\theta_2 + \theta_3) \\
\quad - \{(\nu_z a_1 - \nu_x c_1)\cos\lambda_1 - b_1\sin\lambda_1\}\sin(\theta_2 + \theta_3)\} \\
\quad - \{b_1\cos\lambda_1 + (\nu_z a_1 - \nu_x c_1)\sin\lambda_1\}\{\nu_x\cos(\theta_2 + \theta_3) \\
\quad - \{(\nu_y c_1 - \nu_z b_1)\cos\lambda_1 - a_1\sin\lambda_1\}\sin(\theta_2 + \theta_3)\}\}\sin\lambda_4
\end{array}
\end{bmatrix}
$$

$$
= \begin{bmatrix}
\begin{array}{l}
\{a_1\cos\lambda_1 + (\nu_y c_1 - \nu_z b_1)\sin\lambda_1\}\cos\lambda_4 \\
\quad + \{b_1\nu_z\cos\lambda_1 + (\nu_z^2 a_1 - \nu_x\nu_z c_1)\sin\lambda_1 - c_1\nu_y\cos\lambda_1 \\
\quad - (\nu_x\nu_y b_1 - \nu_y^2 a_1)\sin\lambda_1\}\cos(\theta_2 + \theta_3)\sin\lambda_4 \\
\quad + \{\{b_1\cos\lambda_1 + (\nu_z a_1 - \nu_x c_1)\sin\lambda_1\}\{(\nu_x b_1 \\
\quad - \nu_y a_1)\cos\lambda_1 - c_1\sin\lambda_1\} \\
\quad + \{c_1\cos\lambda_1 + (\nu_x b_1 - \nu_y a_1)\sin\lambda_1\}\{(\nu_z a_1 - \nu_x c_1)\cos\lambda_1 \\
\quad - b_1\sin\lambda_1\}\}\sin(\theta_2 + \theta_3)\sin\lambda_4
\end{array} \\
\begin{array}{l}
\{b_1\cos\lambda_1 + (\nu_z a_1 - \nu_x c_1)\sin\lambda_1\}\cos\lambda_4 \\
\quad + \{-a_1\nu_z\cos\lambda_1 - (\nu_y\nu_z c_1 - \nu_z^2 b_1)\sin\lambda_1 + c_1\nu_x\cos\lambda_1 \\
\quad + (\nu_x^2 b_1 - \nu_x\nu_y a_1)\sin\lambda_1\}\cos(\theta_2 + \theta_3)\sin\lambda_4 \\
\quad + \{\{-a_1\cos\lambda_1 - (\nu_y c_1 - \nu_z b_1)\sin\lambda_1\}\{(\nu_x b_1 \\
\quad - \nu_y a_1)\cos\lambda_1 - c_1\sin\lambda_1\} \\
\quad - \{c_1\cos\lambda_1 + (\nu_x b_1 - \nu_y a_1)\sin\lambda_1\}\{(\nu_y c_1 - \nu_z b_1)\cos\lambda_1 \\
\quad - a_1\sin\lambda_1\}\}\sin(\theta_2 + \theta_3)\sin\lambda_4
\end{array} \\
\begin{array}{l}
\{c_1\cos\lambda_1 + (\nu_x b_1 - \nu_y a_1)\sin\lambda_1\}\cos\lambda_4 \\
\quad + \{a_1\nu_y\cos\lambda_1 + (\nu_y^2 c_1 - \nu_y\nu_z b_1)\sin\lambda_1 - b_1\nu_x\cos\lambda_1 \\
\quad - (\nu_x\nu_z a_1 - \nu_x^2 c_1)\sin\lambda_1\}\cos(\theta_2 + \theta_3)\sin\lambda_4
\end{array}
\end{bmatrix}
$$

$$
\begin{bmatrix}
\quad + \{\{a_1\cos\lambda_1 + (v_y c_1 - v_z b_1)\sin\lambda_1\}\{(v_z a_1 \\
\quad - v_x c_1)\cos\lambda_1 - b_1\sin\lambda_1\} \\
\quad + \{b_1\cos\lambda_1 + (v_z a_1 - v_x c_1)\sin\lambda_1\}\{(v_y c_1 - v_z b_1)\cos\lambda_1 \\
\quad - a_1\sin\lambda_1\}\}\sin(\theta_2 + \theta_3)\sin\lambda_4
\end{bmatrix}
$$

$$
= \begin{bmatrix}
\{a_1\cos\lambda_1 + (v_y c_1 - v_z b_1)\sin\lambda_1\}\cos\lambda_4 \\
+ \{(b_1 v_z - c_1 v_y)\cos\lambda_1 + (v_z^2 a_1 - v_x v_z c_1 - v_x v_y b_1 \\
+ v_y^2 a_1)\sin\lambda_1\}\cos(\theta_2 + \theta_3)\sin\lambda_4 \\
-\{\{b_1(v_x b_1 - v_y a_1) - c_1(v_z a_1 - v_x c_1)\sin^2\lambda_1\}\}\sin(\theta_2 + \theta_3)\sin\lambda_4 \\
\{b_1\cos\lambda_1 + (v_z a_1 - v_x c_1)\sin\lambda_1\}\cos\lambda_4 \\
+ \{(-a_1 v_z + c_1 v_x)\cos\lambda_1 + \{-v_y v_z c_1 + v_z^2 b_1 + v_x^2 b_1 \\
- v_x v_y a_1\}\sin\lambda_1\}\cos(\theta_2 + \theta_3)\sin\lambda_4 \\
- \{-a_1(v_x b_1 - v_y a_1) + c_1(v_y c_1 - v_z b_1)\}\sin(\theta_2 + \theta_3)\sin\lambda_4 \\
\{c_1\cos\lambda_1 + (v_x b_1 - v_y a_1)\sin\lambda_1\}\cos\lambda_4 \\
+ \{(a_1 v_y - b_1 v_x)\cos\lambda_1 + (v_y^2 c_1 - v_y v_z b_1 - v_x v_z a_1 \\
+ v_x^2 c_1)\sin\lambda_1\}\cos(\theta_2 + \theta_3)\sin\lambda_4 \\
- \{a_1(v_z a_1 - v_x c_1) - b_1(v_y c_1 - v_z b_1)\sin\lambda_1^2\}\}\sin(\theta_2 + \theta_3)\sin\lambda_4
\end{bmatrix}
$$

so

$$
\begin{bmatrix} a_3 \\ b_3 \\ c_3 \end{bmatrix} = \begin{bmatrix}
\{a_1\cos\lambda_1 + (v_y c_1 - v_z b_1)\sin\lambda_1\}\cos\lambda_4 \\
\quad + \{\{(b_1 v_z - c_1 v_y)\cos\lambda_1 + a_1\sin\lambda_1\}\cos(\theta_2 + \theta_3) \\
\quad - v_x\sin(\theta_2 + \theta_3)\}\sin\lambda_4 \\
\{b_1\cos\lambda_1 + (v_z a_1 - v_x c_1)\sin\lambda_1\}\cos\lambda_4 \\
\quad + \{\{(-a_1 v_z + c_1 v_x)\cos\lambda_1 + b_1\sin\lambda_1\}\cos(\theta_2 + \theta_3) \\
\quad - v_y\sin(\theta_2 + \theta_3)\}\sin\lambda_4 \\
\{c_1\cos\lambda_1 + (v_x b_1 - v_y a_1)\sin\lambda_1\}\cos\lambda_4 \\
\quad + \{\{(a_1 v_y - b_1 v_x)\cos\lambda_1 + c_1\sin\lambda_1\}\cos(\theta_2 + \theta_3) \\
\quad - v_z\sin(\theta_2 + \theta_3)\}\sin\lambda_4
\end{bmatrix} \tag{B.7}
$$

A 5 *dof* directable robot $|\,\lambda\theta\theta\lambda\theta\,|$ has two planes of motion, $\bar{\mathbf{p}}_2^2$ and $\bar{\mathbf{p}}_3^1$, and the same $\mathbf{u}_5$ as robot $|\,\lambda\theta\theta\lambda\,|$. $\bar{\mathbf{p}}_3^1$ is given by equation (B.7). From equation (3.1)

$$
\mathbf{u}_6 = \mathbf{u}_5 + \frac{l_5}{l_4}\begin{bmatrix}
\cos\theta_5 & -c_3\sin\theta_5 & b_3\sin\theta_5 & 0 \\
c_3\sin\theta_5 & \cos\theta_5 & -a_3\sin\theta_5 & 0 \\
-b_3\sin\theta_5 & a_3\sin\theta_5 & \cos\theta_5 & 0 \\
0 & 0 & 0 & 1
\end{bmatrix}(\mathbf{u}_5 - \mathbf{u}_4)
$$

and $(\mathbf{u}_5 - \mathbf{u}_4)l_5/l_4$ on the right-hand side is given by l_5 times equation (B.4), so $\mathbf{u}_6 - \mathbf{u}_5$ is given by l_5 times

$$
\begin{bmatrix}
\cos\theta_5 & -c_3\sin\theta_5 & b_3\sin\theta_5 & 0 \\
c_3\sin\theta_5 & \cos\theta_5 & -a_3\sin\theta_5 & 0 \\
-b_3\sin\theta_5 & a_3\sin\theta_5 & \cos\theta_5 & 0 \\
0 & 0 & 0 & 1
\end{bmatrix} \times
$$

$$\begin{bmatrix} \nu_x\cos(\theta_2+\theta_3) - \{(\nu_y c_1 - \nu_z b_1)\cos\lambda_1 - a_1\sin\lambda_1\}\sin(\theta_2+\theta_3) \\ \nu_y\cos(\theta_2+\theta_3) - \{(\nu_z a_1 - \nu_x c_1)\cos\lambda_1 - b_1\sin\lambda_1\}\sin(\theta_2+\theta_3) \\ \nu_z\cos(\theta_2+\theta_3) - \{(\nu_x b_1 - \nu_y a_1)\cos\lambda_1 - c_1\sin\lambda_1\}\sin(\theta_2+\theta_3) \\ 0 \end{bmatrix} = \begin{bmatrix} \xi_x \\ \xi_y \\ \xi_z \\ 0 \end{bmatrix}$$

so

$$\begin{aligned}
\xi_x = {} & \cos\theta_5\{\nu_x\cos(\theta_2+\theta_3) - \{(\nu_y c_1 - \nu_z b_1)\cos\lambda_1 - a_1\sin\lambda_1\}\sin(\theta_2+\theta_3)\} \\
& - \{\{c_1\cos\lambda_1 + (\nu_x b_1 - \nu_y a_1)\sin\lambda_1\}\cos\lambda_4 \\
& + \{\{(a_1\nu_y - b_1\nu_x)\cos\lambda_1 + c_1\sin\lambda_1\}\cos(\theta_2+\theta_3) \\
& - \nu_z\sin(\theta_2+\theta_3)\}\sin\lambda_4\}\sin\theta_5\{\nu_y\cos(\theta_2+\theta_3) \\
& \qquad - \{(\nu_z a_1 - \nu_x c_1)\cos\lambda_1 - b_1\sin\lambda\}\sin(\theta_2+\theta_3)\} \\
& + \{\{b_1\cos\lambda_1 + (\nu_z a_1 - \nu_x c_1)\sin\lambda_1\}\cos\lambda_4 \\
& + \{\{(-a_1\nu_z + c_1\nu_x)\cos\lambda_1 + b_1\sin\lambda_1\}\cos(\theta_2+\theta_3) \\
& - \nu_y\sin(\theta_2+\theta_3)\}\sin\lambda_4\}\sin\theta_5 \\
& \qquad \{\nu_z\cos(\theta_2+\theta_3) - \{(\nu_x b_1 - \nu_y a_1)\cos\lambda_1 - c_1\sin\lambda_1\}\sin(\theta_2+\theta_3\} \\
= {} & \nu_x\cos(\theta_2+\theta_3)\cos\theta_5 - \{(\nu_y c_1 - \nu_z b_1)\cos\lambda_1 \\
& - a_1\sin\lambda_1\}\sin(\theta_2+\theta_3)\cos\theta_5 \\
& - \{\{c_1\cos\lambda_1 + (\nu_x b_1 - \nu_y a_1)\sin\lambda_1\}\{\nu_y\cos(\theta_2+\theta_3) \\
& - \{(\nu_z a_1 - \nu_x c_1)\cos\lambda_1 - b_1\sin\lambda_1\}\sin(\theta_2+\theta_3)\} \\
& - \{b_1\cos\lambda_1 + (\nu_z a_1 - \nu_x c_1)\sin\lambda_1\}\{\nu_z\cos(\theta_2+\theta_3) \\
& - \{(\nu_x b_1 - \nu_y a_1)\cos\lambda_1 - c_1\sin\lambda_1\}\sin(\theta_2+\theta_3)\}\}\cos\lambda_4\sin\theta_5 \\
& - \{\{(a_1\nu_y - b_1\nu_x)\cos\lambda_1\cos(\theta_2+\theta_3) + c_1\sin\lambda_1\cos(\theta_2+\theta_3) \\
& - \nu_z\sin(\theta_2+\theta_3)\} \\
& \times \{\nu_y\cos(\theta_2+\theta_3) - \{(\nu_z a_1 - \nu_x c_1)\cos\lambda_1 - b_1\sin\lambda_1\}\sin(\theta_2+\theta_3)\} \\
& - \{(-a_1\nu_z + c_1\nu_x)\cos\lambda_1\cos(\theta_2+\theta_3) \\
& + b_1\sin\lambda_1\cos(\theta_2+\theta_3) - \nu_y\sin(\theta_2+\theta_3)\} \\
& \times \{\nu_z\cos(\theta_2+\theta_3) - \{(\nu_x b_1 - \nu_y a_1)\cos\lambda_1 \\
& - c_1\sin\lambda_1\}\sin(\theta_2+\theta_3)\}\}\sin\lambda_4\sin\theta_5 \\
= {} & \nu_x\cos(\theta_2+\theta_3)\cos\theta_5 - \{(\nu_y c_1 - \nu_z b_1)\cos\lambda_1 \\
& - a_1\sin\lambda_1\}\sin(\theta_2+\theta_3)\cos\theta_5 \\
& + \{-c_1\nu_y\cos\lambda_1\cos(\theta_2+\theta_3) - (\nu_x\nu_y b_1 - \nu_y^2 a_1)\sin\lambda_1\cos(\theta_2+\theta_3)
\end{aligned}$$

$$
\begin{aligned}
&+ \{c_1\cos\lambda_1 + (\nu_x b_1 - \nu_y a_1)\sin\lambda_1\}\{(\nu_z a_1 - \nu_x c_1)\cos\lambda_1 \\
&- b_1\sin\lambda_1\}\sin(\theta_2+\theta_3) \\
&+ b_1\nu_z\cos\lambda_1\cos(\theta_2+\theta_3) + (\nu_z^2 a_1 - \nu_x\nu_z c_1)\sin\lambda_1\cos(\theta_2+\theta_3) \\
&- \{b_1\cos\lambda_1 + (\nu_z a_1 - \nu_x c_1)\sin\lambda_1\}\{(\nu_x b_1 - \nu_y a_1)\cos\lambda_1 \\
&- c_1\sin\lambda_1\}\sin(\theta_2+\theta_3)\}\cos\lambda_4\sin\theta_5 \\
&- \{(a_1\nu_y^2 - b_1\nu_x\nu_y)\cos\lambda_1\cos^2(\theta_2+\theta_3) + c_1\nu_y\sin\lambda_1\cos^2(\theta_2+\theta_3) \\
&- \{(a_1\nu_y - b_1\nu_x)\cos\lambda_1\cos(\theta_2+\theta_3) \\
&+ c_1\sin\lambda_1\cos(\theta_2+\theta_3) - \nu_z\sin(\theta_2+\theta_3)\} \\
&\times \{(\nu_z a_1 - \nu_x c_1)\cos\lambda_1 - b_1\sin\lambda_1\}\sin(\theta_2+\theta_3) \\
&- (-a_1\nu_z^2 + c_1\nu_x\nu_z)\cos\lambda_1\cos^2(\theta_2+\theta_3) - b_1\nu_z\sin\lambda_1\cos^2(\theta_2+\theta_3) \\
&+ \{(-a_1\nu_z + c_1\nu_x)\cos\lambda_1\cos(\theta_2+\theta_3) \\
&+ b_1\sin\lambda_1\cos(\theta_2+\theta_3) - \nu_y\sin(\theta_2+\theta_3)\} \\
&\qquad\qquad \{(\nu_x b_1 - \nu_y a_1)\cos\lambda_1 - c_1\sin\lambda_1\}\sin(\theta_2+\theta_3)\}\sin\theta_4\sin\theta_5 \\
=\ &\nu_x\cos(\theta_2+\theta_3)\cos\theta_5 \\
&- \{(\nu_y c_1 - \nu_z b_1)\cos\lambda_1 - a_1\sin\lambda_1\}\sin(\theta_2+\theta_3)\cos\theta_5 \\
&- \{\{(c_1\nu_y - b_1\nu_z)\cos\lambda_1 + (\nu_x\nu_y b_1 - \nu_y^2 a_1 + \nu_x\nu_z c_1)\sin\lambda_1\}\cos(\theta_2+\theta_3) \\
&- \{(\nu_z a_1 c_1 - \nu_x c_1^2 - \nu_x b_1^2 + \nu_y a_1 b_1)\}\sin(\theta_2+\theta_3)\}\cos\lambda_4\sin\theta_5 \\
&- \{(a_1\nu_y^2 - b_1\nu_x\nu_y + a_1\nu_z^2 - c_1\nu_x\nu_z)\cos\lambda_1\cos^2(\theta_2+\theta_3) \\
&+ (c_1\nu_y - b_1\nu_z)\sin\lambda_1\cos^2(\theta_2+\theta_3) \\
&- \{\{(a_1\nu_y - b_1\nu_x)\cos\lambda_1\cos(\theta_2+\theta_3) + c_1\sin\lambda_1\cos(\theta_2+\theta_3) \\
&+ \nu_z\sin(\theta_2+\theta_3)\}(\nu_z a_1 - \nu_x c_1)\cos\lambda_1 \\
&- \{(a_1\nu_y - b_1\nu_x)\cos\lambda_1\cos(\theta_2+\theta_3) + c_1\sin\lambda_1\cos(\theta_2+\theta_3) \\
&- \nu_z\sin(\theta_2+\theta_3)\}b_1\sin\lambda_1 \\
&- \{(-a_1\nu_z + c_1\nu_x)\cos\lambda_1\cos(\theta_2+\theta_3) + b_1\sin\lambda_1\cos(\theta_2+\theta_3) \\
&- \nu_y\sin(\theta_2+\theta_3)\}(\nu_x b_1 - \nu_y a_1)\cos\lambda_1 \\
&+ \{(-a_1 c_1\nu_z + c_1^2\nu_x)\cos\lambda_1\cos(\theta_2+\theta_3) + b_1 c_1\sin\lambda_1\cos(\theta_2+\theta_3) \\
&- c_1\nu_y\sin(\theta_2+\theta_3)\}\sin\lambda_1\}\sin(\theta_2+\theta_3)\}\sin\lambda_4\sin\theta_5 \\
=\ &\nu_x\cos(\theta_2+\theta_3)\cos\theta_5 - \{(\nu_y c_1 - \nu_z b_1)\cos\lambda_1 \\
&- a_1\sin\lambda_1\}\sin(\theta_2+\theta_3)\cos\theta_5 \\
&- \{\{(c_1\nu_y - b_1\nu_z)\cos\lambda_1 - a_1\sin\lambda_1\}\cos(\theta_2+\theta_3)
\end{aligned}
$$

$$- \nu_x \sin(\theta_2 + \theta_3)\}\cos \lambda_4 \sin \theta_5$$
$$- \{a_1 \cos \lambda_1 \cos^2(\theta_2 + \theta_3) + (c_1 \nu_y - b_1 \nu_z)\sin \lambda_1 \cos^2(\theta_2 + \theta_3)$$
$$- (\nu_z^2 a_1 - \nu_x \nu_z c_1 - \nu_y \nu_x b_1 + \nu_y^2 a_1)\cos \lambda_1 \sin^2(\theta_2 + \theta_3)$$
$$- \nu_z b_1 \sin^2(\theta_2 + \theta_3) + c_1 \nu_y \sin \lambda_1 \sin^2(\theta_2 + \theta_3)\}\sin \lambda_4 \sin \theta_5$$
$$= \nu_x\{\cos(\theta_2 + \theta_3)\cos \theta_5 - \sin(\theta_2 + \theta_3)\cos \lambda_4 \sin \theta_5\}$$
$$- \{(\nu_y c_1 - \nu_z b_1)\sin \lambda_1 + a_1 \cos \lambda_1\}\sin \lambda_4 \sin \theta_5$$
$$- \{(\nu_y c_1 - \nu_z b_1)\cos \lambda_1 - a_1 \sin \lambda_1\}\{\sin(\theta_2 + \theta_3)\cos \theta_5$$
$$+ \cos(\theta_2 + \theta_3)\cos \lambda_4 \sin \theta_5\} \tag{B.8}$$

Similarly, it can be shown that

$$\xi_y = \nu_y\{\cos(\theta_2 + \theta_3)\cos \theta_5 - \sin(\theta_2 + \theta_3)\cos \lambda_4 \sin \theta_5\}$$
$$- \{(a_1 \nu_z - c_1 \nu_x)\sin \lambda_1 + b_1 \cos \lambda_1\}\sin \lambda_4 \sin \theta_5$$
$$- \{(a_1 \nu_z - c_1 \nu_x)\cos \lambda_1 - b_1 \sin \lambda_1\}\{\sin(\theta_2 + \theta_3)\cos \theta_5$$
$$+ \cos(\theta_2 + \theta_3)\cos \lambda_4 \sin \theta_5 \tag{B.9}$$

and

$$\xi_z = \nu_z\{\cos(\theta_2 + \theta_3)\cos \theta_5 - \sin(\theta_2 + \theta_3)\cos \lambda_4 \sin \theta_5\}$$
$$- (b_1 \nu_x - a_1 \nu_y)\sin \lambda_1 + c_1 \cos \lambda_1\}\sin \lambda_4 \sin \theta_5$$
$$- \{(b_1 \nu_x - a_1 \nu_y)\cos \lambda_1 - c_1 \sin \lambda_1\}\{\sin(\theta_2 + \theta_3)\cos \theta_5$$
$$+ \cos(\theta_2 + \theta_3)\cos \lambda_4 \sin \theta_5\} \tag{B.10}$$

APPENDIX C

Common Commercial and Training Robots

In this appendix the schematics for many common commercial and training robots are given. In some cases these include the full configurations, with joint angle limitations, link lengths, and characteristics of the drive systems. Attempts were made to contact every known robot manufacturer—even manufacturers with fewer than ten employees. The cottage-industry size of some manufacturers hints at the volatility and impermanence of a compilation such as this. For example, about 20% of commercial robot manufacturers have gone out of business or reorganized into another corporation in the last 2 years.

In some cases the order of the axes has been changed from the manufacturer's description to conform to a standard type of robot. For example, some manufacturers use the order $|\sigma\phi\sigma|$ to describe a cylindrical axis robot, but without altering the operation in any way, we can describe the same robot as $|\phi\sigma + \sigma|$.

The descriptions given by the manufacturers for the configuration and operation of their robots varies from very good (as with GMF Corporation, which is out of business) to very poor, with the average being poor. The information provided by the manufacturers rarely enables the complete configuration (link lengths, angle limitations, and so on) to be established. Except in a few cases, it would be impossible to program the kinematic equations of the robots from the descriptions given.[1] Of course, having the

[1] The presence of gaps or spaces in specific kinematic descriptions indicates missing information.

physical robot on which to perform measurements would enable the models to be formed. The quantity of information per robot varies from a fraction of a page to many pages, including blueprints (as with Kuka). Extensive use has been made of the compilations of others [Allan 1980; Bowker 1985; Cugy and Page 1984; Hunt 1983; Japan Industrial Robot Assoc. 1986; Simons 1980; Tanner 1981; Tver 1983], of which the most useful has been the work of Cugy and Page [Cugy and Page 1984].

The applications of each robot are as given by the manufacturer, so if none were offered, none are presented here. Most manufacturers quote positional accuracy and/or repeatability. The meanings of these terms are illustrated in Figure C.1. The deviation of the actual from the desired position (usually for the end effector) is called the *positioning accuracy*. The *repeatability* of the actual position is about 30% of the positional accuracy.

A robot may appear more than once in the listings; this is not an error as the manufacturer may permit additional degrees of freedom or different orientation mechanisms to be attached to the robot.

The cost of a robot may be no more than 10% of the overall cost of a working system within which the robot operates. The principal cost is in tooling for fixturing, specialized design, and software support. Further, the time between the purchase of a robot and its operation can be extensive. Some manufacturers such as Seiko in Japan have found it prudent to design and build their own robots, enabling the robot best to fit the work cell. Others purchase a robot and then consider its work-cell environment. The entrée to membership in the automation club is high cost and long delays. By some estimates General Motors has spent upwards of $20 billion on automation, of which most has not been cost-effective. With better planning and management, that cost could have been reduced and the effectiveness of the automation facilities improved.

Industrial managers need a clearer picture of the costs, installation times, and capabilities of an automation work cell. Such a work cell may or may not contain a robot. If engineers could simulate a work cell, they could

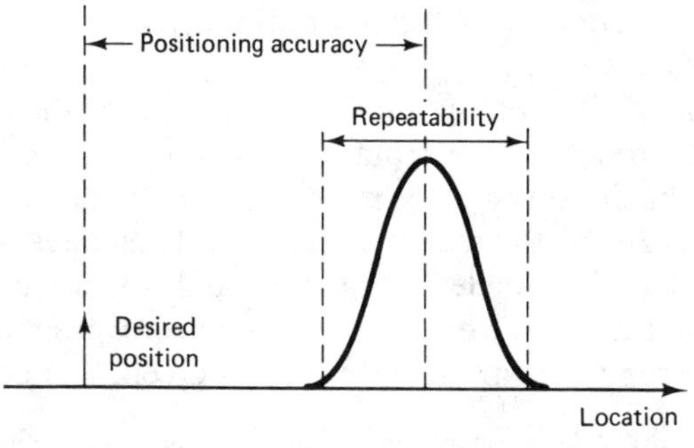

Figure C.1 Accuracy characteristics

optimize layouts and designs at low cost before committing capital resources. A prerequisite for such a simulation that involves one or more robots is the complete kinematic equations of each robot.

C.1 REVOLUTE ROBOTS

$|\lambda\theta\theta\lambda|$, $\overline{\mathbf{p}}^2\overline{\mathbf{p}}^0$, pencilable, with logical position space $\mathbf{u}_5$ (which establishes all joint positions) and the specific plane $\overline{\mathbf{p}}^0$ in the pencil of planes defined by line $\mathbf{u}_5 + \alpha(\mathbf{u}_4 - \mathbf{u}_5)$. Given $\mathbf{u}_5$ and with known $\mathbf{u}_1$ and $\boldsymbol{\nu}$, the plane of motion $\overline{\mathbf{p}}^2$ is found and the IKS is readily completed. Some robots with this configuration are: Accumatic Machinery Co., NUMAN; FMC Corp., A-360 and A-450; Microbot Alpha; Precision Robots, PRI-1100; Reis Machines, V15; Yaskawa Electric Co., L100.

$|\lambda\theta\theta\theta\lambda|$, $\overline{\mathbf{p}}^3\overline{\mathbf{p}}^0$, drillable, with logical position space $\mathbf{u}_6$ (which is on $\overline{\mathbf{p}}^3$, which itself is then established), $\mathbf{u}_5$ (which is on $\overline{\mathbf{p}}^3$ at distance l_5 from $\mathbf{u}_6$), and $\overline{\mathbf{p}}^0$ (which is on the pencil of planes defined by line $\mathbf{u}_5 + \alpha(\mathbf{u}_4 - \mathbf{u}_5)$). Since the plane with area is established, the remainder of the IKS is simple. Robots with this configuration include: Air Technical Industries, Electrobot ET; ASAE Co., IRB L6/2, G6/2, and 60/2; Daewoo, NOVA 10; Daihen, LA IRB-301 and SA IRB-311; Daihen, models LA IRB-301 and SA IRB-311; Eshed Robotec, SCORBOT-ER III; GCA, DKP200V and DKP300V; GEC (of England) PRW-10; GE, P5 and P6; Graco, OM Series; Hitachi, PW10II; International Robomation/Intelligence, IRI M50; Microbot, Alpha II; Mitsubishi, RW-212, RV-133P, and RM-501; Nachi-Fujikoshi Corp., 7501-AE; Rhino XR series; Sapri Earnest One; Syke Robotic Holdings, 600-5 Mk II; Tecquipment (Open University), MA 2000; Thermwood Robotics, Series Three; Unimation, Unimate Puma series 200 and 500; Volkswagon, GP100 and K15; Yaskawa Electric Co., Motoman L10W.

$|\lambda\theta\theta\lambda\theta|$, $\overline{\mathbf{p}}^2\overline{\mathbf{p}}^1$, directable. Thus, the logical position space is $\mathbf{u}_6$ and $\mathbf{u}_5$ (constrained to be at distance l_5 from $\mathbf{u}_6$). The IKS of this robot is given in Section 6.3.1. The only robot found to have this configuration is the Prab G-05.

$|\lambda\theta\theta\phi\theta|$, $\overline{\mathbf{p}}^2\overline{\mathbf{p}}^1$, directable. Thus, the logical position space and IKS are the same as the prior robot. The robots with this configuration include the Daihen Corp. models LK, LA, and SA.

$|\lambda\theta\theta\theta\gamma|$, $\overline{\mathbf{p}}^3\overline{\mathbf{p}}^1$, directable. Thus, the logical position space is $\mathbf{u}_6$ and $\mathbf{u}_5$ (constrained to be at distance l_5 from $\mathbf{u}_6$). $\mathbf{u}_1$, $\boldsymbol{\nu}$, and $\mathbf{u}_5$ establish the plane with area, and $\mathbf{p}^1$ is found from $\overline{\mathbf{p}}^3$, $\mathbf{u}_5$, and $\mathbf{u}_6$. The line of intersection of the two planes determines $\mathbf{u}_4$, and the rest of the IKS is simple. The only robot found to have this configuration is the Blohm + Vovv D1000 series, model IR-D 1252 K.

$|\lambda\theta\theta\lambda\phi|$, $\bar{\mathbf{p}}^2\bar{\mathbf{p}}^0\bar{\mathbf{p}}^0$, loose pencilable, with no logical position space. The only robot found with this configuration is the Mitsubishi model RW-111.

$|\lambda\theta\theta\lambda\theta\lambda|$, $\bar{\mathbf{p}}^2\bar{\mathbf{p}}^1\bar{\mathbf{p}}^0$, orientable, with logical position space $\mathbf{u}_7$, $\mathbf{u}_6$ (constrained so $|\mathbf{u}_7 - \mathbf{u}_6| = l_6$), and $\bar{\mathbf{p}}^0$ (which lies in the pencil of planes defined by line $\mathbf{u}_7 + \alpha(\mathbf{u}_6 - \mathbf{u}_7)$. The IKS of this configuration is given in Section 6.3.1. Robots with this popular configuration include: American Cimflex, Merlin; ASAE IRB 90/2; Binks Finishing Robot and model 88-800; Cybotech, V80; EKE Group, HDS 06; GMF Robotics, S-200, S-300, and S-400; Hispano-Suiza, COBRA; Kuka, IR Series with in-line wrist; Mitsubishi, RW-211, RW-213, RV-242, and RV-133S; Prab, G-06; Reis Machines, extraction robot; Sapri Earnest Zero; Thermwood Robotics, Series Six; Toshiba/Houstan International SR-606V; Unimation, Unimate Puma series 700 and 800; Volkwagon G8 and G60; Yaskawa Electric Co., V6, V12, and L120.

$|\lambda\theta\theta\gamma\lambda|$, $\bar{\mathbf{p}}^3\mathbf{p}^1\bar{\mathbf{p}}^0$, orientable with logical position space $\mathbf{u}_7$, $\mathbf{u}_6$ (constrained so $|\mathbf{u}_7 - \mathbf{u}_6| = l_6$), and $\bar{\mathbf{p}}^0$ (which lies in the pencil of planes defined by line $\mathbf{u}_7 + \alpha(\mathbf{u}_6 - \mathbf{u}_7)$. The IKS proceeds as for configuration $|\lambda\theta\theta\gamma|$ discussed previously. Robots found to have this configuration include: Action Machinery Co., model 460-10, 760-10, 760-12; Air Technical Industries, RMX series; Blohm + Vovv D1000 series, model IR-D 1260 H; Cybotech, model TP15; Hitachi, model M6060.

$|\lambda\theta\theta\theta\lambda\phi|$, $\bar{\mathbf{p}}^3\bar{\mathbf{p}}^0\bar{\mathbf{p}}^0$, orientable, with logical position space $\mathbf{u}_7$, $\mathbf{u}_6$ (constrained so $|\mathbf{u}_7 - \mathbf{u}_6| = l_6$), and $\bar{\mathbf{p}}^0$ (which lies in the pencil of planes defined by line $\mathbf{u}_7 + \alpha(\mathbf{u}_6 - \mathbf{u}_7)$. Since $\mathbf{u}_6$ is on $\bar{\mathbf{p}}^3$, this plane with area is established. $\mathbf{u}_6 - \mathbf{u}_4$ is orthogonal to $\mathbf{u}_6 - \mathbf{u}_7$ and $\mathbf{u}_4$ is on $\bar{\mathbf{p}}^3$, so $\mathbf{u}_4$ can be found, and the rest of the IKS is simple. Robots with this configuration include: GMF Robotics, model S-100; Nachi-Fujikoshi 7601-AE.

$|\lambda\theta\theta\theta\lambda\gamma|$, $\bar{\mathbf{p}}^3\bar{\mathbf{p}}^0\mathbf{p}^1$, loose directable. Robots with this configuration include: Thermwood Robotics Series Seven; Tokico America, Armstar 846RP and 256RP.

$|\lambda\theta\theta\lambda\phi\phi|$, $\bar{\mathbf{p}}^2\bar{\mathbf{p}}^0\bar{\mathbf{p}}^0\bar{\mathbf{p}}^0$, orientable with logical position space $\mathbf{u}_7$, $\mathbf{u}_6$ (constrained so $|\mathbf{u}_7 - \mathbf{u}_6| = l_6$), and $\bar{\mathbf{p}}^0$ (which lies in the pencil of planes defined by line $\mathbf{u}_7 + \alpha(\mathbf{u}_6 - \mathbf{u}_7)$. The IKS of this robot is discussed in Section 6.5. Robots with this configuration include: Kuka IR series with offset wrist; Nachi-Fujikoshi 8000-AK series; Yaskawa Electric, Motoman L15, L30, and L60:.

$|\lambda\theta\theta\gamma\theta\gamma\lambda|$, $\bar{\mathbf{p}}^2\mathbf{p}^2\bar{\mathbf{p}}^0$, loose orientable, with no logical position space. The only robot found with this configuration is the Cybotech model P15.

$|\lambda\theta\theta\theta\lambda\theta\lambda\theta|$, $\bar{\mathbf{p}}^3\bar{\mathbf{p}}^1\bar{\mathbf{p}}^1$, double loose orientable, with no logical position space. The only robot found with this configuration is the Hispano-Suiza ISIS.

The full configuration of some specific revolute robots are given next.

ASAE Company, model IRB L6/2. 5 *dof*, 13 pounds payload. Weight 319 lb. Adding an optional sixth *dof* θ_6 reduces the payload to 9 lb. Speed 114°/s on ϕ_1 and 135°/s on θ_2 and θ_3. Repetition accuracy ±0.008 in.

$$\overset{\bar{\mathbf{p}}^3}{} \qquad\qquad\qquad\qquad \overset{\bar{\mathbf{p}}^0}{}$$

$$\left|\lambda_1\begin{Bmatrix}170\\-170\end{Bmatrix}22.5\text{ in.}\;\middle|\;\theta_2\begin{Bmatrix}40\\-40\end{Bmatrix}22.5\;\middle|\;\theta_3\begin{Bmatrix}25\\-40\end{Bmatrix}26\;\middle|\;\theta_4\begin{Bmatrix}90\\-90\end{Bmatrix}0\;\middle|\;\lambda_5\begin{Bmatrix}180\\-180\end{Bmatrix}8\right|$$

Drillable. Applications: Arc welding, adhesives and sealants, cleaning of castings, inspection and vision, machine tending, material handling, polishing, deburring, spot and stud welding. See Figure C.2.

Binks Manufacturing Co., model 88-800. 6 *dof*. Hydraulic on first three axes, electric on last three axes. Pneumatic end effector. Maximum speeds 60°/s on θ_1, 45°/*s* on θ_2, 32°/s on θ_3, 90°/s on λ_4 and θ_5, and 150°/s of λ_6. Lead-through controller.

$$\overset{\bar{\mathbf{p}}^2}{} \qquad\qquad\qquad\qquad \overset{\bar{\mathbf{p}}^1}{} \qquad\qquad\qquad \overset{\bar{\mathbf{p}}^0}{}$$

$$\left|\lambda_1\begin{Bmatrix}135\\-135\end{Bmatrix}22\text{ in.}\;\middle|\;\theta_2\begin{Bmatrix}37\\-37\end{Bmatrix}39.6\;\middle|\;\theta_3\begin{Bmatrix}130\\40\end{Bmatrix}50\;\middle|\;\lambda_4\begin{Bmatrix}180\\-180\end{Bmatrix}10\;\middle|\;\theta_5\begin{Bmatrix}90\\-90\end{Bmatrix}0\;\middle|\;\lambda_6\begin{Bmatrix}135\\-135\end{Bmatrix}10\right|$$

Orientable. Applications: paint spraying. See Figure C.3.

Daewoo Heavy Industries, NOVA-10, 5 *dof*, with optional sixth *dof*. DC servomotor. Weight 350 kg, payload 10 kg. Repeatability ±0.2 mm. Maximum speeds 96°/s on ϕ_1, 108°/s on θ_2, 72°/s on θ_3, 120°/s on θ_4, and 180°/s on λ_5. Teach pendant control.

$$\overset{\bar{\mathbf{p}}^3}{} \qquad\qquad\qquad\qquad \overset{\bar{\mathbf{p}}^0}{}$$

$$\left|\lambda_1\begin{Bmatrix}150\\-150\end{Bmatrix}800\text{ mm}\;\middle|\;\theta_2\begin{Bmatrix}17\\-77\end{Bmatrix}650\;\middle|\;\theta_3\begin{Bmatrix}25\\-45\end{Bmatrix}850\;\middle|\;\theta_4\begin{Bmatrix}70\\-110\end{Bmatrix}100\;\middle|\;\lambda_5\{360\}0\right|$$

Drillable. Applications: arc welding, material handling, sealing, water jet cutting, machine loading. See Figure C.4.

Kuka model IR 662/100. 6 *dof*. Payload 100 kg. Electro-mechanical drive with DC servomotors. Maximum speed at the end effector at full extension 3.7m/s. Repeatability ±0.8 mm. PTP/CP. Speeds 79°/s for ϕ_1, 1.25 m/s at u_3, 1.45 m/s at u_5, 110°/s for λ_4, 86°/s for ϕ_5, and 120°/s for ϕ_6.

When fitted to an in-line wrist, the configuration is:

$$\overset{\bar{\mathbf{p}}^2}{} \qquad\qquad\qquad\qquad \overset{\bar{\mathbf{p}}^1}{} \qquad\qquad\qquad \overset{\bar{\mathbf{p}}^0}{}$$

$$\left|\lambda_1\begin{Bmatrix}160\\-160\end{Bmatrix}1000\;\middle|\;\theta_2\begin{Bmatrix}45\\-55\end{Bmatrix}l_2\;\middle|\;\theta_3\begin{Bmatrix}49.5\\-48\end{Bmatrix}l_3\;\middle|\;\lambda_4\begin{Bmatrix}225\\-225\end{Bmatrix}l_4\;\middle|\;\theta_5\begin{Bmatrix}113\\-112\end{Bmatrix}l_5\;\middle|\;\lambda_6\begin{Bmatrix}225\\-225\end{Bmatrix}l_6\right|$$

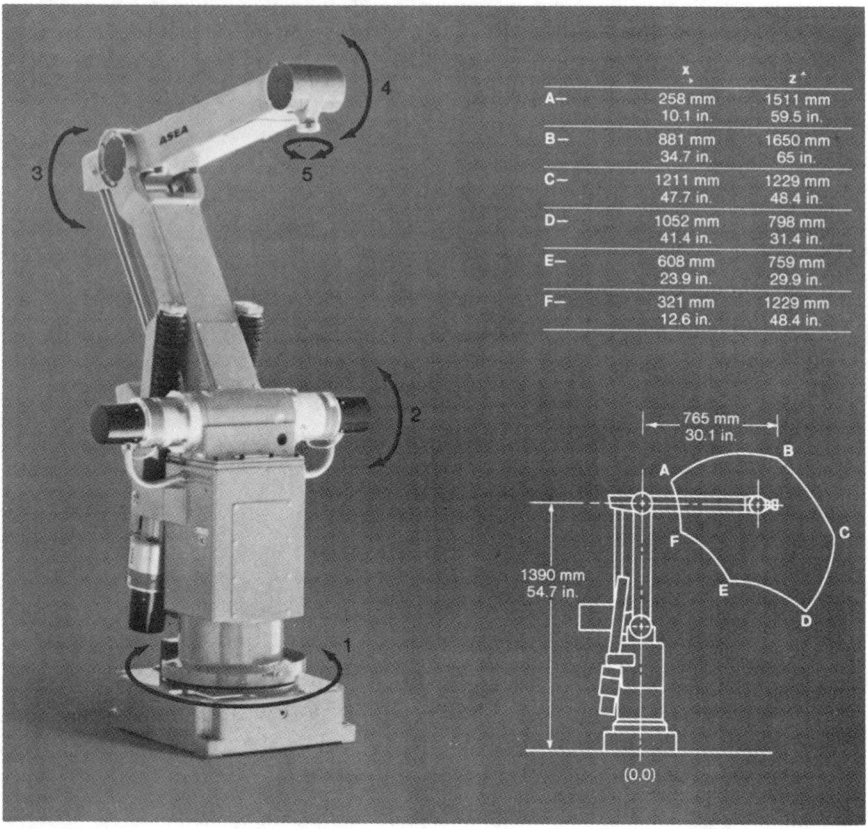
1
2
3
4
5
ASEA
x
z
A— 258 mm 10.1 in. 1511 mm 59.5 in.
B— 881 mm 34.7 in. 1650 mm 65 in.
C— 1211 mm 47.7 in. 1229 mm 48.4 in.
D— 1052 mm 41.4 in. 798 mm 31.4 in.
E— 608 mm 23.9 in. 759 mm 29.9 in.
F— 321 mm 12.6 in. 1229 mm 48.4 in.
765 mm
30.1 in.
1390 mm
54.7 in.
A
B
C
D
E
F
(0,0)

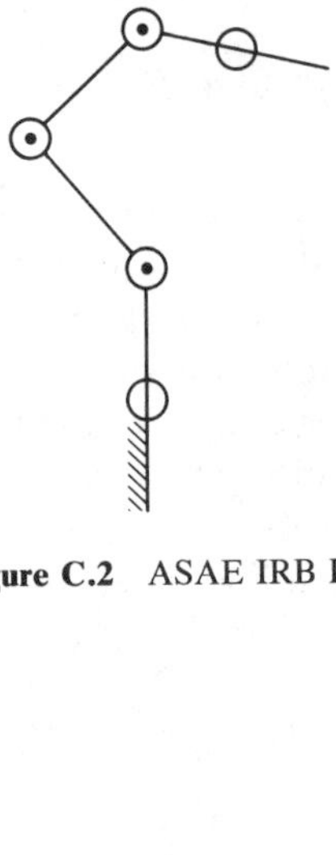

Figure C.2 ASAE IRB L6/2

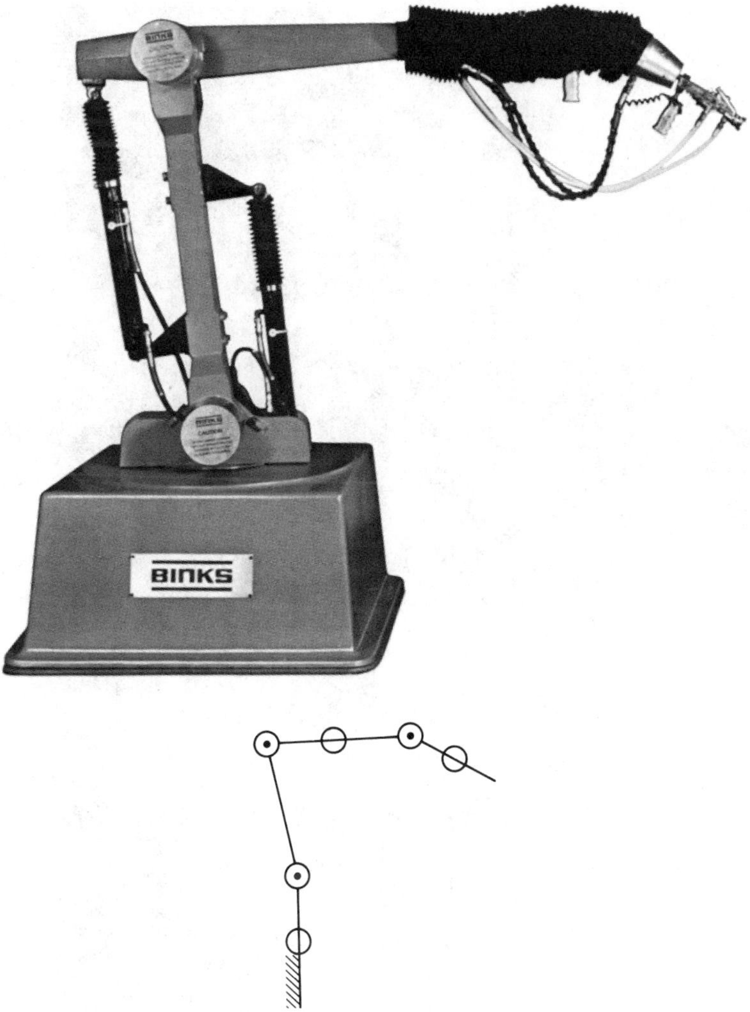

Figure C.3 Binks 88-800

When fitted with an offset wrist the configuration is:

$$\begin{array}{cccccc} \bar{\mathbf{p}}^2 & & & \bar{\mathbf{p}}^1 & \bar{\mathbf{p}}^0 & \bar{\mathbf{p}}^0 \\ \left|\lambda_1\begin{Bmatrix}160\\-160\end{Bmatrix}1000\right. & \left|\theta_2\begin{Bmatrix}45\\-55\end{Bmatrix}l_2\right. & \left|\theta_3\begin{Bmatrix}49.5\\-48\end{Bmatrix}l_3\right. & \left|\lambda_4\begin{Bmatrix}225\\-225\end{Bmatrix}l_4\right. & \left|\phi_5\begin{Bmatrix}113\\-112\end{Bmatrix}l_5\right. & \left|\phi_6\begin{Bmatrix}225\\-225\end{Bmatrix}l_6\right| \end{array}$$

Both configurations are orientable. Applications: machining, material handling, assembly, and spot and arc welding. An optional sliding case can be employed to move complete robot. See Figure C.5.

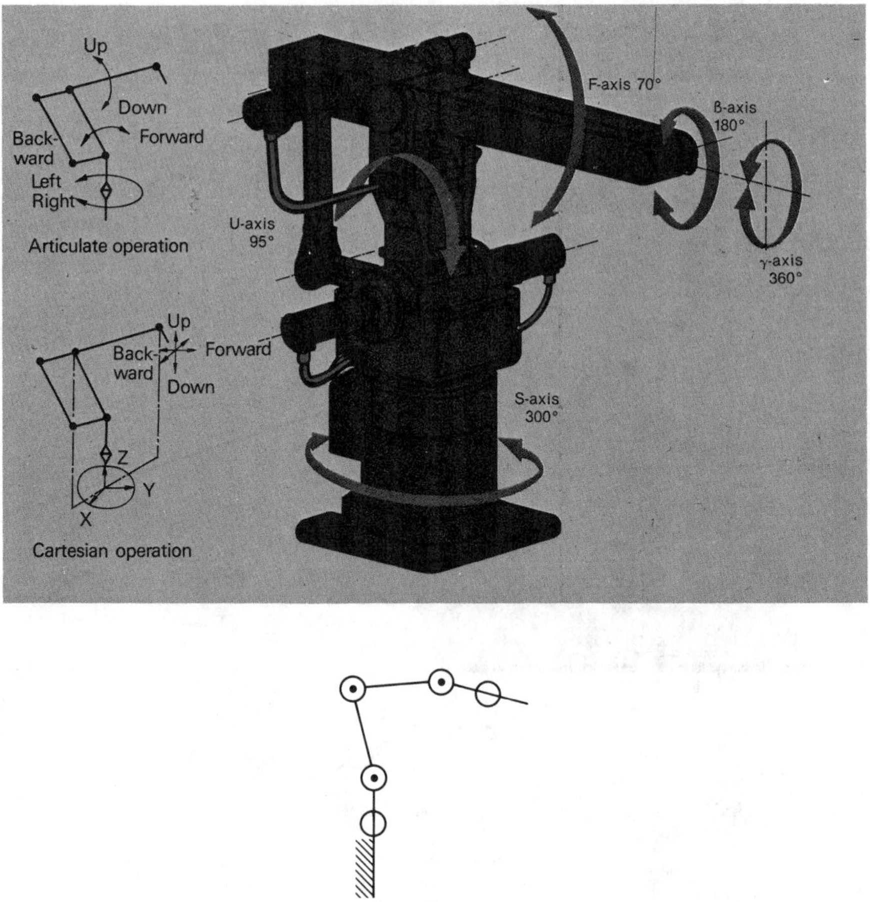

Figure C.4 Daewoo, NOVA-10

Nachi-Fujikoshi Corporation, 8601-AK robot. 6 *dof*. Repeatability ±0.5 mm. Payload 100 kg. Speeds 90°/s on ϕ_1, θ_2, and θ_3, PTP controller.

$\bar{\mathbf{p}}^2$			$\bar{\mathbf{p}}^0$	$\bar{\mathbf{p}}^0$	$\bar{\mathbf{p}}^0$
$\lambda_1\left\{\begin{matrix}135\\-135\end{matrix}\right\}$ 875 mm	$\theta_2\left\{\begin{matrix}60\\-45\end{matrix}\right\}$ 1135	$\theta_3\left\{\begin{matrix}30\\-60\end{matrix}\right\}$ 1500	$\lambda_4\left\{\begin{matrix}190\\-190\end{matrix}\right\}$ 0	$\phi_5\left\{\begin{matrix}240\\-240\end{matrix}\right\}$ 220	$\phi_6\left\{\begin{matrix}190\\-190\end{matrix}\right\}$ 135

Orientable. Applications: spot welding, machine loading, material handling, and assembly. The IKS of this robot is discussed in example 3 of Section 6.6. See Figure C.6.

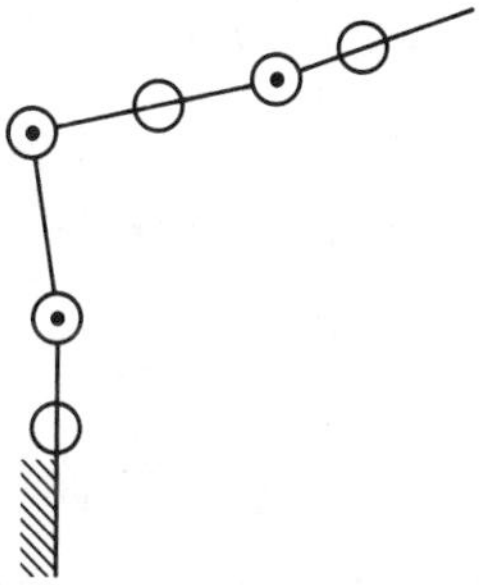

Figure C.5 Kuka IR 662/100

Syke Robotic Holdings, model 600-5 Mk II. 5 *dof*. DC servomotors. Payload 2 kg. Weight 26 kg. Accuracy ±0.25 mm. Speed 2.1 m/s at end effector. Teach pendant control. PTP controller.

$$\overset{\bar{\mathbf{p}}^3}{\left|\lambda_1\begin{Bmatrix}165\\-165\end{Bmatrix}0.63\text{ m}\right.}\left|\theta_2\begin{Bmatrix}135\\-135\end{Bmatrix}0.31\right|\theta_3\begin{Bmatrix}120\\-120\end{Bmatrix}0.31\left|\theta_4\begin{Bmatrix}90\\-90\end{Bmatrix}0.05\right|\overset{\bar{\mathbf{p}}^0}{\lambda_5\begin{Bmatrix}270\\-270\end{Bmatrix}0\left.\right|}$$

Drillable. Applications: This is a fast, very lightweight robot used for pick and place. See Figure C.7.

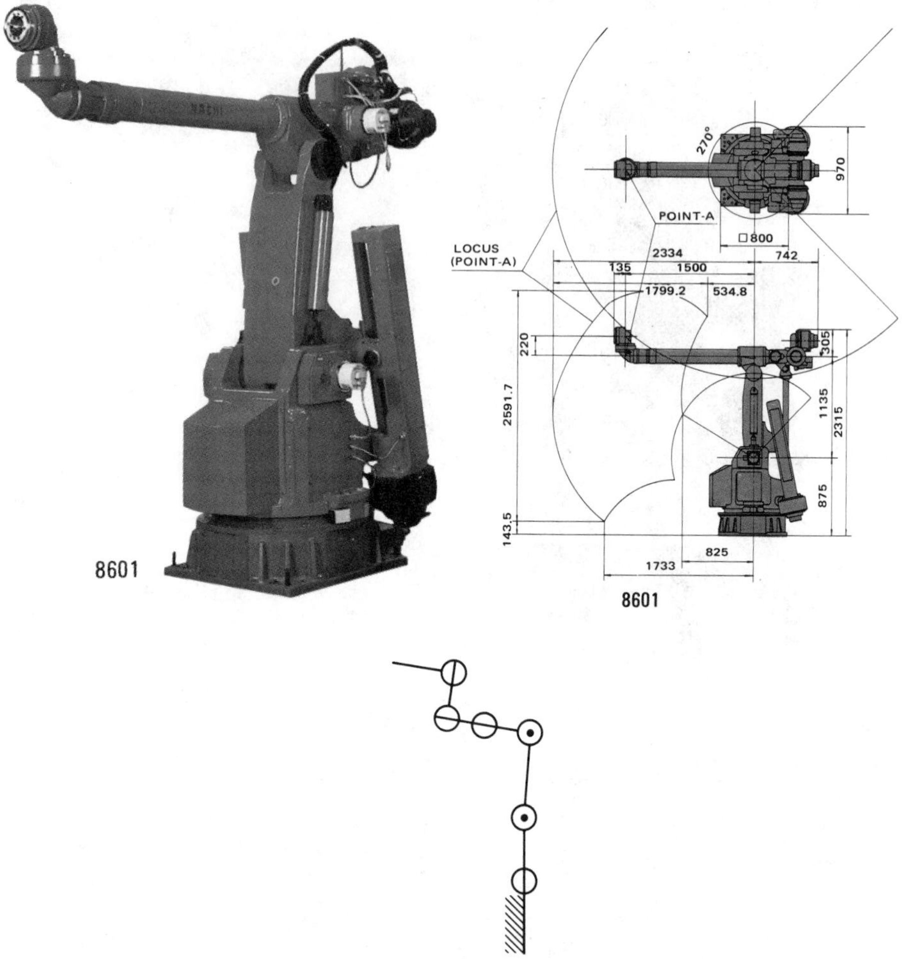

Figure C.6 Nachi Robot, 8601

Thermwood Robotics, Series Six. 6 *dof.* Hydraulic/mechanical drives. Payload 18 lb. Weight 500 lb. PTP and continuous-path controller. Teach pendant control. Repeatability ±0.125 in. Speeds 30 in./s horizontally and 30 in./s vertically at the end effector. Has a manually adjusted pivot between joints θ_2 and θ_3.

$$\bar{\mathbf{p}}^3 \qquad\qquad\qquad\qquad \bar{\mathbf{p}}^1$$

$$\left|\lambda_1\begin{Bmatrix}67\\-67\end{Bmatrix}21\text{ in.}\right|\theta_2\begin{Bmatrix}\\ \end{Bmatrix}42\left|\theta_3\begin{Bmatrix}\\ \end{Bmatrix}28\right|\lambda_4\begin{Bmatrix}90\\-90\end{Bmatrix}l_4\left|\theta_5\begin{Bmatrix}225\\-225\end{Bmatrix}l_5\right|\lambda_6\begin{Bmatrix}90\\-90\end{Bmatrix}0\Bigg|$$

Orientable. Applications: spray-painting robot. See Figure C.8.

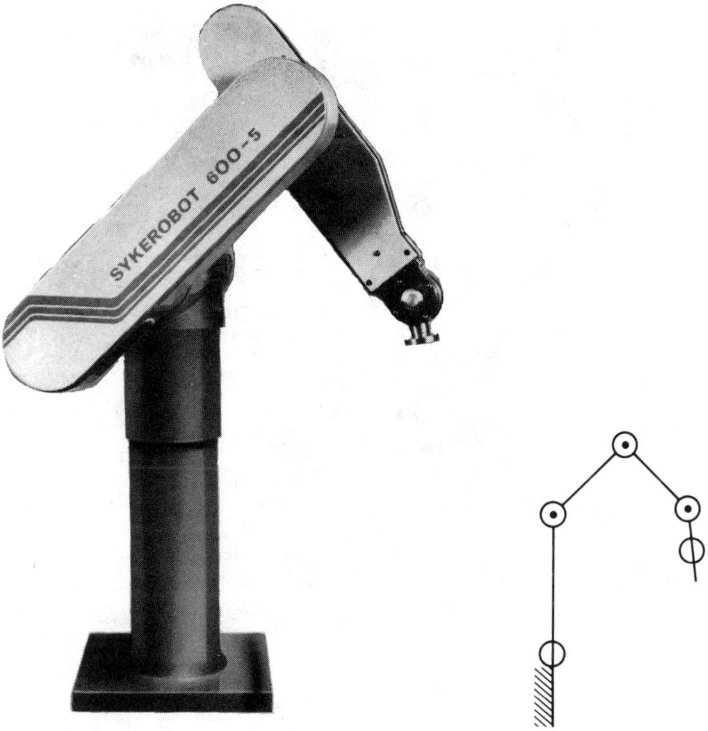

Figure C.7 Syke 600-5

C.2 CYLINDRICAL ROBOTS

$| \lambda\sigma+\sigma |$, $\bar{\mathbf{p}}^2$, placeable, with logical position space $\mathbf{u}_4$. Robots with this configuration include: Air technical Industries, Hydro-Arm HAX, Electro-Arm EAX, and RT Series; American Monarch AMR-7; Automation Equipment Co., Pacesetter 4; Cincinnati Milacron, T^3 363; GMF Robotics, E-101, E-201, E-310, A-200,M-100, M-300; Lamberton Robotics (with up to 7 *dof*); Mitsubishi Electric, RC-136P.

$| \lambda\sigma+\sigma\lambda |$, $\bar{\mathbf{p}}^2\bar{\mathbf{p}}^0$, pencilable, with logical position space $\mathbf{u}_5$ (which enables all joint location to be found) and the specific plane $\bar{\mathbf{p}}^0$ in the pencil of planes defined by line $\mathbf{u}_5 + \alpha(\mathbf{u}_4 - \mathbf{u}_5)$. Robots with this configuration include: Amatrol, Mercury; Cincinnati Milacron, T^3 373; GMF Robotics, E101; Pearse-Pearson, Probaut ARC-66; Positech, Probot CC-1A; R-2000 Corp., Botnek; RIMROCK Copperweld; Scrader-Bellows, MotionMate MM-I and MM-II;

$| \lambda\sigma+\sigma\phi |$, $\bar{\mathbf{p}}^2\bar{\mathbf{p}}_0$, pencilable, with logical position space $\mathbf{u}_5$ (which enables all joint location to be found) and the specific plane $\bar{\mathbf{p}}_0$ in the

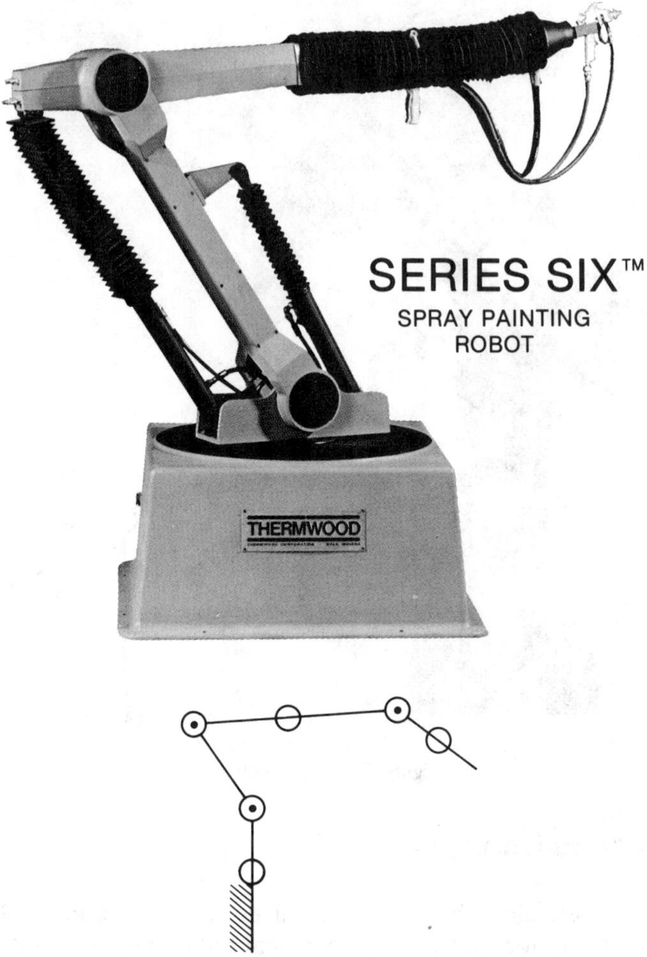

Figure C.8 Thermwood

pencil of planes defined by line $\mathbf{u}_5 + \alpha(\mathbf{u}_4 - \mathbf{u}_5)$. The only robot found with this configuration is the Mitsibishi Electric RC-136P.

$| \lambda\sigma+\sigma\theta\lambda |$, $\bar{\mathbf{p}}^3\bar{\mathbf{p}}^0$, drillable: Air Technical Industries, HAX-596, HAX-5108; GMF Robotics, E-101, E-201, E-310, A-200, M-100, M-300;

$| \lambda\sigma+\sigma\lambda\phi |$, $\bar{\mathbf{p}}^2\bar{\mathbf{p}}^1$, loose drillable, with no logical position space. GMF Robotics has models E-101, E-201, E-310, A-200, M-100, and M-300 with this configuration.

The configurations of some specific cylindrical axis robots are given next.

(X)

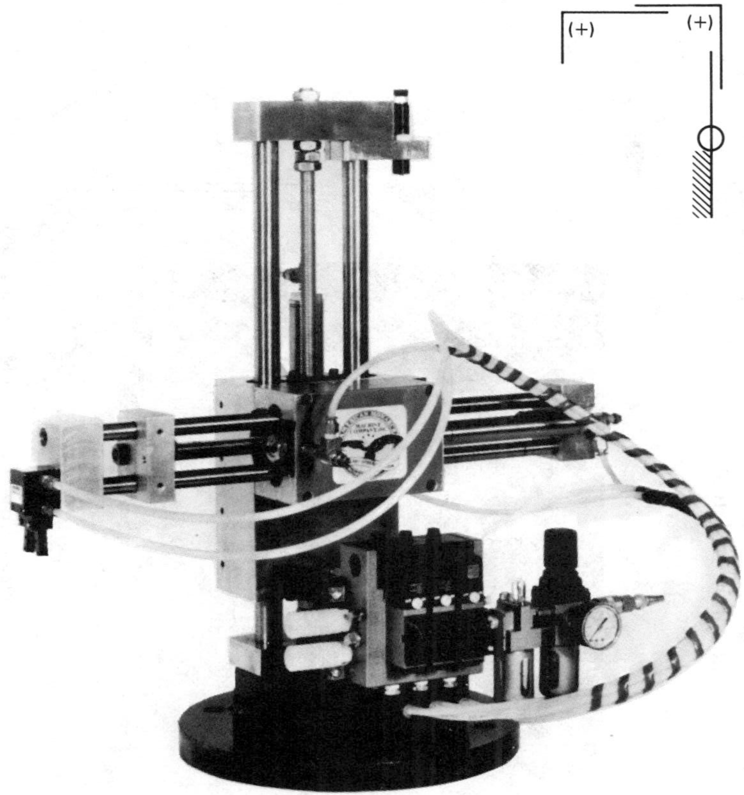

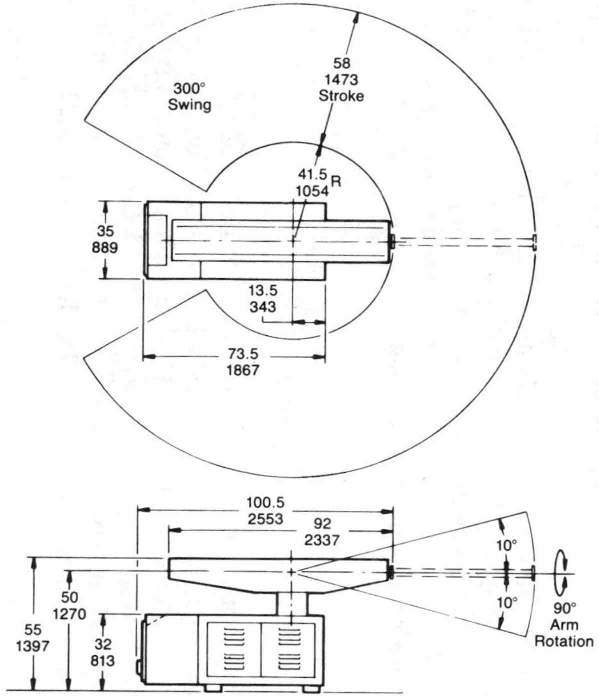

Note: 5800 HD (heavy duty) does not have 90° arm rotation.

Red = Millimeters
Black = Inches

Figure C.9 American Monarch AMR-7

American Monarch AMR-7. 3 *dof* with optional fourth roll axis. Pneumatic drive. Payload 5 lb. Repeatability 0.003 in. Maximum speeds 90°/s on ϕ_1, 21 in./s on σ_2, and 21 in./s on σ_3. Applications: secondary press operations, machine loading and unloading, assembly, packaging, palletizing and material handling.

$$\bar{\mathbf{p}}^2$$
$$\left| \lambda_1 \begin{Bmatrix} 90 \\ -90 \end{Bmatrix} \begin{matrix} 15 \text{ in.} \\ 22 \end{matrix} \middle| (+) \begin{matrix} 9 \\ 17 \end{matrix} \right|$$

Placeable. Applications: pick and place. See Figure C.9.

Cincinnati Milacron, model T^3 363. T^3 stands for Tomorrow's Tool Today. 3 *dof* (shown in Figure C.10) with optional pitch or yaw axis to make 4 *dof*. Weight 1550 lb and payload 75 lb, DC servo. Speeds up to 90°/s on ϕ_1, 20 in./s on σ_2 and 40 in./s on σ_3. Position repeatability 40 mils. Serial communication controller (RS232C and RS422).

$$\bar{\mathbf{p}}^2$$
$$\left| \lambda_1 \begin{Bmatrix} 150 \\ -150 \end{Bmatrix} \begin{matrix} 24 \\ 41 \end{matrix} (+) \begin{matrix} 31 \\ 57 \end{matrix} \right|$$

Placeable. Applications: parts handling, package handling, machine loading/unloading, palletizing.

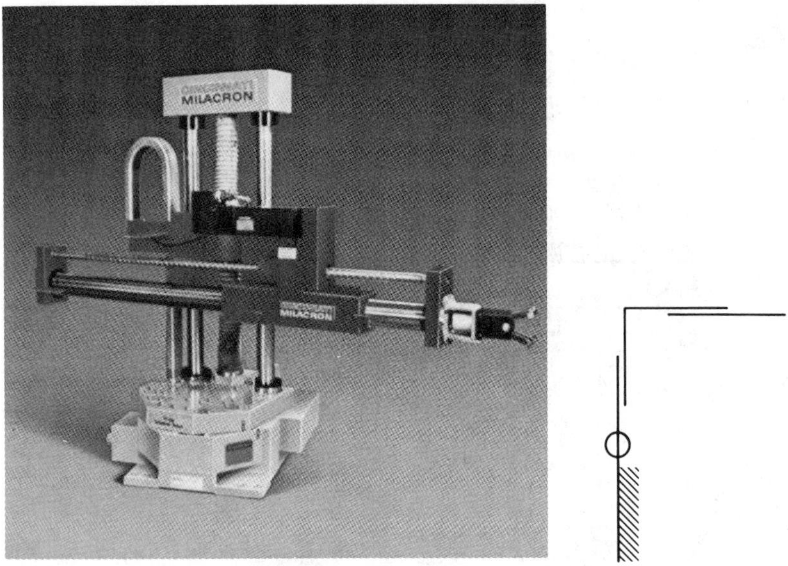

Figure C.10 Cincinnati Milacron T^3 363

C.3 CARTESIAN ROBOTS

$|\ \sigma+\sigma\ |$, $\mathbf{p}^2$, where this plane is usually $[0\ \ 0\ \ 1\ \ d]$. Robots with this configuration include: Pickomatic, various models; Sculer, Inc., Handler II-2; Toshiba/Houstan Int. Corp., SR-L series, type C.

$|\ \sigma+\sigma\lambda\ |$, $\mathbf{p}^2\bar{\mathbf{p}}^0$, with no logical position space. A robot with this configuration is the Kuka IR 460/100.

$|\ \sigma+\sigma(\times)\sigma\ |$, $\mathbf{p}^2\mathbf{p}^1$, placeable, with logical position space $\mathbf{u}_4$. Robots with this configuration include: Anorod Co., 3A-1 and 3A-3; Durr, gantry robots; FMC Corp., models 250, 500, 00; Kuka, models IR 460/100 and PU 800; Lamson Corp., Robotrol; Liebherr Machine Tool Co., LPR gantry robots; MACK Corp., modular robots from 3 *dof* upward in almost any configuration; Reis, LR series and GR70 gantry robot; Toshiba/Houstan International Corporation, SR-L series, type 2C; Towa, Transman-II; Unimation, Unimate 6000 series gantry; Volkswagon, P30 and P50; VSI Automation Assembly, Archie series MT.

$|\ \sigma+\sigma\psi\sigma\ |$, $\mathbf{p}^2\bar{\mathbf{p}}^1$, pencilable, with logical position space $\mathbf{u}_4$ and the specific $\bar{\mathbf{p}}^1$ in the pencil of planes defined by line $\mathbf{u}_4 + \alpha[a_2\ \ b_2\ \ c_2\ \ 0]^\#$. The point where this line intersects $\mathbf{p}^2$ gives $\mathbf{u}_3$, and the rest of the IKS is simple. Robots with this configuration include: American Monarch AMR-4; Anorod Co., models 4A-1 and 4A-3; Bosch modular Cartesian; Durr, gantry robots; Epple Buxbaum; GMB Robotics, G-200 gantry; Towa, Transman-II.

$|\ \sigma+\sigma\psi\phi\ |$, $\mathbf{p}^2\bar{\mathbf{p}}^0\bar{\mathbf{p}}^0$, pencilable. Camco Auto-Load L-2200 and L-4400.

$|\ \sigma+\sigma\gamma\psi\ |$, $\mathbf{p}^2\mathbf{p}^1\bar{\mathbf{p}}^0$, pencilable. N/C Concepts, model L-Robo.

$|\ \sigma+\sigma\psi\sigma\theta\ |$, $\mathbf{p}^2\bar{\mathbf{p}}^2$, directable, with logical position space $\mathbf{u}_5$ and $\mathbf{u}_4$ (constrained to be at distance l_5 from $\mathbf{u}_5$). With $\mathbf{u}_4$ known, the IKS proceeds as in the pencilable robot $|\ \sigma+\sigma\psi\sigma\ |$ just discussed. Two robots found with this configuration are: Durr, Gantry Robots; Kuka, PU 800.

$|\ \sigma+\sigma\psi\sigma\phi\ |$, $\mathbf{p}^2\bar{\mathbf{p}}^1\bar{\mathbf{p}}^0$, loose pencilable, with no logical position space. Robots found with this configuration are: Anorod, models 5A-1 and 5A-3; General Electric Co., model A12.

$|\ \sigma+\sigma\psi\sigma\theta\lambda\ |$, $\mathbf{p}^2\bar{\mathbf{p}}^2\bar{\mathbf{p}}^0$, directable with logical position space $\mathbf{u}_7$, $\mathbf{u}_6$ (constrained so $|\ \mathbf{u}_7 - \mathbf{u}_6\ | = l_6$), and $\bar{\mathbf{p}}_0$ (which lies in the pencil of planes defined by line $\mathbf{u}_7 + \alpha(\mathbf{u}_6 - \mathbf{u}_7)$. American Cimflex, GM-2600; Anorod Co., Anorobot model 1700; GCA Corp., models XR50 and XR100.

$|\ \sigma+\sigma\psi\sigma\phi\phi\ |$, $\mathbf{p}^2\bar{\mathbf{p}}^1\bar{\mathbf{p}}^0\bar{\mathbf{p}}^0$, orientable. Kuka 460/100.

$|\ \sigma+\sigma\gamma\sigma\theta\lambda\ |$, $\mathbf{p}^2\mathbf{p}^3\bar{\mathbf{p}}^0$, loose directable. Volkswagon model P200.

$|\ \sigma+\sigma\psi\theta\theta\phi\phi\ |$, . Mitsubishi Electric, models RW-251 and 253/253W.

$|\ \sigma+\sigma\psi\sigma\theta\psi\psi\psi\ |$, . Cybotech, model WCX90.

The full configuration of a specific Cartesian robot follows.

Liebherr Machine Tool Co., series LPR gantry robots. Basic 3 *dof xyz* gantry-style robots with one, two, or three additional revolute axes. AC servomotor drives. Payloads 110 lb. for LPR 200 and 1100 lb. for LPR 2000.

$$\overset{\mathbf{p}^2}{\left|\begin{matrix}6\text{ m}\\0\end{matrix}\right|}\;\overset{\mathbf{p}^1}{\left|(+)^{6}_{0}\right|}\left|(\times)^{1}_{0}\right|$$

Placeable. Applications: loading/unloading, parts transfer. See Figure C.11.

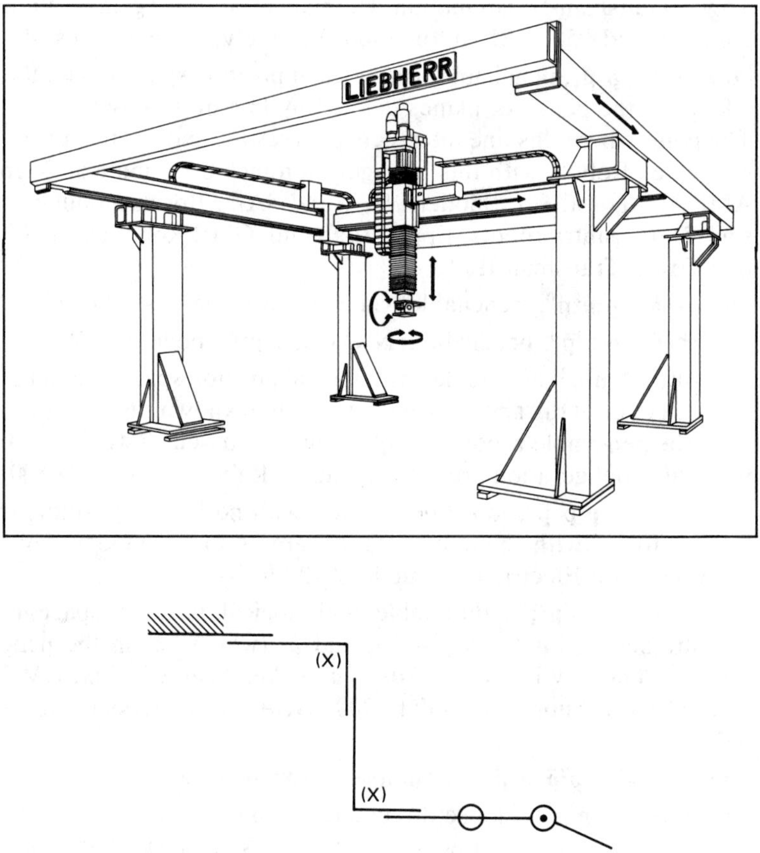

Figure C.11 Liebherr

C.4 POLAR ROBOTS

$|\ \lambda\theta\sigma\ |$, $\bar{\mathbf{p}}^2$,placeable, with logical position space $\mathbf{u}_4$. The robots found with this configuration are the Prab models 4200HD and 5800HD.

$|\ \lambda\theta\sigma\lambda\ |$, $\bar{\mathbf{p}}^2\bar{\mathbf{p}}^0$, pencilable. Hardinge, model unspecified; Prab, models 4200 and 5800.

$|\ \lambda\theta\sigma\theta\ |$, $\bar{\mathbf{p}}^3$, loose placeable. GEC (of England), Little Giant.

$|\ \lambda\theta\sigma\theta\lambda\ |$, $\bar{\mathbf{p}}^3\bar{\mathbf{p}}^0$, drillable. Air Technical Industries, RAX series; United States Robots, Maker 110.

$|\ \lambda\theta\sigma\lambda\theta\ |$, $\bar{\mathbf{p}}^2\bar{\mathbf{p}}^1$, directable. Kuka IR 260/500.2, 260/500.4 and 260/60.5

$|\ \lambda\theta\sigma\gamma\theta\theta\ |$, $\bar{\mathbf{p}}^2\mathbf{p}^3$, and the position space of this robot is not evident. GEC Robot Systems, RAMP 2000.

$|\ \lambda\theta\sigma\theta\lambda\theta\ |$, $\bar{\mathbf{p}}^3\bar{\mathbf{p}}^1$, loose directable. Volkswagon, R30.

The full configuration of one polar robot follows.

Prab, models 4200 and 5800. 4 *dof*. Payload 75 lb for 4200 and 50 lb for 5800. Weight 2600 lb. Electro-hydraulic. Repeatability ±0.008 in. Applications: die casting, injection moulding, forging, machine tool loading, heat treating, glass handling, parts cleaning, dip coating, press loading, material transfer. For 4200 the mathematical expression is

$$\overbrace{\left|\lambda_1\begin{Bmatrix}300\\0\end{Bmatrix}50\text{ in.}\right|\theta_2\begin{Bmatrix}20\\0\end{Bmatrix}\begin{matrix}41.5\\83.5\end{matrix}}^{\bar{\mathbf{p}}_2^2}\ \overbrace{\left|\lambda_4\begin{Bmatrix}90\\0\end{Bmatrix}0\right|}^{\bar{\mathbf{p}}_3^0}$$

The expression is the same for 5800 except σ_3 is 41.5 in. to 99.5 in. See Figure C.12.

C.5 SCARA ROBOTS

Several manufacturers offer $|\ \theta\theta\ |$ as a configuration that is expandable into a SCARA. We have not listed these.

$|\ \theta\theta\times\sigma\ |$, $\mathbf{p}^2\mathbf{p}^1$, placeable. American Monarch AMR series; Bosch, SR series. Hirata Industrial Machineries Co., PPR-270; Hitachi America, A4010H; Toshiba/Houstan, model 1053H; United States Robots, Maker Series.

$|\ \theta\theta\psi\sigma\ |$, $\mathbf{p}^2\bar{\mathbf{p}}^1$, or $|\ \theta\theta\times\sigma\lambda\ |$, $\mathbf{p}^2\mathbf{p}^1\bar{\mathbf{p}}^0$; drillable: Adept Technology, AdeptOne; Bosch, SR series; ESAB North America, MAC 500 and MAC 2000; GCA Corp., DKP200H; GEC Electrical Products, A4010H, A3020, A4020 (made by Hitachi), A3100 and A4100; Hirata

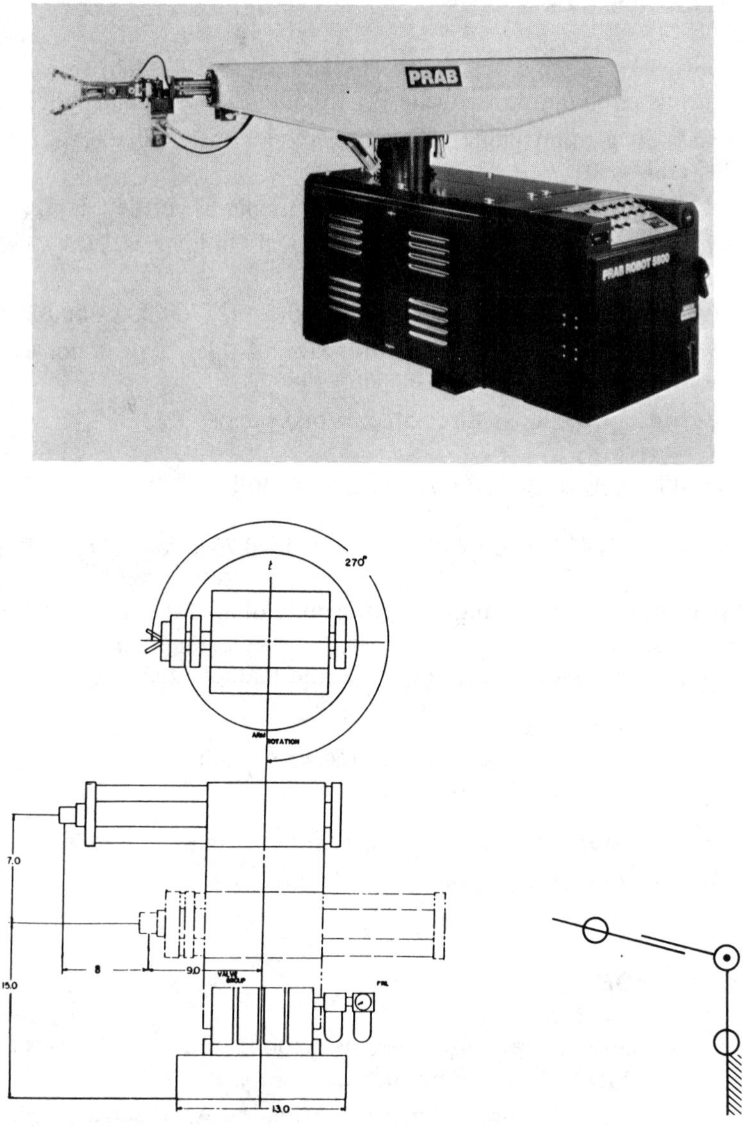

Figure C.12 Prab, models 5800 and 5800 HD

Industrial Machineries, PPR-270, AR-i350, and AR-i550; Hitachi America, A3020, A4020, A3100L, A4100L, and A4010S; IBM, models 7535, 7540, 7545, 7545-800, 7547, and 7575; Intelledex, Microsmooth 400 and model 1400; Nachi-Fujikoshi, models 160-AL, 200-AL, 400-AL, and 600-AL; Pentel, models GL-50 and GL-63; Toshiba/Houstan, model 1054H; United States Robots, Maker Series; VSI Automation Assembly, Charlie-Screwdriver Series, types F and S and Series 2, 4, and 6.

$| \theta\theta\underline{\psi}\sigma\theta |$, $\mathbf{p}^2\bar{\mathbf{p}}^2$, directable. Bosch, models SR-450, SR-600, and SR-800.

The full configurations of some specific SCARA robots follow.

Adept Technology, AdeptOne Manipulator. 4 *dof.* Weight 400 lb. Maximum payload 13.2 lb. Maximum velocity 30 ft./s at end effector.

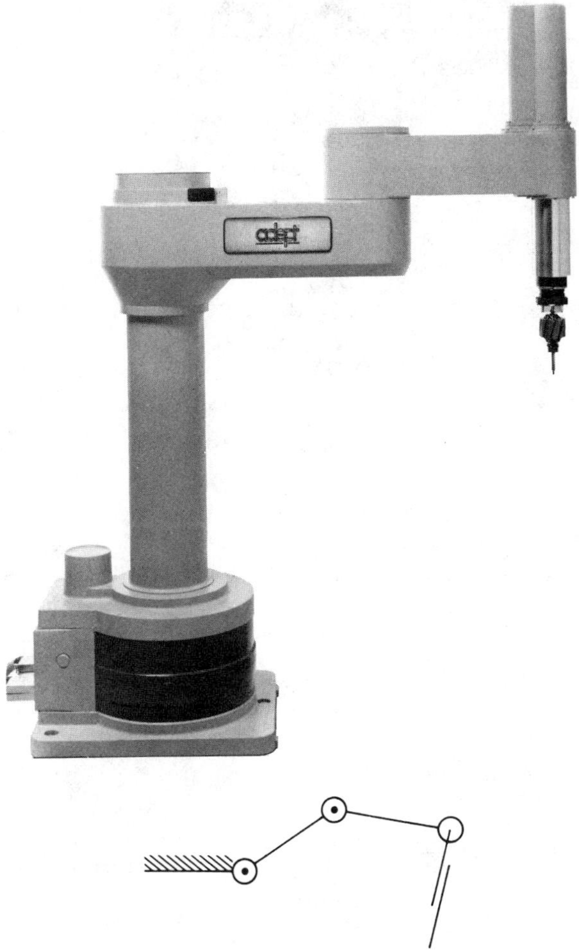

Figure C.13 Adept, AdeptOne

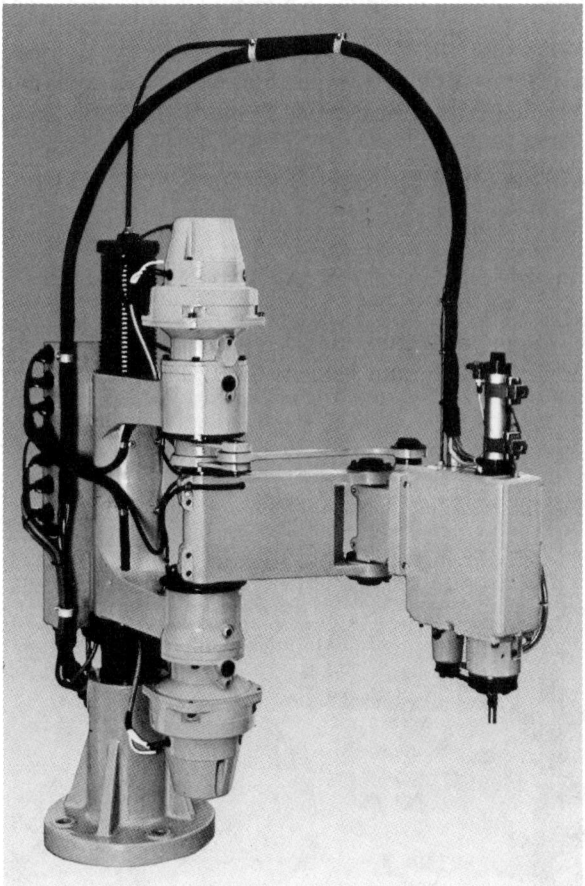

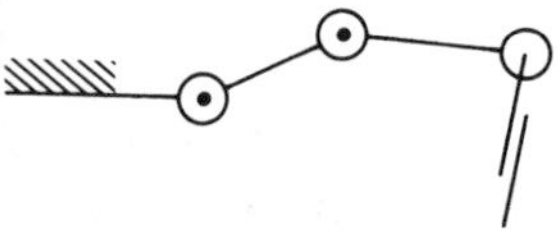

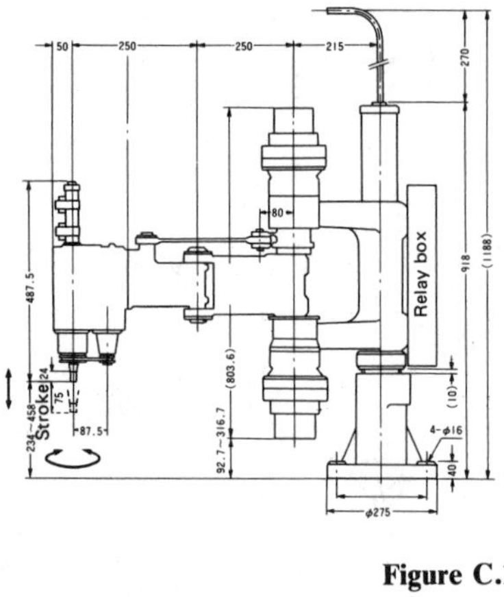

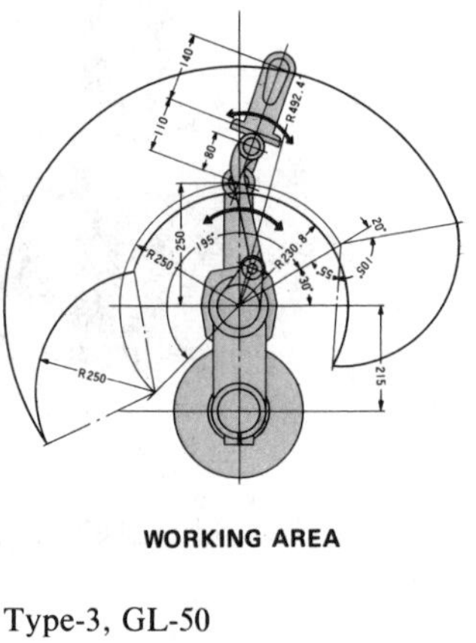

WORKING AREA

Figure C.14 Pentel, Type-3, GL-50

Torque at end effector 60 lb-in. Resolution ±0.0005 in., repeatability ±0.001 in. in *xy*-plane, ±0.002 in. in *z*-direction. Cycle time 0.9 s under no load and 1.3 s with 13 lb payload. Optional pitch axis available. Two $\frac{5}{32}$ in. air lines to end effector. Motorola 6800–based controller operating under VAL-II. Specific applications not given by manufacturer. 34.52-in. pedestal. The stroke on σ_4 will change to 11.7 in. when the pedestal is 30.52 in. See Figure C.13.

$$\mathbf{p}^2 \qquad\qquad \bar{\mathbf{p}}^1$$
$$\left|\theta_1\begin{Bmatrix}150\\-150\end{Bmatrix}16.73\text{ in.}\left|\theta_2\begin{Bmatrix}147\\-147\end{Bmatrix}14.76\right|\psi_3\begin{Bmatrix}554\end{Bmatrix}^{7.7}_{0}\right|$$

Pentel type-3 (GL-50). 4 *dof.* DC servo with hydraulic on σ_4. 1-kg payload. Weight 90 kg (with tool unit). Tool speed 1.3 m/s. Repeatability ±0.03 mm. Applications: accurate assembly (insertion, sealing, fitting), loading/unloading, moulding inserts, palletizing, inspection, automatic gauging, fine-part soldering. See Figure C.14.

$$\mathbf{p}^2 \qquad\qquad \bar{\mathbf{p}}^1$$
$$\left|\theta_1\begin{Bmatrix}195\\0\end{Bmatrix}250\text{ mm}\left|\theta_2\begin{Bmatrix}105\\0\end{Bmatrix}250\right|\psi_3\begin{Bmatrix}360\end{Bmatrix}^{75}_{0}\right|$$

Terms Used in the Text

Accuracy of a Robot: Difference between the taught position and the actual position of the position space of a robot at a particular point in the operating cycle; contrast with repeatability.

Actuator: The motor or driver that provides motive force to move a robotic joint.

AGV: Automatic guided vehicle.

Analog: Measurement or control where the information is a continuous function of time; a glass thermometer is an example.

Android: A robot that resembles a human.

Anthropomorphic: A revolute robot configured to resemble the human arm.

Bang-bang: A binary control in which the device passes rapidly from one state to the other; can be used to describe maximum acceleration and deceleration.

Bang-bang Robot: A robot having binary states, as in a pneumatic robot.

Base of a Robot: The platform on which the joints and links of the robot are mounted.

Baud: The unit of measure of signaling speed in transferring electronic data, where 1 baud = 1 bit of information per second.

CAD/CAM/CIM: Computer-aided design/computer-aided manufacturing/computer-integrated manufacturing.

CIKS: Closed-form inverse kinematic solution.

Closed-form Solution: A solution that can be obtained without error in a preassigned number of steps.

CNC: Computer numerical control.

Compliance: Displacement of the position space of the robot due to external forces; used as an aid to alignment.

Controller: The computer dedicated to controlling the robotic actuators.

Curved Link: A link with a curve or twist such that it is not considered to lie in a single plane of motion.

Cycle Time: The time from starting one operation to starting the next. Generally considered the best measure of the effective speed of a robot.

Defective Joint System: A kinematic chain of links and joints that can be replaced by an equivalent set of links and joints with fewer joints.

Dependent Plane: A plane of motion whose values are dependent on the preceding (downlink side) plane and that is unaffected by the status of the downlink joint.

Dextrous Volume: The working volume of a robot comprised of the totality of points that can be reached by the end effector where at each point the end effector can assume every possible orientation.

Digital: Measurement, processing, or control in which information is processed as a series of timed pulses—the mode of operation in a digital computer.

Directable: A placeable robot whose last link can lie in any assigned direction.

Distal: Toward the end effector (same as uplink).

DOF: Degree of freedom; *m dof* means *m* degrees of freedom.

Downlink: The link connected to a joint closer to the base.

Drillable: A placeable robot capable of placing its end link in any orientation in the pencil of planes containing its first link.

Dynamics: The study of position and motion considering forces and masses as well as kinematics.

Encoder: A transducer used to convert from rotational position to digital information.

Fixturing: The accurate positioning and holding of workpieces so that operations can be performed in these pieces.

FKS: Forward kinematic solution.

FMS: Flexible manufacturing system.

Fold Line: The line of intersection between adjacent planes of motion.

Gantry Robot: An overhead mounted robot with Cartesian configuration.

Gripper: The end effector of a robot with movable jaws that can grip an object.

Hierarchical Control: A distributed processing system comprising two or more controllers where one or more controllers are subordinated to the commands of another.

IKS: Inverse kinematic solution.

Independent Plane: A plane of motion where the vector partially defining the plane can rotate as the downlink joint rotates.

Iteration: A cycle of calculations in an open form solution. If the iteration is convergent the set of values at the end of the iteration is closer to the required solution than the set of values at the beginning of the iteration.

Joint Space: The ordered set of joint angles, given symbol ξ.

Kinematically Simple Joints: The restricted set of 1 *dof* joints that characterize all common robots.

Kinematics: The study of position and motion without considering mass.

Lead-Through Control: The method of teaching a robot a sequence of operations by a human operator physically moving the robot through these operations.

Link: The rigid link that connects two joints or a joint to the end effector.

Loose: A robot with more degrees of freedom than the order of the position space.

Mobile Robot: A robot mounted on a movable base.

OIKS: Open-form inverse kinematic solution.

Open-Form Solution: A solution that cannot be made in a preassigned number of steps, characterized by guessing at the solution and refining the result until an acceptable (within an acceptable tolerance) solution is obtained.

Optical Encoder: A transducer that converts rotational position to electrical information using flashes of light passing through slits in an opaque disc—the most common method of determining joint position in a robot.

Orientable: A robot whose end effector is placeable in any orientation.

Payload: The maximum weight that the end effector of the robot can handle during normal operation. This is often a function of speed at the end effector and/or extension of robot—that is, the payload may be reduced when the end effector is close to its working envelope.

Pencilable: A placeable robot whose last link can rotate while its position is unaltered.

Pick-and-Place Robot: A simple robot that transfers objects from one location to the next without control of the path followed. The pneumatic bang-bang robot is an example.

Pixel: Picture-cell, a digital picture element.

Placeable: A robot whose end effector can be placed at any point in the neighborhood (within an incremental distance) of any point within its working volume.

Plane of Motion: The plane within which a kinematic chain of joints and links move, assuming the link before the downlink joint is fixed.

Plane with Area: A plane of motion whose linkages have 2 *dof* or more within the plane.

Position Space: The set of positional and orientational requirements of the end effector, plus other designated joints, given the symbol ζ.

Prehension: Act of gripping or holding an object.

Proximity Sensor: A transducer that senses when an object is close.

Repeatability of a Robot: The maximum error expected between positions when operating from cycle to cycle; contrast with accuracy.

Resolver: A rotary or linear transducer that converts mechanical motion into analog electrical signals.

SCARA: Selective Compliant Articulative Robot for Assembly, or Selective Compliant Assembly Robot Arm.

Slew: The simultaneous motion of all joints in which each joint moves linearly and all joints start and stop together.

Spline Function: A function in which the position and its first derivative are matched to some desired function at the end points.

Stepper Motor: Sometimes known as stepping motor. A bidirectional electric motor that turns a set number of degrees for each input pulse.

Stiff Robot: A robot that does not have a plane with area. The robot can assume certain position spaces that require massive changes in joint space to move to some proximate position spaces.

Structured Light: Light projected in narrow rays or bands to produce patterns on an object so that the object and/or its location can be identified.

Teach Pendant: Also known as teach box. A programming device attached to a robot controller and held by an operator physically close to a robot, where the operator teaches the robot to move through a sequence of motions. The complete sequence can be edited and speeds modified as desired.

Torque Space: The set of applied torques required to move the robot through the required joint space (position space) time function; given the symbol τ.

Transducer: A device that converts from one type of energy, position, or motion to another.

Uplink: The link connected to a joint closer to the end effector.

Vacuum Cups: Pneumatically activated cups, usually with soft perimeters to decrease air leakage, typically for picking up objects with a flat face, such as printed circuit boards.

Voids: The places inside the working envelope of a robot that cannot be reached by any feasible joint space.

Working Envelope: The outer limit of the work area of the position space of a robot.

Working Volume: The totality of points that can be reached by the end effector of the robot regardless of end effector orientation.

References

At the end of each citation is the chapter(s) and section(s) for which the citation is appropriate. If the citation is generally applicable to a chapter, only the chapter number is given.

ADAMS, J. A. 1974. "Cubic Spline Curve Fitting with Controlled End Conditions." Computer Aided Design 6:1–9. 7.1, 8.3.2

ALLAN, J. J., III. 1980. ed. "A Survey of Industrial Robots." 2nd ed. Dallas, Tex.: Leading Edge. App. C

ANGELES, J. 1985. *Spatial Kinematic Chains*. New York: Springer-Verlag. 3 (intro.)

ANGELES, J. 1985. "On the Numerical Solution of the Inverse Kinematic Problem." *Int. J. Robotics Research* 4(2):21–37. 6.4

ANGELES, J. 1986. "Iterative Kinematic Analysis of General Five-Axis Robot Manipulators." *Int. J. Robotics Research* 4(4):59–70. 6.4

ARMSTRONG, W. W. 1979. "A Recursive Solution to the Equations of Motion of an *N*-Link Manipulator." Proc. 5th World Cong. on Theory of Mach. and Mech., July, Montreal, Canada, pp. 1343–6. 6.4

ASADA, H., AND A. B. BY. 1985. "Kinematic Analysis of Workpart Fixturing for Flexible Assembly with Automatically Reconfigurable Fixtures," *IEEE J. Robotics and Automation* RA-1, no. 2 (June):86–94, 8.8

ASADA, H., AND J. E. SLOTINE. 1986. *Robot Analysis and Control*. New York: John Wiley. 9.2

BALAFOUTIS, C. A., R. V. PATEL, AND P. MISRA. 1988. "Efficient Modeling and Computation of Manipulator Dynamics Using Orthogonal Cartesian Tensors," *J. Robotics and Automation* 4, no. 6 (Dec.):665–76, 9.3

BEZIER, B. E. 1971. "Example of an Existing System in the Motor Industry: the Unisurf System." Proc. Royal Soc. London, vol. A321, pp. 207–18. 7.3

BEZIER, B. E. 1972. *Numerical Control Mathematics and Applications*. London: John Wiley. 7.3

DEBOOR, C. 1972. "On Calculating with B-Splines." *J. Approx. Theory* 6:50–62. 7.3

DEBOOR, C. 1978. *A Practical Guide to Splines*. New York: Springer-Verlag. 7.3

BOTTEMA, O., AND B. ROTH. 1979. *Theoretical Kinematics*. Amsterdam: North-Holland, 2.1

BOWKER COMPANY. 1985. "Robotics, CAD/CAM Market Place, 1985." New York. App. C

BRADY, M., ET AL. 1983. *Robot Motion: Planning and Control*. Cambridge, Mass.: MIT Press. Mass, 6 (intro.), 8 (intro.)

CHASEN, S. H. 1978. *Geometric Principles and Procedures for Computer Graphics*. Prentice Hall. 2.8

CHURCHILL, R. V. 1963. *Fourier Series and Boundary Value Problems*. New York: McGraw-Hill. 7 (intro.), 8.4

CRAIG, J. J. 1986. *Introduction to Robotics, Mechanics and Control*. Reading, Mass.: Addison-Wesley. 2.1.3, 3.4, 5.1.3, 6 (intro.), 6.4, 6.5, 9 (intro.), 10.8.2

CUGY, A., AND K. PAGE. 1984. *Industrial Robot Specifications*. rev. ed. London: Kogan Page, App C

CUTKOSKY, M. R. 1985. *Robotic Grasping and Fine Manipulation*. Boston, Mass.: Kluwer Acad. Publishers. 8.7

DENAVIT, J., AND R. S. HARTENBERG. 1955. "A Kinematic Notation for Lower Pair Mechanisms Based on Matrices." *J. App. Mechanics* 22 (June):215–21. 3 (intro.), 3.1, 3.6

DENMAN, K. H. 1971. "Smooth Cubic Spline Interpolation Functions." *Ind. Math., J. Ind. Math. Soc.* 21 part 2:55–75. 7.1

DUFFY, J. 1980. *Analysis of Mechanisms and Robot Manipulators*. New York: John Wiley. 2.1, 3.1

ERSÜ, E., AND D. NUNGESSER. 1984. "A Numerical Solution of the General Kinematic Problem." Proc. Int. Conf. on Robotics, March, Atlanta, Georgia, pp. 162–75. 6 (intro.)

FEATHERSTONE, R. 1983. "Position and Velocity Transformations Between Robot End Effector Coordinates and Joint Angles." *Int. J. Robotics Research* 2. 8 (intro.), 8.4

FITZGERALD, A. E., C. KINGSLEY, JR., AND A. KUSKO. 1971. *Electric Machinery*. 3d. ed. New York: McGraw-Hill. 10.4, 10.5, 10.6

FOLEY, J. D. 1982. *Fundamentals of Interactive Computer Graphics*. Reading, Mass.: Addison-Wesley, 2.8

FREUND, E. 1977. "Path Control for a Redundant Type of Industrial Robot." Proc. 7th Int. Symp. on Industrial Robots, Oct. Tokyo, pp. 107–14, 8(int, VIII)

FU, K. S., R. C. GOZALVES, AND C. S. G. LEE. 1987. "Robotics: Control, Sensing, Vision and Intelligence." New York: McGraw-Hill. 8.8, 9 (intro.), 10.8.2

GANTMACHER, F. R., 1959. *The Theory of Matrices*. New York: Chelsea Publishing.

GELLERT, W., H. KÜSTNER, M. HELLWICH, AND H. KÄSTNER. 1977. *The CNR Concise Encyclopedia of Mathematics*. New York: Van Nostrand Reinhold. 1.4.5

GOLDENBERG, A. A., B. BENHABIB, AND R. G. FENTON. 1985. "A Complete Generalized Solution to the Inverse Kinematics of Robots." *IEEE J. Robotics and Automation* RA-1, no. 1 (March):14–20. 6.4

GORLA, B., AND M. RENAUD. 1984. *Robots Manipulateurs*. Toulouse, Fr.: Cepadues-Editions. France. 6 (intro.)

GREENWOOD, D. T. 1965. *Principles of Dynamics*. Englewood Cliffs, N.J.: Prentice Hall. 9.2, 9.4

GROOVER, M. P., M. WEISS, R. N. NAGEL, AND N. G. ODREY. 1986. *Industrial Robotics: Technology, Programing and Applications*. New York: McGraw-Hill. 10.10

GUPTA, K. C., AND B. ROTH. 1982. "Design Considerations for Manipulator Workspace." *ASME Journal of Mechanical Design* 104:704–11, 5.3, 8 (intro.)

HANAFUSA, H., T. YOSHIKAWA, AND Y. NAKAMURA. 1978. "Control of Multi-Articulated Robot Arms with Redundancy," pp. 237–38. "Part 1: Analysis and Determination of Input Considering Task Priority," pp. 319–20. "Part 2: Application to Obstacle Avoidance Problem," pp. 421–26. 21st Joint Aut. Control Conf., Japan. 4.2, 8.8

HANAFUSA, H., T. YOSHIKAWA, AND Y. NAKAMURA. 1983. "Redundancy Analysis of Articulated Robot Arms and Its Utilization for Tasks with Priority." *Trans J. Soc. Inst. and Control Eng*. 19, no. 5:421–26. 4.2, 8.8

HARTENBERG, R. S., AND J. DENAVIT. 1964. *Kinematic Synthesis of Linkages*. New York: McGraw-Hill. 2.1, 3 (intro.), 3.1, 3.6

HUANG, H.-P., AND N. H. MCCLAMROCH. 1988. "Time-Optimal Control for a Robotic Contour Following Problem." *J. Robotics and Automation* 4, no. 2 (April):140–9. 10.9

HUNT, K. 1978. *Kinematic Geometry of Mechanisms*. Fair Lawn, N.J.: Oxford Univ. press. 3.1

HUNT, V. A. 1983. *Industrial Robotics Handbook*. New York: Industrial Press. App. C

JAPAN INDUSTRIAL ROBOT ASSOC., "The Specifications and Applications of Industrial Robots in Japan." 1986. App. C

KAFRISSEN, E., AND M. STEPHANS. 1984. *Industrial Robots and Robotics*. Reston, Va.: Reston Publishing. 10.10

KAHN, M. E., AND B. ROTH. 1971. "The Near-Minimum Time Control of Open-Loop Articulated Kinematic Chains." *Trans. ASME, J. Dynamic Systems, Meas. and Control* 93, no. 3:164–72. 8.4

KHERADPIR, S., AND J. S. THORP. 1988. "Real-Time Control of Robot Manipulators in the Presence of Obstacles." *J. Robotics and Automation* 4, no. 6(Dec.):687–98. 8.8

KINOSHITA, G. 1981. "The Maneuverability of a Manipulator with Multi-Joints." Proc. 11th Int. Symp. on Industrial Robots, Japan Ind. Robot Soc., pp. 325–32. 3.1

KRAUSE, P. C. 1986. "Analysis of Electric Machinery." New York: McGraw-Hill. 10.5

KUMAR, A. AND K. J. WALDRON. 1981. "The Workspace of a Mechancial Manipulator," ASME Journal of Mechanical Design 103:665–72. 5.3, 6.5

KUO, B. C. 1987. "Automatic Control Systems." 5th ed. Englewood Cliffs, N.J.: Prentice Hall, 10 (intro.), 10.4, 10.8, 10.8.2

LAI, Z.-C., AND C.-H. MENQ. 1988. "The Dextrous Workspace of Simple Manipulators." *J. Robotics and Automation* 4, no. 1(Feb):99–103. 5.3

LAMMINEUR, P., AND O. CORNILLIE. 1984. *Industrial Robots*. Oxford, Eng.: Pergamon Press.

LEE, C. S. G. 1983a "Robot Arm Kinetics," in "Tutorial on Robotics," C. S. G. Lee et al., eds. IEEE Computer Society, pp. 47–65. 2.1, 5 (intro.)

LEE, C. S. G. AND M. ZIEGLER. 1983b. "A Geometric Approach in Solving the Inverse Kinematics of PUMA Robots." Proc. 13th Int. Symp. Indust. Robots, Chicago, Ill., pp. 16–1 to 16–18. 6.4

LEE, T. W. AND D. C. H. YANG. 1983. "On the Evaluation of the Manipulator Workspace." *J. Mechanisms, Transmissions and Aut. in Design.* 105, No. 1 (March): 70–77. 5.3

LITVIN, F. L., AND V. P. CASTELI. 1984. "Robot's Manipulators: Simulation and Identification of Configurations, Execution of Prescribed Trajectories." Proc. Int. Conf. on Robotics, March, Atlanta, pp. 34–44. 8 (intro.)

LOZANO-PEREZ, T., AND M. WESLEY. 1979. "An Algorithm for Planning Collision Free Paths among Polyhedral Obstacles." *Comm. ACM* 22, no. 10(Oct.):560–70. 8.7

LOZANO-PEREZ, T. 1981. "Automatic Planning of Manipulator Transfer Movements." *IEEE Trans. Syst., Man and Cybern.* SMC-11(Oct.):681–98. 8 (intro.), 8.8

LOZANO-PEREZ, T., M. M. MASON, AND R. H. TAYLOR. 1984. "Automatic Synthesis of Fine-Motion Strategies for Robots," in Robotics Research, 1st Int. Symp., ed. M. Brady and R. Paul. Cambridge, Mass.: MIT Press. 8.7, 8.8

LUH, J. Y. S., AND M. W. WALKER. 1977. "Minimum-time Along the Path for a Mechanical Arm." *Proc. 1977 IEEE Conf. Decision and Control* 1, pp. 755–59 New Orleans, La. 8.4

LUH, J. Y. S., M. W. WALKER, AND R. P. C. PAUL. 1980a. "Resolved Acceleration Control of Mechanical Manipulators." *IEEE Trans. Automatic Control* 25, no. 3:468–74, 8 (intro.), 8.4

LUH, J. Y. S., M. W. WALKER, AND R. P. C. PAUL. 1980b. "On-Line Computational Scheme for Mechanical Manipulators." *ASME J. Dyn Syst. Meas. and Control* 102 (June):69–76. 9.3

LUH, J. Y. S., AND C. S. LIN. 1981. "Optimum Path Planning for Mechanical Manipulators." *Trans. ASME, J. Dynamic Syst. Meas. Cont.* 103, no. 2 (June):142–51. 8 (intro.), 8.4

LUH, J. Y. S., AND C. E. CAMPBELL. 1982. "Collision Free Path Planning for Industrial Robots." Proc. 21st IEEE Conf. on Decision Contr., Dec., Orlando, Fl. pp. 84–8. 8 (intro.), 8.8

MAKINO, H. 1976. "A Kinematical Classification of Robot Manipulators." Proc. 3rd Conf. on Industrial Robot Technology, March, Nottingham. 5 (intro.)

MARDEN, M. 1966. "Geometry of Polynomials." American Math Soc., Providence, R.I. 9.8

MASON, M. T., AND J. K. SALISBURY, JR. 1985. *Robot Hands and the Mechanics of Manipulation*. Cambridge, Mass.: MIT Press. 5.1

MCCARTHY, J. M. "Dual Orthogonal Matrices in Manipulator Kinematics." *Int. J. Robotics Research* 5, no. 2:45–51. 3.6

MCINNIS, B. C., AND C. F. LIU. "Kinematics and Dynamics in Robotics: A Tutorial Based Upon Classical Concepts of Vectorial Mechanics." *IEEE J. Robotics and Automation* RA-2, no. 4 (Dec.):181–87. 5 (intro.)

NAKAMURA, Y., AND H. HANAFUSA. 1985. "Task Priority Based Redundancy Control of Robot Manipulators." Robotics Research, 2d Int Symp., ed. H. Hanafusa and H. Inoue, Cambridge, Mass.: MIT Press. 4.2

NAKAGAWA, Y., AND T. NINOMIYA. 1985. "The Structured Light Method for Inspection of Solder Joints and Assembly Robot Vision Systems." Robotics Research, 2d. Int Symp., ed by H. Hasafusa and H. Inoue. Cambridge, Mass.: MIT Press. 2.6

NEWMAN, W. M. 1979. *Principles of Interactive Computer Graphics*. New York: McGraw-Hill. 2.8

OGATA, K. 1970. *Modern Control Engineering*. Englewood Cliffs, N.J.: Prentice Hall. 10 (intro.), 10.2, 10.4, 10.5, 10.7, 10.8.1

OVERHAUSER, A. W. 1968. "Analytic Definition of Curves and Surfaces by Parabolic Blending." Ford Motor Co. Tech. Rept. #SL68-40, May. 7.1

PARKIN, R. E. AND HUTCHINSON. 1985a. "A Compliant Mechanical Gripper." *Robotics Age* 7, no. 5 (May). 8 (intro.)

PARKIN, R. E. 1985b. "Inverse Robotic Kinematics Derived from Planes of Movement." *Robotics Age* 7, no. 8 (Aug.):20–29. 3.9, 4.3, 5.2, 6 (intro.), 6.2–6.4

PARKIN, R. E. 1985c. "Simplified Determination of Robotic Movement Using Planes." ASME Int. Conf. on Computers in Engineering, Aug., Boston, vol. 1, pp. 59–64. 4 (intro.)

PARKIN, R. E. 1986. "Geometry of the Robotic Workplace and the Homogeneous Representation." *Robotics Engineering* 8, no. 1 (Jan.):16–22. 1.1–1.4, 2.1

PAUL, R. P. 1975. "Manipulator Path Control," Proc. IEEE Int. Conf. Cybernetics and Society, Sept., New York, pp. 147–52. 8.4

PAUL, R. P. 1979. "Manipulator Cartesian Path Control." *IEEE Trans. on Systems, Man. and Cybernetics* SMC-9, no. 11 (Nov.):702–11. 8 (intro.), 8.2, 8.4

PAUL, R. P., B. SHIMANO, AND G. E. MAYER. 1981a. "Kinematic Control Equations for Simple Manipulators." *IEEE Trans. Systems, Man, and Cybernetics* SMC-11, no. 6 (June):449–55. 3.1, 3.6, 6 (intro.)

PAUL, R. P. 1981b. *Robot Manipulators—Mathematics, Programming and Control*. Cambridge, Mass.: MIT Press. 1.1, 2.1, 5.1, 5.2, 6 (intro.), 8 (intro.), 9.2

PAUL, R. P., AND C. N. STEVENSON. 1983. "Kinematics of Robot Wrists." *Int. J. Robotics Research* 2, no. 1:31–38. 3.2

PAUL, R. P., AND H. ZHANG. 1986. "Computationally Efficient Kinematics for Manipulators with Spherical Wrists Based on the Homogeneous Transformation Representation." *Int. J. Robotics Research* 5, no. 2:32–44. 5 (intro.), 6 (intro.)

PAVLIDIS, T. 1982. "Algorithms for Graphics and Image Processing." Rockville, Md.: Computer Science Press. 2.8

PIEPER, D. 1968. "The Kinematics of Manipulators under Computer Control," Ph.D. thesis, Stanford University. 2.1, 6 (intro.)

PIEPER, D. L., AND B. ROTH. 1969. "The Kinetics of Manipulators Under Computer Control." Proc. 2nd Int. Conf. Theory of Machines and Mechanisms, Sept., Warsaw. 3.1, 6 (intro.)

RAIBERT, M. H., AND B. K. P. HORN. 1978. "Manipulator Control Using the Configuration Space Method." *Industrial Robot* 5, no. 2(June):69–73. 8 (intro.)

RALSTON A., AND P. RABINOWITZ. 1978. *A First Course in Numerical Analysis*. 2d ed. New York: McGraw-Hill. 2.7, 7 (intro.), 7.2, 7.3, 8.3.2, 9.9

REULEAUX, F. 1963. *Kinematics of Machinery*. Mineola, N.Y.: Dover Publications (reprint of original 1876 work published by Macmillan Company). 3.1

ROTH, B. 1967a. "The Kinematics of Motion Through Finitely Separated Positions." *J. App. Mechanics* 34 (Sept.):591–98. 3 (intro.), 3.1

ROTH, B. 1967b. "Finite Position Theory Applied to Mechanism Synthesis." *J. App. Mechanics* 34 (Sept.):599–605. 3 (intro.), 3.1

ROUTH, E. J. 1877. *A Treatise On The Dynamics Of A Rigid Body*. London: Macmillan. 10.7

ROUTH, E. J. 1959. *The Advanced Part Of A Treatise On The Dynamics Of A Rigid Body*. London: Macmillan. 1905; repr. Dover, New York. 10.7

SHETH, P. N., AND J. J. UICKER. 1971. "A Generalized Symbolic Notation for Mechanisms." *ASME Journal of Engineering for Industry* 93, no. 1:102–12. 3.1

SIMONS, G. L. 1980. "Robots in Industry." Manchester, Eng.: N.C.C. Publications. App. C

SNYDER, W. 1985. *Industrial Robots: Computer Interfacing and Control*. Englewood Cliffs, N.J.: Prentice Hall. 10 (intro.)

STANISIC, M. M., AND G. R. PENNOCK. 1985. "A Nondegenerate Kinematic Solution of a Seven-Jointed Robot Manipulator." *Int J. Robotics Research* 4, no. 2:10–20. 3.1, 3.2, 6 (intro.)

SUGIMOTO, K., AND Y. MATSUMOTO. 1984. "Kinematic Analysis of Manipulators by Means of the Projective Transformation of Screw Coordinates." Robotics Research, 1st Int. Symp., ed. M. Brady and R. Paul, Cambridge, Mass.: MIT Press. 6 (intro.)

SUH, S.-H., AND K. G. SHIN. 1988. "A Variational Dynamic Programming Approach to Robot Planning with a Distance-Safety Criterion." *J. Robotics and Automation* 4, no. 3 (June):334–49. 8.8

TANNER, W. R. 1981. "Industrial Robots." 2d edition, Soc. Manuf. Engineers. App. C

TAYLOR, R. H. 1979. "Planning and Execution of Straight Line Manipulator Trajectories." *IBM J. of Res. and Devel.* 23, no. 4 (July):424–36. 8 (intro.), 8.1, 8.2

TSAI, Y. C. AND A. H. SONI. 1983. "An Algorithm for the Workspace of a General n-R Robot." *Trans. ASME J. of Mechanisms, Transmissions and Automation in Design* 105, no. 1 (March):52–57. 5.3

TSAI, L., AND A. MORGAN. 1984. "Solving the Kinematics of the Most General Six and Five-Degree-of-Freedom Manipulators by Continuation Methods." ASME Mechanisms Conf., Oct., Boston, Mass. 6 (intro.)

TVER, D. F. 1983. *Robotics Source Book and Dictionary*. New York: Industrial Press. App. C

VEITSCHEGGER, W. K. AND C. H. WU. 1986. "Robot Accuracy Analysis Based on Kinematics." *IEEE J. Robotics and Automation* RA-2, no. 3 (Sept):171–9. 5.4, 10.8.3

WALDRON, K. 1982. "Geometrically Based Manipulator Rate Control Algorithms." *Mechanism and Machine Theory* 17, no. 6:379–85. 8 (intro.), 8.4

WHITNEY, D. E. 1969. "Resolved Motion Rate Control of Manipulators and Human Prostheses." *IEEE Trans. Man-Mach Syst.* MMS-10 (June):47–53. 8 (intro.), 8.4

WHITNEY, D. E., AND A. C. EDSALL. 1985. "Modelling Robot Contour Processes." Robotics Research, 2nd Int. Symp., ed. H. Hanafusa and H. Inoue. Cambridge, Mass.: MIT Press. 3.7

WOLOVICH, W. A. 1987. "Robotics: Basic Analysis and Design," Holt, Rinehart and Winston, New York, 2(1.3), 3(VI), 6(int)

YOSHIKAWA, T. 1983. "Analysis and Control of Robot Manipulators with Redundancy." Int. Symp. Robotics Research, pp. 735–48. 4.2, 8.8

YOSHIMOTO, K., AND K. WAKATSUKI. 1984. "Application of the Preview Tracking Algorithm to Servoing a Robot Manipulator," Robotics Research, 1st Int. Symp., ed. M. Brady and R. Paul, Cambridge, Mass.: MIT Press, 10.8.3

YOSHIMOTO, K., AND H. SULIUCHI. 1985. "Trajectory Control of Robot Manipulator Based on the Preview Tracking Control Algorithm." Robotics Research, 2d Int. Symp., Cambridge, Mass.: MIT Press. 10.8.3

Additional Reading

Beni, G., and S. Hackwood, eds. 1985. *Recent Advances in Robotics*. New York: John Wiley.

Coiffet, P. 1983. *Robot Technology*. Vol I–III, London: Kogan Page.

Colson, J. C., and N. D. Perreira. 1983. "Kinematic Arrangements Used in Industrial Robots." Proc. 13th Int. Symp. on Industrial Robots, April, pp. 20–1 to 20–18.

Critchlow, A. 1985. *Introduction to Robotics*. New York: Macmillan.

Dorf, R. 1983. *Robotics and Flexible Automation*. Reston, Va.: Reston.

Dubowski, S., and D. T. DesForges. 1979. "The Application of Model-Referenced Adaptive Control to Robotic Manipulations." *ASME J. Dyn. Systems, Measurement and Control* 101:193–200.

Goertz, R. C. 1952. "Fundamentals of General Purpose Remote Manipulators." *Nucleonics* 10(Nov.):36–42.

Hansen, J. A., K. C. Gupta, and S. M. K. Kazerounian. 1983. "Generation and Evaluation of the Workspace of a Manipulator." Proc. Sixth IFToMM Congress on Theory of Machines and Mechanisms, New Delhi; also *Int. Journal of Robotics Research*.

Hayati, S. 1983. "Robot Arm Geometric Link Parameter Estimation." Proc. 22d IEEE Conf. on Decision and Control, December.

Koren, Y. 1983. *Computer Control of Manufacturing Systems*. New York: McGraw-Hill.

Koren, Y. 1985. *Robotics for Engineers*. New York: McGraw-Hill.

PAUL, R. P., B. SHIMANO, AND G. E. MAYER. 1981. "Differential Kinematic Control Equations for Simple Manipulators." *IEEE Trans. Syst., Man and Cybern.* SMC-11(June):456–60.

RENAUD, M. 1981. "Geometric and Kinematic Models of a Robot Manipulator: Calculation of the Jacobian Matrix and its Inverse." Proc. of Eleventh Int. Symp. on Ind. Robots, Oct., Tokyo, Japan. 1984.

SADRE, A., R. SMITH, AND W. CARTWRIGHT. 1984. "Coordinate Transformations for Two Industrial Robots." Proc. Int. Conf. on Robotics, March, Atlanta, pp. 45–61.

SUGIMOTO, K., J. DUFFY, AND K. H. HUNT. 1982. "Special Configurations of Spatial Mechanisms and Robot Arms." *Mechanism and Machine Theory* 17, no. 2:119–32.

ULLRICH, R. A. 1983. *The Robotics Primer: the What, Why and How of Robots in the Workplace*. Englewood Cliffs, N.J.: Prentice Hall.

YANG, D. C. H., AND T. W. LEE, 1983. "On the Workspace on Mechanical Manipulators." *ASME Journal of Mechanisms, Transmissions,and Automation in Design* 105:62–70.

Index